Edited by
Dmitri A. Molodov

**Microstructural Design of
Advanced Engineering
Materials**

Related Titles

Riedel, Ralf / Chen, I-Wei (eds.)

Ceramics Science and Technology

4 Volume Set

2014

ISBN: 978-3-527-31149-1

Plasticity, Failure and Fatigue in Structural Materials-From Macro to Nano

Proceedings of the HaelMughrabi Honorary Symposium

2008

Print ISBN: 978-0-873-39714-8

Pfeiler, W. (ed.)

Alloy Physics

A Comprehensive Reference

2007

Print ISBN: 978-3-527-31321-1, also available in electronic formats

Levitin, V.

High Temperature Strain of Metals and Alloys

Physical Fundamentals

2006

Print ISBN: 978-3-527-31338-9, also available in electronic formats

Scheel, H.J., Capper, P. (eds.)

Crystal Growth Technology

From Fundamentals and Simulation to Large-scale Production

2008

Print ISBN: 978-3-527-31762-2, also available in electronic formats

Köhler, M., Fritzsche, W.

Nanotechnology

An Introduction to Nanostructuring Techniques

2nd Edition

2007

Print ISBN: 978-3-527-31871-1, also available in electronic formats

Herlach, D.M. (ed.)

Phase Transformations in Multicomponent Melts

2009

Print ISBN: 978-3-527-31994-7, also available in electronic formats

Herlach, Dieter M. / Matson, Douglas M. (eds.)

Solidification of Containerless Undercooled Melts

2012

ISBN: 978-3-527-33122-2

Levitin, V., Loskutov, S.

Strained Metallic Surfaces

Theory, Nanostructuring and Fatigue Strength

2009

Print ISBN: 978-3-527-32344-9, also available in electronic formats

Dubois, J., Belin-Ferré, E. (eds.)

Complex Metallic Alloys

Fundamentals and Applications

2011

Print ISBN: 978-3-527-32523-8, also available in electronic formats

Jackson, K.A.

Kinetic Processes

Crystal Growth, Diffusion, and Phase Transitions in Materials

Second Edition

2010

Print ISBN: 978-3-527-32736-2

Edited by Dmitri A. Molodov

Microstructural Design of Advanced Engineering Materials

Verlag GmbH & Co. KGaA

The Editor

Dmitri A. Molodov
RWTH Aachen
Institut für Metallkunde und Metallphysik
Kopernikusstr. 14
52074 Aachen
Germany

All books published by **Wiley-VCH** are carefully produced. Nevertheless, authors, editors, and publisher do not warrant the information contained in these books, including this book, to be free of errors. Readers are advised to keep in mind that statements, data, illustrations, procedural details or other items may inadvertently be inaccurate.

Library of Congress Card No.: applied for

British Library Cataloguing-in-Publication Data
A catalogue record for this book is available from the British Library.

Bibliographic information published by the Deutsche Nationalbibliothek
The Deutsche Nationalbibliothek lists this publication in the Deutsche Nationalbibliografie; detailed bibliographic data are available on the Internet at <http://dnb.d-nb.de>.

© 2013 Wiley-VCH Verlag GmbH & Co. KGaA, Boschstr. 12, 69469 Weinheim, Germany

All rights reserved (including those of translation into other languages). No part of this book may be reproduced in any form – by photoprinting, microfilm, or any other means – nor transmitted or translated into a machine language without written permission from the publishers. Registered names, trademarks, etc. used in this book, even when not specifically marked as such, are not to be considered unprotected by law.

Print ISBN: 978-3-527-33269-4
ePDF ISBN: 978-3-527-65284-6
ePub ISBN: 978-3-527-65283-9
mobi ISBN: 978-3-527-65282-2
oBook ISBN: 978-3-527-65281-5

Composition   Thomson Digital, Noida, India
Printing and Binding   Markono Print Media Pte Ltd, Singapore
Cover Design   Adam Design, Weinheim

Printed in Singapore
Printed on acid-free paper

Contents

Preface *XV*
List of Contributors *XVII*

Part I   Materials Modeling and Simulation: Crystal Plasticity, Deformation, and Recrystallization *1*

1   Through-Process Modeling of Materials Fabrication: Philosophy, Current State, and Future Directions *3*
Günter Gottstein
1.1   Introduction *3*
1.2   Microstructure Evolution *5*
1.3   Microstructural Processes *6*
1.4   Through-Process Modeling *10*
1.5   Future Directions *14*
   References *16*

2   Application of the Generalized Schmid Law in Multiscale Models: Opportunities and Limitations *19*
Paul Van Houtte
2.1   Introduction *19*
2.2   Crystal Plasticity *20*
2.2.1   Generalized Schmid Law *22*
2.2.2   Calculation of Slip Rates, Lattice Rotation, and Stress from a Prescribed Deformation *23*
2.2.3   Taylor Factor *26*
2.3   Polycrystal Plasticity Models for Single-Phase Materials *27*
2.3.1   Sachs Model *28*
2.3.2   Taylor Theory *28*
2.3.3   Relaxed Constraints Taylor Theory *29*
2.3.4   Grain Interaction Models *30*
2.4   Plastic Anisotropy of Polycrystalline Materials *33*
2.5   Experimental Validation *34*

2.5.1	Prediction of Rolling Textures 34
2.5.2	Prediction of Cup Drawing Textures 36
2.6	Conclusions 37
	References 38

3	**Crystal Plasticity Modeling** 41
	Franz Roters, Martin Diehl, Philip Eisenlohr, and Dierk Raabe
3.1	Introduction 41
3.2	Fundamentals 45
3.2.1	Constitutive Models 46
3.2.1.1	Dislocation Slip 47
3.2.1.2	Displacive Transformations 47
3.2.2	Homogenization 48
3.2.3	Boundary Value Solvers 49
3.3	Application Examples 49
3.3.1	Texture and Anisotropy 49
3.3.1.1	Prediction of Texture Evolution 49
3.3.1.2	Prediction of Earing Behavior 50
3.3.1.3	Optimization of Earing Behavior 51
3.3.2	Effective Material Properties 55
3.3.2.1	Direct Transfer of Microstructures 55
3.3.2.2	Representative Volume Elements 57
3.3.2.3	The *Virtual Laboratory* 59
3.4	Conclusions and Outlook 61
	References 62

4	**Modeling of Severe Plastic Deformation: Time-Proven Recipes and New Results** 69
	Yuri Estrin and Alexei Vinogradov
4.1	Introduction 69
4.2	One-Internal Variable Models 70
4.3	Two-Internal Variable Models 77
4.4	Three-Internal Variable Models 81
4.5	Numerical Simulations of SPD Processes 82
4.6	Concluding Remarks 86
	References 87

5	**Plastic Anisotropy in Magnesium Alloys – Phenomena and Modeling** 91
	Bevis Hutchinson and Matthew Barnett
5.1	Deformation Modes and Textures 91
5.2	Anisotropy of Stress and Strain 92
5.3	Modeling Anisotropic Stress and Strain 103
5.4	Concluding Remarks 114
	References 115

6	**Application of Stochastic Geometry to Nucleation and Growth Transformations** *119*	
	Paulo R. Rios and Elena Villa	
6.1	Introduction *119*	
6.2	Mathematical Background and Basic Notation *121*	
6.2.1	Modeling Birth-and-Growth Processes *121*	
6.2.2	Mean Densities Associated to a Birth-and-Growth Process *123*	
6.2.3	Causal Cone *124*	
6.3	Revisiting JMAK *126*	
6.4	Nucleation in Clusters *130*	
6.4.1	The Matérn Cluster Process *130*	
6.4.2	Evaluation of the Integral in Eq. (6.17) *131*	
6.4.3	Numerical Examples *133*	
6.4.3.1	Influence of Cluster Radius *133*	
6.4.3.2	Influence of Number of Nuclei per Cluster *135*	
6.5	Nucleation on Lower Dimensional Surfaces *136*	
6.5.1	Derivation of General Expressions for Surface and Bulk Nucleation *136*	
6.5.1.1	Surface Nucleation *136*	
6.5.1.2	Bulk Nucleation *137*	
6.5.2	Numerical Examples *138*	
6.5.2.1	Surface Nucleation *138*	
6.5.2.2	Bulk Nucleation *138*	
6.5.2.3	Simultaneous Bulk and Surface Nucleation *140*	
6.6	Analytical Expressions for Transformations Nucleated on Random Planes *141*	
6.6.1	General Results for Nucleation on Random Planes *141*	
6.6.2	Behavior at the Origin as a Model for the Behavior in an "Unbounded" Specimen *142*	
6.6.3	Nucleation on Random Parallel Planes Located Within a Specimen of Finite Thickness *143*	
6.6.4	Nucleation on Random Parallel Planes Located Within an "Unbounded Specimens" *143*	
6.6.5	Computer Simulation Results *145*	
6.7	Random Velocity *145*	
6.7.1	Time-Dependent, Random Velocity *145*	
6.7.2	Particular Cases *147*	
6.7.3	Computer Simulation *148*	
6.8	Simultaneous and Sequential Transformations *150*	
6.8.1	Simultaneous Transformations *151*	
6.8.2	Sequential Transformations *152*	
6.8.3	Application to Recrystallization of an IF Steel *153*	
6.9	Final Remarks *157*	
	References *157*	

7		**Implementation of Anisotropic Grain Boundary Properties in Mesoscopic Simulations** *161*
		Anthony D. Rollett
7.1		Introduction *161*
7.2		Overview of Simulation Methods *161*
7.3		Anisotropy of Grain Boundaries *162*
7.3.1		Energy *162*
7.3.2		Mobility *163*
7.4		Simulation Approaches *164*
7.4.1		Potts Model *164*
7.4.2		Cellular Automata *169*
7.4.3		Phase Field *170*
7.4.4		Cusps in Grain Boundary Energy *174*
7.4.5		Level Set *174*
7.4.6		Vertex *175*
7.4.7		Moving Finite Element *176*
7.4.8		Particle Pinning of Boundaries *179*
7.5		Summary *180*
		References *180*
Part II		**Interfacial Phenomena and their Role in Microstructure Control** *187*
8		**Grain Boundary Junctions: Their Effect on Interfacial Phenomena** *189*
		Lasar S. Shvindlerman and Günter Gottstein
8.1		Introduction *189*
8.2		Experimental Measurement of Grain Boundary Triple Line Energy *190*
8.3		Impact of Triple Line Tension on the Thermodynamics and Kinetics in Solids *192*
8.3.1		Grain Boundary Triple Line Contribution to the Driving Force for Grain Growth *192*
8.3.2		Effect of the Triple Junction Line Tension on the Zener Force *193*
8.3.3		Effect of Triple Junction Line Tension on the Gibbs–Thompson Relation *195*
8.4		Why do Crystalline Nanoparticles Agglomerate with Low Misorientations? *196*
8.5		Concluding Remarks *198*
		References *199*
9		**Plastic Deformation by Grain Boundary Motion: Experiments and Simulations** *201*
		Dmitri A. Molodov and Yuri Mishin
9.1		Introduction *201*
9.2		What is the Coupled Grain Boundary Motion? *202*
9.3		Computer Simulation Methodology *204*

9.4	Experimental Methodology	206
9.5	Multiplicity of Coupling Factors	208
9.6	Dynamics of Coupled GB Motion	212
9.7	Coupled Motion of Asymmetrical Grain Boundaries	216
9.8	Coupled Grain Boundary Motion and Grain Rotation	221
9.9	Concluding Remarks	227
	References	229

10 Grain Boundary Migration Induced by a Magnetic Field: Fundamentals and Implications for Microstructure Evolution 235
Dmitri A. Molodov

10.1	Introduction	235
10.2	Driving Forces for Grain Boundary Migration	236
10.3	Magnetically Driven Grain Boundary Motion in Bicrystals	237
10.3.1	Specimens and Applied Methods to Measure Grain Boundary Migration	237
10.3.2	Measurements of Absolute Grain Boundary Mobility	239
10.3.3	Misorientation Dependence of Grain Boundary Mobility	243
10.3.4	Effect of Boundary Plane Inclination on Tilt Boundary Mobility	245
10.4	Selective Grain Growth in Locally Deformed Zn Single Crystals under a Magnetic Driving Force	246
10.5	Impact of a Magnetic Driving Force on Texture and Grain Structure Development in Magnetically Anisotropic Polycrystals	248
10.5.1	Texture Evolution during Grain Growth	248
10.5.2	Microstructure Evolution and Growth Kinetics	253
10.6	Magnetic Field Influence on Texture and Microstructure Evolution in Polycrystals Due to Enhanced Grain Boundary Motion	258
10.7	Concluding Remarks	261
	References	262

11 Interface Segregation in Advanced Steels Studied at the Atomic Scale 267
Dierk Raabe, Dirk Ponge, Reiner Kirchheim, Hamid Assadi, Yujiao Li, Shoji Goto, Aleksander Kostka, Michael Herbig, Stefanie Sandlöbes, Margarita Kuzmina, Julio Millán, Lei Yuan, and Pyuck-Pa Choi

11.1	Motivation for Analyzing Grain and Phase Boundaries in High-Strength Steels	267
11.2	Theory of Equilibrium Grain Boundary Segregation	271
11.2.1	Gibbs Adsorption Isotherm Applied to Grain Boundaries	271
11.2.2	Langmuir–McLean Isotherm Equations for Grain Boundary Segregation	272
11.2.3	Phase-Field Modeling of Grain Boundary Segregation and Phase Transformation at Grain Boundaries	274
11.2.4	Interface Complexions at Grain Boundaries	278

11.3	Atom Probe Tomography and Correlated Electron Microscopy on Interfaces in Steels *280*	
11.4	Atomic-Scale Experimental Observation of Grain Boundary Segregation in the Ferrite Phase of Pearlitic Steel *282*	
11.5	Phase Transformation and Nucleation on Chemically Decorated Grain Boundaries *288*	
11.5.1	Introduction to Phase Transformation at Grain Boundaries *288*	
11.5.2	Grain Boundary Segregation and Associated Local Phase Transformation in Martensitic Fe-C Steels *290*	
11.6	Conclusions and Outlook *295*	
	References *295*	

12 Interface Structure-Dependent Grain Growth Behavior in Polycrystals *299*

Suk-Joong L. Kang, Yang-Il Jung, Sang-Hyun Jung, and John G. Fisher

12.1	Introduction *299*	
12.2	Fundamentals: Equilibrium Shape of the Interface *300*	
12.2.1	Equilibrium Crystal Shape *300*	
12.2.2	Equilibrium Boundary Shape *301*	
12.3	Grain Growth in Solid–Liquid Two-Phase Systems *302*	
12.3.1	Growth Mechanisms and Kinetics of a Single Crystal in a Liquid *302*	
12.3.1.1	Diffusion-Controlled Crystal Growth *302*	
12.3.1.2	Interface Reaction-Controlled Crystal Growth *303*	
12.3.1.3	Mixed Controlled Growth of a Faceted Crystal *306*	
12.3.2	Grain Growth Behavior *307*	
12.3.2.1	Stationary Grain Growth in Systems with Spherical Grains *308*	
12.3.2.2	Nonstationary Grain Growth in Systems with Faceted Grains *309*	
12.4	Grain Growth in Solid-State Single-Phase Systems *312*	
12.4.1	Migration Mechanisms and Kinetics of the Grain Boundary *312*	
12.4.2	Grain Growth Behavior *315*	
12.5	Concluding Remarks *317*	
	References *318*	

13 Capillary-Mediated Interface Energy Fields: Deterministic Dendritic Branching *323*

Martin E. Glicksman

13.1	Introduction *323*	
13.2	Capillary Energy Fields *324*	
13.2.1	Background *324*	
13.2.2	Melting Experiments *325*	
13.2.3	Self-Similar Melting *326*	
13.2.4	Influence of Capillarity on Melting *327*	
13.3	Capillarity-Mediated Branching *329*	

13.3.1	Local Equilibrium	*329*
13.3.2	Gibbs–Thomson–Herring Interface Potential	*329*
13.3.3	Tangential Gradients and Fluxes	*330*
13.3.4	Capillary-Mediated Energy	*331*
13.4	Branching	*333*
13.4.1	Stefan Balance	*333*
13.4.2	Zeros of the Surface Laplacian	*333*
13.5	Dynamic Solver Results	*334*
13.6	Conclusions	*336*
	References	*337*

Part III Advanced Experimental Approaches for Microstructure Characterization *339*

14 High Angular Resolution EBSD and Its Materials Applications *341*
Claire Maurice, Romain Quey, Roland Fortunier, and Julian H. Driver

14.1	Introduction: Some History of HR-EBSD	*341*
14.2	HR-EBSD Methods	*342*
14.2.1	Basic Geometry of HR-EBSD	*342*
14.2.2	EBSP Numerical Simulations	*345*
14.2.3	Practical Problems	*345*
14.2.3.1	Lens Distortion	*346*
14.2.3.2	Orientation Gradients	*347*
14.2.3.3	Pattern Center	*349*
14.3	Applications	*351*
14.3.1	Low-Angle Grain Boundaries	*351*
14.3.2	Pure Elastic Strains	*351*
14.3.3	Crystal Defects and Lattice Misorientations: GND Analysis	*354*
14.3.4	Crystal Defects and Elastic Strains	*359*
14.4	Discussion	*359*
14.5	Conclusions	*362*
	References	*363*

15 4D Characterization of Metal Microstructures *367*
Dorte Juul Jensen

15.1	Introduction	*367*
15.2	4D Characterizations by 3DXRD – From Idea to Implementation	*368*
15.2.1	The 3DXRD Microscope	*369*
15.2.2	Another Approach for 3D Mapping of Crystallographic Contrast Microstructure by Synchrotron x-rays	*371*
15.3	Examples of Applications	*372*
15.3.1	Recrystallization Studies	*373*
15.3.1.1	Nucleation	*373*

15.3.1.2	Grain Boundary Migration *374*
15.3.1.3	Recrystallization Kinetics *376*
15.3.2	Other 4D Studies *377*
15.3.2.1	Plastic Deformation *377*
15.3.2.2	Grain Growth *378*
15.3.2.3	Phase Transformation *379*
15.4	Challenges and Suggestions for the Future Success of 3D Materials Science *379*
15.5	Concluding Remarks *381*
	References *382*

16 Crystallographic Textures and a Magnifying Glass to Investigate Materials *387*
Jürgen Hirsch

16.1	Introduction *387*
16.2	Texture Evolution and Exploitation of Related Information in Metal Processing *388*
16.2.1	Casting *389*
16.2.2	Deformation *389*
16.2.2.1	Textures Revealing Details on the Mechanisms of Deformation *391*
16.2.2.2	Stacking Fault Energy Analysis in Rolling Texture Formation *395*
16.2.2.3	Surface Shearing and Lubrication Effects *395*
16.2.3	Recrystallization *397*
16.2.3.1	Oriented Nucleation, Oriented Growth and in situ Recrystallization in Aluminum Alloys *397*
16.2.3.2	Recrystallization in Particle Containing Al Alloys *399*
16.3	Summary *399*
	References *400*

Part IV Applications: Grain Boundary Engineering and Microstructural Design for Advanced Properties *403*

17 The Advent and Recent Progress of Grain Boundary Engineering (GBE): In Focus on GBE for Fracture Control through Texturing *405*
Tadao Watanabe

17.1	Introduction *405*
17.2	Historical Background *406*
17.2.1	Demand for Strong and Tough Structural Materials *406*
17.2.2	Long Pending Materials Problem and Dilemma *407*
17.2.3	Origin of Heterogeneity of Grain Boundary Phenomena *409*
17.3	Basic Concept of Grain Boundary Engineering *410*
17.3.1	Beneficial and Detrimental Effects of Grain Boundaries *410*
17.3.2	Basic Concept of Grain Boundary Engineering *411*
17.3.3	Structure-Dependent Grain Boundary Properties in Bicrystals *413*

17.3.4	Structure-Dependent Fracture Processes in Polycrystals	*417*
17.3.5	Parameters to Characterize Grain Boundary Microstructure	*419*
17.4	Characteristic Features of Grain Boundary Microstructures	*420*
17.4.1	Grain Size Dependence of Grain Boundary Microstructure	*420*
17.4.2	Change in Grain Boundary Energy during Grain Growth	*420*
17.4.3	Relation between Grain Boundary Character Distribution and Grain Size	*421*
17.5	Relation between Texture and Grain Boundary Microstructure	*426*
17.5.1	Theoretical Basis of GB Microstructure in Textured Polycrystals	*426*
17.5.2	Characteristic Features of GB Microstructures in Textured Polycrystals	*427*
17.6	Grain Boundary Engineering for Fracture Control through Texturing	*434*
17.6.1	The Control of Microscale Texture and GB Microstructure	*434*
17.6.2	GB Microstructure-Dependent Brittle Fracture in Molybdenum	*437*
17.6.3	Percolation-Controlled and GB Microstructure-Dependent Fracture	*438*
17.6.4	Fracture Control Based on Fractal Analysis of GB Microstructure	*439*
17.7	Conclusion	*441*
	References	*441*

18 Microstructure and Texture Design of NiAl via Thermomechanical Processing *447*

Werner Skrotzki

18.1	Introduction	*447*
18.2	Experimental	*447*
18.3	Microstructure and Texture Development	*450*
18.4	Texture Simulations	*457*
18.5	Mechanical Anisotropy	*459*
18.6	Conclusions	*463*
	References	*463*

19 Development of Novel Metallic High Temperature Materials by Microstructural Design *467*

Martin Heilmaier, Joachim Rösler, Debashis Mukherji, and Manja Krüger

19.1	Introduction	*467*
19.2	Alloy System Mo–Si–B	*468*
19.2.1	Manufacturing and Microstructure	*468*
19.2.2	Strength and Ductility at Ambient Temperatures	*471*
19.2.2.1	Single-phase Mo Solid Solutions	*471*
19.2.2.2	Multiphase Mo–Si–B Alloys	*475*
19.2.3	Oxidation Resistance	*476*
19.2.4	Creep Resistance	*478*
19.3	Alloy System Co–Re–Cr	*480*

	19.3.1	Oxidation *480*
	19.3.2	Strength and Ductility at Ambient Temperatures *483*
	19.3.3	Creep Resistance *486*
	19.4	Conclusions *489*
		References *490*

Index *495*

Preface

Properties of crystalline engineering materials are directly related to their microstructure, defined as the spatial distribution of elements, phases, defects, and orientations. In view of the dramatically increased specific property material requirements during the past decades, the efforts to understand how the granular microstructure of polycrystals develops and how it can be influenced and predicted became extremely important, since microstructure control is crucial, both for improvement of materials performance and design of advanced materials with tailored properties.

The topic of microstructural design of advanced materials was recently the focus of a special symposium,[1] in honor of Professor Dr rer. nat. Dr h.c. Günter Gottstein (*Günter Gottstein Honorary Symposium on Characterization and Design of Microstructure for Advanced Materials*), which was held in the frame of the MSE 2012 (Materials Science and Engineering) Congress in Darmstadt, Germany, September 25–27, 2012, organized by the Deutsche Gesellschaft für Materialkunde (DGM).

This book represents a collection of manuscripts written by leading scientists in the field of microstructural design of engineering materials, who were invited to deliver keynote lectures at the Günter Gottstein Honorary Symposium. This provided a unique opportunity to bring together experts in various aspects of microstructure design and to address a wide range of topics, which are crucial for predicting and controlling the microstructure evolution, including crystal plasticity due to slip, twinning, and grain boundary motion; nucleation during recrystallization; grain boundary migration under various forces; impact of boundary junctions; interfacial anisotropy and solute segregation, interaction between interfaces and particles, and so on. As obvious from the reviews comprising this book, an interaction between various research approaches – experiment, microstructural modeling, computation, and theory – is indispensable for successful and effective microstructural design of advanced engineering materials.

The book is subdivided into four parts, beginning with the modeling of the basic processes of microstructure development, that is, crystal plasticity, deformation, and

---

1) Sponsored by the Deutsche Forschungsgemeinschaft (DFG), Deutsche Gesellschaft für Materialkunde (DGM), Aleris Rolled Products Germany GmbH, Hydro Aluminium Deutschland GmbH, ThyssenKrupp VDM GmbH, Wieland-Werke AG.

recrystallization in different metallic materials subjected to various processing routes including severe plastic deformation. The second part addresses grain boundaries and interfaces, their kinetics and thermodynamics, and their effects on microstructure evolution. The third part is dedicated to advanced experimental methods to characterize the microstructure and to elucidate the underlying mechanisms of its development. The final chapters comprise various applications – grain boundary engineering for improving fracture resistance of various metals and alloys and microstructural design of advanced high temperature materials.

The editor is grateful to all authors for their engagement and cooperation as well as the Wiley-VCH editorial team for the enthusiasm and help to prepare and publish this book and in such a way to celebrate Professor Günter Gottstein and his unique contributions to Materials Science.

Aachen, January 2013   *Dmitri A. Molodov*

List of Contributors

Hamid Assadi
MPI für Eisenforschung
Max-Planck-Str. 1
40237 Düsseldorf
Germany

Matthew Barnett
Deakin University
Institute for Frontier Materials
Pigdons Rd
Geelong, VIC 3217
Australia

Pyuck-Pa Choi
MPI für Eisenforschung
Max-Planck-Str. 1
40237 Düsseldorf
Germany

Martin Diehl
MPI für Eisenforschung
Max-Planck-Str. 1
40237 Düsseldorf
Germany

Julian H. Driver
SMS Centre
Ecole des Mines de Saint Etienne
158 Cours Fauriel
42032 Saint Etienne
France

Philip Eisenlohr
MPI für Eisenforschung
Max-Planck-Str. 1
40237 Düsseldorf
Germany

Yuri Estrin
Monash University
Centre for Advanced Hybrid Materials
Department of Materials Engineering
Clayton, VIC 3800
Australia

John G. Fisher
Chonnam National University
School of Materials Science and Engineering
77 Yongbong-ro, Buk-gu
Gwangju 500-757
Republic of Korea

Roland Fortunier
SMS Centre
Ecole des Mines de Saint Etienne
158 Cours Fauriel
42032 Saint Etienne
France

Martin E. Glicksman
Florida Institute of Technology
Mechanical & Aerospace
Engineering Department
150 West University Blvd.
Melbourne, FL 32901
USA

Shoji Goto
MPI für Eisenforschung
Max-Planck-Str. 1
40237 Düsseldorf
Germany

Günter Gottstein
RWTH Aachen University
Institute of Physical Metallurgy and
Metal Physics (IMM)
Kopernikusstr. 14
52056 Aachen
Germany

Martin Heilmaier
Karlsruher Institut für Technologie
Institut für Angewandte Materialien
Engelbert-Arnold-Str. 4
76131 Karlsruhe
Germany

Michael Herbig
MPI für Eisenforschung
Max-Planck-Str. 1
40237 Düsseldorf
Germany

Jürgen Hirsch
Research & Development Bonn
Hydro Aluminium Rolled
Products GmbH
Georg-von-Boeselager-Str. 21
53117 Bonn
Germany

Bevis Hutchinson
Swerea-KIMAB
Box 7047
164 07 Stockholm
Sweden

Dorte Juul Jensen
Technical University of Denmark
Materials Science and Advanced
Characterization Section
Department of Wind Energy
Risø Campus
4000 Roskilde
Denmark

Sang-Hyun Jung
Korea Advanced Institute of Science
and Technology
Department of Materials Science and
Engineering
291 Daehak-ro, Yuseong-gu
Daejeon 305-701
Republic of Korea

Yang-Il Jung
Korea Atomic Energy Research
Institute
111 Daedeok-daero, 989beon-gil
Yuseong-gu
Daejeon 305-353
Republic of Korea

Suk-Joong L. Kang
Korea Advanced Institute of Science
and Technology
Department of Materials Science and
Engineering
291 Daehak-ro, Yuseong-gu
Daejeon 305-701
Republic of Korea

Reiner Kirchheim
MPI für Eisenforschung
Max-Planck-Str. 1
40237 Düsseldorf
Germany

Aleksander Kostka
MPI für Eisenforschung
Max-Planck-Str. 1
40237 Düsseldorf
Germany

Manja Krüger
Otto-von-Guericke Universität
Magdeburg
Institut für Werkstoff- und
Fügetechnik
Große Steinernetischstr. 6
39104 Magdeburg
Germany

Margarita Kuzmina
MPI für Eisenforschung
Max-Planck-Str. 1
40237 Düsseldorf
Germany

Yujiao Li
MPI für Eisenforschung
Max-Planck-Str. 1
40237 Düsseldorf
Germany

Claire Maurice
SMS Centre
Ecole des Mines de Saint Etienne
158 Cours Fauriel
42032 Saint Etienne
France

Julio Millán
MPI für Eisenforschung
Max-Planck-Str. 1
40237 Düsseldorf
Germany

Yuri Mishin
George Mason University
School of Physics
Astronomy and Computational
Sciences
MSN 3F3
4400 University Drive
Fairfax, VA 22030
USA

Dmitri A. Molodov
RWTH Aachen University
Institute of Physical Metallurgy and
Metal Physics
Kopernikusstr. 14
52056 Aachen
Germany

Debashis Mukherji
Technische Universitat
Braunschweig
Institut für Werkstoffe
Langer Kamp 8
38106 Braunschweig
Germany

Dirk Ponge
MPI für Eisenforschung
Max-Planck-Str. 1
40237 Düsseldorf
Germany

Romain Quey
SMS Centre
Ecole des Mines de Saint Etienne
158 Cours Fauriel
42032 Saint Etienne
France

Dierk Raabe
MPI für Eisenforschung
Max-Planck-Str. 1
40237 Düsseldorf
Germany

Paulo R. Rios
Universidade Federal Fluminense
Escola de Engenharia Industrial
Metalúrgica de Volta Redonda
Av. dos Trabalhadores 420
Volta Redonda, RJ 27255-125
Brazil

Anthony D. Rollett
Carnegie Mellon University
Department of Materials Science and Engineering
5000 Forbes Avenue
Pittsburgh, PA 15213-3890
USA

Joachim Rösler
Technische Universitat Braunschweig
Institut für Werkstoffe
Langer Kamp 8
38106 Braunschweig
Germany

Franz Roters
MPI für Eisenforschung
Max-Planck-Str. 1
40237 Düsseldorf
Germany

Stefanie Sandlöbes
MPI für Eisenforschung
Max-Planck-Str. 1
40237 Düsseldorf
Germany

Lasar S. Shvindlerman
RWTH Aachen University
Institute of Physical Metallurgy and Metal Physics
Kopernikusstr. 14
52056 Aachen
Germany

and

Russian Academy of Sciences
Institute of Solid State Physics
Academician Ossipyan Str. 2
Chernogolovka
142432 Moscow
Russia

Werner Skrotzki
Dresden University of Technology
Institute of Structural Physics
01062 Dresden
Germany

Paul Van Houtte
Katholieke Universiteit Leuven
Department MTM
Kasteelpark Arenberg 44
3001 Leuven
Belgium

Elena Villa
University of Milan
Department of Mathematics
via Saldini 50
20133 Milano
Italy

Alexei Vinogradov
Togliatti State University
Laboratory of Materials and Intelligent Diagnostic Systems
14 Belorusskaya St.
445667 Togliatti
Russia

and

Osaka State University
Department of Intelligent Materials Engineering
Osaka 558-8585
Japan

Tadao Watanabe
Northeastern University
Key Laboratory of Anisotropy and
Texture of Materials
Shenyang 110004
China

Formerly, Tohoku University
Graduate School of Engineering

Permanent address:
4-29-18, Yurigaoka
Natori, Miyagi 981-1245
Japan

Lei Yuan
MPI für Eisenforschung
Max-Planck-Str. 1
40237 Düsseldorf
Germany

Part I
Materials Modeling and Simulation: Crystal Plasticity, Deformation, and Recrystallization

1
Through-Process Modeling of Materials Fabrication: Philosophy, Current State, and Future Directions
Günter Gottstein

1.1
Introduction

Mathematical modeling of physical phenomena is not new science or a recent development but a fundamental ingredient of physical sciences. In fact, mathematics is the language of natural sciences, and it is the objective of physical research to extract from observed phenomena the general behavior in terms of mathematical relations that will allow making quantitative predictions. Physical phenomena are usually described in terms of respective equations of state (thermodynamic considerations) and equations of motion (kinetic considerations). Both types of relations are typically expressed in terms of differential equations, mostly partial differential equations (PDEs). The solution of these equations for usually complex boundary conditions can most commonly not be obtained in closed form, and therefore, the behavior of respective thermodynamic or kinetic systems can only be determined for very special conditions, for example, at the limits of time and space. Forty years ago, owing to the lack of easy to handle closed form solutions particularly engineers refrained from utilizing physics-based concepts, but instead they developed empirical models by fitting simple mathematical functions to obtained data, mostly power law relations for monotonic dependencies, since a power law could still be handled by a slide rule, the typical personal computational tool at that time. Such empirical approaches were actually very accurate as long as the same material was processed the same way, but beyond measured regimes they lacked any predictive power.

With the advent of powerful computers, the situation changed dramatically. Besides the fact that complicated PDEs and complex boundary conditions could now be solved numerically, simulation tools became available to probe virtual materials behavior at any length and time scale. On the macroscopic scale the finite element method (FEM) became the predominant numerical tool for engineers; on the mesoscopic level, the phase-field theory besides Monte Carlo (MC) methods, cellular automata (CA), and front tracking algorithms such as vertex models or level-set methods advanced to established modeling approaches for microstructural evolution of materials. On the atomistic level molecular dynamics

Microstructural Design of Advanced Engineering Materials, First Edition. Edited by Dmitri A. Molodov.
© 2013 Wiley-VCH Verlag GmbH & Co. KGaA. Published 2013 by Wiley-VCH Verlag GmbH & Co. KGaA.

(MD) simulations enabled a large variety of atomistic phenomena to be explored, and eventually density functional theory allowed *ab-initio* quantum mechanical studies of complex atomistic configurations, to name only the most popular approaches. With these computational tools at hand, one does not generate new physics, since the models and tools essentially reflect our understanding of physical phenomena and the underlying mechanisms. Instead, available computational power allows us to address complex phenomena, mutual interaction of different physical processes, and nonsteady-state behavior of physical systems. If we confine our consideration to materials, in particular to crystalline solids, specifically commercial metallic materials, we have now the option to utilize computer power, sophisticated simulation approaches, and advanced numerical algorithms for the prediction of material properties and therefore, for an optimization of materials processing and materials performance in service, in other words we are now able to put 50 years of physical metallurgy to work.

To make reliable predictions of materials behavior one has to understand that, contrary to common believe of engineers, the properties of a material are not controlled by the processing conditions but by chemical composition and microstructure. In other words, there are no processing–property relationships that can be utilized for the prediction of material properties; rather the only state variable of properties is, besides the unchanged overall composition, the microstructure, which is liable to change by thermal and mechanical processing (Figure 1.1). Hence, the prediction of final properties of a material requires to pursue the development of microstructure along the entire processing chain, in principle from solidification through the semifinished product and eventually to a part in service. The simulation of microstructure evolution and therefore of materials properties along the processing chain is referred to as through-process modeling (TPM) in Europe or, more

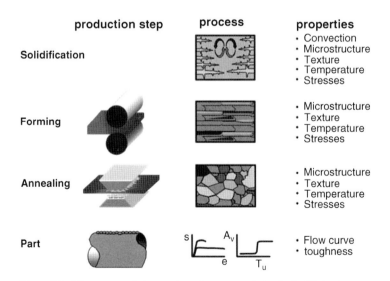

Figure 1.1 Microstructural change along the process chain of tube fabrication.

recently, integrated computational materials engineering (ICME) in the United States. In the following, we will use throughout the term TPM, keeping the identity with ICME in mind.

1.2 Microstructure Evolution

In view of the observed microstructural complexity in commercial materials the prediction of microstructure evolution during processing seems to be an intractable problem. Hence, it seems surprising at first glance that physical metallurgy research of the past 80 years has shown that there are only three processes involved that have to be considered for microstructural change, that is, crystal plasticity, recrystallization and related processes, and phase transformations. Admittedly each of these three microstructural processes is very complex and their mutual interaction can lead to widely different microstructures; the principles have been laid out by metal physics research in the recent past and mathematical concepts exist to address microstructure evolution quantitatively. Also, specific relations have been derived that associate microstructure with material properties.

Most of concepts of microstructural phenomena are formulated in a continuum approach on the mesoscopic level, which is typically of the order of micrometer (Figure 1.2).

The underlying mechanisms proceed on an atomistic level and determine continuum properties like diffusivities, mobilities, and enthalpies. The mesoscopic approach is attached to the macroscopic world by microstructure–property relationships that can be used in FEM simulations of materials processing. The most germane approach to TPM is therefore modeling of microstructure on a mesoscopic scale under processing conditions delivered by FEM simulations. On demand, atomistic simulations are engaged to generate intrinsic material data that are needed in the mesoscopic approach. In the following, we will shortly introduce the essentials of current modeling approaches of respective microstructural processes.

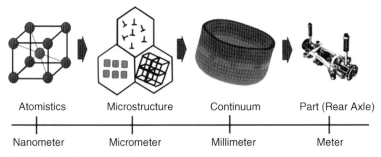

Figure 1.2 Multiscale modeling, macroscopic properties, microstructure, and atomistic processes are defined on different length scales.

1.3
Microstructural Processes

Crystalline solids can deform plastically by crystallographic slip, mechanical twinning, diffusion controlled plasticity, or transformation plasticity. The dominant mechanism of *crystal plasticity* in most commercial materials is crystallographic slip by dislocation motion. During deformation a material undergoes work hardening, that is, dislocation storage, and a change of crystal orientation, that is, texture [1]. The introduced dislocation structure is the microstructural variable of the mechanical properties of the material after deformation, for example, strength. The orientation change leads to the formation of a nonrandom orientation distribution. Whereas modeling of deformation texture is already well advanced and yields reasonable predictions that compare well with experiments, modeling of the deformation microstructure is much more complicated. Even if we neglect the microstructural inhomogeneities that usually accompany cold forming, a prediction of large strain work hardening and dislocation arrangement on the basis of 3D discrete dislocation dynamics is not yet feasible. More powerful in this context are statistical deformation models such as the Kocks–Mecking approach [2] with the total dislocation density as single microstructural state parameter or the 3IVM+ with three internal variables [3] (Figure 1.3).

The latter considers the evolution (production + and reduction −) of dislocation density ρ or even several (n) types of dislocation densities ρ_i with strain ε:

$$d\rho_i = d\rho_i^+ + d\rho_i^- \qquad (i = 1, n) \tag{1.1}$$

and utilize the classical kinetic equation of state for dislocation plasticity (Orowan equation)

$$\dot{\varepsilon} = \rho_m b v \tag{1.2}$$

where $\dot{\varepsilon}$ is the strain rate, ρ_m is the mobile dislocation density, b is the Burgers vector, and v is the dislocation velocity.

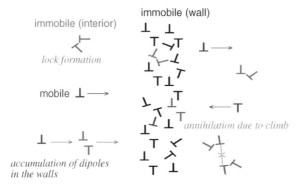

Figure 1.3 The 3IVM+ work hardening model distinguishes three different dislocation densities and their mutual interaction.

The strength σ is then obtained via the Taylor equation

$$\sigma = \alpha G b \sum f_i \sqrt{\rho_i} \qquad (1.3)$$

f_i represents the respective volume fraction associated with the dislocation density ρ_i.

Respective models can approximate the measured flow curves quite well and are able to make reasonable predictions for different deformation conditions and changing alloy composition after optimizing a significant number of unknown physical parameters. While this development is promising, more detailed investigations are necessary to make the models easier to handle and to anticipate the model parameters by theoretical concepts.

Crystallographic slip by dislocation motion proceeds by pure shear and therefore is accompanied by a rotation, that is, a change of crystallographic texture of a polycrystal. The respective orientation changes in a polycrystal can be calculated by a variety of methods, notably by the Taylor approach [4], where it is assumed that each crystal undergoes the same deformation as the macroscopic specimen. More refined models take also the interaction of the grains into account, for example, the Lamel code [5], viscoplastic self-consistent approaches [6] or the grain interaction model (GIA) [7,8]. In the latter, an eight grain aggregate is considered that has grain boundaries in all three spatial directions (Figure 1.4). The entire cluster is forced to comply with the Taylor conditions but the grains in the cluster are allowed to deform freely and impose shears on next neighbor grains across these internal boundaries; however, this incompatibility has to be compensated for by the introduction of geometrically necessary dislocations. An energy minimization yields the activated slip systems and the respective shears. These models effectively reproduce the experimentally obtained crystallographic textures.

Softening phenomena are caused by *recovery, recrystallization*, and *grain growth* [1]. Whereas *recovery* and related phenomena such as continuous recrystallization or recrystallization *in situ* are simply dislocation controlled processes and can be addressed by means of crystal plasticity concepts outlined earlier, recrystallization and grain growth require different modeling approaches.

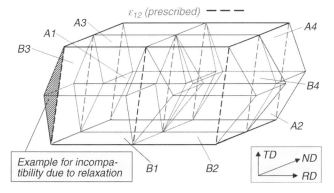

Figure 1.4 The grain cluster model GIA considers an arrangement of eight grains and allows for interaction across their boundaries.

Recrystallization proceeds during annealing of deformed crystals and consists of the nucleation of strain free grains and their growth into the deformed matrix until the entire deformed volume has been consumed. The growth process can be modeled reasonably well with a variety of mesoscopic simulation tools, like Monte Carlo simulations, cellular automata, or phase field methods. The crucial issue in recrystallization modeling is a physical approach of nucleation and, correspondingly, the prediction of nucleus locations and orientations and nucleation kinetics. The latter determines incubation time and nucleation frequency. Owing to the dramatic change of microstructure and thus properties of a material during recrystallization, an inaccuracy in the selection of nucleation kinetics and/or mechanisms is liable to engender a totally wrong prediction of the state of the material during processing. Therefore, substantial efforts have been dedicated to this problem with obvious success but far from a satisfactory solution. While for specific alloys, a prediction of distinct properties such as texture or grain size can be accomplished, it is still beyond our capabilities to predict the recrystallization behavior of a new and untested alloy. For real time and real-space simulations, deterministic modeling tools like cellular automata are most appropriate if modified to accommodate the specific circumstances of recrystallization. For this purpose, the sample volume is discretized and each volume elements can assume discrete states depending on the local environment; with regard to recrystallization the state can switch from "deformed" to "recrystallized." The transformation occurs when a volume element in a deformed state is touched by the grain boundary of a growing recrystallization nucleus, so that the speed of transformation is determined by the growth rate of a grain, that is, the grain boundary velocity

$$v = mp \qquad (1.4)$$

where m is the grain boundary mobility that will depend on misorientation, local chemistry, temperature, and so on, and p is the local driving force, essentially the local stored dislocation energy. To accommodate a locally varying growth rate, advanced CA codes like the smart cellular automata code CORe [9] are capable of adapting its grid size to the local and temporal environment and are also referred to as cellular operators (Figure 1.5).

Like all recrystallization simulation tools CORe is a pure growth model that has to be merged with a nucleation model like the ReNuc [10] code that has recently been

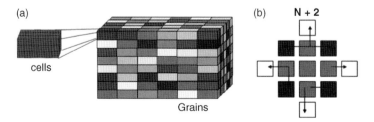

Figure 1.5 Cellular automata modeling of recrystallization; a grain in a polycrystal is subdivided into volume elements (a) that can change their state once touched by a growth front (b).

developed for aluminum alloys to output nucleation rates for typical nucleation mechanisms such as recrystallization at grain boundaries, transition bands, shear bands or at large particles [11]. Given the nucleation frequency and nucleus texture the kinetics of recrystallization, grain size distribution and texture development can be computed.

Recrystallization is succeeded by *grain growth*, which is observed by a coarsening of the grain structure. It is driven by grain boundary curvature and proceeds by grain boundary motion with local migration rate

$$v = m \frac{2\gamma}{R} \tag{1.5}$$

where γ is the specific grain boundary energy and R is the local radius of the boundary curvature [1].

Grain growth can be simulated by a variety of methods such as the phase field method, cellular automata, Monte Carlo simulations, or front-tracking algorithms like the network models. The latter have the advantage that they are deterministic and simulate grain growth in real time and real space [12]. Moreover, they most closely reflect the physical nature of the process in terms of curvature-driven grain boundary motion. A typical vertex model discretizes the grain boundary network into linear segments (2D) or triangular elements (3D) that are bound by virtual vertices (on the boundaries and triple lines in 3D) or real vertices (quadruple points in 3D) for which the displacement in a specific time increment is computed according to Eq. (1.5).

This setup allows flexibility in driving force p and mobility m assignment to the various constituents of a grain boundary network in a polycrystal (Figure 1.6), namely grain boundaries (p_b, m_b), triple lines (p_t, m_t), and quadruple points (p_q, m_q). Since only the boundaries are considered whereas the grain interior remains unconsidered, respective codes are very fast and allow us to address systems of substantial size and thus, cover an essential change of microstructure during grain growth. An example is the 3D virtual vertex model developed at the Institute of Physical Metallurgy and Metal Physics (IMM) of RWTH Aachen University.

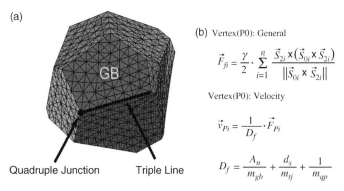

Figure 1.6 Three-dimensional vertex modeling of grain growth; grain boundaries (GB), triple lines, and quadruple points are discretized by vertices (a) that are displaced according to the equation of motion (b) under local conditions.

Nucleation energy: $\Delta G(r_c) = \frac{4}{3} \cdot \pi \cdot r_c^2 \cdot \sigma$

Critical radius: $r_c = \frac{2 \cdot \sigma}{\Delta g_T}$

Growth rate (Zener): $\frac{dr}{dt} = \frac{c(t) - c^\alpha(r)}{c^\beta - c^\alpha(r)} \cdot 0.5 \cdot \left| \frac{c(t) - c^\alpha(r)}{c^\beta - c(t)} \right| \frac{D}{r}$

Nucleation rate (Becker/Döring): $\dot{N} = N_0 \cdot Z \cdot \beta \cdot \exp\left(-\frac{\Delta G(r_c)}{k_B \cdot T}\right)$

$$\frac{\partial f(r,t)}{\partial t} + \frac{\partial}{\partial r}\left(\frac{dr}{dt} \cdot f(r,t)\right) = \dot{N}$$

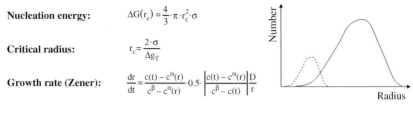

Figure 1.7 Fundamental relations of the ClaNG model; the grain size distribution $f(r,t)$ obeys the continuity equation.

Phase transformations like the α–γ-transformation in steels or precipitation and dissolution phenomena in aluminum alloys are most adequately addressed by the phase field method. However, this method is computationally demanding and requires substantial computer power with long computation times. Moreover, for many, in particular industrial applications, there is no need to obtain highly detailed information on space resolved microstructure evolution, since the relations between microstructure and properties are generally based on statistical average values that can be obtained from much simpler approaches. For instance, precipitation, particle coarsening, or dissolution in aluminum alloys can be addressed successfully by classical nucleation and growth theory as laid out by Becker and Döring 80 years ago. Coupled to a thermodynamic database for a computation of the equilibrium phase diagram and supplemented by the Kampmann–Wagner approach for the evolution of particle size distribution, it renders solute content, particle size, and particle volume fractions as function of annealing time for a given temperature or temperature profile. An example is the ClaNG code [13] (Figure 1.7) that can be readily interfaced to Calphad databases, and its fast computation speed lends itself for interfacing with concurrent processes such as recrystallization and grain growth.[1]

Work hardening, recrystallization, and related phenomena and phase transformations constitute the basis for following and predicting microstructure evolution during materials processing, if connected to a process model that is capable of predicting temporally and locally temperature and strain as usually provided by FEM.

1.4
Through-Process Modeling

In the following, we shall present an example of through-process modeling for the fabrication of the aluminum sheet [14,15]. To begin with, one has to specify which particular property or properties are to be determined. In this example, we want to

---

[1] All the models outlined in this text have been published and respective codes are accessible through SimWeb on the IMM homepage (www.imm.rwth-aachen.de).

1.4 Through-Process Modeling | 11

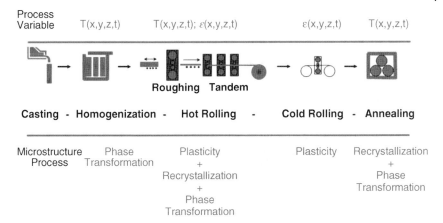

Figure 1.8 Through-process modeling of aluminum sheet fabrication; the process variables are obtained by FEM simulations, the microstructure undergoes changes by three microstructural processes.

predict the earing behavior during deep drawing of a processed sheet. For this, one needs to know the microstructure of the rolled and annealed sheet in terms of crystallographic texture and dislocation structure to predict the behavior during sheet forming. We start out with the cast and homogenized ingot, and the process chain consists of hot rolling, coiling, and multiple cold rolling and annealing until the final sheet thickness is obtained (Figure 1.8).

The process data are given by streamlines from FEM simulations of deformation and temperature for select characteristic sheet locations such as center plane and surface. With the temperature and displacement gradient known as function of time and space along the entire process chain, microstructure evolution in each volume element can be simulated by considering deformation, recrystallization, and precipitation–dissolution in a given volume element for each time increment (Figure 1.9).

During hot rolling the material will undergo deformation in the rolls as well as recrystallization and precipitation during interstand times and during coiling. It is stressed that deformation and recrystallization will impact texture so that texture evolution needs to be followed along the processing chain. During cold rolling work hardening and deformation texture development will occur and subsequent annealing will cause recrystallization and potentially precipitation, particle coarsening, or dissolution that will also affect texture. Eventually, after the last annealing step the final texture and dislocation structure are established.

This information can then serve as input to compute the in-plane strength anisotropy and yield surface as needed for simulation of the deep drawing process. For the aluminum alloy AA5182, this exercise was conducted in cooperation with the Institute of Metal Forming (IBF) of RWTH Aachen University and industrial partners, notably Hydro Aluminium Germany [16]. At defined processing steps,

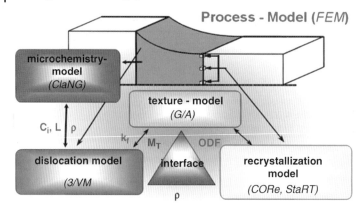

Figure 1.9 In each increment of time, the change of state of an FEM volume element has to be updated with respect to its microstructural changes.

samples could be taken to compare predictions and experimental results. It is emphasized that in this study the computational prediction was made prior to the measurements in order to evaluate the true predictive power of the simulation tools. The results are given in terms of the volume fractions of the major texture components (Figure 1.10).

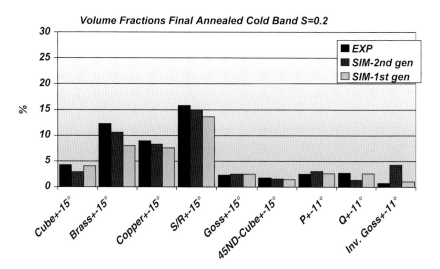

Figure 1.10 Volume fractions of major texture components in the center plane of finally annealed AA5182 sheet after 13 processing steps. Predictions prior to measurements (light gray), refined predictions with experimental results known (dark grey), and experimental results (black).

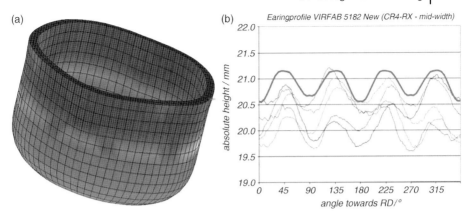

Figure 1.11 Result of FEM simulation of deep drawn cup (a) measured on several samples (thin lines) and computed (bold line) earing profiles of AA5182 sheet. (courtesy IBF)

Apparently, the predictions after 13 processing steps are quite reasonable, although a second simulation, after the experimental results were known, allowed to even better optimize the parameters and rendered some quantitative improvement.

If the computed final texture is input in a subsequent sheet forming FEM simulation, the experimentally observed earing behavior can apparently be predicted with acceptable accuracy (Figure 1.11).

At this point, it is important to realize how the predictive power or the accuracy of a simulation has to be evaluated. Typically, the quality of a simulation is obtained by benchmarking with experimental results, so that the experiment always constitutes the reference. However, one should be aware that also experimental measurements and in particular industrial processes are subject to a certain scatter that has to be taken into account for an assessment of the quality and reliability of a model. In principle, an optimized simulation tool does not lend itself to an error analysis, since the prediction is made exactly according to the assumptions implemented in the code. However, the used fit parameters may be subject to scatter and frequently a multivariate regression analysis of a multiparameter model may yield more than a single set of optimum parameters, which however may render different predictions for different conditions. With respect to the predictive power of the work hardening model 3IVM+, the variance of the predicted flow curve was compared to the expected fluctuation of temperature in an industrial hot rolling process. As obvious from Figure 1.12, the potential variation of the predicted work hardening curve is still within the accepted inaccuracy due to temperature fluctuations, which means that the predictions are in an acceptable range. Of course, for multistep through-process modeling it is important to know, how the permissible deviations in each step affect the final result, but this analysis has always to be related to the potential variations of the simulated industrial process that will still generate an acceptable final product.

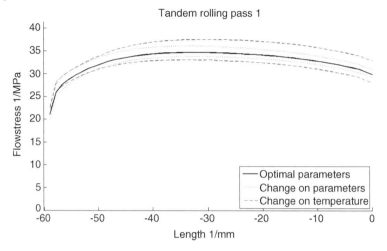

Figure 1.12 Scatter of predicted flow curve (three center curves) and variation due to acceptable temperature fluctuations (top and bottom curve) (courtesy IBF).

1.5
Future Directions

Although the results of the simulation trial were very promising, there is still a long way to go, if more complex processing schemes or more advanced alloys are to be addressed. To correctly predict work hardening with strength contributions of various alloying elements and several constitutive components in an alloy is still a challenge. The effect of particles, on both nucleation of recrystallization and nucleus growth, the problem of orientation-dependent recovery and thus, incubation time of recrystallization, and local fluctuations of Zener drag, need to be refined by respective theoretical concepts to obtain better predictions of recrystallization kinetics and texture.

Most importantly, one should not underrate that all models rely on the knowledge of specific material properties, such as elastic constants, diffusivities, boundary mobilities and boundary energies, and so on that are difficult to obtain experimentally and are likely to depend sensitively on composition since they are known to be seriously affected by alloying elements and misorientation distribution. It is very unlikely that such data can be generated experimentally so that in the future it will be indispensable to employ atomistic simulations for providing these data. While *ab-initio* simulations are promising for computing elastic properties and for providing energetic information [17], today's predictions of grain boundary mobility by MD are still orders of magnitude different from experimental results [18,19]. The same is true for diffusivities, in particular for interdiffusion in multicomponent alloys.

For commercial applications, it is also necessary to develop adaptive interfaces for automation of the computational procedure and for the handling of more complex microstructural features, for example, partially recrystallized microstructures or concurrent precipitation and recrystallization [20].

Finally, it is mentioned that computation times are still by far too long to address large and complex systems. Despite of the availability of seemingly ever increasing computer power simulations that mimic real world conditions take too long. Correspondingly, the developed simulation tools will have only limited acceptance in an industrial environment. While they lend themselves for alloy development in a research laboratory, for process control one would need orders of magnitude higher computational speed. In this context, it is helpful to remember that virtually most of the simulation tools have been designed for single CPU operations. The future of high-performance computing, however, calls for codes that can make use of massive parallel computer architectures. It is also not of much gain to optimize a serial code for parallel computation. If really a quantum leap in computational speed is to be obtained it will be necessary to redevelop basic physical models and to utilize modern mathematical tools to completely separate individual computational procedures and still extract from the separately obtained output the desired physical information.

It is important to realize that through-process modeling is designed to predict material development and the terminal state of a material for a given process chain. This is highly valuable information since computer simulations do speed up alloy and process development and are much more cost-effective than traditional empirical approaches. However, true materials design calls for the inversion of the procedure, that is, poses the question: What material and process one has to choose for obtaining a material for optimum performance for a specific application? This so-called inverse modeling is certainly much more difficult to perform than the current – already highly demanding – through-process modeling activities. However, there is no time and reason to wait until all open questions of microstructure evolution and through-process modeling have been solved, since substantial efforts for a development of inverse simulation approaches and adequate mathematical tools will be necessary to take the next step.

Computer simulations have long faced skepticism and criticism of scientists who prefer the classical approach of analytical modeling. It is certainly true that one cannot generate new physics on the computer since the computer can only work in the frame of the encoded physical models. Computational materials science however, means that one can handle complexity that cannot be addressed analytically. This is the true benefit of computer simulation of materials, and with increasing computational modeling power there is hope and expectation that we shall be able to make virtual materials design and virtual materials engineering eventually come true.

Acknowledgments

The author is indebted to the Deutsche Forschungsgemeinschaft for continued funding of through-process modeling activities at IMM in the framework of the collaborative research center SFB 370 "Integral Materials Modeling" and transfer program TFB 63 "Industrially Relevant Modeling Tools." I am grateful to Hydro

Aluminium Germany for continued support and encouragement as well as for providing processed material. Special thanks go to the Institute of Metal Forming (IBF) of RWTH Aachen University for conducting the FEM computations and for many years of excellent cooperation. Finally I like to acknowledge all my doctoral students and post-docs who were engaged in through-process modeling programs at IMM, notably Volker Mohles, Luis Barrales-Mora, Mischa Crumbach, Matthias Goerdeler, Carmen Schäfer and many more. Finally, I want to thank Matthias Loeck, head of our computer department for his personal engagement in code development, SimWeb design and his valuable support of computational procedures at IMM.

References

1 Gottstein, G. (2004) *Physical Foundations of Materials Science*, Springer, Berlin.
2 Mecking, H. and Kocks, U.F. (1981) Kinetics of flow and strain hardening. *Acta Metallurgica*, **29**, 1865–1875.
3 Roters, F., Raabe, D., and Gottstein, G. (2000) Work hardening in heterogeneous alloys: A microstructural approach based on three internal state variables. *Acta Materialia*, **48**, 4181–4189.
4 Taylor, G.I. (1938) Plastic strain in metals. *Institute of Metals*, **662**, 307–324.
5 Van Houtte, P., Delannay, L., and Samajdar, I. (1999) Quantitative prediction of cold rolling textures in low-carbon steel by means of the LAMEL model. *Textures and Microstructures*, **31**, 109–149.
6 Molinari, A., Canova, G.R., and Ahzi, S. (1987) A self consistent approach of the large deformation polycrystal viscoplasticity. *Acta Metallurgica*, **35**, 2983–2994.
7 Crumbach, M., Pomana, G., Wagner, P., and Gottstein, G. (2001) A Taylor type deformation texture model considering grain interaction and material properties: Part I. fundamentals, in *Recrystallization and Grain Growth* (eds G. Gottstein and D. Molodov), Springer, Berlin, pp. 1053–1060.
8 Crumbach, M., Pomana, G., Wagner, P., and Gottstein, G. (2001) A Taylor type deformation texture model considering grain interaction and material properties: Part II. Experimental validation and coupling to FEM, in *Recrystallization and Grain Growth* (eds G. Gottstein and D. Molodov), Springer, Berlin, pp. 1061–1068.
9 Mukhopadhyay, P., Loeck, M., and Gottstein, G. (2007) A cellular operator model for the simulation of static recrystallization. *Acta Materialia*, **55**, 551–564.
10 Crumbach, M., Goerdeler, M., and Gottstein, G. (2006) Modelling of recrystallization textures in aluminium alloys: I. Model set-up and integration. *Acta Materialia*, **54**, 3275–3289.
11 Schäfer, C., Song, J., and Gottstein, G. (2009) Modeling of texture evolution in the deformation zone of second-phase particles. *Acta Materialia*, **57**, 1026–1034.
12 Barrales Mora, L.A., Gottstein, G., and Shvindlerman, L.S. (2008) Three-dimensional grain growth: Analytical approaches and computer simulations. *Acta Materialia*, **56**, 5915–5926.
13 Schneider, M., Gottstein, G., Löchte, L., and Hirsch, J. (2002) A statistical model for precipitation – applications to commercial Al–Mn–Mg–Fe–Si alloys, in *Aluminium Alloys. Their Physical and Mechanical Properties* (eds P.J. Gregson and S.J. Harris), Trans Tech Publications, Switzerland, pp. 637–642.
14 Gottstein, G. (ed.) (2007) *Integral Materials Modeling: Toward Physics Based Through-Process Models*, Wiley-VCH, Weinheim.
15 Hirsch, J. (ed.) (2006) *Virtual Fabrication of Aluminum Products*, Wiley-VCH, Weinheim
16 Crumbach, M., Goerdeler, M., Gottstein, G., Neumann, L., Aretz, H., and Kopp, R. (2004) Through-process texture modelling

of aluminium alloys. *Modelling and Simulation in Materials Science and Engineering*, **12**, S1–S18.
17 Counts, W.A., Friák, M., Raabe, D., and Neugebauer, J. (2009) Using *ab initio* calculations in designing bcc Mg–Li alloys for ultra-lightweight applications. *Acta Materialia*, **57**, 69–76.
18 Schönfelder, B., Gottstein, G., and Svindlerman, L.S. (2006) Atomistic simulations of grain boundary migration in copper. *Metallurgical and Materials Transactions*, **37**, 1757–1771.
19 Zhou, J. and Mohles, V. (2011) Towards realistic molecular dynamics simulations of grain boundary mobility. *Acta Materialia*, **59**, 5997–6006.
20 Schäfer, C., Mohles, V., and Gottstein, G. (2011) Modeling of non-isothermal annealing: Interaction of recrystallization, recovery, and precipitation. *Acta Materialia*, **59**, 6574–6587.

2
Application of the Generalized Schmid Law in Multiscale Models: Opportunities and Limitations

Paul Van Houtte

2.1
Introduction

Crystallographic textures [1–3] play an important role in understanding processing and final properties of metallic products. The latter can partly or entirely be understood from the chemical and physical properties of the material. One of these is the crystallographic texture of each of the phases of the material, that is, the set of orientations of the crystallographic lattices of the grains. The texture of the final product is a major source of anisotropy of physical and mechanical properties of the material. The texture of an intermediate product affect events during further processing steps such as annealing or hot or cold forming. Last but not least the textures change during processing steps. Combining these insights leads to a conclusion that there may be a two-way interaction between textures and processing; the course of the process depends on the current texture, but at the same time modifies the texture. The study of the interaction between texture, processing, and final material properties is a subject that has always been close to the heart of Professor Günter Gottstein.

In this chapter, there is no room for an introduction to crystallographic textures in metals. A nice general introduction can be found in the book by Kocks, Tomé, and Wenk [2]. Bunge's book [1] gives the classical mathematical methods for texture analysis. For an introduction in the recent developments in local texture analysis, the book by Engler and Randle [3] is recommended. In the following, the study of the interaction between texture, processing, and final material properties will be discussed, but only for the case of metal forming at room temperature. In the next section, it will be reminded how the plastic response of a single grain can be understood and modeled. This then enables the discussion of two-scale models for the plastic deformation of polycrystalline materials in the following section. This will include a short section about modeling the plastic anisotropy of polycrystalline materials. Finally a section will be devoted to the experimental validation of predicted deformation textures and associated mechanical anisotropy. Furthermore, note that models like these can also be used as starting basis to model other features of the microstructure, such as local textures and/or the heterogeneity at different length scales of lattice orientations (e.g., lattice curvature) and defect densities (e.g., density of geometrically necessary dislocations).

2.2
Crystal Plasticity

Plastic deformation of metallic crystals requires re-arrangement of the atoms. This means that after a large plastic deformation, most atoms will have different neighbors. The main mechanism that makes this possible is dislocation glide. The book by Professor Gottstein [4] offers a very good introduction to this. Some highlights will be reminded here. It was concluded from early study of deformed crystals that under applied shear stress, one half of a crystal could slide over the other half along certain crystallographic planes, called "slip planes." Attempts to calculate the stress required for this failed in the sense that the stress levels found were orders of magnitude larger than the experimentally measured ones. This was explained later by the theory that two crystallographic planes do not slide upon each other as a whole. This happens step-by-step with the help of a large number of line defects, called "dislocations," which travel through the crystal. The combined effect of all these events effectively results in a displacement of one crystallographic plane along the adjacent one in a particular direction that is fixed, called "slip direction." The process is called "dislocation glide." A combination of slip direction (characterized by a unit vector **b**) and slip plane (characterized by its normal **m**, also a unit vector) is called a "slip system" (Figure 2.1). In this figure, **b** is parallel to the axis x_2 and **m** to the axis x_3. The component τ_{23} of the local stress is the driving force to move the dislocation, called "resolved shear

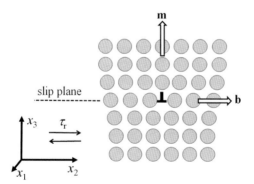

Figure 2.1 An "edge" dislocation indicated by the symbol ⊥ is shown. It travels along a *slip plane* under the action of a local shear stress. In a given material, this can only happen on slip planes that belong to certain families of crystallographic planes characterized by their Miller indices (*hkl*) [4]. The vector **m** shown in this figure is a unit vector normal to such family of parallel planes. The *Burgers vector* is another physical quantity used to characterize the line defect along the dislocation [4]. The vector **b** is a vector with unit length that has the same direction as the Burgers vector. It will be called the *slip direction* in this paper, and it must also belong to certain crystal orientation characterized by Miller indices [*uvw*]. The vectors **m** and **u** must be orthogonal to each other.

stress" τ_r. When it is positive and sufficiently large (it must reach τ_c, which is called the "critical resolved shear stress" or CRSS), the dislocation will start moving to the right until it reaches either an obstacle or the end of the crystal. One then gets the impression that the top half of the crystal has slipped one atomic spacing to the right with respect to the lower half. Suppose that during a time interval Δt many dislocations have slipped over a series of slip planes that are parallel to each other. When contemplating a much larger portion of the crystal (e.g., with a size of $(1\,\mu m)^3$), one would get the impression that a homogeneous shear has taken place.

Note that the dislocation shown in Figure 2.1 is a so-called edge dislocation. Other types of dislocations are also introduced in the book of Professor Gottstein [4]. However, the following model remains valid for all these types.

- In each type of metallic crystals, there is a limited number of combinations of slip planes **m** and slip directions **b** that define a "slip system."
- The movement of any type of dislocations along these slip systems causes a simple shear solely described by **m** and **b** and by the "amount of slip" γ.
- It is then convenient to work with the shear (or slip) rate, that is, the shear per unit time. The plastic shear rate caused by dislocations moving along a certain slip system (the one with the sequence number s) is mathematically described by a velocity gradient tensor $\mathbf{l}^s$.

$$l^s_{ij} = b^s_i m^s_j \dot\gamma^s \qquad (2.1)$$

in which $\dot\gamma^s$ is called the slip rate acting on the slip system.[1] Note that l^s_{ij}, the description of $\mathbf{l}^s$ used here, is expressed in a reference frame that is fixed to the crystal lattice. Figure 2.2 shows the slip system families that are agreed upon by many researchers for cubic metals, namely 12 in fcc metals with four slip planes of the $\{1\,1\,1\}$ type, each combined with three of the six available $\langle 1\,1\,0\rangle$ type slip directions. In bcc metals there are 24 slip systems, with four slip directions of the $\langle 1\,1\,1\rangle$ type, combined with either slip planes of the $\{1\,1\,0\}$-type or of the $\{1\,1\,2\}$-type.

Note that the symmetric part of a velocity gradient tensor l_{ij} is called the "strain rate tensor" d_{ij}.

$$d_{ij} = \frac{1}{2}(l_{ij} + l_{ji}) \qquad (2.2)$$

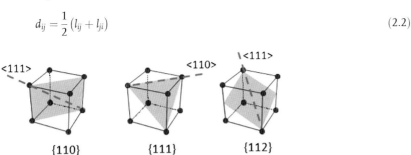

Figure 2.2 Slip plane families $\{hkl\}$ (gray surfaces) and slip direction families $\langle uvw\rangle$ in cubic crystals.

---

1) Note that the Einstein summation convention is NOT used in this paper.

2.2.1
Generalized Schmid Law

Consider a crystal on which an increasing tensile stress σ is applied. The "resolved shear stress" τ_r on the slip system with sequence number s will be equal to $(\cos \kappa^s \cos \lambda^s)\sigma$, in which κ^s and λ^s are the angles between the tensile axis and the vectors **m** and **b** belonging to that slip system, respectively (see Figure 6.30 in [4]). In the beginning, when τ_r is small, its absolute value will not reach the CRSS τ_c on any of the slip systems; however, after some time, this value will be reached on a first slip system, on which the dislocations will then begin to move. This first slip system will be the one for which $m = \cos \kappa \cos \lambda$ has the highest value, at least when the CRSS are equal on all available slip systems. This is called the Schmid law [4]. It is possible that at the time when a first slip system is activated, by coincidence the CRSS is also reached on other slip systems. In such case, several slip systems are activated simultaneously.

A similar analysis is possible for multiaxial loading of the crystal. The applied stress tensor is then more complicated and so is the formula for the resolved shear stress. Let us derive this formula here. A reference frame $\langle x_i^s \rangle$ is defined for the slip system s, with x_1^s being parallel to $\mathbf{m}^s$ and x_2^s to $\mathbf{b}^s$. Let the co-ordinate transformation from the sample reference frame to the slip system reference frame $\langle x_i^s \rangle$ be described by the matrix $\left[a_{ij}^s\right]$. Note that a_{ij}^s is the cosine of the angle between the axis i of the reference frame of the slip system s and the axis j of the sample reference frame [5]. The local stress is now expressed as follows in the slip system reference frame:

$$\sigma_{ij}^s = \sum_k \sum_l a_{ik}^s a_{jl}^s \sigma_{kl} \tag{2.3}$$

Taking the definition of the axes of the slip system reference frame into account, we now find for the resolved shear stress.

$$\tau_r^s = \sigma_{12}^s = \sum_k \sum_l a_{1k}^s a_{2l}^s \sigma_{kl} \tag{2.4}$$

Because of the definition of the unit vectors $\mathbf{m}^s$ and $\mathbf{b}^s$, we also have $m_k^s = a_{1k}^s$ and $b_k^s = a_{2k}^s$. This combined with the symmetry of the stress tensor leads to the classical formula

$$\tau_r^s = \sum_k \sum_l M_{kl}^s \sigma_{kl} \tag{2.5}$$

in which the M_{kl}^s are the components of the so-called Schmid tensor **M** associated with the slip system s.

$$M_{kl} = \frac{1}{2}(m_k b_l + b_k m_l) \quad \text{or, in tensor notation}: \quad \mathbf{M} = \frac{1}{2}(\mathbf{m} \otimes \mathbf{b} + \mathbf{b} \otimes \mathbf{m}) \tag{2.6}$$

As described earlier, the first active slip system will be the one on which the CRSS will be reached first by the absolute value of the resolved shear stress during an increase of loading. This is called the generalized Schmid law (see also [2]).

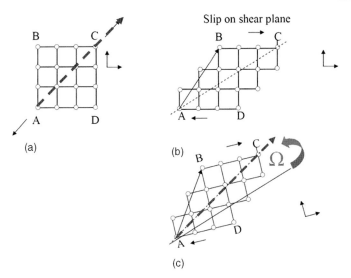

Figure 2.3 (a) A hypothetical crystal with a primitive cubic lattice is subjected to a tensile load along the axis AC. Suppose that the available slip planes are parallel with AD or BC. (b) Slip has taken place, and the crystal underwent a simple shear deformation. So far, it is assumed that the orientation of the crystal lattice has not changed. Note that this implies that the tensile axis AC has rotated. (c) However, the external boundary conditions, which are dictated by the tensile testing machine, do not allow a rotation of the tensile axis. The rotation Ω compensates for it. It is seen that it changes the orientation of the crystal lattice.

The parameter $\dot{\gamma}^s$, that is, the slip rate on the slip system s, will now be used to characterize the amount of moving dislocations on the slip system s multiplied by the distance traveled by them per unit time. Equations (2.1) and (2.2) give the contribution of such slip rate to the velocity gradient tensor and the strain rate tensor of the crystal, respectively.

After some time, such slip rate will cause a finite shear deformation. Figure 2.3a and b illustrate this for a tensile test in which only one slip system is activated. Without rotation of the crystal lattice, the line connecting the points AC (which are lying on the axis of the tensile testing machine, i.e., the connection between the clamps) seem to rotate, which is conflicting with the external boundary conditions, as in reality, the axis of the machine is not rotating. Hence, this external boundary condition can be used to capture the lattice rotation that results from the plastic deformation (Figure 2.3c).

2.2.2
Calculation of Slip Rates, Lattice Rotation, and Stress from a Prescribed Deformation

Let **l** be the velocity gradient tensor that describes the deformation applied on the material per unit time. Its components are l_{ij} in a reference frame fixed to the crystal lattice. d_{ij} is its symmetrical part, called the strain rate tensor.

Equation (2.1) gives the contribution of an active slip system, the velocity gradient of the crystal. All these contributions have to be added:

$$l_{ij} = \sum_{s \in \text{all slip systems}} b_i^s m_j^s \dot{\gamma}^s \tag{2.7}$$

in which it is taken care of that $\dot{\gamma}^s = 0$ for the nonactive slip systems. The symmetrical part of this equation is

$$d_{ij} = \sum_{s \in \text{all slip systems}} M_{ij}^s \dot{\gamma}^s \tag{2.8}$$

Note that the elastic part of the strain is usually neglected in this paper unless explicitly stated otherwise.

Equation (2.8) formally stands for nine equations (all combinations of i and j), but this is redundant, as the equations for ij and ji are numerically identical because of the symmetry of the strain rate tensor and the Schmid tensors. Hence, there are only six independent equations. However, there is more redundancy. d_{ij} are plastic strain rates, for which

$$d_{11} + d_{22} + d_{33} = 0 \tag{2.9}$$

holds, because in nonporous metals, the volume is conserved upon plastic deformation. On the other hand, the fact that for a given slip system the vectors **m** and **b** are perpendicular to each other leads to

$$\sum_{i=1}^{3} b_i^s m_i^s = \sum_{i=1}^{3} M_{ii}^s = 0 \tag{2.10}$$

Because of Eqs. (2.9) and (2.10), the equations for $ij = 11, 22,$ and 33 in Eq. (2.8) are not independent of each other. One could deal with this by, for example, leaving the equation for $ij = 33$ out, with only five independent equations left over. Instead, it is better to contract the double indices ij used in these equations to a single index p in order to remediate the two types of redundancy. There are several ways to do this. In his polycrystal deformation software, the present author has used the method explained in Appendix A1.2 of [6]. p has a range from 1 to 5. The contraction is such that the following property holds:

$$\sum_{k=1}^{3} \sum_{l=1}^{3} A_{kl} B_{kl} = \sum_{p=1}^{5} A_p B_p \tag{2.11}$$

in which **A** and **B** are symmetric tensors of the type that have the property given by Eq. (2.9), such as strain rates, stress deviators, and Schmid tensors. Equation (2.8) now becomes

$$d_p = \sum_{s \in \text{all slip systems}} M_p^s \dot{\gamma}^s \quad \text{with} \quad p = 1, 5 \tag{2.12}$$

If d_p is prescribed, this is a set of five linear equations in which $\dot{\gamma}^s$ is unknown. In the previous section, it was indicated that there are 12 slip systems available in fcc metals and 24 in bcc metals. So it would be easy to find the slip rates $\dot{\gamma}^s$ by solving the system of five linear equations given by Eq. (2.12) if one would know in advance which five slip systems would be active. Unfortunately, this is usually not known in advance. The so-called rate insensitive methods find this out by using the generalized Schmid law. It requires that the absolute value of τ_r, the resolved shear on a slip system, must never be higher than τ_c, its CRSS. The latter depends on the internal state of the material. However, on active slip systems, the absolute values of τ_r must be *equal* to the CRSS.[2] The problem is then reduced to finding a stress tensor for the crystal that activates at least five slip systems but overstresses none. Bishop and Hill [7,8] have proposed a solution method for this problem based on the Maximum Work Principle. This method makes use of a prescribed strain rate tensor d_{ij}. A stress tensor can then be proposed that will activate one or several equivalent sets of five slip systems. The slip rates of all other slip systems are then forced to zero, which makes the solution of Eqs. (Eqs. (2.8)–(2.12) easier in order to find the slip rates on the active slip systems. Bishop and Hill [7,8] have also demonstrated that their method is strictly equivalent to the Taylor [9] model that would solve Eq. (2.12) for all combinations of 5 out of 12 slip systems for fcc metals or 24 for bcc metals to obtain the slip rates $\dot{\gamma}^s$. A set of slip rates is found for each of these combinations (792 in the case of fcc metals). The one that gives the smallest value for the dissipated frictional work per unit time ($\dot{W}$) is selected.

$$\dot{W} = \sum_{s \in \{\text{active slip systems}\}} \tau_c^s |\dot{\gamma}^s| = \text{Min} \qquad (2.13)$$

in which τ_c^s is the CRSS of the slip system s. Van Houtte *et al.* [10] have given a short overview of these methods. Equations (2.12) and (2.13) constitute a problem of minimization subject to linear constraints, which can be solved by means of linear programming [11]. It is found that the conditions that must be satisfied in order to have a minimum are in fact identical to the generalized Schmid law [10].

The so-called rate sensitive or visco-plastic methods offer an alternative to these solution methods. The generalized Schmid law is not used as such. Instead, a viscoplastic law relating to the resolved shear stress on a slip system to the slip rate on the same slip system is adopted [12]. The results of these methods are nevertheless not very different from those obtained by the "rate insensitive methods" [10].

Suppose that only the strain rate tensor d_{ij} is prescribed, and that it is intentionally chosen in such way that the result of the methods discussed earlier consists of only one nonzero slip rate $\dot{\gamma}^s$. As described earlier, the velocity gradient tensor $\mathbf{l}_s$ can be expressed in a reference frame that is fixed to the crystal lattice by the matrix l_{ij}^s. Since both s and $\dot{\gamma}^s$ are now known, $\mathbf{l}^s$ can be obtained by means of Eq. (2.1). This is graphically illustrated by Figure 2.3a and b for a time increment Δt: a simple shear strain is achieved, and the crystal lattice is not rotated. In fact, the boundary conditions of the tensile test considered in the figure cannot be satisfied unless it is assumed that there may also be a nonzero lattice rotation Ω (Figure 2.3c).

---

2) It is possible that on a given slip system, the value of the τ_c depends on the direction of slip. In that case, the two directions of slip of every slip system should somehow be treated as separate slip systems.

In modeling practice, this is implemented by working with a velocity gradient tensor (prescribed by the boundary conditions) $\mathbf{l}$ described by a matrix l_{ij} which is not expressed in a frame that is co-rotational with the crystal lattice. Rather it would be expressed in some frame that is co-rotational with the frame $\langle x_i \rangle$ in which the boundary conditions that are imposed on the crystal by its surroundings are known. As a result, Eq. (2.7) must be replaced by

$$l_{ij} = \sum_{s \in \text{all slip systems}} b_i^s m_j^s \dot{\gamma}^s + \dot{w}_{ij} \tag{2.14}$$

in which $\dot{w}_{ij}$ is a skew-symmetrical velocity gradient tensor that expresses the rate of rotation of the crystal lattice. Its symmetrical part is zero; hence, as before, Eq. (2.8) is the symmetrical part of Eq. (2.14).

Assuming that l_{ij}, the velocity gradient tensor of the crystal, is known from the boundary conditions, Eq. (2.14) makes it possible to calculate $\dot{w}_{ij}$ once the slip rates have been obtained by one of the methods described earlier. Time integration of $\dot{w}_{ij}$ can then be used to calculate the evolution of the crystal orientation with strain.

A problem is that rate-insensitive methods often do not find unique solutions for the slip rates and for $\dot{w}_{ij}$. The rate sensitive method chooses one of the possible solutions without rock-hard guarantee that it is the best choice according to physics. The problem occurs when the CRSS have the same value on all available slip systems, as in annealed and not yet deformed material. But it is surprising that this issue, which is pertinent for the classical Taylor theory, has been found to be much less problematic when more advanced theories are adopted [10].

Once five or more active slip systems have been identified, the generalized Schmid law (Eq. (2.5)) can be used to find the stress components. Note that σ'', the hydrostatic component of the stress, cannot be found in this way for plastically incompressible materials; hence, there is an additional equation to Eq. (2.5).

$$\sigma_{11} + \sigma_{22} + \sigma_{33} = 3\sigma'' \tag{2.15}$$

in which the hydrostatic component σ'' can be chosen arbitrarily. One could use the value 0, and then the σ_{ij} would represent the stress deviator. So, if five slips systems are active, one has five equations (2.5) plus Eq. (2.15), from which the six σ_{ij} can be obtained. This method is not ruined in the case of nonuniqueness of the slip rates: each of the possible solutions would yield the same value for the stress.

2.2.3
Taylor Factor

The CRSS may be different for every slip system. For annealed material, one might adopt the following simplification:

$$\tau_c^s = \alpha^s \tau_c \tag{2.16}$$

in which the α^s represents "variations" upon the reference critical resolved shear stress τ_c. These α^s are also called "critical resolved stress ratios" (CRSSR). They may all be equal to 1, or they may depend on the slip system family to which they belong.

In bcc metals, two families are often considered: those with {1 1 0} slip planes and those with {1 1 2} slip planes. For the first family, crystal symmetry arguments can be used to demonstrate that α^s must also be equal for slip in a positive or negative direction on a given slip system.

Anyway, if all α^s are equal to unity, according to Eq. (2.13), $\dot{W}$, the plastic work dissipated per unit time and unit volume, scales with the reference critical resolved shear stress τ_c. Moreover, if one multiplies the prescribed strain rate tensor with a factor η then all the slip rates $\dot{\gamma}^s$ will be increased likewise and so will $\dot{W}$. It is then customary to introduce a scalar measure (or "magnitude") for the strain rate, namely the von Mises equivalent strain rate.

$$d_{vM} = \sqrt{\frac{2}{3}\sum_i\sum_j (d_{ij})^2} \tag{2.17}$$

The so-called Taylor factor of a single grain is now defined as follows:

$$M = \frac{\dot{W}}{\tau_c d_{vM}} \tag{2.18}$$

in order to neutralize the scaling of $\dot{W}$ with τ_c and d_{vM}. It then follows that

$$M = \sum_{s\in\{\text{active slip systems}\}} \alpha^s \left|\frac{\dot{\gamma}^s}{d_{vM}}\right| \tag{2.19}$$

in which the α^s are often set to unity. The Taylor factor neither scales with the magnitude of the prescribed strain rate d_{vM} nor with the value of the reference CRSS τ_c. It solely represents the plastic anisotropy, that is, the effect of the crystal orientation upon the work needed to deform the crystal. In simple crystal deformation models, work hardening due to dislocation accumulation is incorporated into the critical resolved shear stress τ_c, whereas plastic anisotropy is implemented through the Taylor factor. In a given material, the Taylor factor may be treated as a function of

1) the crystal orientation g [1];
2) the ratio $\frac{d_{ij}}{d_{vM}}$.

2.3
Polycrystal Plasticity Models for Single-Phase Materials

Consider a "representative volume element" (RVE) of the macroscopic length scale in a single-phase polycrystalline material. It consists of a few hundreds or thousands of grains. However, it is only a "point" at the engineering length scale, that is, the length scale used in engineering finite element models for metal forming. The crystallographic texture of the material inside the RVE ("macroscopic texture") must be representative for the texture of the material at the point considered. In this section, models will be discussed for the plastic deformation of such RVE. It is assumed that boundary conditions (applicable to the entire macroscopic RVE) are known, and that they take the form of a prescribed velocity gradient tensor **L**, called the "macroscopic velocity gradient tensor." The average stress of the RVE will be called "macroscopic stress."

2.3.1
Sachs Model

Bishop and Hill [7] have identified two limit models for polycrystalline materials. One of them assumes that the stress is homogeneously distributed throughout the polycrystal. Further analysis would show that except in special cases, this model would predict that only one grain deforms plastically, namely the one with the highest Schmid factor, that is, with the most favorable crystal orientation for plastic deformation. This model, to which no more attention will be paid in this paper, is called "modèle statique" in the French scientific literature. Note that this model is different from the Sachs model [13], which *does not* assume a uniform stress field in the polycrystal. Sachs considered a tensile test on a polycrystalline material. His intention was to use the Schmid law to find the first activated slip system in every grain, as if they were stand-alone crystals. However, he observed that every grain needs a different stress level to achieve plastic deformation because they usually have different Schmid factors because of different crystal orientations. He then assumed the existence of stress interactions between the grains, leading to local fluctuations as to obtain the right stress to activate a slip system in every one of the crystallites. Hence, he introduced the notion of stress interactions between grains many years before anyone else. However, the lateral strains as predicted by this model are too heterogeneous. The displacement fields of neighboring grains would be unrealistically incompatible when observed in a transverse section of the tensile test sample. This explains the usual poor quality of the deformation texture predictions by the Sachs model.

2.3.2
Taylor Theory

Taylor [9] proposed to assume that the plastic strains of all grains are the same and equal to the macroscopic plastic strain. Today, this is usually formulated as follows.

- Elastic strains are neglected; the velocity gradient tensor l_{ij} describing the rate of plastic deformation in a given grain is equal to the macroscopic velocity gradient tensor L_{ij}.
- The latter is prescribed, which in practice means: derived from the boundary conditions at the macroscopic length scale, that is, the boundary conditions valid for the macroscopic RVE:

$$l_{ij} = L_{ij} \tag{2.20}$$

One of the procedures explained earlier in Section 2.3 can now be applied to each of the grains of the RVE, see for example [10] for more details. The stress σ_{ij} in all these grains is obtained as well as the velocity gradient tensor $\dot{w}_{ij}$ that describes the rate of rotation of the crystal lattice. This makes it possible to estimate the change of the lattice orientation during a strain increment, and hence predict the texture evolution during the strain increment. By repeatedly doing this for a series of such increments, a deformation texture can be predicted. The stresses predicted by this model are not in equilibrium at the grain boundaries. Yet the agreement between measured and deformation textures predicted by this method is better than for the Sachs model. The preferred orientations are reasonably well predicted, but their intensities are not correct [10]. It seems nevertheless that neglecting the stress equilibrium is less detrimental than neglecting strain compatibility, as has been discussed by Kocks [14].

2.3.3
Relaxed Constraints Taylor Theory

The classical Taylor model described earlier is often called the *full-constraints (FC) Taylor theory* because it requires all components of the velocity gradient tensor to be homogeneously distributed throughout the macroscopic RVE. The so-called relaxed constraints (RCs) Taylor theories are less strict: not all nine conditions represented by Eq. (2.20) must be satisfied in each grain. The condition for $ij = 13$ ($1 = $ RD and $3 = $ ND in case of rolling) might, for example, be dropped (Figure 2.4). Now, l_{13} is a simple shear. Let $\mathbf{b}^{RLX}$ be a unit vector in the shear direction (RD in Figure 2.4) and $\mathbf{m}^{RLX}$ a unit vector that is normal to the shear plane (ND in Figure 2.4). A mathematically convenient way to implement this in Taylor software is to introduce an additional (fictitious slip) system. By analogy to Eq. (2.1), it would have the following contribution to the velocity gradient of the crystal:

$$l_{ij}^{RLX} = b_i^{RLX} m_j^{RLX} \dot{\gamma}^{RLX} \tag{2.21}$$

Similarly, it would have a Schmid tensor as defined by Eq. (2.6). Index contraction as applied to the "regular" Schmid tensors is also possible. Equation (2.12) would become

$$d_p = M_p^{RLX} \dot{\gamma}^{RLX} + \sum_{s \in \text{all slip systems}} M_p^s \dot{\gamma}^s \quad \text{with} \quad p = 1, 5 \tag{2.22}$$

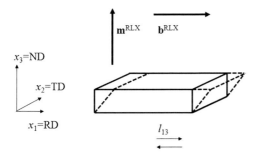

Figure 2.4 Schematic representation of a flattened and elongated grain in a rolled sample. The shear l_{13} is "relaxed."

As before, d_p is equal to the macroscopic strain rate of the RVE. Let τ_c^{RLX} be the critical resolved shear stress that might be associated with this pseudoslip system. Equation (2.13) would now become

$$\dot{W} = \tau_c^{RLX}|\dot{\gamma}^{RLX}| + \sum_{s \in \{\text{active slip systems}\}} \tau_c^s|\dot{\gamma}^s| = \text{Min} \qquad (2.23)$$

Now suppose that one sets τ_c^{RLX} to zero. It is seen in Eq. (2.23) that in that case, $\dot{\gamma}^{RLX}$ would have no direct contribution to the plastic work $\dot{W}$ that is to be minimized. The linear programming software would find a solution to Eqs. (2.22) and (2.23) with a smaller value for $\dot{W}$ than would be found by the full-constraints Taylor theory (Eqs. (2.12) and (2.13)). Indeed, the addition of $\dot{\gamma}^{RLX}$ to the right-hand side of Eq. (2.22) gives the latter more versatility to comply with the (prescribed) left-hand side, and this without "cost," as $\dot{\gamma}^{RLX}$ does not contribute to $\dot{W}$ in Eq. (2.23). Another implication of this is most interesting. In a case like this, the pseudoslip system will always be active, since it is the "softest" of all slip systems in terms of plastic work. As explained in Section 2.3, the generalized Schmid law holds for all active slip systems, hence, also for the pseudoslip system. This is a mere mathematical matter. But it means in practice that the resolved shear stress, τ_r^{RLX}, obtained from the stress as given by Eq. (2.5), needs to be equal to its critical value τ_c^{RLX}. The latter had been set to zero; hence, τ_r^{RLX} needs to be zero too. Now, this shear stress is nothing else as the shear stress corresponding to the shear component represented by the pseudo-slip system (Figure 2.4). So, relaxing a shear strain as is done here is equivalent to setting the corresponding shear stress to zero.

The "pancake model" (see, e.g. [11]) used to be a popular relaxed constraints model for rolling. Not only the ND-RD shear was relaxed (l_{13} in Figure 2.4), but also the ND-TD shear (which would be l_{23} according to the axis definitions in Figure 2.4). A more detailed overview including references to relevant literature is given in [10]. The paper also discussed the fact that conventional RC Taylor models can capture certain features of deformation textures better than the FC Taylor theory (e.g., the precise orientation), but, just as the FC Taylor theory, perform insufficiently when it comes to quantitative texture predictions.

2.3.4
Grain Interaction Models

The Sachs and the Taylor models both allow us to treat every grain independently from the rest of the polycrystal. In addition, they do not consider the fact that stresses and stain rates may also be heterogeneous *inside* the grains. These over-simplifications have a price, as these models cannot achieve *both* stress equilibrium *and* strain compatibility at grain boundaries. Note that the velocity field offered by the Taylor model may be simplistic but it at least offers a geometrically consistent field over the entire macroscopic RVE. The stress field offered by it is, however, not in equilibrium in the entire RVE, but only inside every given grain, where it is assumed to be homogeneous. In the last two decades, other models have emerged that try to do better.

The highest accuracy can probably be achieved by the crystal plasticity finite element method (CPFEM), because it offers a solution that achieves stress equilibrium at the boundaries of the elements and maintains geometrically consistency throughout the entire RVE [15]. The price to pay is a considerable computational cost. The very new crystal plasticity fast Fourier (CP-FFT) model can possibly replace the CPFEM models for many applications at a much smaller computational effort [16]. However, there are also some less ambitious and less computationally intensive methods that try to achieve some (but not all) of the advantages of the CPFEM method as compared to the Taylor theory.

Elastoplastic or viscoplastic self-consistent models [17], a sort of mean-field models, implement a certain stress–strain interaction between an individual grain and the average behavior of the RVE. As this is done for every grain separately, without direct accounting of short range interactions between directly neighboring grains, neither single stress field in equilibrium nor geometrically compatible velocity is offered for the entire RVE. Lebensohn *et al.* [17] have published an interesting comparison between the viscoplastic self-consistent method and the CP-FFT method.

Grain interaction models try to achieve strain compatibility and stress equilibrium between directly neighboring grains (GIA model [18] or ALAMEL model [10]). Instead of, such as the Taylor model, solving the crystal plasticity equations for every grain separately, they do it simultaneously for all grains of a small clusters (eight grains in case of GIA and two in case of ALAMEL). The initial crystal orientations of the grains of such clusters are chosen at random from the initial ODF of the macroscopic RVE.[3] These models do not assume that the nominal velocity gradients of the grains that belong to a particular clusters are the same. However, they require that within a cluster, the average of these nominal velocity gradients is equal to the average velocity gradient of the macroscopic RVE of the polycrystal. They also assure stress equilibrium at the internal grain boundaries of the cluster. Let us illustrate this for the ALAMEL model. It makes random selections of grain boundary segments in a polycrystal (Figure 2.5 a). For each grain boundary segment, two sizable parts of the grains at both sides of the grain boundary segment are considered (Figure 2.5b). Together they form the "cluster." It is not assumed that the velocity gradient is homogeneous in each of these parts. However, inside each of these two parts, a very shallow region just at the grain boundary segment is assumed to have a homogeneous velocity gradient, which will be called "nominal velocity gradient." The nominal velocity gradients of the two grains of the cluster are different, as each of them features a few shear relaxations. The model however enforces the *average* of the two nominal velocity gradients of the grains in the cluster to be equal to the prescribed velocity gradient tensor of the macroscopic RVE. A first shear relaxation of the type l_{13} is allowed in one of the shallow regions, and an opposing shear relaxation $-l_{13}$ in the other one (Figure 2.5b). l_{13} has the same meaning as in the

---

3) The meaning of the word "random" as used here is the one which it has in the statistical literature. All crystal orientations of all clusters taken together must constitute a set of discrete orientations that is representative for the macroscopic RVE of the material.

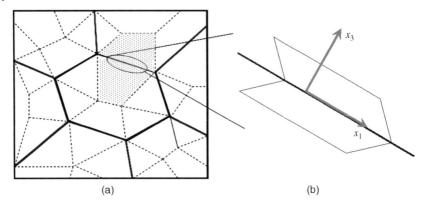

Figure 2.5 (After [19]) (a) Schematic representation of the microstructure of a polycrystalline material. A grain boundary segment is randomly selected. It is supposed to be representative for part of the volumes of the grains that it separates. (b) A relaxation is allowed in a shallow region at both sides of the grain boundary segment. It consists of two opposing shears of the l_{13} type (axis definitions as given in the figure).

pancake model, except that the axes x_1 and x_3 shown in Figure 2.4 do not stand for RD and ND now. Instead, the vector $\mathbf{m}^{RLX}$ (see the previous section) will now be normal to the grain boundary (axis x_3 in Figure 2.5b) segment and $\mathbf{b}^{RLX}$ will be parallel with axis x_1. The mathematical treatment goes as follows. Equation (2.21) is still valid. Equation (2.22) is now written separately for each of the two parts of the shallow region.

$$d_p = M_p^{RLX}\dot{\gamma}^{RLX} + \sum_{s \in \text{all slip systems}} M_p^{As}\dot{\gamma}^{As} \quad \text{with} \quad p = 1, 5$$

$$d_p = -M_p^{RLX}\dot{\gamma}^{RLX} + \sum_{s \in \text{all slip systems}} M_p^{Bs}\dot{\gamma}^{Bs} \quad \text{with} \quad p = 1, 5$$

(2.24)

As before, d_p is equal to the macroscopic strain rate of the RVE. The symbols A and B denote the top and bottom grain in Figure 2.5b, respectively. Equation (2.23) now becomes

$$\dot{W} = \tau_c^{RLX}\left|\dot{\gamma}^{RLX}\right| + \sum_{G=A,B}\sum_{s \in \{\text{active slip systems}\}} \tau_c^{Gs}\left|\dot{\gamma}^{Gs}\right| = \text{Min}$$

(2.25)

that is, the plastic work is now minimized for the two parts of the shallow zone taken together.

The ALAMEL model not only updates the crystal orientations of the shallow zones at both sides of a grain boundary segment after each strain increment, but also updates the orientation and the length of the grain boundary segments themselves. The conventional formulas for affine transformations are used. The transformation is described by the deformation gradient tensor of the strain increment, itself obtained by integration over the strain increment of the prescribed velocity gradient tensor of the macroscopic RVE. Hence, an ALAMEL simulation actually updates the set of orientations and lengths of the grain boundary segments that together constitute a model of the microstructure.

The set of grain orientations that is obtained in this way by the ALAMEL model actually describes merely the nominal orientations of the shallow regions at both sides of the grain boundary segments. Since any particular grain has different grain boundary segments, not just one, it can be expected that the evolution of the crystal orientation at the grain boundary segments of this grain will start to diverge according to this model. This also implies the development of gradients in the distribution of stresses, strains, and lattice orientation. It is however assumed that the set of nominal crystal orientations (two at each grain boundary segment) obtained by the model after each strain increment remains representative for the global texture of the macroscopic RVE. This is a strong assumption that is not supported by a direct theoretical proof; hence, experimental validation is necessary. This will be further discussed in Section 2.5.

2.4
Plastic Anisotropy of Polycrystalline Materials

Consider a polycrystalline material (= macroscopic RVE) subjected to a deformation described by a strain rate tensor **D**. Any of the polycrystal deformation models described earlier can be used to obtain the rate of plastic work, denoted by $\dot{W}$, dissipated in a particular crystal. The average over the entire macroscopic RVE is denoted by $\dot{\bar{W}}$. The latter will be a function of the macroscopic strain rate tensor **D**. This function can be used to describe the plastic anisotropy of the macroscopic RVE, at least according to the polycrystal model used. In practice, it is preferred to use a sort of "polar co-ordinates" for strain rate space, namely

$$D_0 = \sqrt{\sum_{p=1}^{5} (D_p)^2} \qquad (2.26)$$

and

$$a_p = \frac{D_p}{D_0} \qquad (2.27)$$

representing the components of **D** and **a** by means of contracted indices, as introduced in Section 2.2, just after Eq. (2.10). It is seen that

$$\mathbf{D} = D_0\,\mathbf{a} \qquad (2.28)$$

The vector **a** (components a_p) is a unit vector in strain rate space (called "strain mode" hereafter), and hence an angular coordinate, whereas D_0 is a sort of "radius." Note also that the von Mises equivalent strain rate is given by

$$D_{vM} = \sqrt{\frac{2}{3}} D_0 \qquad (2.29)$$

Let us now use one of the rate insensitive polycrystal models (Taylor, GIA or ALAMEL) to calculate $\dot{W}$ (of a particular macroscopic RVE with known texture) for a given value of the "strain modes" **a** and for a fixed value of D_0. Note that for such models, $\dot{W}$ simply scales with D_0 if **a** is kept fixed. Such a calculation can be called a "virtual mechanical test," **a** representing, for example, a uniaxial tensile test (with a prescribed Lankford ratio!), or a torsion test. The resulting values of $\dot{W}$ would only be valid for the beginning of plastic flow (work hardening is not modeled in this way), but depend on the texture and on the strain mode and hence describe the plastic anisotropy. Imagine now that a very large number of virtual mechanical tests are performed for various strain modes. Many values $\dot{W}(\mathbf{D})$ would be found, **D** given by Eq. (2.28). Such result could be implemented in a FE software for the engineering length scale if one manages to fit the data for $\dot{W}(\mathbf{D})$ to a relatively simple analytical function $\Psi(\mathbf{D})$. An approximate model for the anisotropic stress–strain rate relationship of the material can then be obtained [20]:

$$\bar{\sigma}_p = \frac{\partial \Psi}{\partial D_p} \tag{2.30}$$

This can be used in a constitutive model of a finite element code for metal forming. The recent so-called hierarchical multiscale model (HMS) [21] is an example. It makes use of a particular analytical form for $\Psi(\mathbf{D})$ called the "Facet model" [20].

2.5
Experimental Validation

2.5.1
Prediction of Rolling Textures

All important texture components of the rolling texture of a bcc metal can be seen in the $\varphi_2 = 45°$ section of the ODF (Figure 2.6). It is therefore extensively used in texture

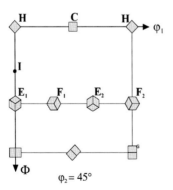

Figure 2.6 Overview of the principal components of steel textures, as they can be seen in a $\varphi_2 = 45°$ section.

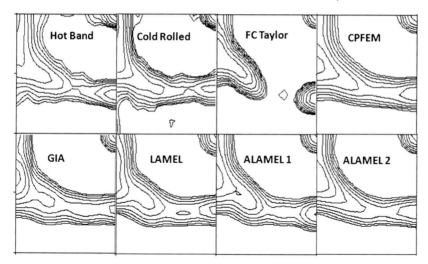

Figure 2.7 $\varphi_2 = 45°$ sections of the ODF of eight textures of IF steel: hot band texture, texture after 70% cold rolling reduction, simulated textures according to six models. CPFEM by Bate [23]. For LAMEL: see [24]. ALAMEL 1: starting from equiaxed grains; ALAMEL 2: starting from elongated grains (as if the grains of the hot rolled microstructure already had 40% rolling reduction) and using additional $\{1\,2\,3\}\langle 1\,1\,1\rangle$ slip systems. Levels: 1; 1.4; 2; 2.8; 4; 5.6; 8; 11; 16; 22.

research on bcc metals. Figure 2.7 shows such sections of the hot band and the cold band texture (after 70% cold rolling reduction) of a particular type of IF steel. In this case, the hot band texture actually looks as if it were a cold rolling texture, although it is less sharp than the true cold rolling texture. So these two textures are the start and the end textures of the cold rolling operation. The hot band texture had been converted into 3000 discrete orientations using the "statistical" method [22]. This discrete texture was used as a starting texture for the simulation of the cold rolling texture by means of six different polycrystal models, as indicated in the subtitle of Figure 2.7 [10]. All models have considered the $\{1\,1\,0\}\langle 1\,1\,1\rangle$ and $\{1\,1\,2\}\langle 1\,1\,1\rangle$ slip systems, except ALAMEL2, which has also considered the $\{1\,2\,3\}\langle 1\,1\,1\rangle$ slip systems family. Visual inspection of Figure 2.7 reveals that the prediction of the cold rolling texture by the FC Taylor model is really bad; even the "zero" model (i.e., *leave the texture as it is*) would have performed better. The results of the CPFEM, GIA, LAMEL, and the two versions of the ALAMEL model all give quite satisfactory results. These are all models that take interactions between directly neighboring grains into account. It must however be remarked that the CPFEM method used here [23] had only assigned one element per grain. A CPFEM method that would use 27 or more elements per grain would probably perform noticeably better than the GIA, LAMEL [24] or ALAMEL models, but would require an even larger calculation time.

Similar validation efforts have been done for aluminum alloys. Van Houtte *et al.* [25] have reported a study on the prediction of rolling textures of AA1200, AA5005, and AA5182 after thickness reductions ranging from 40% to 98%. The GIA, ALAMEL, and CPFEM (one element per grain) models had been selected to do the simulations, because it was already known from earlier work that the results of other models, such as FC and RC Taylor, were not satisfactory. It was found that GIA and ALAMEL gave quite good results for AA1200, less satisfactory ones for AA5005, and not satisfactory at all for AA 5182. In this alloy, the dominating deformation mode is suspected to be severe shear banding, which is not captured by these models. However, Engler [26] demonstrated that a modified version of the GIA model that includes work hardening could do a better job in this case.

2.5.2
Prediction of Cup Drawing Textures

Cups have been drawn from low carbon steel sheet. The blank diameter was 100 mm, the die had an inner diameter of 52 mm, and the punch an outer diameter of 50 mm. A HMS was used to simulate the drawing of low carbon steel cups [21]. This model is in fact a finite element model for metal forming. In each integration point, the material is represented by a 5000-grain RVE at the macroscale. The FC Taylor or the ALAMEL model was applied on it for two purposes: (i) generate the local anisotropic plastic response of the material (as explained in Section 2.4) to be used by the FE model and (ii) to simulate the evolution of the deformation texture of the material. These actions are not necessarily performed after each deformation step (see [21] for more details). Figure 2.8 shows the deformation texture in the cup wall close to the rim, at 90° from the rolling direction of the sheet material. The

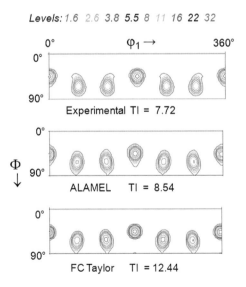

Figure 2.8 Deformation texture in the cup wall close to the rim, at 90° from the rolling direction of the sheet material. Experimental texture: average over several samples. Predicted texture: using the ALAMEL and the FC Taylor hierarchical multiscale model (HMS). The sharpness of the texture is characterized by the TI. (After [21]).

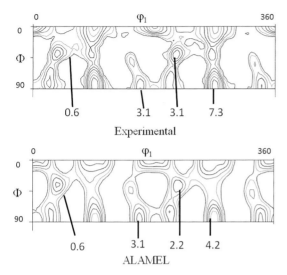

Figure 2.9 (After [27]) $\varphi_2 = 45°$ sections of the ODF of measured and simulated textures for AA3103-O at the rim of drawn cups, 45° from RD. Texture evolution simulated with ALAMEL.

experimental texture is shown as well as the results of the ALAMEL and the Taylor models. The texture index (TI) introduced by Bunge [1] is also given. It is seen that the experimental and simulated textures are quite similar, the simulated textures being sharper than the experimental one. The texture index (a quadratic criterion) of the ALAMEL and FC Taylor result was about 10% and 60% higher than the experimental one, respectively.

Similar work has also been done for aluminum alloys [27]. Figure 2.9 shows a result for a cup made from an AA3103-O alloy. It is seen that the texture predicted by the ALAMEL model agrees well with the experimentally observed texture. It is however admitted that in the case of fcc metals, the $\varphi_2 = 45°$ section of the ODF does not reveal all texture components.

2.6 Conclusions

The generalized Schmid law is in the first place a constitutive model for the plastic deformation of a crystal with mobile dislocations. It can be used to establish a link between the local stress and the strain rate (or vice-versa) as well as to obtain the lattice spin due to the plastic deformation. Its predecessor, the Schmid law, has been used from 1929 [13] with an attempt to at least explain the development of deformation textures in polycrystalline metals. This requires a two-scale model, connecting the stress and strain rate felt or known at the macroscopic length scale to events taking place at the mesocopic length scale (grain scale). A milestone in this was the work of Taylor [9] who in 1938 proposed a much better model than the Sachs model however without making use of the generalized Schmid law. This was done in the early 1950s by Bishop and Hill [7,8] who also demonstrated that the Taylor theory was not in conflict with their own. In the 1970s, Asaro and Needleman [12] proposed to replace the generalized Schmid law by a visco-plastic model, which however also makes use of the resolved shear stress, and does not lead to fundamentally different results than the older methods that are nowadays called "rate-insensitive methods." Extensive study showed later that the Taylor–Bishop theory could give good qualitative predictions of deformation textures, but was not capable to make quantitatively accurate predictions of the ODF of a deformation texture. Such predictions are required when calculations of the anisotropy of mechanical or physical properties should be based on them. The search for the perfect method is still going on. The best and most general method probably is the CPFEM model [15]. Very promising (less computationally intensive) is the recent CP-FFT model [16]. But it remains surprising that excellent results have also been obtained by much simpler (and often faster) models such as GIA [18] and ALAMEL [10]. All these successful new models have one thing in common: they somehow try to take the short-range stress–strain interaction between directly neighboring grains into account. In all these models, the generalized Schmid law or the related visco-plastic model remains the cornerstone.

Acknowledgments

The author wishes to thank many colleagues who have co-authored papers on this subject: E. Aernoudt, J. Gil Sevillano, D. Roose, R. Wenk, F. Wagner, O. Engler, P. Bate, M. Seefeldt, A. Van Bael, L. Delannay, L.S. Toth, I. Samajdar, S. Li, S. He, A. K. Kanjarla, S. K. Yerra, J. Gawad, G. Samaey, P. Eyckens, and Q. Xie. He has moreover very much enjoyed numerous exciting discussions on related topics with K. Lücke, G. Gottstein, U.F. Kocks, J. Hirsch, C. Tomé, R. Lebensohn, F. Roters, P. Wagner, and M. Crumbach. The companies ArcelorMittal and TataSteel have supported this work as well as the Belgian and Flemish governments through FWO grants and grants from Belgian Science Policy (present contract: P7/21).

References

1 Bunge, H.J. (1993) *Texture Analysis in Material Science*, Cuvillier Verlag, Göttingen, ISBN 3–928815-81-4.
2 Kocks, U.F., Tomé, C.N., and Wenk, H.-R. (1998) *Texture and Anisotropy: Preferred Orientations and their Effect on Materials Properties*, Cambridge University Press, Cambridge, UK, ISBN 0 521 46516 8.
3 Engler, O. and Randle, V. (2009) *Introduction to Texture Analysis: Macrotexture, Microtexture and Orientation Mapping*, 2nd edn, CRC Press, London, ISBN 10: 1420063650.
4 Gottstein, G. (2004) *Physical Foundations of Materials Science*, Springer-Verlag, Berlin, Heidelberg, New York, (ISBN 3–540–40139-3).
5 Nye, J.F. (1990) *Physical Properties of Crystals*, Clarendon Press, Oxford, ISBN 0–19–851165-5.

6 Van Houtte, P. (1995) Heterogeneity of plastic strain around an ellipsoidal inclusion in an ideal plastic matrix. *Acta Metallurgica*, **43**, 2859–2879.

7 Bishop, J.F.W. and Hill, R. (1951) A theory of the plastic distortion of a polycrystalline aggregate under combined stress. *Philosophical Magazine*, **42**, 414–427.

8 Bishop, J.F.W. and Hill, R. (1951) A theoretical derivation of the plastic properties of a polycrystalline face-centred metal. *Philosophical Magazine*, **42**, 1298–1307.

9 Taylor, G.I. (1938) Plastic strain in metals. *Institute of Metals*, **62**, 307–324.

10 Van Houtte, P., Li, S., Seefeldt, M., and Delannay, L. (2005) *International Journal of Plasticity*, **21**, 589–624.

11 Van Houtte, P. (1988) A comprehensive mathematical formulation of an extended Taylor–Bishop–Hill model featuring relaxed constraints, the Renouard–Wintenberger theory and a strain rate sensitivity model. *Textures and Microstructures*, **8–9**, 313–350.

12 Asaro, R.J. and Needleman, A. (1985) Overview No. 42: Texture development and strain hardening in rate dependent polycrystals. *Acta Metallurgica*, **33**, 923–953.

13 Sachs, G. (1928) Zur Ableitung einer Fließbedingung. *Zeitschrift des Vereins Deutscher Ingenieure*, **72**, 734–736.

14 Kocks, U.F. (1970) The relation between polycrystal deformation and single crystal deformation. *Metallurgical Transactions*, **1**, 1121–1143.

15 Roters, F., Eisenlohr, P., Hantcherli, L., Tjahjanto, D.D., Bieler, T.R., and Raabe, D. (2010) Overview of constitutive laws, kinematics, homogenization and multiscale methods in crystal plasticity finite-element modeling: Theory, experiments, applications. *Acta Materialia*, **58**, 1152–1211.

16 Lebensohn, R.A., Rollet, A., and Suquet, P. (2011) Fast Fourier transform-based modeling for the determination of micromechanical fields in polycrystals. *Journal of Metals*, **63/3**, 13–18.

17 Lebensohn, R.A., Liu, Y., and Ponte Castañeda, P. (2004) On the accuracy of the self-consistent approximation for polycrystals: Comparison with full-field numerical simulations. *Acta Materialia*, **52**, 5347–5361.

18 Engler, O., Crumbach, M., and Li, S. (2005) Alloy-dependent rolling texture simulation of aluminium alloys with a grain-interaction model. *Acta Materialia*, **53**, 2241–2257.

19 Kanjarla, A.K., Van Houtte, P., and Delannay, L. (2010) Assessment of plastic heterogeneity in grain interaction models using crystal plasticity finite element method. *International Journal of Plasticity*, **26**, 1220–1233.

20 Van Houtte, P., Yerra, S.K., and Van Bael, A. (2009) The Facet method: A hierarchical multilevel modelling scheme for anisotropic convex plastic potentials. *International Journal of Plasticity*, **25**, 332–360.

21 Van Houtte, P., Gawad, J., Eyckens, P., Van Bael, B., Samaey, G., and Roose, D. (2011) A full-field strategy to take texture induced anisotropy into account during FE simulations of metal forming processes. *Journal of Metals*, **63/11**, 37–43.

22 Toth, L.S. and Van Houtte, P. (1992) Discretization techniques for orientation distribution functions. *Textures and Microstructures*, **19**, 229–244.

23 Bate, P. (1999) Modelling deformation microstructure with the crystal plasticity finite-element method. *Philosophical Transactions of the Royal Society of London Series A – Mathematical Physical and Engineering Sciences*, **357**, 1589–1601.

24 Van Houtte, P., Delannay, L., and Samajdar, I. (1999) Quantitative prediction of cold rolling textures in low-carbon steel by means of the LAMEL model. *Textures and Microstructures*, **31**, 109–149.

25 Van Houtte, P., Li, S., and Engler, O. (2004) Modelling deformation texture of aluminium alloys using grain interaction models. *Aluminium*, **80/6**, 702–706.

26 Engler, O. (2003) Through-process modelling of the impact of intermediate annealing on texture evolution in aluminium alloy AA 5182. *Modelling and*

Simulation in Materials Science and Engineering, **11**, 863–882.

27 Van Houtte, P., Gawad, J., Eyckens, P., Van Bael, B., Samaey, Giovanni, and Roose, D. (2012) Multi-scale modelling of the development of heterogeneous distributions of stress, strain, deformation texture and anisotropy in sheet metal forming, in *Linking Scales in Computations: From Microstructure to Macro-scale Properties* (ed. O. Cazacu), Procedia IUTAM 3, Elsevier, pp. 67–75 (Available online at http://www.sciencedirect.com/science/journal/22109838 , accessed on October 29, 2012).

3
Crystal Plasticity Modeling

Franz Roters, Martin Diehl, Philip Eisenlohr, and Dierk Raabe

3.1
Introduction

Among all material properties, the mechanical properties can, to date, probably be regarded as the most important ones. This not only holds for structural materials, but it is also true for many functional materials, which are in general also subjected to certain mechanical loads. The tailoring of mechanical properties is, therefore, a key aspect of this book. Within this process, the modeling of mechanical properties based on the microstructure of the material has gained a lot of importance during the past 50 years. With the increase of computational power, more and more physical aspects could be included in these modeling efforts. As the vast majority of structural materials is of crystalline nature, modeling of crystal elastoplasticity became a key topic.

The elastic–plastic deformation of crystalline aggregates depends on the direction of loading, that is, crystals are mechanically anisotropic. This phenomenon is due to the anisotropy of the elastic behavior and the orientation dependence of the activation of the crystallographic deformation mechanisms (dislocations, twins, martensitic transformations). A consequence of crystalline anisotropy is that the associated mechanical phenomena such as shape change, crystallographic texture, strength, strain hardening, deformation-induced surface roughening, and damage also depend on the orientation. This is not a trivial statement as it implies that the mechanical parameters of crystalline matter are tensor quantities. Another major consequence of the single crystalline elastic–plastic anisotropy is that it adds up to produce macroscopically directional properties as well when the orientation distribution (crystallographic texture) of the grains in a polycrystal is not random. Figure 3.1 shows such an example of a plain carbon steel sheet with a preferred crystal orientation (here high probability for a crystallographic {1 1 1} plane being parallel to the sheet surface) after cup drawing. Plastic anisotropy leads to the formation of an uneven rim (referred to as *ears* or *earing*) and a heterogeneous distribution of material thinning during forming. It must be emphasized in this context that a random texture is not the rule but a rare exception in real materials. In other words, practically all crystalline materials reveal macroscopic anisotropy.

3 Crystal Plasticity Modeling

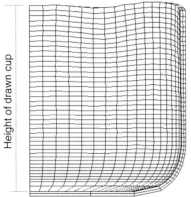

Figure 3.1 Consequence of plastic anisotropy when drawing a textured sheet into a cup. The orientation distribution before deformation exhibits a high-volume fraction of grains with a crystallographic [1 1 1] axis parallel to the sheet normal. The arrows on the left mark six ears resulting from preferential material flow. On the right, the corresponding crystal plasticity finite element simulation is shown.

Therefore, a theory of the mechanical properties of crystals must include, first, the crystallographic and anisotropic nature of those mechanisms that create shear and, second, the orientation(s) of the crystal(s) studied relative to the applied boundary conditions (e.g., loading axis and rolling plane).

Early approaches to describe anisotropic plasticity under simple boundary conditions have considered these aspects, such as the Sachs [1], Taylor [2], Bishop–Hill [3,4], or Kröner [5] formulations. However, these approaches were designed neither for considering explicitly the mechanical interactions among the crystals in a polycrystal nor for responding to complex internal or external boundary conditions, see Figure 3.2. Instead, they are built on certain simplifying assumptions of strain or stress homogeneity to cope with the intricate interactions within a polycrystal.

The finite element method (FEM) [8–10] is today the most used simulation technique in the field of continuum mechanics. This is mainly due to its great flexibility in the treatment of (mechanical) boundary conditions. However, the FEM can be used to solve almost any kind of boundary value problem, of which mainly the simulation of thermal fields is of interest also in conjunction with mechanical simulations. While the state-of-the-art continuum simulations are in many cases coupled thermo-mechanical calculations, as plastic deformation is always producing some heat, in the field of crystal plasticity FEM (CPFEM) most simulations are still purely mechanical and ignore any temperature effects, that is, assume constant temperature.

Even though it has been known since 1934 [11–14] that crystalline materials deform plastically by the slip of dislocations on discrete slip systems, for a long time continuum mechanical FE simulations have used isotropic material models that were based on empirical equations. The first CPFE simulations were performed by Peirce, Asaro, and Needleman in 1982 [15]. Owing to computational restrictions,

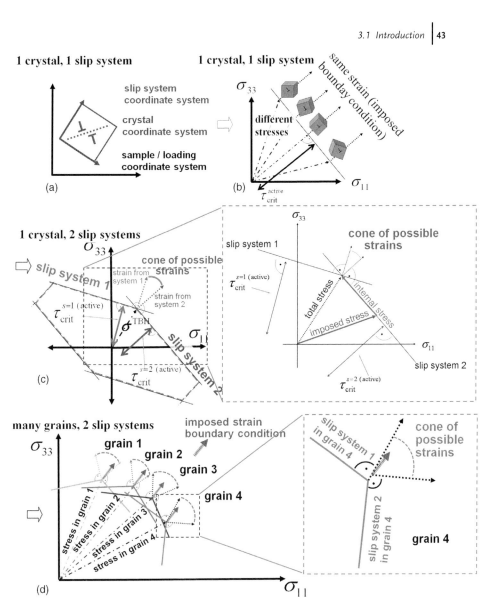

Figure 3.2 Schematical presentation of the increasing complexity of crystal-scale micromechanics with respect to the equilibrium of the forces and the compatibility of the displacements for different situations: (a, b) Single-slip situation in a single crystal presented in stress space. (c) Portion of a single crystal yield surface with three-slip systems. (d) Multislip situation in a polycrystal where all different crystals have to satisfy an assumed imposed strain in their respective yield corners. If the strain is homogeneous this situation leads to different stresses in each crystal [6,7]. τ_{crit}: critical shear stress; σ^{TBH}: Taylor–Bishop–Hill stress state (stress required to reach a yield corner).

they used a simplified setup of two symmetric slip systems in order to study the tensile behavior of a single crystal. These simulations were later extended to a polycrystalline arrangement by Harren *et al.* [16,17] using a 2-D setup with two or three slip systems. In 1991, Becker was the first to perform simulations on the basis of the 12 slip systems of a face-centered cubic (fcc) crystal. Using a 3-D model for the crystallographic degrees of freedom, he simulated channel-die deformation of a single crystal [18] and of a columnar polycrystal aggregate [19].

One main advantage of CPFE models lies in their capability of solving crystal mechanical problems under complicated internal and/or external boundary conditions. This aspect is not a mere computational advantage, but it is an inherent part of the physics of crystal mechanics since it enables one to tackle those boundary conditions that are imposed by inter- and intragrain micromechanical interactions, Figure 3.3 [20]. This is essential not only to study in-grain or grain cluster mechanical problems, but also to better understand the often quite abrupt mechanical transitions at interfaces [21].

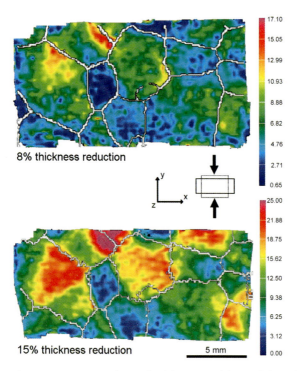

Figure 3.3 Experimental example of the heterogeneity of plastic deformation at the grain and subgrain scale using an aluminum oligocrystal with large columnar grains [20]. The images show the distribution of the accumulated von Mises equivalent strain in a specimen after $\Delta y/y_0 = 8\%$ and 15% thickness reduction in plane strain (y_0: initial sample height). The experiment was conducted in a lubricated channel-die setup. White lines indicate high-angle grain boundaries derived from EBSD microtexture measurements. The equivalent strains (determined using digital image correlation) differ across some of the grain boundaries by a factor of 4–5 giving evidence of the enormous orientation-dependent heterogeneity of plasticity even in pure metals.

However, the success of CPFE methods is not only built on their efficiency in dealing with complicated boundary conditions, but they also offer high flexibility with respect to including various constitutive formulations for plastic flow and hardening at the elementary shear system level. The constitutive flow laws that were suggested during the past decades have gradually developed from empirical viscoplastic formulations [22,23] into microstructure-based multiscale models of plasticity, including a variety of deformation mechanisms, size-dependent effects, and interface mechanisms [24–43] (see Section 3.2.1). In this context it should be emphasized that the finite element method itself is not the actual model but the variational solver for the underlying constitutive equations.

Since its first introduction by Peirce et al. [15] the CPFE method has matured into a whole family of constitutive and numerical formulations, which has been applied to a broad variety of crystal mechanical problems, see Roters et al. [44] for a recent review. In the field of direct or one-to-one crystal plasticity models numerous grain- and subgrain-scale problems have been tackled using meshes with sub-grain resolutions and, in part, complex 2-D and 3-D grain arrangements [20,45–51]. Alternatively, advanced homogenization schemes were developed for the application of the CPFE method to large-scale problems (see Section 3.2.2). In this case, one main problem was the correct representation of the (statistical) crystallographic texture of the material in the CPFE mesh. This can be achieved in different ways using, for example, texture components [52,53] or direct sampling of single orientations from the orientation distribution function (ODF) [54–56] (see Section 3.2.2).

Finally, as mentioned before, the FEM is just one way of solving mechanical boundary value problems. Alternative methods, such as spectral methods introduced first by Moulinec and Suquet [57,58] in the field of mechanics, can be used as well.

3.2 Fundamentals

Most crystal plasticity models discussed in the literature are formulated in a finite strain framework. The total deformation gradient, $F = \frac{dy}{dx}$, can therefore be multiplicatively decomposed ([59], Figure 3.4)

$$F = F_e F_p \tag{3.1}$$

Here F_e is the elastic part of the deformation gradient that describes both the elastic distortion and the rigid-body rotation. F_p is the plastic part of the total deformation that is due to crystallographic slip. It is noteworthy that crystallographic slip does not change the orientation of the crystal lattice even though F_p in general does include a rotational part, that is, can be decomposed as $F_p = R_p U_p$.

As shown in Figure 3.4, the decomposition of the total deformation gradient can be illustrated by introducing a reference configuration, an intermediate (relaxed) configuration, and a current configuration. The plastic deformation gradient F_p maps a material point from the reference configuration to the intermediate configuration. Respectively, the elastic deformation gradient F_e projects the point

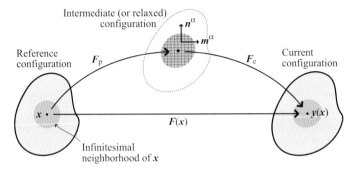

Figure 3.4 Schematic representation of the decomposition of deformation gradient $F = \frac{dy}{dx}$. Vectors m^α and n^α are, respectively, the slip direction and the slip plane normal of the crystallographic lattice in the intermediate (relaxed) configuration.

from the intermediate configuration to the current configuration. It is, however, important to realize that, in general, the order of the decomposition does not correspond to the actual deformation sequence.

The velocity gradient in the current configuration, denoted as $\tilde{L}$, can be written as

$$\tilde{L} := \dot{F}F^{-1} = \dot{F}_e F_e^{-1} + F_e \dot{F}_p F_p^{-1} F_e^{-1} = \tilde{L}_e + \tilde{L}_p \tag{3.2}$$

In case of dislocation slip being the only plastic deformation mechanism, the plastic velocity gradient L_p (in the intermediate (relaxed) configuration) is determined by the slip rates $\dot{\gamma}^\alpha$ on all N_{slip} active slip systems α

$$L_p := \dot{F}_p F_p^{-1} = \sum_{\alpha=1}^{N_{\text{slip}}} \dot{\gamma}^\alpha m^\alpha \otimes n^\alpha \tag{3.3}$$

where m^α is the slip direction and n^α the slip plane normal of slip system α. The magnitude of the shear rates $\dot{\gamma}^\alpha$ is determined by the constitutive model used, as described in the following section.

3.2.1
Constitutive Models

The constitutive model determines the character of the simulation. Strictly speaking, two constitutive models are needed, namely one for the elastic response and the other for the plastic deformation. However, practically all implementations of crystal plasticity use Hooke's law for the description of the elastic part. In contrast, the models developed for the plastic deformation range from purely phenomenological descriptions to models incorporating many aspects of the physics of dislocation motion and other deformation mechanisms.

In general, constitutive models for plasticity consist out of two (sets of) equations. (1) The *kinetic equation of state* determines the stress σ necessary to reach a certain strain rate $\dot{\varepsilon}$ at a given temperature T for the current material state S. (2) The *state evolution equation(s)* describe(s) the rate of change of the state $\dot{S}$ as a function of strain rate $\dot{\varepsilon}$, stress σ, temperature T, and the current state S. The choice of appropriate state variables is left to the user. Either phenomenological quantities such as critical stresses or physical ones, for example, dislocation densities or twin volume fractions, can be used.

3.2.1.1 Dislocation Slip

The calculation of the shear rates in Eq (3.3) can be done purely phenomenologically or rather based on physics. One variant of a phenomenological description is the so-called power law formulation [15,23,60,61]

$$\dot{\gamma}^\alpha = \dot{\gamma}_0 \left|\frac{\tau^\alpha}{\tau_c^\alpha}\right|^n \mathrm{sgn}(\tau^\alpha) \tag{3.4}$$

where $\dot{\gamma}^\alpha$ is the shear rate on slip system α subjected to the resolved shear stress τ^α at a critical stress τ_c^α; $\dot{\gamma}_0$ and n are material parameters that determine the reference shear rate and the stress sensitivity of slip, respectively.

In case of a physical model based on dislocation densities, it is straightforward to use the Orowan equation [11] for this purpose:

$$\dot{\gamma}^\alpha = \rho_m^\alpha b v^\alpha \tag{3.5}$$

where ρ_m^α is the density of mobile dislocations, b the magnitude of the Burgers vector, and v^α the average velocity of the mobile dislocations.

In addition to these kinetic equations, state evolution equations are needed that describe the evolution of τ_c^α or ρ_m^α for the above examples.

3.2.1.2 Displacive Transformations

The preceding sections focused on dislocations as carriers of plastic shear. However, materials such as austenitic steels, transformation-induced plasticity (TRIP) steels, brass, twinning-induced plasticity (TWIP) steels, and shape memory alloys deform not only by dislocation slip, but also by displacive deformation mechanisms (also referred to as displacive transformations).

Two such mechanisms and their incorporation into the CPFE framework will be discussed here, namely martensite formation [62,63] and mechanical twinning [35,41,64,65]. A martensitic transformation changes the lattice structure of a crystal. The resulting shape change involves, as a rule, also a change in the unit cell volume, that is, a volume dilatation or contraction. Mechanical twinning proceeds by a shear mechanism, which reorients the affected volume into a mirror orientation relative to the surrounding matrix.

Due to their different characters, both mechanisms are incorporated in different ways into the constitutive formulations. In case of the TRIP effect, Eq (3.1) is modified by introducing a third deformation gradient F_{tr}:

$$F = F_e F_p F_{tr} \tag{3.6}$$

The transformation part, F_{tr}, of the deformation gradient is given by

$$F_{tr} = I + \sum_{i=1}^{M} \xi^i b^i \otimes d^i \qquad (3.7)$$

I is the second-order identity tensor, vectors b^i and d^i denote, respectively, the transformation shape strain vector and the unit normal to the habit plane of transformation system i and ξ^i represents the fraction of crystal volume, which underwent transformation on system i. The evolution of ξ^i has to be described by an additional state evolution equation.

As twinning is a shearing mechanism, it can be included into crystal plasticity by modifying Eq (3.3) [35]:

$$L_p = \left(1 - \sum_{\beta=1}^{N_{twin}} f^\beta \right) \sum_{\alpha=1}^{N_{slip}} \dot{\gamma}^\alpha m^\alpha \otimes n^\alpha + \sum_{\beta=1}^{N_{twin}} \gamma_{twin} \dot{f}^\beta m^\beta_{twin} \otimes n^\beta_{twin}, \qquad (3.8)$$

where f^β is the twin volume fraction on twin system β, and γ_{twin} is the characteristic twin shear ($\sqrt{2}/2$ for fcc and bcc crystal structures). Again an additional state evolution equation is needed for calculating $\dot{f}^\beta$.

3.2.2
Homogenization

The constitutive equations determine the deformation behavior of a single grain. However, when crystal plasticity is applied to large-scale problems, each material point (e.g., integration point in an FE mesh) represents a potentially large number of crystals. In this case, numerical schemes are required that homogenize the multi-grain response into an average response of the grain aggregate. Figure 3.5 illustrates four commonly used homogenization schemes, namely the isostrain ([66] or full constraints Taylor [2]) scheme, Taylor-based schemes allowing for relaxation (e.g., [67–69]), cluster models (e.g., LAMEL [70–72], GIA [73–76], or RGC [77]), and full-field homogenization by FEM [78–83] or spectral methods using fast Fourier transforms, for example, [57,58].

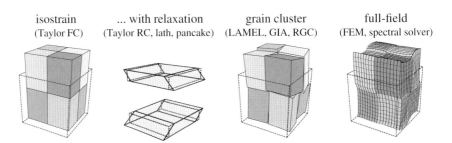

Figure 3.5 Four commonly used homogenization schemes.

3.2.3
Boundary Value Solvers

Crystal plasticity modeling finally relies on a tool to solve for mechanical equilibrium under given boundary conditions. Up to now, the FEM, called CPFEM in this case, is mostly used as it can handle arbitrary boundary conditions, see [44,84]. However, for special purposes, other solution techniques can be numerically superior. Spectral methods using fast Fourier transforms are such an example if one is interested in the response of so called *Representative Volume Elements* under periodic boundary conditions [57,85,86].

3.3
Application Examples

As already mentioned in the introduction, crystal plasticity (FEM) has been applied to a broad variety of crystal mechanical problems, see Roters *et al.* [44]. Therefore, in the spirit of this book, this section presents application examples that illustrate how crystal plasticity modeling can actually contribute to this field of material science.

3.3.1
Texture and Anisotropy

3.3.1.1 Prediction of Texture Evolution
One key aspect of crystal plasticity models is that they naturally account for crystal anisotropy (both elastic and plastic anisotropy). Moreover, they do account not only for the starting orientation of the crystals, but also for the change in orientation during deformation. As a consequence of this, CPFEM was from its very beginning used for texture prediction (e.g., [19,87,88]).

In this field, CPFEM competes with other models such as the Taylor models [2,67–69], cluster models [72,75,77] (see Section 3.2.2), and self-consistent models [89]. Li and Van Houtte [7,90] compared several of these models for the prediction of texture evolution. As a measure for the quality of the predicted textures, they introduced the so-called texture index $I(\Delta f)$. The texture index is the integral of the square of the difference between the ODF $f(g)$ of the experimental texture and the one simulated by a model:

$$I(\Delta f) = \int [f_{\text{model}}(g) - f_{\text{exp}}(g)]^2 \, dg \qquad (3.9)$$

where g is a crystal orientation. The integral is taken over the entire orientation space. The value of $I(\Delta f)$ is zero iff the two ODFs are identical.

While for some cases the prediction quality of one of the other models was comparable or even better than the CPFEM predictions, CPFEM turned out to be the most flexible method, that is, among the best for all cases studied, however, at the price of the highest computational cost, Table 3.1.

Table 3.1 Texture index of difference ODFs for two aluminum alloys and IF steel (Reproduced from Raabe et al. [7, Chapter 22].

Material	AA1200		AA5182							IF steel
Reduction %	40	63	86	95	98	40	63	86	95	98	70
FC	0.54	1.43	2.87	1.87	3.58	0.29	1.43	2.58	2.84	4.53	4.65
RCP	4.25	4.03	4.60	6.69	9.57	1.89	0.84	2.60	3.37	3.89	1.40
CPFEM[1]	–	–	–	–	–	–	–	–	–	–	1.22
CPFEM[2]	0.42	0.92	1.62	1.48	2.52	0.16	0.76	1.41	2.51	3.72	0.90
VPSC	1.11	2.02	2.26	2.81	4.29	0.57	1.33	1.92	2.18	2.68	–
GIA	0.57	0.87	1.16	1.35	2.34	0.12	0.35	0.90	1.67	2.51	0.76
Lamel[1]	0.89	1.60	1.56	1.26	1.81	0.46	1.15	1.66	2.41	2.83	0.59
Lamel[2]	0.49	0.99	1.45	2.25	3.35	0.32	0.81	0.96	1.19	1.49	–
Alamel[1]	0.49	1.06	1.38	1.07	1.58	0.27	1.09	1.29	1.70	2.08	1.08
Alamel[2]	–	–	–	–	–	–	–	–	–	–	0.65
I_{exp}	2.59	2.45	4.51	6.87	9.23	1.96	2.64	3.17	3.99	5.10	6.7

FC: Taylor theory ([2]); RCP: Pancake variant of the RC Taylor theory; CPFEM[1]: developed by Bate [91]; CPFEM[2]: developed by Kalidindi et al. [92]; VPSC: Visco-Plastic Self-Consistent Model (Lebensohn and Tomé [89]); GIA: developed by Crumbach et al. [76]; Lamel[1]: standard Lamel model; Lamel[2]: Lamel with Type I, II and III-relaxation; Alamel[1]: Advanced Lamel Model, starting from equiaxed grains; Alamel[2]: starting from elongated grains (as if the grains of the hot rolled microstructure already had 40% rolling reduction) and using additional {123} ⟨111⟩ slip systems ([70–72] see also Raabe [93,94]).

3.3.1.2 Prediction of Earing Behavior

The presence of nonrandom textures and their evolution is the main reason for anisotropy of structural materials during forming and plays an important role for the precise prediction of the final shape of the parts. Cup drawing is a standard material test for the characterization of this material anisotropy. The cup drawing

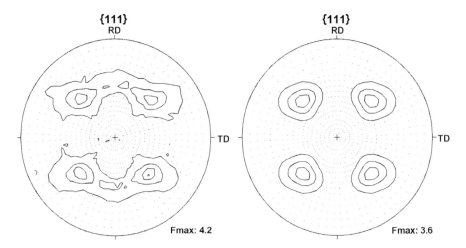

Figure 3.6 Experimental (left) and recalculated (right) {1 1 1} pole figure of AA3104 hot band.

Table 3.2 Texture components used for fitting the texture of AA3104 hot band [52,95,96].

Euler angles (°)			Scatter (°)	Intensity
φ_1	ϕ	φ_2	b^c	I^c
197.9	6.5	245.0	15.2	0.29
	Random		–	0.71

simulation shown here is for aluminum AA3104 hot band. Figure 3.6 shows on the left-hand side the experimental {1 1 1} pole figure of the material. It shows a cube texture typical for hot rolled aluminum alloys. On the right-hand side, the pole figure recalculated from a texture component fit is shown [52,95,96]. Besides the random portion of the texture, only one spherical component was used to fit this rather pronounced cube texture (Table 3.2). Owing to the orthorhombic sample symmetry, the single orientation has to be balanced by three additional symmetrically equivalent orientations to correctly reproduce the response of the material in the CPFE calculations. The resulting earing profile is shown in Figure 3.7, together with the experimentally measured earing profile. There is a very good agreement of simulation and experiment. Figure 3.8 shows the relative wall thickness distribution for the drawn cup (only one-quarter has been simulated because of sample symmetry). It can be seen that the bottom of the cup is thinner whereas the upper part of its side is thicker than the original sheet. The thickening is most pronounced in the valley of the earing profile, that is, at the 45° position.

3.3.1.3 Optimization of Earing Behavior

In the early industrial practice, texture was regarded as a property of polycrystals that was simply inherited from the preceding processing steps without conducting particular anisotropy optimization. This means that textures were known as an inevitable side effect of materials processing, which was hard to avoid and often accepted as it was. In contrast, modern industrial process design gradually aims at optimizing microstructures and properties during production, that is, its goal consists in considering metallurgical mechanisms such as crystal plasticity, recrystallization, grain growth, and phase transformation for the design of well-tailored crystallographic textures with respect to certain desired anisotropy properties of the final product.

The most recent phase in the advancement of quantitative texture and anisotropy engineering consists in the introduction of inverse texture simulation methods. Such approaches are designed for the physically based tailoring of optimum textures for final products under consideration of prescribed processing and materials conditions on an inverse basis. This means that variational texture optimization can nowadays be conducted in a way to match some desired final anisotropy and can

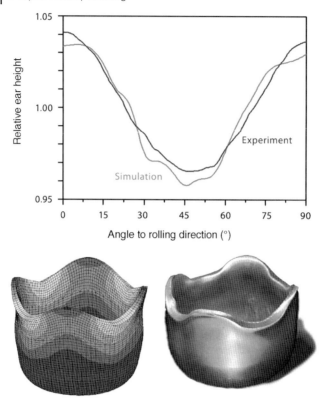

Figure 3.7 Simulated and experimental earing profile for AA3104 hot band.

help to identify beneficial corresponding processing parameters. This amounts to a tenet change, in the sense that the process should no longer determine the textures but that the desired textures should determine the process.

As shown in the previous section, elastic–plastic anisotropy during deep drawing may entail earing. One important consequence of that is – besides the irregular shape of the drawn specimen – an inhomogeneous distribution of the mechanical properties and of the wall thickness due to volume conservation (Figure 3.8) and the kinematically necessary strain rate variation. The trivial solution for the control and minimization of earing would be the presence of a random crystallographic texture prior to loading. However, such a supposed obvious approach is prevented because of two reasons. First, random starting textures generally do not remain random when the material is plastically deformed. This applies, in particular, to sheet forming operations. Second, complete spherical and spatial randomization of textures is very difficult. Most metallurgical and mechanical processes promote, rather than reduce, orientation distributions. This applies to most fcc metals, in particular to those without bulk phase transformation during forming such as aluminum.

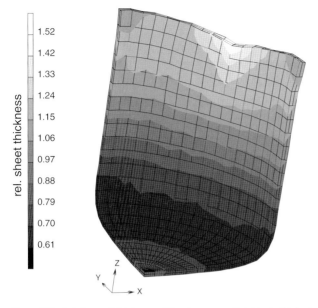

Figure 3.8 Relative wall thickness of the drawn cup (brighter is thicker).

Therefore, a more practical approach for reducing shape anisotropy lies in combining the texture components constituting the initial sheet in such a way that the resulting ear profile – accounting also for texture changes during forming – can be minimized due to the mutual compensation of the shape anisotropy contributions introduced by the individual texture components during forming.

For example, a combination of S ($\varphi_1 = 50°, \phi = 35°, \varphi_2 = 70°$) and Cube ($\varphi_1 = 0°, \phi = 0°, \varphi_2 = 0°$) texture components for aluminum, which are typical rolling and annealing components of this material and produce opposite earing, are chosen. Ear profiles were computed for different combinations of the two components. The aim of this simulation series is the prediction of the optimum texture composition for minimum earing. Earing is quantified in terms of the ear height and ear area (Figure 3.9). The ear height describes the difference between the highest and the lowest points on the profile of the drawn cup. The ear area integrates the entire surface between these two extremal values.

Figure 3.9 Earing is quantified in terms of the ear height and the ear area. The ear height describes the difference between the highest and the lowest point on the profile of the drawn cup. The ear area integrates the entire surface between these two extremal values.

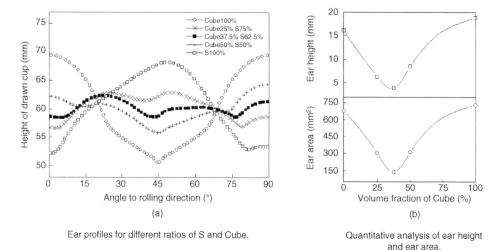

(a) Ear profiles for different ratios of S and Cube.

(b) Quantitative analysis of ear height and ear area.

Figure 3.10 Minimization of ear formation by mixing the S and Cube texture components. The simulation results show that the ear behavior can be minimized by mixing S and Cube texture components at an optimum ratio of about 62.75 vol%/37.25 vol% (S/Cube).

Figure 3.10 presents the ear profiles for different ratios of the volume fractions of the two texture components. Figure 3.10a shows that an optimum profile is obtained for a combination of 62.75 vol% of the S component and 37.25 vol% of the Cube component in case that the orientational scatter width amounts to 0° prior to elastic–plastic loading (combination of two originally perfect single crystals). Figure 3.10b reveals that this result applies to both the ear height and the ear area. Figure 3.10b also substantiates that the dependence of earing in terms of height and area on the texture composition reveals a steep change in the vicinity of the earing minimum, that is, small modifications in the ratio of the two texture component volumes entail a strong change in earing. This means that even minor texture changes can lead to remarkable optimization or degradation of the ear profile. It is important to note in this context that the ideal components with 0° orientation scatter width, which characterize the starting texture, develop during the simulation into an array of similar orientations, each of which may undergo individual reorientations according to the local boundary conditions. It is the special advantage of the (texture component) crystal plasticity finite element method to take these individual local reorientation- and strain-paths properly into account.

It is of some interest in this context to observe that the optimum volume ratio between the S orientation and the Cube orientation amounts to about 1.67 : 1 (62.75 vol%: 37.25 vol%) rather than to 1 : 1. The initial assumption that an optimum ratio between the two texture components might amount to 1 : 1 was suggested by Figure 3.10a, which shows that the ear profiles created by the S or the Cube component alone are opposite indicating the possibility of mutual compensation. The fact that 1.67 : 1 and not 1 : 1 is the optimum ratio underlines that the interaction of different texture components is highly nonlinear. It is also important to learn from this ratio that the volume fraction of the S component must obviously be much larger than that of the Cube component in order to compensate the anisotropy. This means that the Cube orientation has a much larger effect on the overall anisotropy during deep drawing than the S orientation.

3.3.2
Effective Material Properties

The demands on the properties of structural materials are ever increasing. The properties to be optimized include, but are not limited to, strength, ductility, stiffness, and so on. This trend leads to the development of structural materials with more and more complex microstructures. Examples are high-strength steels such as dual-phase or complex multiphase steels. This complexity makes it difficult to predict effective material properties, as most of the homogenization schemes mentioned in Section 3.2.2 fail to give correct results. However, without correct material models describing their properties new materials cannot be introduced into the market, as FEM engineers cannot perform decent product simulations. In particular, not being able to predict effective properties implies not being able to optimize properties.

Full-field crystal plasticity simulations (sometimes referred to as direct crystal plasticity) are a very promising way of obtaining effective material properties of complex microstructures. However, as the simulated volume of material needs to be relatively large and at the same time it requires high mesh resolution, these simulations are computationally demanding. Moreover, for optimum results, sophisticated constitutive models are needed, which not only describe all individual phases but also the interface properties need to be correctly incorporated.

3.3.2.1 Direct Transfer of Microstructures

Microstructure characterization has made tremendous progress in the last 20 years, for example, the use of electron backscatter diffraction (EBSD) to measure orientation maps. While 2-D measurements [97–99] can be regarded standard, 3-D measurements using serial sectioning [100] become more and more common. The main disadvantage of serial sectioning is certainly the fact that it is a destructive technique, that is, it is impossible to perform experiments with the characterized sample. Therefore, other nondestructive methods such as 3-D X-ray tomography [101–103] gain importance as well. Direct comparison of experiments and simulations is possible in this case even in 3-D; however, to date, in most cases only surface data is compared using 2-D EBSD measurements.

The data provided by these methods need to be transferred into simulation models. As it usually comes in the form of point (or voxel) data, this is most conveniently done for simulation methods also using point grids, for example, spectral methods. While a direct transfer is also possible for the FEM, it is in many cases desirable to use shape-conforming meshes.

Figure 3.11 shows an orientation map of a dual-phase steel. The orientation data connected to each pixel (measurement resolution 1618×1154 pixel) are used as

Figure 3.11 Orientation map of a dual-phase steel obtained by EBSD. The gray shades indicate the crystal orientation while the dark shaded areas indicate Martensite.

input for a crystal plasticity simulation using a recently developed spectral solver [85] included in the general crystal plasticity framework DAMASK [104,105]. Figure 3.12 shows the distribution of the stress component P_{11} after 3.1% strain in the 1-direction (the horizontal direction). While the average stress level is about 0.8 GPa, it can be well seen that within the Martensite regions stresses of up to 2.8 GPa can be found, whereas in the other regions of the sample even compressive stresses prevail.

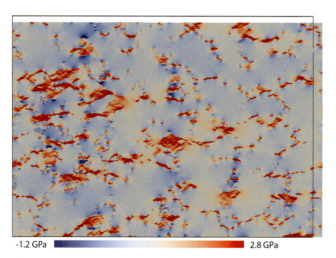

Figure 3.12 Stress component P_{11} after straining the microstructure shown in Figure 3.11 3.1% in the 1-direction (horizontal). Black frame indicates the original shape.

3.3.2.2 Representative Volume Elements

Direct transfer of microstructures is useful for comparing simulations results with experiments, that is, model validation. However, it only provides the properties of a special microstructure. Therefore, for microstructure optimization, other simulation tools are needed. Representative volume elements (RVEs) are suited to obtain properties of *typical* microstructures. They are artificial volumes that reproduce some statistical measures of real microstructures, such as grain size, shape, orientation, and so on. The more of such measures are taken into account the more representative the volume becomes. Representative in this context means that the homogenized properties of the volume element do not depend on the exact realization of the RVE, but that any RVE built using the same microstructure measures has the same properties. To avoid surface effects usually periodic boundary conditions are used in the simulation of RVEs. While it is desirable to model an RVE with as large volume as possible, at the same time the spatial resolution should also be maximized, especially when modeling microstructures where large gradients are to be expected. As computer resources, that is, the total number of points that can be handled, are limited, a compromise between both aspects has to be found.

Figure 3.13 shows an (R)VE containing 50 grains of arbitrary orientation. It is constructed using a Voronoi tessellation [106], that is, the average grain size is matched by choosing the appropriate number of grains. When additional microstructure measures are to be taken into account the process of building the RVE quickly becomes rather complex and more sophisticated approaches are needed [107–109]. The DREAM.3D software [110] is one example of free software for automatically creating RVEs respecting a number of statistical microstructure measures at the same time.

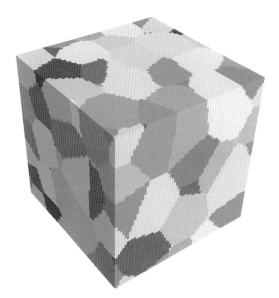

Figure 3.13 (Representative) Volume Element containing 50 randomly oriented grains.

3 Crystal Plasticity Modeling

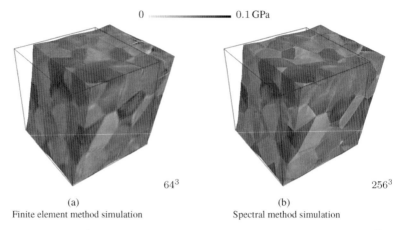

(a) Finite element method simulation — 64^3

(b) Spectral method simulation — 256^3

Figure 3.14 Local first Piola–Kirchhoff shear stress P_{yz} at average shear deformation of $\bar{F}_{yz} = 0.2$ mapped onto the deformed configuration.

The homogenized properties of the RVE are then found by applying some load to it. Figure 3.14 shows the RVE from Figure 3.13 after a simple shear deformation up to 20%. The left subfigure shows the resulting stress distribution obtained using the FEM, while the right subfigure shows the result using a spectral method. As the spectral method is much faster and more memory efficient than the FEM, a much higher mesh resolution could be used. It is, therefore, no surprise that the result from the spectral method shows greater details than that obtained using the FEM. Figure 3.15 shows the stress–strain curves resulting from the RVE simulations. Here another advantage of the spectral method becomes visible. While the FEM results do show systematic mesh dependence, the spectral method results in almost the same stress–strain curve for all mesh resolutions tested (16^3 to 256^3).

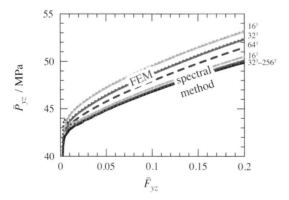

Figure 3.15 Volume-averaged shear stress $\bar{P}_{yz}$ as function of shear deformation $\bar{F}_{yz}$. Mesh/grid resolution increases from light to dark colors: 16^3 to 64^3 (finite element method, upper curves) and 16^3 to 256^3 (spectral method, lower curves). Convergence tolerance decreases from dotted to dashed to solid lines; however, results essentially coincide.

As the RVEs are built on the basis of statistical descriptions of the microstructure, they can easily be used to find optimized microstructures. In this process, the respective microstructure measures are varied until a microstructure with the desired properties is found.

3.3.2.3 The *Virtual Laboratory*

As mentioned in Section 3.2.2, full-field simulation of RVEs can serve as a homogenization scheme for large-scale forming simulations [111]. However, the computational effort is extremely high, which makes the approach prohibitive up to now at least for 3-D simulations. Nevertheless, RVE simulations can be of great value for large-scale simulations. In an approach often called a *Virtual Laboratory*, RVE simulations can be used to calibrate analytical yield surfaces commonly used in large-scale FEM forming simulations. These yield surface descriptions are normally calibrated using experimental data, for example, yield stresses determined from tests with different loading conditions. In the *Virtual Laboratory* these tests are replaced by respective RVE simulations (see Figure 3.16). The advantage in this approach is that some of the experiments necessary are difficult to impossible to perform while RVE simulations are easy to carry out with almost any kind of boundary condition.

Kraska *et al.* [113] have used the *Virtual Laboratory* to calibrate the material model as proposed by Vegter *et al.* [112] and implemented in PAM-STAMP 2G [114]. The interpolation requires at least the following deformation tests:

- A stack compression test with measurement of ovalization.
- Three uniaxial tensile tests (0°, 45°, 90°) with lateral strain measurement.
- Three tensile tests with constrained lateral strain (plane strain, 0°, 45°, 90°). Contrary to real tests, the lateral stress can be identified in virtual tests.
- Three shear tests, providing yield locus data in the lower right quadrant for different principal stress directions with respect to the rolling direction.

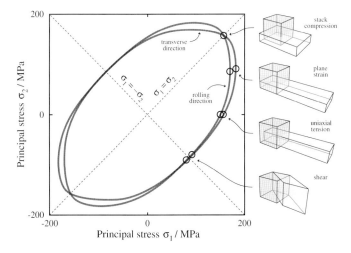

Figure 3.16 The Vegter *et al.* [112] yield locus derived from virtual test data. Corresponding individual deformation tests are schematically depicted at right.

Table 3.3 Parameters for the Vegter yield locus (PAM-STAMP 2G [114]) obtained from virtual deformation tests on H320LA.

Parameter	Angle to rolling direction (°)		
	0	45	90
σ-uniaxial	1	1.01	1.029
r-uniaxial	0.624	0.798	0.950
σ-plane strain	1.1	1.14	1.16
α-plane strain	0.5	0.5	0.5
σ-pure shear	0.5615	0.5743	0.603
r-biaxial	0.75	–	–
σ-biaxial	1.004	–	–

The required parameters (see Table 3.3) were determined from all virtual deformation experiments at 1% accumulated shear deformation on the slip systems, which corresponds to approximately 0.4% plastic strain.

With the aforementioned yield locus parametrization, stamping, trimming, and springback of a car boot made of H320LA was simulated (Figure 3.17). For comparison, two additional simulations were carried out. One using a simple yield locus (Hill48 [115]) fitted to tensile data from experiments and another in which the Vegter parametrization relied on experimentally determined r-values instead of those resulting from the virtual deformation tests.

The differences between optically scanned and simulated geometries are illustrated in Table 3.4. The maximum deviation of the simulation using purely virtual test data (center row) is larger than that of the simple yield locus. If the r-values of the Vegter model fitted to virtually derived data are replaced by measured ones, the discrepancy is reduced. However, the remaining disagreement is only marginally less than that resulting from the Hill48 yield locus.

Springback simulation and compensation generate an increasing demand for precise material models. Microstructural approaches such as crystal plasticity require, however, still too many resources in terms of memory and computational power for a direct simulation of industrial sheet metal forming. Therefore, more

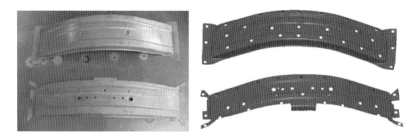

Figure 3.17 Stamping part of car boot. Photographs and visualized measured data. After stamping and after trimming.

Table 3.4 Shape deviation between measured and simulated geometry for different material models and different types of test data for parameter fitting (Reproduced from Roters et al. [84, Section 10.4]).

Yield criteria	Test data	r0	r45	r90	Maximum difference (mm)
Hill48	Real[a]	0.815	1.245	1.087	5.7
Vegter	Virtual[b]	0.624	0.798	0.950	6.9
Vegter	Mixed[c]	0.815	1.245	1.087	5.6

a) Corresponding to DIN.
b) At 0.4% plastic strain.
c) Virtual and real (r-values).

sophisticated empirical models are likely to emerge. The resulting increased flexibility (and accuracy) of those models entails a higher experimental effort for parameter identification. In this context, virtual specimens with microstructure-based constitutive laws may be used to move the effort from real-world mechanical testing to computer simulation. It was shown that material parameters obtained from texture data and tensile tests using the virtual test program can compete in simulation quality with the full parameter set obtained experimentally. Finally, the simulations can be used in the future to not only predict the behavior of given microstructures but also to optimize the microstructure in view of the sheet-forming process.

3.4
Conclusions and Outlook

Crystal plasticity simulations offer a very high potential for modeling the effective properties of complex multiphase microstructures based on the properties of single crystals/phases. However, a number of conditions need to be satisfied to explore this potential:

Constitutive models: Crystal plasticity simulations rely on the fact that the single crystal/phase behavior can be correctly modeled. Real microstructures contain an almost infinite wealth of building blocks: interstitials, precipitates, second phases, and so on. It is the constitutive model that has to describe all these aspects. To reach the goal of really predicting properties – rather than reproducing them – sophisticated constitutive models need to be developed.

Interface models: Besides the single crystal behavior, the effect of interfaces (e.g., grain/phase boundaries) needs to be correctly modeled by the constitutive laws.

Homogenization models: For application of crystal plasticity at component scale advanced homogenization techniques are needed.

Numerical efficiency: At present, the computational demands are prohibitive for the direct incorporation of crystal plasticity models in component scale forming simulations. However, concepts such as the *Virtual Laboratory* overcome this problem by combining classical yield surface concepts with crystal plasticity RVE simulations. At the same time, new solution techniques like the spectral methods enable direct crystal plasticity simulations with a so far unknown level of resolution. Together with the ever-increasing computer power direct coupling of crystal plasticity and forming simulations might become possible in the not-too-distant future.

Once all these ingredients are in place, crystal plasticity simulations will be undoubtedly among the most powerful tools for *Microstructural Design of Advanced Engineering Materials*. Crystal plasticity simulations can then be used for optimizing the microstructure with respect to whatever property desired by the user.

References

1 Sachs, G. (1928) Zur Ableitung einer Fliessbedingung. *Zeitschrift des Vereines deutscher Ingenieure*, **72**, 734–736.

2 Taylor, G.I. (1938) Plastic strain in metals. *Journal of Institute of Metals*, **62**, 307–324.

3 Bishop, J.F.W. and Hill, R. (1951) A theoretical derivation of the plastic properties of a polycrystalline face centered metal. *Philosophical Magazine*, **42**, 1298–1307.

4 Bishop, J.F.W. and Hill, R. (1951) A theory of the plastic distortion of a polycrystalline aggregate under combined stresses. *Philosophical Magazine*, **42**, 414–427.

5 Kröner, E. (1961) On the plastic deformation of polycrystals. *Acta Metallurgica*, **9**, 155–161.

6 Raabe, D., Klose, P., Engl, B., Imlau, K.P., Friedel, F., and Roters, F. (2002) Concepts for integrating plastic anisotropy into metal forming simulations. *Advanced Engineering Materials*, **4**, 169–180.

7 Raabe, D. Roters, F. Barlat, F. and Chen, L.Q. (eds) (2004) *Continuum Scale Simulation of Engineering Materials, Fundamentals – Microstructures – Process Applications*, Wiley-VCH, Weinheim, Germany, ISBN: 3-527-30760-5.

8 Zienkiewicz, O.C. and Taylor, R.L. (2005) *The Finite Element Method for Solid and Structural Mechanics*, 6th edn, Butterworth-Heinemann, Oxford, UK.

9 Zienkiewicz, O.C., Taylor, R.L., and Nithiarasu, P. (2005) *The Finite Element Method for Fluid Dynamics*, 6th edn, Butterworth-Heinemann, Oxford, UK

10 Zienkiewicz, O.C., Taylor, R.L., and Zhu, J.Z. (2005) *The Finite Element Method: Its Basis and Fundamentals*, 6th edn, Butterworth-Heinemann. Oxford, UK

11 Orowan, E. (1934) Zur Krsitallplastizität I. –III. *Zeitschrift fuer Physik*, **89**, 605–659.

12 Polanyi, M. (1934) Über eine Art Gitterstörung, die einen Kristall plastisch machen könnte. *Zeitschrift Fur Physik*, **89**, 660–664.

13 Taylor, G.I. (1934) The mechanism of plastic deformation of crystals. Part I. – theoretical. *Proceedings of the Royal Society of London Series A*, **145**, 362–387.

14 Taylor, G.I. (1934) The mechanism of plastic deformation of crystals. Part II. – comparison with observations. *Proceedings of the Royal Society of London Series A*, **145**, 388–404.

15 Peirce, D., Asaro, R.J., and Needleman, A. (1982) An analysis of nonuniform and localized deformation in ductile single crystals. *Acta Metallurgica*, **30**, 1087–1119.

16 Harren, S.V. and Asaro, R.J. (1989) Nonuniform deformations in polycrystals and aspects of the validity of the Taylor model. *Journal of the Mechanics and Physics of Solids*, **37**, 191–232.

17 Harren, S.V., Dève, H., and Asaro, R.J. (1988) Shear band formation in plane strain compression. *Acta Metallurgica*, **36**, 2435–2480.

18 Becker, R., Butler, J.F., Hu, H., and Lalli, L.A. (1991) Analysis of an aluminum single crystal with unstable initial orientation (001) [111] in channel die compression. *Metallurgical Transactions A*, **22**, 45–58.

19 Becker, R. (1991) Analysis of texture evolution in channel die compression – I. Effects of grain interaction. *Acta Metallurgica*, **39**, 1211–1230.

20 Sachtleber, M., Zhao, Z., and Raabe, D. (2002) Experimental investigation of plastic grain interaction. *Materials Science and Engineering A*, **336**, 81–87.

21 Raabe, D., Sachtleber, M., Weiland, H., Scheele, G., and Zhao, Z. (2003) Grain-scale micromechanics of polycrystal surfaces during plastic straining. *Acta Materialia*, **51**, 1539–1560.

22 Asaro, R.J. and Rice, J.R. (1977) Strain localization in ductile single crystals. *Journal of the Mechanics and Physics of Solids*, **25**, 309–338.

23 Rice, J.R. (1971) Inelastic constitutive relations for solids: an internal variable theory and its application to metal plasticity. *Journal of the Mechanics and Physics of Solids*, **19**, 433–455.

24 Arsenlis, A. and Parks, D.M. (1999) Crystallographic aspects of geometrically-necessary and statistically-stored dislocation density. *Acta Materialia*, **47**, 1597–1611.

25 Arsenlis, A. and Parks, D.M. (2002) Modeling the evolution of crystallographic dislocation density in crystal plasticity. *Journal of the Mechanics and Physics of Solids*, **50**, 1979–2009.

26 Arsenlis, A., Parks, D.M., Becker, R., and Bulatov, V.V. (2004) On the evolution of crystallographic dislocation density in non-homogeneously deforming crystals. *Journal of the Mechanics and Physics of Solids*, **52**, 1213–1246.

27 Cheong, K.S. and Busso, E.P. (2004) Discrete dislocation density modelling of single phase FCC polycrystal aggregates. *Acta Materialia*, **52**, 5665–5675.

28 Evers, L.P., Brekelmans, W.A.M., and Geers, M.G.D. (2004) Non-local crystal plasticity model with intrinsic SSD and GND effects. *Journal of the Mechanics and Physics of Solids*, **52**, 2379–2401.

29 Evers, L.P., Brekelmans, W.A.M., and Geers, M.G.D. (2004) Scale dependent crystal plasticity framework with dislocation density and grain boundary effects. *International Journal of Solids and Structures*, **41**, 5209–5230.

30 Evers, L.P., Parks, D.M., Brekelmans, W.A.M., and Geers, M.G.D. (2002) Crystal plasticity model with enhanced hardening by geometrically necessary dislocation accumulation. *Journal of the Mechanics and Physics of Solids*, **50**, 2403–2424.

31 Fleck, N.A. and Hutchinson, J.W. (1997) Strain gradient plasticity, *Advances in Applied Mechanics*, vol. 33, Academic Press, New York, pp. 1825–1857.

32 Fleck, N.A., Muller, G.M., Ashby, M.F., and Hutchinson, J.W. (1994) Strain gradient plasticity: Theory and experiment. *Acta Metallurgica*, **42**, 475–487.

33 Gao, H. and Huang, Y. (2003) Geometrically necessary dislocation and size-dependent plasticity. *Scripta Materialia*, **48** (2), 113–118.

34 Gao, H., Huang, Y., Nix, W.D., and Hutchinson, J.W. (1999) Mechanism-based strain gradient plasticity – I. Theory. *Journal of the Mechanics and Physics of Solids*, **47**, 1239–1263.

35 Kalidindi, S.R. (1998) Incorporation of deformation twinning in crystal plasticity models. *Journal of the Mechanics and Physics of Solids*, **46**, 267–290.

36 Kalidindi, S.R. (2001) Modeling anisotropic strain hardening and deformation textures in low stacking fault energy fcc metals. *International Journal of Plasticity*, **17**, 837–860.

37 Ma, A. and Roters, F. (2004) A constitutive model for fcc single crystals based on dislocation densities and its application to uniaxial compression of aluminium single crystals. *Acta Materialia*, **52**, 3603–3612.

38 Ma, A., Roters, F., and Raabe, D. (2006) A dislocation density based constitutive model for crystal plasticity FEM including

geometrically necessary dislocations. *Acta Materialia*, **54**, 2169–2179.
39 Ma, A., Roters, F., and Raabe, D. (2006) On the consideration of interactions between dislocations and grain boundaries in crystal plasticity finite element modeling – theory, experiments, and simulations. *Acta Materialia*, **54**, 2181–2194.
40 Nix, W.D. and Gao, H. (1998) Indentation size effects in crystalline materials: A law of strain gradient plasticity. *Journal of the Mechanics and Physics of Solids*, **46**, 411–425.
41 Salem, A.A., Kalidindi, S.R., and Semiatin, S.L. (2005) Strain hardening due to deformation twinning in α-titanium: Constitutive relations and crystal-plasticity modeling. *Acta Materialia*, **53**, 3495–3502.
42 Staroselskya, A. and Anand, L. (2003) A constitutive model for hcp materials deforming by slip and twinning: application to magnesium alloy AZ31B. *International Journal of Plasticity*, **19**, 1843–1864.
43 Suiker, A.S.J. and Turteltaub, S. (2005) Computational modelling of plasticity induced by martensitic phase transformations. *International Journal for Numerical Methods in Engineering*, **63**, 1655–1693.
44 Roters, F., Eisenlohr, P., Hantcherli, L., Tjahjanto, D.D., Bieler, T.R., and Raabe, D. (2010) Overview of constitutive laws, kinematics, homogenization and multiscale methods in crystal plasticity finite-element modeling: Theory, experiments, applications. *Acta Materialia*, **58**, 1152–1211.
45 Bachu, V. and Kalidindi, S.R. (1998) On the accuracy of the predictions of texture evolution by the finite element technique for fcc polycrystals. *Materials Science and Engineering A*, **257**, 108–117.
46 Beaudoin, A.J., Mecking, H., and Kocks, U.F. (1996) Development of localized orientation gradients in fcc polycrystals. *Philosophical Magazine A – Physics of Condensed Matter Structure Defects and Mechanical Properties*, **73**, 1503–1517.
47 Mika, D.P. and Dawson, P.R. (1998) Effects of grain interaction on deformation in polycrystals. *Materials Science and Engineering A*, **257**, 62–76.
48 Sarma, G.B. and Dawson, P.R. (1996) Texture predictions using a polycrystal plasticity model incorporating neighbor interactions. *International Journal of Plasticity*, **12**, 1023–1054.
49 Sarma, G.B., Radhakrishnan, B., and Zacharia, T. (1998) Finite element simulations of cold deformation at the mesoscale. *Computational Materials Science*, **12**, 105–123.
50 Zhao, Z., Kuchnicki, S., Radovitzky, R., and Cuitiño, A. (2007) Influence of in-grain mesh resolution on the prediction of deformation textures in fcc polycrystals by crystal plasticity FEM. *Acta Materialia*, **55**, 2361–2373.
51 Zhao, Z., Ramesh, M., Raabe, D., Cuitiño, A., and Radovitzky, R. (2008) Investigation of three-dimensional aspects of grain-scale plastic surface deformation of an aluminum oligocrystal. *International Journal of Plasticity*, **24**, 2278–2297.
52 Raabe, D. and Roters, F. (2004) Using texture components in crystal plasticity finite element simulations. *International Journal of Plasticity*, **20**, 339–361.
53 Zhao, Z., Mao, W., Roters, F., and Raabe, D. (2001) Introduction of a texture component crystal plasticity finite element method for anisotropy simulations. *Advanced Engineering Materials*, **3**, 984–990.
54 Eisenlohr, P. and Roters, F. (2008) Selecting a set of discrete orientations for accurate texture reconstruction. *Computational Materials Science*, **42**, 670–678.
55 Melchior, M.A. and Delannay, L. (2006) A texture discretization technique adapted to polycrystalline aggregates with non-uniform grain size. *Computational Materials Science*, **37**, 557–564.
56 Tóth, L.S. and Van Houtte, P. (1992) Discretization techniques for orientation distribution functions. *Textures and Microstructures*, **19**, 229–244.
57 Lebensohn, R.A. (2001) N-site modeling of a 3D viscoplastic polycrystal using fast Fourier transform. *Acta Materialia*, **49**, 2723–2737.
58 Moulinec, H. and Suquet, P. (1998) A numerical method for computing the overall response of nonlinear composites

with complex microstructure. *Computer Methods in Applied Mechanics and Engineering*, **157**, 69–94.

59 Lee, E.H. (1969) Elastic–plastic deformation at finite strains. *Journal of Applied Mechanics of the ASME*, **36**, 1–6.

60 Hutchinson, J.W. (1976) Bounds and self-consistent estimates for creep of polycrystalline materials. *Proceedings of the Royal Society of London Series A*, **348**, 101–127.

61 Peirce, D., Asaro, R.J., and Needleman, A. (1983) Material rate dependence and localized deformation in crystalline solids. *Acta Metallurgica*, **31**, 1951–1976.

62 Lan, Y.J., Xiao, N.M., Li, D.Z., and Li, Y.Y. (2005) Mesoscale simulation of deformed austenite decomposition into ferrite by coupling a cellular automaton method with a crystal plasticity finite element model. *Acta Materialia*, **53**, 991–1003.

63 Thamburaja, P. and Anand, L. (2001) Polycrystalline shape-memory materials: effect of crystallographic texture. *Journal of the Mechanics and Physics of Solids*, **49**, 709–737.

64 Marketz, W.T., Fischer, F.D., Kauffmann, F., Dehm, G., Bidlingmaier, T., Wanner, A., and Clemens, H. (2002) On the role of twinning during room temperature deformation of TiAl based alloys. *Materials Science and Engineering A*, **329–331**, 177–183.

65 Staroselsky, A. and Anand, L. (1998) Inelastic deformation of polycrystalline face centered cubic materials by slip and twinning. *Journal of the Mechanics and Physics of Solids*, **46**, 671–696.

66 Voigt, W. (1889) Über die Beziehung zwischen den beiden Elastizitätskonstanten isotroper Körper. *Wiedmanns Annalen der Physik und Chemie*, **38**, 573–587.

67 Honeff, H. and Mecking, H. (1981) Analysis of the deformation texture at different rolling conditions, in *Proc. ICOTOM 6*, vol. 1 (ed. S. Nagashima), The Iron and Steel Institute of Japan, Tokyo, Japan, pp. 347–355.

68 Kocks, U.F. and Chandra, H. (1982) Slip geometry in partially constrained deformation. *Acta Metallurgica*, **30**, 695–709.

69 Van Houtte, P. (1982) On the equivalence of the relaxed Taylor theory and the Bishop–Hill theory for partially constrained plastic deformation of crystals. *Materials Science and Engineering*, **55** (1), 69–77.

70 Delannay, L., Kalidindi, S.R., and Van Houtte, P. (2002) Quantitative prediction of textures in aluminium cold rolled to moderate strains. *Materials Science and Engineering A*, **336**, 233–244.

71 Van Houtte, P., Delannay, L., and Kalidindi, S.R. (2002) Comparison of two grain interaction models for polycrystal plasticity and deformation texture prediction. *International Journal of Plasticity*, **18**, 359–377.

72 Van Houtte, P., Delannay, L., and Samajdar, I. (1999) Quantitative prediction of cold rolling textures in low-carbon steel by means of the LAMEL model. *Textures and Microstructures*, **31**, 109–149.

73 Crumbach, M., Goerdeler, M., and Gottstein, G. (2006) Modelling of recrystallisation textures in aluminium alloys: I. Model set-up and integration. *Acta Materialia*, **54**, 3275–3289.

74 Crumbach, M., Goerdeler, M., and Gottstein, G. (2006) Modelling of recrystallisation textures in aluminium alloys: II. Model performance and experimental validation. *Acta Materialia*, **54**, 3291–3306.

75 Crumbach, M., Goerdeler, M., Gottstein, G., Neumann, L., Aretz, H., and Kopp, R. (2004) Through-process texture modelling of aluminium alloys. *Modelling and Simulation in Materials Science and Engineering*, **12**, S1–S18.

76 Crumbach, M., Pomana, G., Wagner, P., and Gottstein, G. (2001) A Taylor type deformation texture model considering grain interaction and material properties. Part I – fundamentals, in *Recrystallisation and Grain Growth, Proc. First Joint Conference* (eds. G. Gottstein and D.A. Molodov), Springer, Berlin, pp. 1053–1060.

77 Tjahjanto, D.D., Eisenlohr, P., and Roters, F. (2010) A novel grain cluster-based homogenization scheme. *Modelling and Simulation in Materials Science and Engineering*, **18**, 015006.

78 Feyel, F. and Chaboche, J.L. (2000) FE^2 multiscale approach for modelling the elastoviscoplastic behaviour of long fibre SiC/Ti composite materials. *Computer Methods in Applied Mechanics and Engineering*, **183**, 309–330.

79 Kouznetsova, V., Brekelmans, W.A.M., and Baaijens, F.P.T. (2001) An approach to micro-macro modeling of heterogeneous materials. *Computational Mechanics*, **27**, 37–48.

80 Kouznetsova, V., Geers, M.G.D., and Brekelmans, W.A.M. (2002) Multi-scale constitutive modelling of heterogeneous materials with a gradient-enhanced computational homogenization scheme. *International Journal for Numerical Methods in Engineering*, **54**, 1235–1260.

81 Miehe, C., Schotte, J., and Lambrecht, M. (2002) Homogenization of inelastic solid materials at finite strains based on incremental minimization principles. Application to the texture analysis of polycrystals. *Journal of the Mechanics and Physics of Solids*, **50**, 2123–2167.

82 Miehe, C., Schröder, J., and Schotte, J. (1999) Computational homogenization analysis in finite plasticity Simulation of texture development in polycrystalline materials. *Computer Methods in Applied Mechanics and Engineering*, **171**, 387–418.

83 Smit, R.J.M., Brekelmans, W.A.M., and Meijer, H.E.H. (1998) Prediction of the mechanical behavior of nonlinear heterogeneous systems by multi-level finite element modeling. *Computer Methods in Applied Mechanics and Engineering*, **155**, 181–192.

84 Roters, F., Eisenlohr, P., Bieler, T.R., and Raabe, D. (2010) *Crystal Plasticity Finite Element Methods in Materials Science and Engineering*, WILEY-VCH, Weinheim.

85 Eisenlohr, P., Diehl, M., Lebensohn, R., and Roters, F. (2013) A spectral method solution to crystal elasto-viscoplasticity at finite strains. *International Journal of Plasticity*, **46**, 37–53.

86 Prakash, A. and Lebensohn, R. (2009) Simulation of micromechanical behavior of polycrystals: Finite elements versus fast Fourier transforms. *Modelling and Simulation in Materials Science and Engineering*, **17**, 064010–064016.

87 Asaro, R.J. and Needleman, A. (1985) Texture development and strain hardening in rate dependent polycrystals. *Acta Metallurgica*, **33**, 923–953.

88 Bronkhorst, C.A., and Kalidindi, S.R., and Anand, L. (1992) Polycrystalline plasticity and the evolution of crystallographic texture in FCC metals. *Philosophical Transactions of the Royal Society of London Series A – Mathematical Physical and Engineering Sciences*, **341** (1662), 443–477.

89 Lebensohn, R.A. and Tomé, C.N. (1993) A self-consistent anisotropic approach for the simulation of plastic deformation and texture development of polycrystals: Application to zirconium alloys. *Acta Metallurgica*, **41**, 2611–2624.

90 Li, S. and Van Houtte, P. (2002) Performance of statistical (Taylor, Lamel) and CP-FE models in texture prediction of aluminium alloys in cold rolling. *Aluminium*, **78**, 918–922.

91 Bate, P. (1999) Modelling deformation microstructure with the crystal plasticity finite-element method. *Philosophical Transactions of the Royal Society of London Series A – Mathematical Physical and Engineering Sciences*, **357**, 1589–1601.

92 Kalidindi, S.R., Bronkhorst, C.A., and Anand, L. (1992) Crystallographic texture evolution in bulk deformation processing of fcc metals. *Journal of the Mechanics and Physics of Solids*, **40**, 537–569.

93 Raabe, D. (1995) Investigation of contribution of 123 slip planes to development of rolling textures in bcc metals by use of Taylor models. *Materials Science and Technology*, **11**, 455–460.

94 Raabe, D. (1995) Simulation of rolling textures of bcc metals under consideration of grain interactions and 110, 112 and 123 slip planes. *Materials Science and Engineering A*, **197**, 31–37.

95 Helming, K. (1996) *Texturapproximation durch Modellkomponenten*, Cuvillier Verlag, Göttingen.

96 Helming, K., Schwarzer, R.A., Rauschenbach, B., Geier, S., Leiss, B., Wenk, H., Ullemeier, K., and Heinitz, J.

(1994) Texture estimates by means of components. *Zeitschrift Fur Metallkunde*, **85**, 545–553.

97 Adams, B.L., Wright, S.I., and Kunze, K. (1993) Orientation imaging: The emergence of a new microscopy. *Metallurgical Transactions A*, **24**, 819–831.

98 Dingley, D. (1984) On-line determination of crystal orientation and texture determination in an SEM. *Proceedings of the Royal Microscopical Society*, **19**, 74–75.

99 Dingley, D. (2004) Progressive steps in the development of electron backscatter diffraction and orientation imaging microscopy. *Journal of the Minerals Metals and Materials Society*, **213**, 214–224.

100 Zaefferer, S., Wright, S.I., and Raabe, D. (2008) 3D-orientation microscopy in a FIB SEM: A new dimension of microstructure characterisation. *Metallurgical and Materials Transactions A – Physical Metallurgy and Materials Science*, **39**, 374–389.

101 Ice, G.E. and Larson, B.C. (2000) 3D X-ray crystal microscope. *Advanced Engineering Materials*, **2**, 643–646.

102 Lienert, U., Brandes, M.C., Bernier, J.V., Weiss, J., Shastri, S., Mills, M.J., and Miller, M.P. (2009) In situ single-grain peak profile measurements on Ti-7Al during tensile deformation. *Materials Science and Engineering A*, **524**, 46–54.

103 Lienert, U., Li, S.F., Hefferan, C.M., Lind, J., Suter, R.M., Bernier, J.V., Barton, N.R., Brandes, M.C., Mills, M.J., Miller, M.P., Jakobsen, B., and Pantleon, W. (2008) High-energy diffraction microscopy at the advanced photon source. *JOM*, **63**, 70–77.

104 DAMASK (2013) *Düsseldorf Advanced Material Simulation Kit*. http://damask.mpie.de. accessed 22.04.2013.

105 Roters, F., Eisenlohr, P., Kords, C., Tjahjanto, D., Diehl, M., and Raabe, D. (2012) DAMASK: The Düsseldorf Advanced MAterial Simulation Kit for studying crystal plasticity using an FE based or a spectral numerical solver, in *Procedia IUTAM: IUTAM Symposium on Linking Scales in Computation: From Microstructure to Macroscale Properties*, vol. 3 (ed. O. Cazacu), Elsevier, vol. 3, pp. 3–10.

106 Okabe, A., Boots, B., Sugihara, K., and Chiu, S.N. (2000) *Spatial Tessellations: Concepts and Applications of Voronoi Diagrams*, 2nd edn, John Wiley & Sons, Inc., Hoboken, NJ

107 Brands, D., Balzani, D., and Schröder, J. (2010) On the construction of statistically similar representative volume elements based on the lineal-path function. *Proceedings in Applied Mathematics and Mechanics*, **10** (1), 399–400.

108 Groeber, M. (2007) Development of an automated characterization - representation framework for the modeling of polycrystalline materials in 3D, Ph.D. thesis, Ohio State University.

109 Lyckegaard, A., Lauridsen, E., Ludwig, W., Fonda, R., and Poulsen, H. (2011) On the use of Laguerre tessellations for representations of 3D grain structures. *Advanced Engineering Materials*, **13** (3), 165–170.

110 DREAM3D (2013) Digital Representation Environment for Analyzing Microstructure in 3D. http://dream3d.bluequartz.net accessed 22.04.2013.

111 Schröder, J., Balzani, D., and Brands, D. (2010) A FE^2-Homogenization Technique for Two-Phase Steels based on Statistically Similar Representative Volume Elements, in Proceedings of the International Symposium of Plasticity 2010, Neat press.

112 Vegter, H., Horn, C.t., An, Y., Atzema, E., Pijlman, H., and Boogaard, A.v.d., and Huétink, J. (2003) Characterisation and modelling of the plastic material behaviour and its application in sheet metal forming simulation., in Proceedings of COMPLAS VII, CIMNE (eds E. Oñate and D.R.J. Owen).

113 Kraska, M., Doig, M., Tikhomirov, D., Raabe, D., and Roters, F. (2009) Virtual material testing for stamping simulations based on polycrystal plasticity. *Computational Materials Science*, **46**, 383–392.

114 PAM-STAMP, 2G (2004) *1 Standard. Stamping Solutions Manual*, ESI Group, Paris.

115 Hill, R. (1948) A theory of the yielding and plastic flow of anisotropic metals. *Proceedings of the Royal Society of London Series A*, **193**, 281–297.

4
Modeling of Severe Plastic Deformation: Time-Proven Recipes and New Results
Yuri Estrin and Alexei Vinogradov

4.1
Introduction

Microstructure-based constitutive models are an indispensable tool to describe strength and plasticity of metallic materials. In terms of their architecture and the number of fit parameters involved, such models are commonly much more economical than most phenomenological models that are still in use in the mechanics community. Besides, the parameters in microstructure-based constitutive model have a clear physical interpretation, so that it can be predicted, at least in principle, how they would change with alloy composition and heat treatment. The dislocation theory has provided solid foundations for such microstructure-related constitutive models, and their use has advanced the development of new processes and products in industries as diverse as automotive, aerospace, and electronics. The use of integrated computational tools for through-process modeling, which employ dislocation theory based approaches, has become a reality in large-scale manufacturing domain [1]. However, despite the immense efforts that have been put in the development of dislocation-theory-based constitutive modeling, some unresolved problems still remain. This confirms the prophetic words of Cottrell [2] who wrote 60 years ago that the modeling of strain hardening "was the first problem to be attempted by dislocation theory and may be the last to be solved."

In this chapter, we shall be focusing on the arsenal of models currently available for description and, more importantly, prediction of the behavior of metals and alloys under severe plastic deformation (SPD). A group of metal processing methods, collectively defined as SPD techniques [3], provide ways to deform metallic materials to giant strains hardly achievable with conventional processing routes. Not only do these techniques offer an interesting test ground for validation of large deformation models, but they also open up new possibilities to produce bulk materials with exceptionally small grain size – down to the deep submicron range – and remarkable strength [3–6]. After two decades of intensive but largely empirical research, process optimization in this area relies on computational modeling, and microstructure-based models capturing the physics of the deformation processes, and yet robust enough to be used in practical finite element codes, are in demand.

Microstructural Design of Advanced Engineering Materials, First Edition. Edited by Dmitri A. Molodov.
© 2013 Wiley-VCH Verlag GmbH & Co. KGaA. Published 2013 by Wiley-VCH Verlag GmbH & Co. KGaA.

We shall describe the modeling approaches that fulfill this demand. Some insights into the specifics of modeling of SPD processes were provided in an earlier review [7]. In our exposé, we shall start from the classical modeling recipes that go back to the work of Kocks [8] (see also [9–11]) and introduce more complexity where it is needed to capture the specifics of SPD. Examples of calculations of the material behavior during SPD processes will be given and outstanding problems to be resolved by future model development will be outlined.

4.2
One-Internal Variable Models

A seminal model, which indisputably has set the scene for future modeling efforts and which dominates the constitutive modeling philosophy to the present day, is the Kocks–Mecking model [8–11]. The structure of the model is footed on the fact that the dislocation contribution to flow stress, σ_d scales with the square root of the total dislocation density ρ:

$$\sigma_d = M\alpha Gb\sqrt{\rho} \qquad (4.1)$$

where M is the Taylor factor reflecting the texture of the material, G is the shear modulus, and b is the magnitude of the dislocation Burgers vector. The scaling law prescribed by Eq. (4.1) holds quite generally, regardless of the geometrical arrangement or the detail of the dislocation–dislocation interaction mechanism. The thermally activated character of dislocation glide underlying Eq. (4.1) can be represented by a power-law expression for the factor α:

$$\alpha = \alpha_o \left(\frac{\dot{\varepsilon}}{\dot{\varepsilon}_0}\right)^{1/m} \qquad (4.2)$$

which takes its origin in an Arrhenius equation for the plastic strain rate $\dot{\varepsilon}_0$. Here α_o is a constant, which represents the mechanical resistance to dislocation glide due to dislocation–dislocation interaction at absolute zero temperature. The temperature dependence is associated with thermally activated dislocation glide and resides in the parameters $\dot{\varepsilon}_0$ and m. Thus, at low temperatures (below half the melting point, say) $\dot{\varepsilon}_0$ may be considered to be constant, while m is inversely proportional to the absolute temperature. Conversely, at sufficiently high temperatures, m can be regarded as a constant, while $\dot{\varepsilon}_0$ is temperature dependent (and given by an Arrhenius equation) [8–12]. As the dislocation density evolves in the process of straining, an additional equation is required to describe this evolution. A general structure of this equation is given by

$$d\rho/d\varepsilon = M(1/(bL) - k_2\rho) \qquad (4.3)$$

where L represents the dislocation mean free path, that is, the average distance a dislocation travels before it gets immobilized at an unsurmountable obstacle. The coefficient k_2 is associated with the dynamic recovery leading to loss of dislocation density due to annihilation by processes such as cross-slip of screw dislocations (low-temperature case) or climb of edge dislocations (high-temperature case). Accordingly, k_2 is strain rate and temperature dependent. Kocks and Mecking associated the dislocation mean free path L with the average spacing between the dislocations (or, for the case dislocations are organized in a dislocation-cell pattern, with the dislocation cell size). As both these characteristic lengths scale inversely with $\sqrt{\rho}$, the evolution equation for ρ assumes the form

$$d\rho/d\varepsilon = M(k_1\sqrt{\rho} - k_2\rho) \tag{4.4}$$

where k_1 is a constant. Together, Eqs. (4.1, 4.3,) and (4.4) represent the original Kocks–Mecking model of strain hardening. While this type of model is commonly associated with these names, some other authors also deserve credit for introducing the dislocation density evolution approaches. In particular, a tribute should be paid to Janusz Klepaczko who has been developing similar concepts independently [13]. Yngve Bergström has also been promoting the use of Eq. (4.3), but mainly by considering the case of a constant mean free path L [14]. Gottstein and Argon [15] have extended Eq. (4.4) by considering the effect of subgrain boundary motion on the dislocation density evolution.

The Kocks–Mecking model was further advanced by the introduction of an additional dislocation storage term in Eq. (4.4) [9]:

$$d\rho/d\varepsilon = M(1/(b\Lambda) + k_1\sqrt{\rho} - k_2\rho) \tag{4.5}$$

The dislocation mean free path Λ in this equation can be governed by obstacles of various kinds at which gliding dislocations get immobilized. A lot of physical metallurgy can be captured by this term. Thus, Λ can be determined by the average spacing between nonshearable particles, the average distance between twins, the lamellae spacing in a lamellar structure, or the grain size. The latter case is of particular interest in the context of deformation of ultrafine grained materials produced by SPD. However, identifying Λ with the average grain size in modeling SPD processing is only sensible if the latter is already saturated and does not vary in the process of deformation or if its variation with strain is known or can be modeled [16].

Examples of successful application of the modified Kocks–Mecking model that employs the dislocation density evolution equation (4.5), sometimes referred to as the Kocks–Mecking–Estrin (KME) model [17], can be found in [8–12]. Here we present results showing that despite its simplicity the one-internal variable model can account for the material behavior under cyclic loading. Indeed, despite the complexity of the microstructures created in the course of SPD, the cyclic behavior of UFG metals allows a more simple description than that of ordinary poly- and single-crystals because dislocation patterning within the UFG structures does not occur. The characteristic dimensions of the major structural elements – grains or cells – are smaller than the characteristic length scale of dislocation structures self-organization during cyclic loading would induce. Vinogradov et al. [18,19] suggested that in light of this argument it is sensible to describe the shape of a

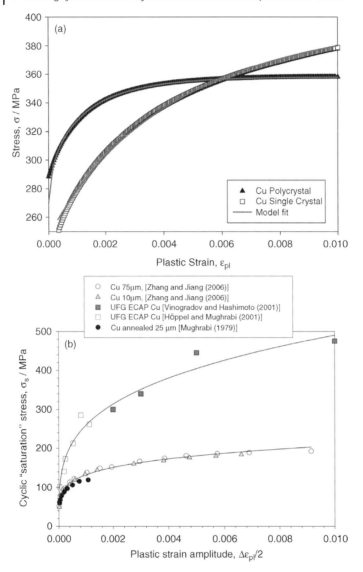

Figure 4.1 Typical ascending part of the stable hysteresis loop of SPD processed poly- and single-crystals of 99.96% copper subjected to one ECAP pass through a 90° die at room temperature. Curve fits by Eq. (4.7) are shown by thin solid lines.

stable hysteresis loop and the cyclic stress–strain curve in terms of the one-internal variable approach. Elementary integration of Eq. (4.5) with $k_1 = 0$ and $\rho(0) = \rho_0$ initial condition yields

$$\rho(\varepsilon) = \frac{k_0}{k_2 b \Lambda} \left(1 - e^{-Mk_2 \varepsilon}\right) + \rho_0 e^{-Mk_2 \varepsilon} \tag{4.6}$$

which, using again the Taylor relationship (4.1), provides a simple analytical expression for the flow stress as a function of strain:

$$\sigma(\varepsilon) = M\alpha Gb\sqrt{\frac{k_0}{k_2 b \Lambda}(1 - e^{-Mk_2\varepsilon}) + \rho_0 e^{-Mk_2\varepsilon}} \quad (4.7)$$

To fit the cyclic hysteresis loop, Vinogradov et al. [18,19] applied a slightly different variant of the KME model, Eq. (4.3), originally proposed by Essmann and Mughrabi [20] for dislocation annihilation in the form

$$\tau = \alpha Gb\sqrt{\frac{1 - \exp(-2y\gamma/b)}{\Lambda \gamma}} \quad (4.8)$$

where τ and γ are the shear stress and shear strain, respectively, Λ is the slip path of dislocations, which is assumed to be constant and approximately equal to the ultra-fine grain (or cell) size, and y is the so-called annihilation length for the dislocation density evolution during a stable cyclic deformation. A tacit assumption here is that mobile dislocations generated at a grain boundary pass through the grain and disappear at the opposite grain boundary, that is, the grain boundaries act as effective sources and sinks for dislocations. Since TEM observations do not reveal substantial differences between the initial and the postfatigue structures (unless fatigue-induced dynamic grain coarsening occurs in pure fcc metals [21,22]), it is plausible that dislocations are not accumulated inside the fine grains during cycling. The model was tested on UFG Al–Mg alloy AA5056 [18,19] and commercial purity (CP) Grade 2 titanium [18,19] after ECAP, and good agreement with experiment was found, cf. Figures 4.1 and 4.2. This is

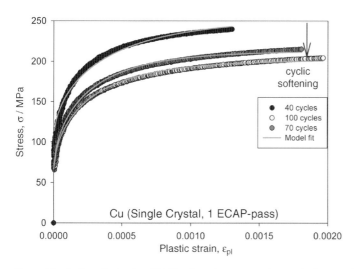

Figure 4.2 Cyclic softening in a {1 1 0}-oriented copper single crystal after one ECAP pass: reduction in the stress amplitude due to continual cyclic softening is seen; curve fits by Eq. (4.7) are shown by thin solid lines.

surprising in view of the simplifications made, particularly in that back stresses that arise due to the two-phase character of the system were neglected. Similarly to results reported in Refs. [18,19], a typical ascending part of the stable hysteresis loop of the ECAP-processed poly- and monocrystalline copper was found, Figure 4.1a. A nonlinear curve fit of data points by the function given by Eq. (4.7) or (4.8) provides an excellent agreement between the experimental and the calculated hysteresis loops. Moreover, the cyclic stress–strain curve for conventional polycrystals and UFG specimens manufactured by SPD can also be approximated by an equation of the same type as Eq. (4.7), Figure 4.1b.

Later the same model was successfully applied by Klemm [23] for cyclic deformation of UFG nickel after different processing schedules (see also [24]). Both groups of researchers arrived at essentially the same conclusions and at very reasonable values of the slip distance Λ (of the order of the grain size) entering Eq. (4.8) explicitly and the annihilation distance y, which is implicitly hidden in k_2. Another example of a successful interpretation of the stable hysteresis loops in terms of the KME approach has been reported recently for UFG titanium created under different SPD processing schedules [6] (after [25]). A good agreement between the model fit and the experimental data is seen, cf. [18,19].

It is interesting to note that in the low strain limit, Eq. (4.7) can be rewritten as

$$\sigma \approx M\alpha Gb \sqrt{\frac{(k_0 - k_2 \rho_0) M\varepsilon}{b\Lambda} + \rho_0} \qquad (4.9)$$

which represents the Hall–Petch-type hardening behavior: the $d^{-1/2}$ dependence of the flow stress is recovered theoretically from Eq. (4.7), in agreement with the experimentally observed behavior of the cyclic stress–strain curve and the fatigue limit of UFG metals manufactured by SPD.

More rigorous modeling would require accounting for the composite-like structure and distinguishing between at least two kinds of dislocations involved: "mobile" and "immobile" (or "cell" and "wall") dislocations, as outlined earlier, with a proper account of back stresses (see also a model by Estrin et al. [26] for ordinary polycrystals).

The constitutive model outlined in this section is applicable only if the Peierls stress, which is always present in a crystalline material [27], is sufficiently small, which is the case for fcc metals and alloys. However, the model can easily be adapted to other systems, such as bcc metals where the Peierls stress can no longer be neglected. An example was provided in [28], where the deformation behavior of ferritic steels was considered. In a first approximation, the Peierls stress σ_P and the dislocation-related stress given by Eq. (4.1) were taken as additive contributions to the overall stress:

$$\sigma = \sigma_P + M\alpha_0 Gb\sqrt{\rho}\left(\frac{\dot{\varepsilon}}{\dot{\varepsilon}_0}\right)^{1/m} \qquad (4.10)$$

The strain rate and temperature dependence of the Peierls stress needs to be represented in an adequate way, as does the possible influence of the dislocation density.

An appropriate relation is determined through an implicit equation [27]:

$$\dot{\varepsilon} = \frac{\rho_m b}{M} \nu_D \frac{2bh^2 a}{kT} \frac{\sigma_P}{M} \exp\left(-\frac{F'_k(\sigma_P/M) + W_m}{kT}\right)\left(\frac{\tilde{L}}{\tilde{L}+X}\right) \quad (4.11)$$

Here $F'_k(\sigma_P/M)$ is the stress-dependent energy required to form a kink, W_m is the activation energy for kink migration in the secondary Peierls relief along the dislocation line, X is the average distance between individual kinks on the dislocation line in thermal equilibrium, ν_D is the Debye frequency, and a and h are the kink geometry parameters, which are of the order of the Burgers vector b. The quantity $\tilde{L}$ denotes the distance between the pinning points on the dislocation line assumed here to be associated with dislocation forest junctions. Hence, $\tilde{L}$ is scaled with the inverse square root of the dislocation density, see the aforementioned text. The equilibrium kink distance X is given by the following equation:

$$X = 2a \exp\left[F'_k(\sigma_P/M)/kT\right] \quad (4.12)$$

Accordingly, for small dislocation density, when $\tilde{L} \gg X$ holds, the last factor in Eq. (4.11) reduces to unity and the single kink formation energy enters the exponential. In the opposite limit case, $\tilde{L} \ll X$, the last factor is given by $\tilde{L}/X$, so that the double-kink formation energy turns out to be rate controlling. Using a particular form of the stress dependence of the kink formation energy [29], this model has been recently applied to describing the mechanical behavior of a TRIP steel with co-existing ferritic and austenitic phases [30]. Of particular interest in the context of this chapter is the success of the model in describing the deformation behavior of ultrafine-grained fcc and bcc phases.

Although the aforementioned model accounting for the Peierls stress works pretty well, the additive form of Eq. (4.10) is a simplification. A more complete theory for the case when the "de-convolution" of the contributions of the localized obstacles and the Peierls relief to the overall stress is not possible in this simple way was presented in [31,32].

Hexagonal materials represent a significant challenge to constitutive modeling. Indeed, the scarcity of the available slip systems in hcp metals leads to plastic incompatibilities between the neighboring grains, which need to be considered in the model. An approach suggested in [33,34] addresses this issue by inclusion of an incompatibility-induced back stress σ_B as an additive term on the right-hand side of Eq. (4.10):

$$\sigma = \sigma_P + M\alpha_0 Gb\sqrt{\rho}\left(\frac{\dot{\varepsilon}}{\dot{\varepsilon}_0}\right)^{1/m} + \sigma_B \quad (4.13)$$

The evolution equation for the back stress is written as

$$\frac{d\sigma_B}{d\varepsilon} = \left(K - \left(Q + \frac{R}{\dot{\varepsilon}}\right)\sigma_B\right) \quad (4.14)$$

where the parameter K represents the rate of increase of the incompatibility stress, the parameter Q characterizes its dynamic relaxation, and R defines the

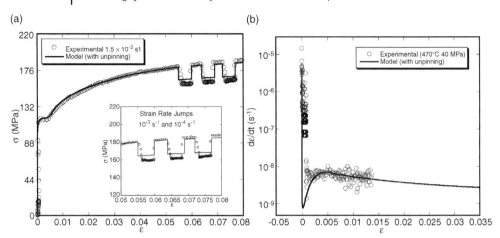

Figure 4.3 Deformation behavior of recrystallized Zircaloy-4 at 470 °C: constant strain rate and strain rate jump response (a) and creep at 40 MPa (b). Model results and experimental data are represented by solid lines and open circles, respectively [33].

rate of time-driven, diffusion-controlled static relaxation. The KME model extended in this way and with a slight modification was shown to predict the deformational behavior of Zircaloy-4 [33] in an excellent way, cf. Figure 4.3.

In order to provide a constitutive description of the deformation behavior of hexagonal α-Ti, the model was extended further, in that a contribution to plastic strain due to twinning was included [34] in the way suggested by [35]. In addition, the evolution equation for the dislocation density was modified to the following form:

$$\frac{d\rho}{d\varepsilon_g} = M\left[k_1\sqrt{\rho} + \frac{1}{bd} + \frac{F}{2eb(1-F)}H(\varepsilon - \varepsilon_{onset}) - k_2\rho\right] \quad (4.15)$$

Here d is the average grain size, e is the twin thickness, and F is the volume fraction of twins; the quantity ε_{onset} entering the Heaviside step function is the onset strain for deformation twinning. The quantity ε_g denotes the part of the strain associated with dislocation glide (as opposed to twinning). Following [35], the twin volume fraction was considered to evolve with strain from an initial value F_0 at the onset strain to a saturation one, F_∞, at large strains:

$$F = F_o + (F_\infty - F_o)\left[1 - \exp\left(-\frac{\varepsilon - \varepsilon_{onset}}{\tilde{\varepsilon}}\right)\right] \quad (4.16)$$

The parameter $\tilde{\varepsilon}$ controls the rate of approach of saturation. We would like to remark that in our current thinking the evolution of twin volume fraction is better represented with respect to stress, rather than strain, variation. This aspect of the role of deformation twinning will be discussed in a forthcoming paper.

An experimental verification of the aforementioned version of the one-internal variable model, which was custom-designed for α-titanium, was provided in [34].

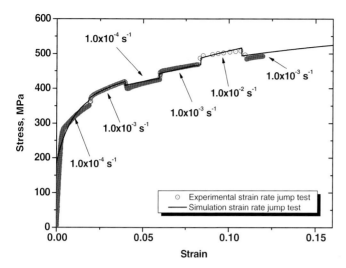

Figure 4.4 A comparison between the calculated and the experimental stress–strain curves for α-Ti deformed in tension at room temperature [34].

An example of good agreement between the calculated and the measured deformation curves (including strain rate jump episodes) is shown in Figure 4.4. It should be mentioned that a provision for accounting for small grain-size effects is included in the model through the second term on the right-hand side of Eq. (4.15). Application of the model to UFG titanium and other fine-grained hcp metals are thus possible.

4.3
Two-Internal Variable Models

While single-internal variable models based on the Kocks–Mecking approach are generally in good agreement with a large body of experimental data on the mechanical response of metals and alloys, they do not provide an explanation of the occurrence to late stages of strain hardening, particularly stages IV and V [36,37]. These stages extend well into the range of very large strains, which are of relevance to SPD processing. Models involving two internal variables potentially possess greater richness and flexibility [38]. For a description of the evolution of the dislocation population in dislocation-cell forming materials, a split of the total dislocation density into the densities of cell interior and cell wall dislocations [39–45] appears to be more appropriate than separation into mobile and immobile dislocation densities considered in Ref. [38]. Motivated by good first-hand experiences with the model [42,43], we present it here in more detail.

As already mentioned, the model refers to dislocation cell forming materials, and it assumes that a cell structure is already formed. Considering the rule of mixture, the shear stress τ is composed of two components:

$$\tau = f_w \tau_w + f_c \tau_c = f_w \alpha G b \sqrt{\rho_w} + f_c \alpha G b \sqrt{\rho_c} \tag{4.17}$$

It is further assumed that the average dislocation cell size, d, scales inversely with the square root of the total dislocation density, ρ as is supported by a wealth of experimental TEM observations:

$$d = K_0/\sqrt{\rho} \tag{4.18}$$

where K_0 is the proportionality constant. The total dislocation density is given by the weighted sum of the dislocation densities in the cell walls, ρ_w and cell interiors, ρ_c:

$$\rho = f \rho_w + (1-f)\rho_c \tag{4.19}$$

Here f is the volume fraction of the cell walls, which can be expressed in terms of the cell size d and the wall thickness w. This quantity depends on grain morphology, but for $w/d \ll 1$ it can be approximated by

$$f \cong 3w/d \tag{4.20}$$

where d can be interpreted as the length of a side of a cube for cube-shaped grains or the grain diameter for spherical grains. According to Eq. (4.18,) the cell size will typically increase in the process of straining leading to growth of the dislocation density, and one might expect that the cell volume fraction f will increase. However, this is not what the experiment shows. The observations for Cu suggest that the volume fraction decreases from an initial value f_o to a saturation value f_∞ (assuming $f_\infty < f_o$). This approach of saturation with growing plastic shear strain γ was described in [42,43] by an empirical equation

$$f = f_\infty + (f_o - f_\infty)\exp(-\gamma/\tilde{\gamma}) \tag{4.21}$$

The parameter $\tilde{\gamma}$ represents the inverse of the rate of this variation. This observation suggests that a decrease of the cell volume fraction f is only possible if the cell wall thickness decreases with strain fast enough. This "thinning" of cell walls can be rationalized in terms of annihilation of dislocations not contributing to misorientation between neighboring cells in the process of straining. In Ref. [6], we suggested to replace the *ad hoc* relation given by Eq. (4.20) by another one, based on the idea that w should scale with $1/\sqrt{\rho_w}$. With $w \cong 1/\sqrt{\rho_w}$ one obtains by combining Eqs. (4.18) and (4.20): $f \cong (3/K)\sqrt{\rho/\rho_w}$, $f = 0.3\sqrt{\rho/\rho_w}$. It should be noted that the ratio of the total dislocation density and the cell wall dislocation density also contains f, so that this implicit equation needs to be solved to obtain the variation with strain of the wall volume fraction. Depending on the parameters of the model, a decrease of f with strain can be predicted for most (but not all) cases.

Assuming that a Taylor-type condition applies to the two "phases" of the material (the cell interiors and the cell walls), we consider the plastic strain to be the same in these "phases." The evolution of the dislocation densities in the cell interiors and the walls is described by a set of coupled differential equations:

$$\frac{d\rho_w}{d\gamma} = \frac{6\beta^*(1-f)^{2/3}}{bdf} + \frac{\sqrt{3}\beta^*(1-f)\sqrt{\rho_w}}{bf} - k_0^*\left(\frac{\dot{\gamma}}{\dot{\gamma}_0}\right)^{-1/n}\rho_w \quad (4.22)$$

$$\frac{d\rho_c}{d\gamma} = \alpha^*\frac{1}{\sqrt{3}}\frac{\sqrt{\rho_w}}{b} - \beta^*\frac{6}{bd(1-f)^{1/3}} - k_0^*\left(\frac{\dot{\gamma}}{\dot{\gamma}_0}\right)^{-1/n}\rho_c \quad (4.23)$$

Here $\dot{\gamma}$ is the plastic shear rate. The individual terms in these equations correspond to the various dislocation reactions involved [42,43]. For example, the first term in Eq. (4.22) and the second term in Eq. (4.23) are related, as the former represents the loss of cell interior dislocations to the walls (and a gain of dislocation density therein). The *dynamic recovery* processes are captured in the last term in both equations. The quantity $\dot{\gamma}_o$ denotes a reference shear rate. As with the exponent m, in the low-temperature regime typical for SPD processing, the exponent n can be taken to be inversely proportional to the absolute temperature T, while the coefficients α^*, β^*, and k_0^* can be considered constant. The latter condition corresponds to dynamic recovery being governed by dislocation cross slip. If, as distinct from cell interiors, the dynamic recovery in cell walls is controlled by the nonconservative process of dislocation climb, as assumed in Zehetbauer's model [36,41,46], $\dot{\gamma}_o$ in Eq. (4.22) is different from that in Eq. (4.23). In that case, it is given by an Arrhenius equation where the activation energy represents that for self-diffusion and the exponent n is a constant. Hydrostatic pressure – a frequent component of SPD processing – has a direct effect on the dislocation recovery in the cell walls by climb [47]. An appropriate modification of the dynamic recovery term in Eq. (4.23) to include the hydrostatic pressure made it possible to account for the effect of back pressure on the ECAP processing of an Al alloy [48].

The applicability of the model outlined earlier hinges on the validity of Eq. (4.18) that establishes a scaling relation between the dislocation cell size and the total dislocation density. This relation, which is suggested by dimensional consideration and stochastic modeling of dislocation ensembles [49], is universally accepted, but a strict proof of its validity for nonsteady state deformation regimes is still a challenge for dislocation theory, cf. [6].

It is interesting to note that by combining the two internal variables approach with that of irreversible thermodynamics applied to dislocation plasticity [50], the original equation for the dislocation density evolution of the Kocks–Mecking–Estrin model, Eq. (4.5), can be obtained. A plastically deforming crystal is an open system that exchanges energy with exterior. The total entropy flux is

$$dS = d_i S + d_e S \quad (4.24)$$

where $d_i S$ is the entropy production due to changes in the internal microstructure and $d_e S$ is the entropy flux due to heat exchange with external heat reservoir. Following a general formalism of nonequilibrium thermodynamics and dissipative systems proposed by Prigogine [51], Huang et al. [52] have related the entropy production d_iS in the bulk of the deforming metal with energy dissipation due to dislocation reactions associated with a shear strain increment $d\gamma$ as

$$d_iS = \frac{1}{T}(dW^+ + dW^= + dW^-) \tag{4.25}$$

where the terms in the parentheses are the energies associated with dislocation generation, glide, and annihilation.

The energy dissipation for the mentioned processes was calculated as [27,52]

$$dW^+ = \frac{1}{2}Gb^2 d\rho^+, \quad dW^- = \frac{1}{2}Gb^2 d\rho^-, \quad dW^= = \tau b \Lambda d\rho^+ \tag{4.26}$$

where $d\rho^+$ and $d\rho^-$ are the increments of dislocation densities for dislocation production and recovery (annihilation), respectively, and τ is the shear stress acting on the dislocations. The mean free path Λ is supposed to be equal to the average interdislocation spacing $\Lambda \sim 1/\sqrt{\rho}$ with the total dislocation density $\rho = \rho^+ - \rho^-$. Using the Taylor expression, Eq. (4.1), and combining Eqs. (4.25) and (4.26), the entropy production can be written as

$$d_iS = \frac{(1+2\alpha)Gb^2}{2T}d\rho^+ + \frac{(2+2\alpha)Gb^2}{2T}d\rho^- \tag{4.27}$$

The entropy flux d_eS associated with heat exchange between the solid undergoing a plastic strain increment $d\gamma$ and the environment is related to plastic work $\tau d\gamma$ [51]:

$$d_eS = -\frac{\tau d\gamma}{T} \tag{4.28}$$

where the shear stress is given by Eq. (4.1) with the appropriate correction for the Taylor factor M. As stated earlier, the deformed dislocation structure can be treated as a "composite" in which the cell walls and cell interiors represent two distinct "phases:" the dislocation-rich regions and the dislocation-depleted ones. Assuming that the dislocation density in cell walls exhibits the same scaling behavior as the total dislocation density ρ, which is inversely proportional to the square of the reciprocal cell size [53], Eq. (4.18), or the mean free path Λ: $\rho \propto 1/\Lambda^2$, $\rho_w \propto 1/\Lambda^2$, it was argued that proportionality between ρ_c and ρ_w can be expected: $\rho_c = \beta \rho_w$ (where β is a constant), cf. [54]. Since for realistic dislocation configurations $\rho_w \gg \rho_c$ holds, one can assume in a first approximation that $\rho_w \cong \rho$ can be used, which leads to the following expression for τ:

$$\tau = \alpha Gb\left(\frac{f_w}{\Lambda} + \sqrt{\beta}f_c\sqrt{\rho}\right) \tag{4.29}$$

Combining Eq. (4.27) with Eqs. (4.28)–(4.29) one obtains

$$\frac{dS}{d\gamma} = \frac{(1+2\alpha)Gb^2}{2T}\frac{d\rho^+}{d\gamma} + \frac{(2+2\alpha)Gb^2}{2T}\frac{d\rho^-}{d\gamma} - \left(\sqrt{\beta}f_c\sqrt{\rho} + \frac{f_w}{\Lambda}\right)\frac{\alpha Gb}{T} \tag{4.30}$$

Assuming further that the rate of dynamic recovery is governed by second-order dislocation annihilation kinetics mediated by climb

$$\frac{d\rho^-}{dt} = \rho^2 b^2 \nu_D e^{-\frac{\Delta G_a}{k_b T}} \qquad (4.31)$$

where ΔG_a is the activation energy for climb and k_b is the Boltzmann constant, and using the Orowan relation $\dot{\gamma} = \rho b \langle v \rangle$, where $\langle v \rangle$ is the average dislocation velocity, the $d\rho^-/d\gamma$ term on the right-hand side of Eq. (4.30) is expressed as

$$\frac{d\rho^-}{d\gamma} = \frac{\rho b \nu_D}{\langle v \rangle} e^{-\frac{\Delta G}{k_b T}} \qquad (4.32)$$

Hence, the explicit relation between the total entropy change during plastic deformation and the total dislocation density reads as

$$\frac{dS}{d\gamma} = \frac{(1+2\alpha)Gb^2}{2T}\frac{d\rho}{d\gamma} + \frac{(2+2\alpha)Gb^2}{2T}\frac{\rho b\nu_D}{\langle v \rangle}e^{-\frac{\Delta G}{k_b T}} - \left(\sqrt{\beta f_c}\sqrt{\rho} + \frac{f_w}{\Lambda}\right)\frac{\alpha Gb}{T} \qquad (4.33)$$

In the steady state determined by the condition $dS = 0$ one obtains a kinetic equation for the average dislocation density:

$$\frac{d\rho}{d\gamma} = \frac{2\alpha f_w}{1+2\alpha} \cdot \frac{1}{b\Lambda} + \frac{2\alpha\sqrt{\beta f_c}}{1+2\alpha} \cdot \frac{\sqrt{\rho}}{b} - \frac{2+2\alpha}{1+2\alpha} \cdot \frac{b\nu_D}{\langle v \rangle} e^{-\frac{\Delta G}{k_b T}} \rho \qquad (4.34)$$

which replicates the familiar evolution equation of the KME model (cf. Eq. (4.5)):

$$\frac{d\rho}{d\gamma} = \frac{k_0}{b\Lambda} + k_1\sqrt{\rho} - k_2 \rho \qquad (4.35)$$

The coefficients k_0, k_1, are k_2 of the KME model (which is a single-internal variable model) can thus be re-interpreted in terms of the quantities pertaining to the two-internal variable model as follows:

$$k_0 = \frac{2\alpha f_c}{1+2\alpha}, \quad k_1 = \frac{2\alpha f_w}{b(1+2\alpha)}, \quad k_2 = \frac{2+2\alpha}{1+2\alpha} \cdot \frac{b\nu_D}{\langle v \rangle} e^{-\frac{\Delta G}{k_b T}} \qquad (4.36)$$

This is a somewhat unexpected outcome of two-internal variable modeling, highlighting the universality of the evolution equation for the total dislocation density in the form represented by Eq. (4.34). A further interesting outcome of the analysis in [50] is the established correlation between the fractal dimension of the dislocation cell structure and the details of the deformation curve for tensile loading.

4.4
Three-Internal Variable Models

A three-internal variable model (3IVM) proposed by Gottstein and co-workers [55,56] is based on a concept similar to that of the above two-internal variable model for dislocation cell forming materials, but discriminates between mobile and immobile dislocations with the cells. Mughrabi's two-phase approach [54] treating the cell walls and cell interiors as two separate "phases" is also adopted. 3IVM has been employed with great success in through-process modeling codes [57] used in numerical simulations of rolling of Al alloys. It should be noted that a simple approximation can reduce the 3IVM to a two-internal variable one. As suggested in [38], one can reasonably assume that the "relaxation" of the mobile dislocation density to its quasi steady state value occurs much faster than steady state is approached by the slowly evolving immobile dislocation density. This intuitive (yet verifiable) approximation makes it possible to eliminate the fast variable – the mobile dislocation density in the cell interiors – "enslaved" by the slow ones, namely the immobile dislocation densities in the cell interiors and the cell walls. This suggests that 3IVM and the two-internal variable model outlined in Section 4.3 are not very far apart.

4.5
Numerical Simulations of SPD Processes

In this section, we turn to numerical simulations of major processes of SPD – a subject that has gained importance in recent years due to great promise of SPD techniques with regard to improvement of properties of metallic materials. Such techniques, which go back to the pioneering work by Bridgman [58], combine severe shear with hydrostatic pressure, which is now known to lead toward extreme grain refinement. Details of SPD processing techniques can be found in several comprehensive reviews [3–5]. The process parameters involved include die geometry, sequence of processing steps, deformation speed and temperature, and so on. Therefore, numerical simulations are an indispensable tool for process optimization with regard to the degree of grain refinement achieved and the uniformity of the microstructure produced. The modeling approaches capable of capturing the specifics of large strain deformation described earlier have provided the necessary ammunition for such computational tasks. As an example of the application of the two-internal variable model [42,43] we show the results for equal channel angular pressing (ECAP) of Al [59], Figure 4.5. In the ECAP process, a billet is forced through an angular die and the material experiences a large amount of shear when it passes through the plane where the entry and the exit channels meet. As the cross-sectional dimensions of the billet are retained in this operation, multiple ECAP passes are possible. The calculated curves representing the evolution of the dislocation cell size and the equivalent stress as a function of the equivalent strain for the ECAP die (which for the channel angle of 90° practically coincides with the number of ECAP passes) are seen to be in good agreement with the measured values. It should be noted that the emerging grain size was identified here with the dislocation cell size. Thus, the assumed scenario of grain refinement infers that the underlying process is the dislocation cell formation followed by cell size evolution according to Eq. (4.18). That is to say, the dislocation cell structure is assumed to be a precursor of the new grain structure that develops with accumulation of misorientations across the cell boundaries.

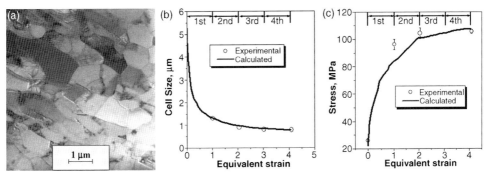

Figure 4.5 Refined grain structure of pure Al after four ECAP passes (a); the evolution of the dislocation cell size (b) and the variation of equivalent stress with equivalent strain for commercial purity Al (c) [59].

The latter process was considered by Pantleon [60] and Estrin et al. [61] using a probabilistic approach. The results of both studies account reasonably well for the misorientation angles associated with the so-called incidental dislocation boundaries [62], but the models used do not apply to high-angle (geometrically necessary) grain boundaries [62] whose fraction grows as a result of SPD processing. The Langevin approach leading to a Fokker–Planck equation for the distribution of misorientation angles [61,62] appears to provide a promising frame for modeling the evolution of grain misorientations. An adequate generalization to a Fokker–Planck equation for the diffusion problem in two-dimensional space covering both the cell size and the misorientation angle is yet to be developed.

While the model predictions based on the two-internal variable model [42,43] have been verified for various materials and SPD processes, there is still an obvious need to improve modeling with regard to grain fragmentation. A recently proposed approach [63] suggests a possible, albeit not a universal, description of the grain fragmentation process. It was suggested that grain subdivision is a result of lattice curvature that develops within a grain due to constraints imposed by the neighboring grains. The rotation of the crystallographic planes in the grain due to dislocation slip was considered to be retarded near the grain boundaries. The grain subdivision was considered to stem from the emergence of geometrically necessary dislocations required to accommodate the resulting lattice curvature. The process of grain subdivision will go on for each generation of newly emerging grains until the dislocation glide mechanism leading to rotation of crystallographic planes ceases to operate to be replaced by diffusion-controlled plasticity. According to [64], this is the case when the dislocation cell size drops below a critical level, d_c, which is determined by

$$d_c = \left(\frac{D_{GB} b}{\dot{\gamma}}\right)^{1/3} \tag{4.37}$$

where D_{GB} is the grain boundary diffusivity. For room temperature SPD processing of common fcc metals, this quantity is estimated to be in the range of several hundred nanometers. For the one-internal variable model, the transition between the two deformation regimes was described by introducing a factor

$$P = \exp\left[-(d_c/d)^3\right] \qquad (4.38)$$

in the storage term associated with grain boundaries, that is, the first term on the right-hand side of Eq. (4.5). Similarly, in the two-internal variable model, this can be done by introducing this factor in the first (storage) term in Eq. (4.22) and the corresponding loss term in Eq. (4.23) – the second term on the right-hand side of that equation.

The aforementioned concept of grain fragmentation was combined with the two-internal variable model [42,43] providing a reliable prediction of strain hardening behavior and the variation of the grain size distribution under large strains [63]. Of particular value is also the prediction of texture evolution. In the past, the effect of grain subdivision was not considered in texture simulations associated with SPD. Inclusion of this effect improved the agreement between the calculated and the experimental texture emerging in Al under SPD processing by simple shear [65].

In several numerical simulation exercises addressing SPD processing, see, for example, [66], it was established that the basic two-internal variable model [42,43] provides a potent and robust tool for predicting the material behavior during processing as well as the properties of the product. For instance, in the aforementioned example of ECAP of Al, such aspects as uniformity of the microstructure produced can be studied by finite element simulations [62]. The results shown in Figure 4.6 demonstrate how the cell size distribution (tantamount to that of the grain

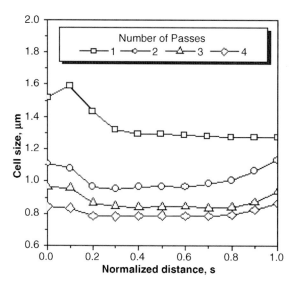

Figure 4.6 Evolution of the dislocation cell size across a billet with the number of ECAP passes. The quantity s denotes the distance from the bottom of the billet normalized with respect to the billet thickness [62].

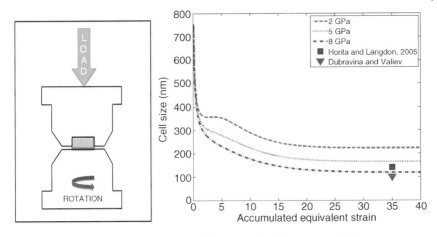

Figure 4.7 Dislocation cell size variation with the accumulated strain during high-pressure torsion of copper [67]. The schematic of the process is shown in the left sketch.

size distribution) across the billet thickness progressively becomes more and more uniform with the number of ECAP passes. The average cell size (which is to become the average grain size) tends to saturate with strain at a level of about 800 nm. For other materials and with different SPD techniques significantly smaller grain sizes are achievable.

An example of more extreme grain refinement is shown in Figure 4.7, which displays the evolution of the calculated dislocation cell size in copper under high pressure torsion (HPT) [67] – a process schematically illustrated in the same figure. For the vertical pressure of 8 GPa the cell/grain size attained a saturation value of about 120 nm, in close agreement with experimental data included in the diagram. The model used in Ref. [67] was based on the two-internal variable approach and additionally included a gradient plasticity term. The provision for gradient plasticity [68] helps explaining the development of nearly uniform, refined microstructure in an inherently nonuniform process such as HPT.

A physically based model, relating higher (second order) gradient terms to plastic strain incompatibility between the adjoining grains [68,69], links the coefficient in the gradient term to the grain size. Another physically sensible variant of gradient plasticity associates a gradient term with mobile dislocation exchange between adjoining regions of the material via dislocation cross slip [70]. Both variants of gradient plasticity were considered in the context of localized strain pattern formation in Mg alloy AZ31 under ECAP deformation [71]. Using linear stability analysis, localized strain patterning in the shear zone of the alloy undergoing ECAP deformation was considered for both variants of gradient plasticity theory in conjunction with the two-internal variable model [42,43]. In such analysis, conditions for growth of small periodic perturbations of a uniform solution of the constitutive equations with different wave numbers are determined. The wave numbers ξ corresponding to the experimentally observed strain localization patterns with a wave length $\lambda = 2\pi/\xi$ are compared with the calculated dispersion curves in

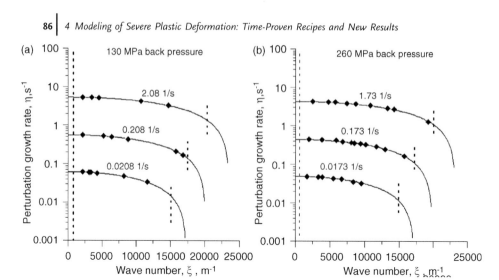

Figure 4.8 Results of linear stability analysis for strain localization under ECAP in alloy AZ31 [71]. Predicted perturbation growth rate is shown as a function of the wave number. The estimated average shear strain rates in the tests are shown for each curve. The diamonds correspond to the experimentally observed wave numbers and are placed on theoretical curves for the respective strain rates. The vertical dotted line on (a) marks the position of the smallest theoretically possible wave number, while the small vertical dashed lines on (b) mark the largest possible wave numbers for the respective strain rates.

Figure 4.8, and very good agreement between the predictions and the measurements is seen for two different back-pressure levels used in ECAP processing.

The two-internal variable formulation can also be used for modeling high-speed severe deformation processes, where heat release plays a substantial role. The model has a provision for considering the thermomechanical coupling through including a temperature dependence in the exponents m and n (or the reference strain rates $\dot{\varepsilon}_0$ and $\dot{\gamma}_0$), as was done recently in modeling impact deformation [72].

4.6
Concluding Remarks

In this exposé, the existing approaches to constitutive modeling of large strain plasticity have been presented. We hope to have been able to demonstrate a very good predictive capability of models based on dislocation density evolution. A particularly suitable modeling frame for describing SPD is provided by the two-internal variable model developed for dislocation cell forming materials [42,43]. The examples of successful application of the model shown above refer to fcc metals. However, their adaptation to bcc or hcp metals is rather straightforward. The recipes for including the Peierls stress and deformation twinning discussed in Section 4.2 for the one-internal variable model can simply be "borrowed." This was done in a recent work [30] on the deformation behavior of medium manganese TRIP steel, which after intercritical annealing contained coarse grained ferrite and two ultrafine-grained phases: ferrite and austenite. The model was shown to perform very well, which can be seen as a further proof of its general applicability – with extra "modules" introduced where necessary. Similarly, scarcity of slip systems in hcp metals can be taken into account in the two-internal variable model through introduction of a back-stress in the same way it was done in the one-internal variable model, as described in Section 4.2. The two-internal variable model was implemented in various finite element packages, including ABAQUS, MARC, LS-DYNA, and meshless codes, such as material point method (MPM). Implementation in the smooth particle hydrodynamics (SPH) code is under way (Lemiale, private communications). The robustness, user-friendliness and predictive capability of the two-internal variable model suggest that it can be used as a broad platform for computational modeling of SPD processing.

Acknowledgments

Partial support from the Ministry of Education and Science of the Russian Federation under the project on Hybrid Nanostructured Materials at the National University of Science and Technology "MISIS" is acknowledged. One of the authors (AV) wishes to thank the Russian Ministry of Education and Science and MEXT, Japan, for partial support of research in this field through grants-in-aid 11. G34.31.0031 and 22102006, respectively. Support of the other author (YE) through an Alexander von Humboldt Award and the hospitality of Prof. G. Gottstein's Institute in Aachen are gratefully acknowledged.

References

1 Gottstein, G. (2007) *Integral Materials Modeling: Towards Physics-Based Through-Process Models*, Wiley-VCH, Weinheim, John Wiley & Sons, Ltd., Chichester, UK, [distributor].

2 Cottrell, A. (1953) *Dislocations and Plastic Flow in Crystals*, Clarendon Press, Oxford.

3 Valiev, R.Z., Estrin, Y., Horita, Z., Langdon, T.G. et al. (2006) Producing bulk ultrafine-grained materials by severe plastic deformation. *Journal of Metals*, **58** (4), 33–39.

4 Valiev, R.Z., Islamgaliev, R.K., and Alexandrov, I.V. (2000) Bulk nanostructured materials from severe plastic deformation. *Progress in Materials Science*, **45** (2), 103–189.

5 Valiev, R.Z. and Langdon, T.G. (2006) Principles of equal-channel angular pressing as a processing tool for grain refinement. *Progress in Materials Science*, **51** (7), 881–981.

6 Estrin, Y. and Vinogradov, A. (2013) Extreme grain refinement by severe plastic deformation: A wealth of challenging science. *Acta Materialia*, **61**, 782–817.

7 Zehetbauer, M.J. and Estrin, Y. (2009) Modeling of strength and strain hardening of bulk nanostructured materials, in *Bulk Nanostructured Materials*, Wiley-VCH Verlag GmbH, Weinheim, pp. 109–136.

8 Kocks, U.F. (1976) Laws for work-hardening and low-temperature creep. *Journal of Engineering Materials and Technology – Transactions of the ASME*, **98** (1), 76–85.

9 Mecking, H. and Kocks, U.F. (1981) Kinetics of flow and strain hardening. *Acta Metallurgica*, **29** (11), 1865–1875.

10 Estrin, Y. and Mecking, H. (1984) A unified phenomenological description of work hardening and creep based on one-parameter models. *Acta Metallurgica*, **32** (1), 57–70.

11 Kocks, U.F. and Mecking, H. (2003) Physics and phenomenology of strain hardening: The FCC case. *Progress in Materials Science*, **48** (3), 171–273.

12 Estrin, Y. (1996) Dislocation-density related constitutive modelling, in *Unified Constitutive Laws of Plastic Deformation* (eds A.S. Krausz and K. Krausz), Academic Press, San Diego, London, pp. 69–106.

13 Klepaczko, J. (1975) Thermally activated flow and strain rate history effects for some polycrystalline fcc metals. *Materials Science and Engineering*, **18** (1), 121–135.

14 Bergström, Y. (1973) *Reviews on Powder Metallurgy and Physical Ceramics*, **2**, 79.

15 Gottstein, G. and Argon, A.S. (1987) Dislocation theory of steady state deformation and its approach in creep and dynamic tests. *Acta Metallurgica*, **35** (6), 1261–1271.

16 Starink, M.J., Qiao, X.G., Zhang, J., and Gao, N. (2009) Predicting grain refinement by cold severe plastic deformation in alloys using volume averaged dislocation generation. *Acta Materialia*, **57** (19), 5796–5811.

17 Alexandrov, I.V. and Chembarisova, R.G. (2007) Development and application of the dislocation model for analysis of the microstructure evolution and deformation behavior of metals subjected to severe plastic deformation. *Reviews on Advanced Materials Science*, **16** (1–2), 51–72.

18 Patlan, V., Vinogradov, A., Higashi, K., and Kitagawa, K. (2001) Overview of fatigue properties of fine grain 5056 Al–Mg alloy processed by equal-channel angular pressing. *Materials Science and Engineering A*, **300** (1–2), 171–182.

19 Vinogradov, A.Y., Stolyarov, V.V., Hashimoto, S., and Valiev, R.Z. (2001) Cyclic behavior of ultrafine-grain titanium produced by severe plastic deformation. *Materials Science and Engineering A*, **318** (1–2), 163–173.

20 Essmann, U. and Mughrabi, H. (1979) Annihilation of dislocations during tensile and cyclic deformation and limits of dislocation densities. *Philosophical Magazine A*, **40** (6), 731–756.

21 Höppel, H.W., Zhou, Z.M., Mughrabi, H., and Valiev, R.Z. (2002) Microstructural study of the parameters governing coarsening and cyclic softening in fatigued ultrafine-grained copper. *Philosophical Magazine A*, **82** (9), 1781–1794.

22 Höppel, H.W., Mughrabi, H., and Vinogradov, A. (2009) Fatigue properties of bulk nanostructured materials, in *Bulk Nanostructured Materials*, Wiley-VCH Verlag GmbH, Weinheim, pp. 481–500.

23 Klemm, R. (2004) *Zyklische Plastizität von mikro- und submikrokristallinem Nickel*. Doctorate thesis. Technische Universität Dresden.

24 Mughrabi, H. and Höppel, H.W. (2010) Cyclic deformation and fatigue properties of very fine-grained metals and alloys. *International Journal of Fatigue*, **32** (9), 1413–1427.

25 Polyakova, V.V., Semenova, I.P., Valiev, R.Z., and Vinogradov, A. *Modelling of cyclic deformation of ultra-fine grain titaninium*, (in preparation).

26 Estrin, Y., Braasch, H., and Brechet, Y. (1996) A dislocation density based constitutive model for cyclic deformation. *Journal of Engineering Materials and Technology ASME*, **118** (4), 441–457.

27 Hirth, J.P. and Lothe, J. (1982) *Theory of Dislocations*, 2nd edn, John Wiley & Sons, Inc., New York.

28 Reichert, B. and Estrin, Y. (2007) Modular modelling of stress–strain behaviour of ferritic steel grades in strain rate ranges relevant for automotive crash situations. *Steel Research International*, **78**, 791–797.

29 Baufeld, B., Petukhov, B.V., Bartsch, M., and Messerschmidt, U. (1998) Transition of mechanisms controlling the dislocation motion in cubic ZrO_2 below 700 °C. *Acta Materialia*, **46** (9), 3077–3085.

30 Lee, S., Estrin, Y., and De Cooman, B.C. (2013) *Constitutive modelling of the mechanical properties of V-added medium manganese TRIP steel*, Metallurgical and Materials Transactions A, DOI: 10.1007/s11661-013-1648-4.

31 Petukhov, B.V. (1996) Phenomenological description of material plasticity in the region of transition from the kink

mechanism of dislocation motion to the mechanism of their local pinning. *Crystallography Reports*, **41**, 181–188.
32. Dour, G. and Estrin, Y. (2002) Dislocation motion in crystals with a high Peierls relief: A unified model incorporating the lattice friction and localized obstacles. *Journal of Engineering Materials and Technology*, **124** (1), 7–12.
33. Dunlop, J.W., Bréchet, Y.J.M., Legras, L., and Estrin, Y. (2007) Dislocation density-based modelling of plastic deformation of Zircaloy-4. *Materials Science and Engineering: A*, **443** (1–2), 77–86.
34. Ahn, D.-H., Kim, H.S., and Estrin, Y. (2012) A semi-phenomenological constitutive model for hcp materials as exemplified by alpha titanium. *Scripta Materialia*, **67** (2), 121–124.
35. Bouaziz, O., Allain, S., Scott, C.P., Cugy, P. et al. (2011) High manganese austenitic twinning induced plasticity steels: A review of the microstructure properties relationships. *Current Opinion in Solid State and Materials Science*, **15** (4), 141–168.
36. Zehetbauer, M. and Seumer, V. (1993) Cold work hardening in stages IV and V of F.C.C. metals: I. Experiments and interpretation. *Acta Metallurgica et Materialia*, **41** (2), 577–588.
37. Gil Sevillano, J., van Houtte, P., and Aernoudt, E. (1980) Large strain work hardening and textures. *Progress in Materials Science*, **25** (2–4), 69–134.
38. Kubin, L.P. and Estrin, Y. (1990) Evolution of dislocation densities and the critical conditions for the Portevin-Le Châtelier effect. *Acta Metallurgica et Materialia*, **38** (5), 697–708.
39. Mughrabi, H. (1983) Dislocation wall and cell structures and long-range internal-stresses in deformed metal crystals. *Acta Metallurgica*, **31** (9), 1367–1379.
40. Prinz, F.B. and Argon, A.S. (1984) The evolution of plastic resistance in large strain plastic flow of single phase subgrain forming metals. *Acta Metallurgica*, **32** (7), 1021–1028.
41. Zehetbauer, M. (1993) Cold work hardening in stages IV and V of fcc metals: II. Model fits and physical results. *Acta Metallurgica et Materialia*, **41** (2), 589–599.
42. Estrin, Y., Tóth, L.S., Molinari, A., and Bréchet, Y. (1998) A dislocation-based model for all hardening stages in large strain deformation. *Acta Materialia*, **46** (15), 5509–5522.
43. Toth, L.S., Molinari, A., and Estrin, Y. (2002) Strain hardening at large strains as predicted by dislocation based polycrystal plasticity model. *Journal of Engineering Materials and Technology*, **124** (1), 71–77.
44. Nes, E. and Marthinsen, K. (2002) Modeling the evolution in microstructure and properties during plastic deformation of fcc metals and alloys: An approach towards a unified model. *Materials Science and Engineering: A*, **322** (1–2), 176–193.
45. Malygin, G.A. (2002) Kinetic mechanism of the formation of fragmented dislocation structures upon large plastic deformations. *Physics of the Solid State*, **44** (11), 2072–2079.
46. Les, P. and Zehetbauer, M.J. (1994) Evolution of microstructural parameters in large strain deformation: description by Zehetbauer's model. *Key Engineering Materials*, **97–98**, 335–340.
47. Zehetbauer, M.J., Stüwe, H.P., Vorhauer, A., Schafler, E., and Kohout, J. (2005) The role of hydrostatic pressure in severe plastic deformation, in *Nanomaterials by Severe Plastic Deformation* (eds M.J. Zehetbauer and R.Z. Valiev), Wiley-VCH Verlag GmbH, Weinheim, pp. 433–446.
48. Mckenzie, P.W.J., Lapovok, R., and Estrin, Y. (2007) The influence of back pressure on ECAP processed. AA 6016: Modeling and experiment. *Acta Materialia*, **55** (9), 2985–2993.
49. Hähner, P. (1996) A theory of dislocation cell formation based on stochastic dislocation dynamics. *Acta Materialia*, **44** (6), 2345–2352.
50. Vinogradov, A., Yasnikov, I.S., and Estrin, Y. (2012) Evolution of fractal structures in dislocation ensembles during plastic deformation. *Physical Review Letters*, **108** (20), 205504.
51. Prigogine, I. (1961) *Introduction to Thermodynamics of Irreversible Processes*, 2nd, rev. edn, Interscience Publishers, New York.
52. Huang, M., Rivera-Díaz-del-Castillo, P.E.J., Bouaziz, O., and van der Zwaag, S. (2008)

Irreversible thermodynamics modelling of plastic deformation of metals. *Materials Science and Technology*, **24** (4), 495–500.

53 Holt, D.L. (1970) Dislocation cell formation in metals. *Journal of Applied Physics*, **41** (8), 3197–3201.

54 Mughrabi, H. (2006) Implications of linear relationships between local and macroscopic flow stresses in the composite model. *International Journal of Materials Research*, **97** (5), 594–596.

55 Roters, F., Raabe, D., and Gottstein, G. (2000) Work hardening in heterogeneous alloys: A microstructural approach based on three internal state variables. *Acta Materialia*, **48** (17), 4181–4189.

56 Goerdeler, M. and Gottstein, G. (2001) A microstructural work hardening model based on three internal state variables. *Materials Science and Engineering A*, **309–310**, 377–381.

57 Gottstein, G., Crumbach, M., Neumann, L., and Kopp, R. (2006) Through process texture simulation for aluminium sheet fabrication. *Materials Science Forum*, **519**, 93–102.

58 Bridgman, P.W. (1946) The tensile properties of several special steels and certain other materials under pressure. *Journal of Applied Physics*, **17** (3), 201–212.

59 Baik, S.C., Estrin, Y., Kim, H.S., and Hellmig, R.J. (2003) Dislocation density-based modeling of deformation behavior of aluminium under equal channel angular pressing. *Materials Science and Engineering A*, **351** (1–2), 86–97.

60 Pantleon, W. (2002) Formation of disorientations in dislocation structures during plastic deformation. *Solid State Phenomena*, **87**, 73–92.

61 Estrin, Y., Tóth, L., Bréchet, Y., and Kim, H.S. (2006) Modelling of the evolution of dislocation cell misorientation under severe plastic deformation. *Materials Science Forum*, **503–504**, 675–680.

62 Hughes, D.A. and Hansen, N. (2000) Microstructure and strength of nickel at large strains. *Acta Materialia*, **48** (11), 2985–3004.

63 Tóth, L.S., Estrin, Y., Lapovok, R., and Gu, C. (2010) A model of grain fragmentation based on lattice curvature. *Acta Materialia*, **58** (5), 1782–1794.

64 Bouaziz, O., Estrin, Y, Bréchet, Y., and Embury J D. (2010) Critical grain size for dislocation storage and consequences for strain hardening of nanocrystalline materials. *Scripta Materialia*, **63**, 477–479.

65 Gu, C.F. and Tóth, L.S. (2012) Texture development and grain refinement in non-equal-channel angular-pressed Al. *Scripta Materialia*, **67** (1), 33–36.

66 Joo, S.-H. and Kim, H. (2010) Comparison of deformation and microstructural evolution between equal channel angular pressing and forward extrusion using the dislocation cell mechanism-based finite element method. *Journal of Materials Science*, **45** (17), 4705–4710.

67 Molotnikov, A. (2008) Application of strain gradient plasticity modelling to high pressure torsion. *Materials Science Forum*, **584–586**, 1051–1056.

68 Estrin, Y., Molotnikov, A., Davies, C.H.J., and Lapovok, R. (2008) Strain gradient plasticity modelling of high-pressure torsion. *Journal of the Mechanics and Physics of Solids*, **56** (4), 1186–1202.

69 Estrin, Y. and Mühlhaus, H.B. (1996) From micro- to macroscale, gradient models of plasticity, in *Proc. of Int. Conf. on Engineering Mathematics* (ed. D. Yuen), IEAust. AEMC Sydney, Australia, pp. 161–165.

70 Estrin, Y. and Kubin, LP. (1991) Plastic instabilities: Phenomenology and theory. *Materials Science and Engineering: A*, **137**, 125–134.

71 Lapovok, R., Toth, L.S., Molinari, A., and Estrin, Y. (2009) Strain localisation patterns under equal-channel angular pressing. *Journal of the Mechanics and Physics of Solids*, **57** (1), 122–136.

72 Lemiale, V., Estrin, Y., Kim, H.S., and O'Donnell, R. (2010) Grain refinement under high strain rate impact: A numerical approach. *Computational Materials Science*, **48** (1), 124–132.

5
Plastic Anisotropy in Magnesium Alloys – Phenomena and Modeling
Bevis Hutchinson and Matthew Barnett

5.1
Deformation Modes and Textures

The mechanical properties of wrought magnesium alloys are dominated by their textures that are both a result of the active modes of plastic deformation and which, together with these modes, dictate the strength and anisotropy of the materials when loaded. Basic understanding of these phenomena is well established and can be found in a number of excellent literature sources [1–3]. The common slip and twinning modes are indicated by the sketches in Figure 5.1. A characteristic of magnesium that sets it apart from other common metals, even many others with hcp crystal structures, is the vast variation in stress levels that are needed to activate shearing on the different systems. Slip is dominated by $\langle a \rangle$ glide in the close packed $\langle 11\bar{2}0 \rangle$ directions on the $\{0001\}$ basal planes at stresses that may be less than 1 MPa. In contrast, the same $\langle a \rangle$ glide on $\{10\bar{1}0\}$ prismatic planes requires almost two orders of magnitude higher stress and pyramidal $\langle c+a \rangle$ slip of the type $\{11\bar{2}2\}\langle 11\bar{2}3 \rangle$ even more. Twinning is similarly diverse, with the so-called tension twinning that provides for extension along the c-axis occurring at shear stress levels of only a few MPa whereas the c-axis "compression twinning", requires more than 50 MPa. The harder deformation modes play important roles in maintaining continuity and inhibiting cracks from forming along grain boundaries, but the textures in wrought products tend to be governed principally by the soft modes and, in particular, by basal slip.

In rolled sheets, basal slip allows rotation of these planes almost into the plane of the sheet. Accordingly, these textures show a high concentration of c-axes in the normal direction of the pole figure and are often referred to as basal textures (see Figure 5.8a). They may approximate to c-axis fiber textures but usually show a preference for $\langle 10\bar{1}0 \rangle$ directions to orient themselves along the rolling direction as a result of the duplex slip along two different $\langle a \rangle$ directions in each grain. These basal textures have the result that further deformation in rolling becomes very difficult since the shear stress on the basal planes is then small. Deformation by shear banding may occur and shear cracking is often a problem. After annealing and recrystallization, the c-axis fiber is usually retained although rotations often occur

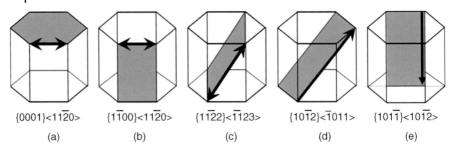

{0001}<11$\bar{2}$0> {1$\bar{1}$00}<11$\bar{2}$0> {11$\bar{2}$2}<$\bar{1}\bar{1}$23> {10$\bar{1}$2}<$\bar{1}$011> {10$\bar{1}\bar{1}$}<10$\bar{1}$2>
(a)   (b)   (c)   (d)   (e)

Figure 5.1 The common deformation modes in magnesium alloys, (a) basal ⟨a⟩ slip, (b) prism ⟨a⟩ slip, (c) pyramidal ⟨c+a⟩ slip, (d) extension twinning, and (e) compression twinning.

such that directions of the type ⟨11$\bar{2}$0⟩ then align themselves with the rolling direction (see Figure 5.15).

The other major forming process for magnesium alloys is hot extrusion. Round sections typically show cylindrical textures with basal plane directions ⟨hk.0⟩ aligned along the extrusion axis and with the c-axes distributed uniformly along the radial directions of the bars, Figure 5.2b. Friction at the dies can cause significant rotations of the texture in the outer subsurface regions. Flat extrusions contain textures that are intermediate between those of rolled sheet and of round bars. These have fiber textures with ⟨10$\bar{1}$0⟩ axes parallel to the extrusion direction and c-axes perpendicular to this and also with a superimposed c-axis peak within the fiber spread perpendicular to the flat surface (see Figure 5.6a).

Descriptions of these textures are broadly true for different magnesium alloys but there are also great variations between them, especially concerning the strength of the textures. Pure magnesium and the traditional alloys containing zinc, aluminum, manganese, and zirconium usually conform to this pattern and produce strong textures after hot working. Recent years have seen much interest in alloys containing rare earth elements such as yttrium, neodymium, gadolinium, or misch metals. These typically have weaker textures and, in some cases at least, may even differ in the types of components that are present. Reasons for this behavior are not fully elucidated but may include both changes in the active deformation modes and in the mechanisms involved in static and/or dynamic recrystallization.

5.2
Anisotropy of Stress and Strain

Although basal slip has a major influence on the formation of texture during processing of magnesium products it tends to play a smaller role in their strength under common states of loading. This is because the textures are sharp and the shear stresses to which the basal planes are exposed are, in most cases, small, that is to say that they have low Schmid factors for basal slip, although there may be exceptions as mentioned later. The major source of strength variability in these products arises as a result of tension twinning {10$\bar{1}$2}⟨$\bar{1}$011⟩. Unlike slip, twinning depends on the sense

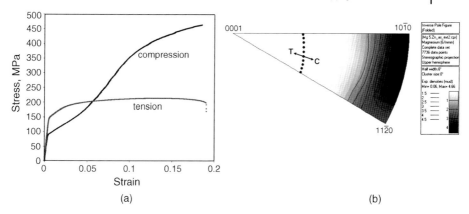

Figure 5.2 (a) Engineering stress–strain curves for extruded rod of magnesium alloy Mg–5% Zn in tension and compression, (b) inverse pole figure for the extrusion direction. Extension twinning occurs in region T in tensile tests and in region C in compression tests.

of shearing and these twins can form when a tensile stress is acting in the vicinity of the c-axis or alternatively when a compression is approximately normal to it. When twinning is favored in these ways, the stress to initiate plastic flow is much reduced.

A good example of how strength is controlled by the availability of twinning is shown in the results by Stanford (Deakin University, private communication) for a round-section extruded magnesium Z5 alloy tested in tension and compression along its length in Figures 5.2 and 5.3. The texture of the alloy is shown as an inverse pole figure in Figure 5.2b. The dotted line in Figure 5.2b marks the boundary between orientations that can undergo twinning when loaded in tension and compression, respectively. It is evident that the compression region is not only larger but also that almost all the crystals are included within this area because of the texture. Since the stress axis lies close to the basal planes of the crystals, slip on these basal planes is very limited in both cases. Schmid factors for the prismatic planes are large, however, and these are the favored slip systems in tension, although giving rise to a high yield stress due to the intrinsic hardness of the prismatic $\langle a \rangle$ slip mechanism. In compression, the situation is totally different since twinning is strongly favored and the yield stress is much lower.

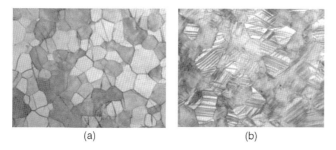

Figure 5.3 Optical micrographs showing structures after 5% deformation (a) in tension and (b) in compression for the material shown in Figure 5.2

Whereas the tension tests show a shape that is normal for most metals, the compression curves are characterized by an inflection and a rapidly rising work hardening rate and flow stress. The possibility for continued twinning decreases rapidly as the original grains are reoriented and their c-axes within the twins become aligned close to the compression direction. Only about 5% strain can be accommodated by the twinning process before the texture is completely transformed. A number of reasons have been proposed for the high strength and work hardening. Barnett [4] concluded that subsequent deformation requires the twinned regions to undergo either $\langle c + a \rangle$ slip or compression twinning, both of which demand much higher stress levels. In Figure 5.2, it can be seen how after about 5% strain the curve for compression rises steeply, later crossing the one for tension where deformation by prismatic $\langle a \rangle$ slip persists.

Micrographs showing the structures of samples deformed to 5% strain in both tension and compression are compared in Figure 5.3. Only a few twins, probably narrow contraction-type twins, are seen in the tensile test specimen where almost all the deformation was accommodated by slip processes. After compression to the same strain level, at least 50% of the volume is converted to a structure consisting of wide-extension twins. It may appear strange that tension is associated with contraction twins while extension twins are found after compression, but it should be remembered that these terms relate to deformations occurring along the directions of the crystal c-axes and not to the specimen loading axes which are orthogonal to one another.

The relatively low strength of magnesium extrusions under compressive loading is recognized as a being a disadvantage and various possibilities for alleviating this have been investigated. Grain refinement is one way in which the situation can be improved. Smaller grains give higher strength for both deformation processes but twinning is especially affected. This can be seen in Figure 5.4 from comparison of the Hall–Petch plots for yield stresses in strongly textured AZ31 alloy [5,6] where the

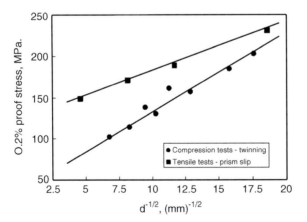

Figure 5.4 Hall–Petch plots for strongly textured magnesium alloy AZ31 from compression and tension tests where deformation is dominated by extension twinning and prism slip, respectively [5,6].

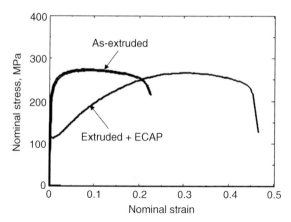

Figure 5.5 Tensile stress–strain curves for an AZ31 alloy in strongly textured extruded condition and weakly textured after ECAP treatment [8].

slope for the twinning case is considerably steeper than for slip, with k_y values of 9.9 and 6.1 MPa $(\text{mm})^{1/2}$, respectively. For conventional grain sizes in the range of 10 to 50 μm, the yield stress in compression lies at least 50 MPa below that for tension. However, the difference reduces with grain refinement, and extrapolation suggests that this type of anisotropy should disappear completely for grain sizes below about 2 μm. In fact, Li et al. [7] have recently demonstrated for the alloy ZK60 having a similar texture to the one in Figure 5.2, that the strength levels in tension and compression become equal to one another when the grain size is reduced to 3 μm. Interestingly, the active deformation modes continued to be slip and twinning, respectively, for these two loading conditions, both operating at the same external stress.

However, it should be noted that grain refinement in itself is not sufficient to ensure high yield stresses. The texture must also be controlled. Several groups, for example, [8,9] have reported that grain refinement of wrought magnesium alloys using equal channel angular pressing (ECAP) actually leads to a reduction in the strength and that this unexpected behavior is attributable to the development of weak and unfavorable textures. Although twinning may be reduced or eliminated, many grains adopt orientations in the diffuse texture where the Schmid factor for basal slip becomes large so that soft modes can be activated during tensile loading. Tensile test curves in Figure 5.5 show examples of this behavior for the alloy AZ31 [8]. Despite the finer grain size after ECAP treatment, the yield stress is reduced by about 50% although ductility is considerably improved.

A different example that demonstrates the important role of twinning on strength is presented in Figure 5.6 for the age-hardenable alloy ZM61 [10]. This had been extruded into flat sections with a sharp texture that is represented by the pole figures shown in Figure 5.6a and b. The yield locus for this material, determined using the method of Lee and Backofen [11] is shown in Figure 5.6c. Strength in the compression quadrant is reduced by the action of twinning, and it is evident that the yielding

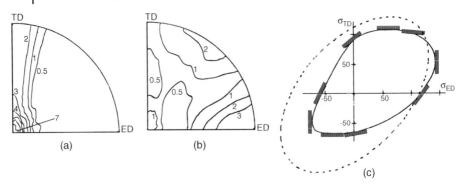

Figure 5.6 Extruded alloy ZM61 (a) (0002) pole figure, (b) (10$\bar{1}$0) pole figure, and (c) plane stress yield locus (units MPa). The dashed line in (c) is a von Mises yielding condition [10].

criterion for this material deviates significantly from the von Mises criterion, which appears as a symmetrical ellipse in this section. While the von Mises criterion is often a good approximation in the case of isotropic metals, it is a poor representation for textured materials, especially hexagonal metals such as magnesium. The dashed curve in Figure 5.6c is the von Mises condition fitted to the tensile strength along the extrusion direction and it can be seen that this greatly overestimates the actual strength for almost all other loading conditions. Engineering designers need to be aware that the standard criteria for calculating strength in complex loading situations may be seriously in error when using wrought magnesium alloys.

An important metallurgical factor in many magnesium alloys is the possibility of strengthening these alloys by precipitation of second phases. Both slip and twinning modes are inhibited by particles and usually the anisotropy is affected by aging treatments. The situation with regard to the effect of particles on twinning is still not fully elucidated, but Robson et al. [12] have shown that the shape and orientation of the particles are likely to be significant factors. Spherical or plate-like particles parallel to the basal planes of the magnesium matrix, as in the AZ91 alloy, raise the yield stress for twinning deformation markedly with the result that the tension/compression asymmetry is greatly reduced. Their analysis suggests that long rod-like particles perpendicular to the basal planes may have the reverse effect by hardening slip more than twinning. Rod-like particles of this type are found in zinc-containing alloys such as Z5, and these alloys do demonstrate slip to be more inhibited than twinning after aging. However, most observations seem to indicate that even when slip hardening exceeds twin hardening in absolute terms, the relative strengthening for twinning is greater. The reason for this is that the preaged strength is much lower in the latter case and, therefore, the overall degree of anisotropy is reduced by the presence of particles.

An example of this behavior was shown for the alloy ZM61, which can be markedly strengthened by an aging treatment at 180 °C for 16 h. Precipitation consists mainly of fine short needles of $MgZn_2$ perpendicular to the basal planes. Although aging raises the strength for all loading conditions, the increments are not

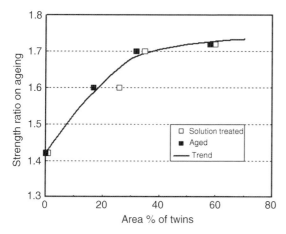

Figure 5.7 Ratio of yield stresses after and before aging for different tension and compression tests on the extruded alloy ZM61. The ratios are shown a function of the area % of twins after 5% strain.

uniform. The increases in yield stress are greater in the tension quadrant of the yield locus ($\Delta Rp \sim 105$ MPa) than in the compression quadrant ($\Delta Rp \sim 85$ MPa) [10]. Despite this, where slip dominates in deformation, aging increases the yield stress by only ~42% whereas in the twinning cases the increase is ~72% so that the degree of anisotropy is reduced. Figure 5.7 shows the ratios of the yield stresses after and before aging plotted together with the area fractions of twins that were measured on metallographic sections after 5% strain. The area fraction of twinned material is somewhat reduced after aging hardening and this is most apparent for intermediate conditions where slip and twinning can to some extent replace one another within the overall deformation process.

As we have seen, the mechanical characteristics and, in particular, the anisotropy of wrought magnesium alloys are frequently controlled by the relative activities of slip and twinning during deformation. Texture, grain size, and precipitation are all factors that influence this balance. Another important factor is the temperature of deformation. Many investigations have led to the conclusion that the various possible deformation modes in magnesium are affected by temperature in different ways. Systematic measurements of critical shear stresses for various modes in pure magnesium at different temperatures have recently been made using plane strain compression tests by Chapuis and Driver [13], who also summarized earlier literature data. The critical shear stress for basal $\langle a \rangle$ glide is small and essentially independent of temperature. It appears that activation of extension twinning is also only weakly affected by temperature. In contrast, many observations show that the critical stress for prismatic $\langle a \rangle$ glide is strongly reduced on raising the deformation temperature and the same is true in the case of pyramidal $\langle c+a \rangle$ slip and compression twinning. The increase in the number of deformation modes, most notably prism slip, that are activated above room temperature has long been associated with the improvement in ductility during hot working [2,3].

A change in the plastic anisotropy is also to be expected at elevated temperatures as slip processes become more favored relative to twinning. In this regard, a particularly sensitive measure of anisotropy is the plastic strain ratio or r-value that is routinely measured in tensile testing of sheet. We will discuss more about the peculiarities of the r-value further, but briefly it is determined as the ratio of width strain to thickness strain within the range of uniform elongation. Typically, the tensile test is interrupted after about 15% elongation when these measurements are made. Equally, plastic strain ratios may be measured on compression test specimens but this is much less common since they are mostly of interest in sheet materials where in-plane compression is difficult to perform. For common engineering materials with cubic crystal structures, the normal deformation modes are symmetrical with respect to their shearing behavior and so the r-values are expected to be identical in tension and compression. This is no longer the case in hexagonal metals when twinning plays a major role. Take, for example, a rolled plate having a basal texture of the type shown in Figure 5.8a. If this is subjected to simple compression along the rolling or transverse directions or, in fact, any of the in-plane directions, the deformation is accommodated principally by twinning that causes expansion along the normal direction and little or none in the width direction. The r-value is therefore close to zero and may even become negative in the early stage of the test [14]. In contrast, tensile straining cannot activate twinning and basal slip is not favored either because of its low Schmid factor. The preferred deformation mode is then prismatic $\langle a \rangle$ glide that produces no change in length along the normal direction (c-axis) with all the extension occurring in the width direction so that the r-value tends toward infinity.

Raising the testing temperature has a similar effect on the r-value as does the change from compression to tension since the critical shear stress for prismatic slip decreases while that for twinning is little affected, as demonstrated by Jain and

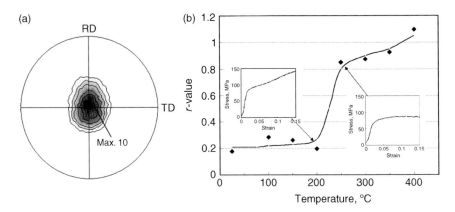

Figure 5.8 (a) Basal pole figure for hot-rolled plate of magnesium alloy similar to the one used in the experiments in Figure 5.8b. (b) Plastic strain ratio measured in compression along the transverse direction of a rolled plate of magnesium alloy AZ31 as a function of the testing temperature.

Agnew [15]. Other results (M.R. Barnett, 2007, Deakin University, unpublished work), shown in Figure 5.8b, confirm that the plastic strain ratio undergoes a rather dramatic change over a narrow range of temperature. The inserted stress–strain curves in Figure 5.8b reveal where the dominant deformation process switches from twinning to slip at the same time. The curve at 200 °C shows the characteristic inflected shape, which is the signature for twinning, whereas at 250 °C the curve has the convex-up shape, which is typical of slip mechanisms. Also, the yield stress is lower at the higher temperature, in conformity with the reduction in the critical shear stress for prismatic slip.

Temperature may also influence the plastic anisotropy of magnesium alloys for other reasons besides its effect on twinning. Agnew and Duygulu [16] were the first to report that the r-value of AZ31 alloy that was tested in tension decreased markedly with increase in test temperature. They investigated rolled plate having an almost perfect basal texture such that no significant effect of twinning could be expected. Agnew *et al.* attributed the fall in r-value and also some change in texture that was observed after straining to an increasing activity of pyramidal $\langle c+a \rangle$ slip at the higher temperatures. This phenomenon was reexamined and confirmed by Hutchinson *et al.* [17], who also used a wide variety of metallographic techniques. No significant enhancement of $\langle c+a \rangle$ slip could be seen by TEM, but there was clear evidence of shearing at or close to the grain boundaries as shown, for example, by scanning probe microscopy carried out on free surfaces, Figure 5.9. Grain boundary sliding (GBS) had been reported separately by Koike *et al.* [18] for AZ31 under similar deformation conditions and this was proposed as the cause of the r-value behavior since the macroscopic strain resulting from GBS should be isotropic with the r-value then tending toward unity, as had been observed. Hutchinson *et al.* reported increasing degrees of GBS with increasing test temperature, although Stanford *et al.* [19] subsequently disputed this, claiming similar grain boundary offsets at both ambient and elevated temperatures and hypothesizing that it was an

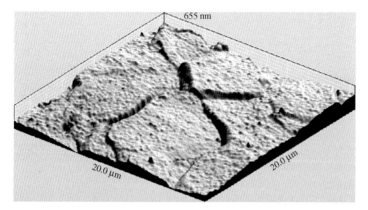

Figure 5.9 Atomic force micrograph of the surface of an AZ31 tensile specimen after straining 5.6% at 250 °C (mean grain size 6 µm) [17].

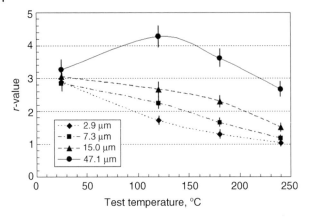

Figure 5.10 Influence of testing temperature on r-values for AZ31 sheets having similar textures but different grain sizes [6].

increase in $\langle c+a \rangle$ slip activity that was responsible for the decrease in r-value. The aforementioned studies were all carried out on fine-grained sheets. In view of the different conclusions, Atwell et al. [6] made a further detailed investigation where a wide range of initial grain sizes was obtained in AZ31 sheets having sharp basal textures. Results of r-value measurements made on these at a range of temperatures are summarized in Figure 5.10.

At room temperature, all the sheets show similar r-values, irrespective of their grain sizes but these values fall with increasing test temperature as was reported earlier and the effect may be large already at 120 °C. The significant difference in this case is the very clear dependence on grain size of the sheets, with more rapid approaches to isotropy as the grain size becomes smaller. There is no other reasonable explanation other than shearing at grain boundaries to explain this behavior. Finer grains imply more boundaries and so larger contributions of shearing to the overall deformation, causing the slip-induced anisotropy to become diluted. Metallographic observations also confirmed that deformation activity at or close to grain boundaries increased steadily with deformation temperature.

The rise in r-value above room temperature for the largest grain size with the peak at nearly 120 °C was confirmed repeatedly and is believed to be due to prismatic slip becoming easier relative to basal slip over this temperature range. Basal slip acts to reduce the r-values while prism slip raises them. Although the same effect will also be present in the finer-grain-sized materials, it is masked by grain boundary shearing in those cases and so does not become apparent in the same way.

The textures and plastic anisotropy in pure magnesium and the traditional alloys containing elements such as aluminum, zinc, zirconium, and manganese all show broadly the same type of behavior. There is presently much interest in developing new alloys containing rare earth (RE) elements such as lanthanum, cerium, neodymium, gadolinium, and also yttrium. These alloys deviate markedly in their mechanical behavior. Probably the main reason for the interest in these alloys is that the ductility is generally improved relative to that of pure magnesium [20–22].

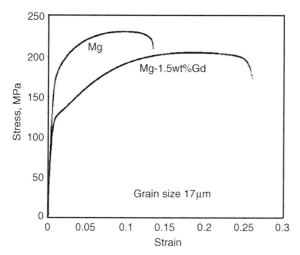

Figure 5.11 Comparison of stress–strain behavior in pure magnesium and a magnesium–1.5% gadolinium alloy processed by extrusion [22].

Figure 5.11 from Stanford and Barnett [22] compares the stress–strain curves of pure magnesium and a magnesium–gadolinium alloy that had been extruded under similar conditions and treated to obtain the same grain size. It is evident that the tensile elongation is greatly increased. However, it may be noted that the yield stress is reduced in the alloy as compared to the pure metal, which runs contrary to usual expectations. The reason for such behavior, which is repeated with other magnesium–rare earth elements, is that their textures are usually much weaker than in pure magnesium or its traditional alloys. Thus, although the metal may be hardened by solid solution effects, the more diffuse textures align more grains into orientations where soft modes, in particular basal slip, may operate. The net effect is to reduce strength, which contributes to greater ductility in the same way as for the ECAP grain refinement treatment described earlier.

Another notable feature of many magnesium alloys containing rare earths is that the usual deviation between tensile and compressive loading conditions is reduced or even eliminated. This was first reported in an extruded magnesium 9 wt%Y alloy by Eckelmeyer and Hertzberg [23], who attributed the effect to inhibition of twinning that removed this directionality. Tension and compression curves for WE54 alloy (Mg-5.2%Y, 1.74%Nd, 0.95%MM, 0.59%Zr) in the extruded condition are presented in Figure 5.12 from the work of Ball and Prangnell [20]. The two curves are quite similar to one another, showing a much different behavior from the usual one that is exemplified in Figure 5.2. In fact, the compression curve lies slightly above the tensile one in the case of WE54. Although the weaker texture resulting from RE-alloying is expected to reduce the strength differential, it cannot account for all of the effect because even a material having a random texture will tend to be weaker in compression since a greater fraction of its grains are suitably loaded to activate twinning than in the case of tension. As can be seen in Figure 5.2b, the solid

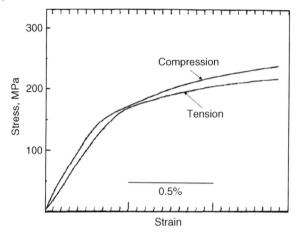

Figure 5.12 Stress–strain curves for bars of magnesium alloy WE54 [20] in tension and compression along the extrusion direction.

angle within which twinning is activated in compression is much greater than the solid angle for tension so that even in the absence of texture, yielding by twinning in compression occurs at lower stresses than in tension.

It seems that the propensity for twinning may be reduced in some RE-alloys at least, although twins are commonly observed after deformation in many alloys. Following the observations of Eckelmeyer and Hertzberg, Safi et al. [10,24] examined an extruded flat magnesium-3%Y alloy and determined its texture and yield locus, which are presented in Figure 5.13. Compare these with the ZM61 alloy results shown in Figure 5.6, which were obtained using the same experimental procedure. The texture is much weaker in the Mg-3%Y and the basal poles tend to be rotated away from the normal direction of the strip about 20° toward the extrusion direction. The yield locus is remarkably symmetrical and corresponds quite closely with the von Mises criterion, which is shown by the dashed line in Figure 5.13c.

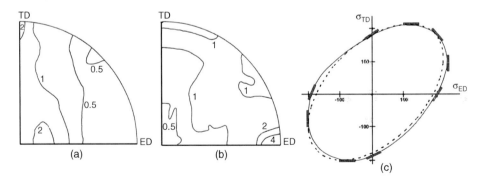

Figure 5.13 Extruded alloy magnesium-3%Y (a) (0002) pole figure, (b) (10$\bar{1}$0) pole figure, and (c) plane stress yield locus (units MPa). The dashed line in (c) is a von Mises yield condition [10].

The results in Figures 5.12 and 5.13 show rather extreme cases and are not typical of all RE-containing alloys. The yttrium alloy in Figure 5.13 had a fine grain size of only 6 μm, which was probably a contributory reason for its isotropic behavior as discussed earlier. Only small contents of rare earth elements are needed to weaken and modify the textures in magnesium alloys. Systematic investigations [25] show that 0.05 wt% or less of RE additions has this effect in extruded rods and also that the texture transition is a function of the processing temperature. Larger RE contents are required to achieve this transition at higher extrusion temperatures. Twinning is not inhibited by these small RE additions but the more diffuse textures act to reduce anisotropy, albeit at the expense of strength. It seems probable that the remarkable isotropy effects shown in Figures 5.12 and 5.13 and also in reference [23] are only associated with high levels of RE alloying. At present, they have no adequate explanation.

5.3
Modeling Anisotropic Stress and Strain

Considerable effort has been expended in recent times in the development of computer models that can explain and predict the stress–strain behavior and plastic anisotropy of textured metals and alloys. Even the common engineering metals with cubic crystal structures, having one single dominant slip mode, present major challenges to this aim although considerable progress has been achieved [26]. Magnesium and other hexagonal metals elicit an even greater complexity because as many as four or even five different deformation modes must be taken into account and these do not respond equally to a given stress field. Taylor-type models have been applied to magnesium but with limited success [27]. The most widely used and most successful approaches to date have been based on viscoplastic self-consistent (VPSC) or elastic–plastic self-consistent (EPSC) models [28,29]. These have allowed quite accurate reproduction of tensile stress–strain curves and also the associated plastic strain ratios (r-values) [30]. Each crystal in the material is considered to reside within a mean field matrix, the characteristics of which are selected to minimize interactions between the individual crystals and itself. Optimized solutions of the mean field then give the predicted plastic properties of the material. Plastic flow for each individual deformation mode is governed by Schmid's law of critical resolved shear stress together with function describing the work hardening, which is usually related to the shear strain through an expression of the Voce type such as

$$\tau = \tau_0 + (\tau_1 + \theta_1 \gamma)\{1 - \exp(\gamma \theta_0 / \tau_1)\} \qquad (5.1)$$

This necessitates four disposable parameters (τ_0, τ_1, θ_0, and θ_1) or three if θ_1 is set to zero, as is often justifiable. In general, it is necessary to invoke at least four different modes of slip and twinning in order to encompass all the deformation processes that metallographic observations have shown to occur and, as a result, the models tend to necessitate up to at least 12 constants that have to be deduced from optimization routines. It would, indeed, be surprising if good agreement were not achieved with the inclusion of so many disposable parameters, but such a situation is clearly not satisfying. This high degree of empiricism, despite the sophistication of the physical models, raises concerns about the possibility of wider predictions, outside the range of conditions where the data for the model were established. Our aim here is, therefore, to look at arguments that can be used to include additional physical principles that should reduce the empirical content of the model parameters. The approach is based on two recent works [31,32]

First, let us consider what is meant by the expression "critical resolved shear stress" or CRSS. This expression has adopted different meanings at different times and for different people. In early literature (e.g., [33]), this was usually taken to mean the minimum shear stress necessary to cause slip and was evaluated from single crystals deformed in Stage I of the generic stress–strain curve, with back-extrapolation to zero strain where necessary. This may also be considered as the best available measure of the critical shear stress necessary to cause movement of a dislocation (in the case of prism slip in magnesium it may alternatively be the stress necessary to activate cross-slip of a dislocation from the basal plane on to the prism plane). It is worth remembering that a segment of dislocation line will move at this CRSS value and only at this CRSS value. At the limiting dislocation level, there is no such thing as work hardening. For clarity, let us refer to this as the "fundamental CRSS." Experimental values of these fundamental CRSSs are available for many metals and deformation modes from single crystal experiments, and it would be most desirable to incorporate their absolute values as independent data within the computer models. At present, this is not a normal practice. If used at all, it is only their relative values that are taken into account as ratios of CRSS values for different modes.

As discussed in a previous paper [34], the effect of increasing deformation (or grain refinement or precipitates, for that matter) is to generate back-stresses resisting the tendency for dislocation movement. Continued plastic flow then requires that the applied stress is large enough to overcome these back-stresses and reach up further to the fundamental CRSS. There is, accordingly, an "external" value of shear stress that must be reached on the slip systems for flow to continue and this value does increase with straining. We will call this the "apparent CRSS". It is this which is referred to in most of the modern literature and which is applicable in the case of modeling. The apparent CRSS is, generally speaking, equal to the fundamental CRSS with an additive term that corresponds to the current degree of hardening. The additional term also includes other contributions from grain boundaries and precipitates and so on. Furthermore, this hardening term should, to a fair approximation, be constant for all the active slip systems. This last point is not rigorously true since specific interactions between different dislocations may not always be identical as, for example, in the cases of self-hardening or of latent hardening between different slip systems or slip modes. However, although such effects are known to occur, they are seldom, if ever, included in calculations because of the complexity involved and the lack of an adequate theoretical basis for treating them. The present argument can therefore provide a great simplification since the apparent CRSSs for different slip modes used in the models should be well approximated by the known fundamental CRSSs together with an additive work hardening term, which is taken to be the same for all the slip modes.

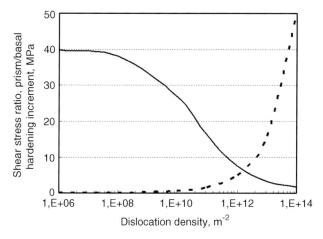

Figure 5.14 Calculated ratio of apparent CRSSs for ⟨a⟩ slip on prismatic and basal planes in magnesium as a function of dislocation density during straining (solid line). Also shown is the increase in shear flow stress (dashed line) [34].

Whereas the fundamental CRSS values may vary considerably in magnitude between different slip modes, the apparent values tend to approach one another rapidly with increasing deformation. As the additive back-stress term becomes increasingly dominant, the ratios between apparent CRSS values for different modes do not remain constant but tend toward unity. This is demonstrated in Figure 5.14 [34] for the case comparing prismatic and basal slip in magnesium where the initial ratio of 40 applies to fundamental CRSS values determined from single crystal data. With only a modest degree of strain hardening, the ratio of apparent CRSSs reduces by more than an order of magnitude, reaching values of only 2 or 3 within the span of a typical mechanical test.

The need to utilize CRSS ratios that deviate significantly from experimental fundamental ones as inputs for computer models has been recognized in numerous cases (e.g., [16,30] and, indeed, the work hardening factors that emerge as outputs from the computer models tend to confirm the same behavior. Recognizing difficulties of this sort, Kocks [35] has suggested that it is inappropriate to quantify work hardening by CRSS values from single crystal experiments carried out under single slip conditions so he proposed that crystals oriented for polyslip deformation should be used instead. Although this viewpoint was specifically concerned with work hardening, it appears to have been extrapolated subsequently to imply that CRSS values from single slip tests are not of practical value. We do not agree with this viewpoint and, in the present treatment, we seek to reinstate the fundamental CRSS values as starting points for polycrystalline modeling.

On the basis of the arguments put forward, we suggest that a physically justifiable approach to modeling is to start out from the experimentally determined fundamental CRSS values and then allow the back-stress term to develop as straining proceeds to quantify the work hardening behavior. This generalized back-stress describes not only kinematic hardening effects but also isotropic contributions arising from both deformation and microstructure. The work hardening part should be approximately identical for all deformation modes and ought to be limited to no more than three parameters, which should be sufficient to permit the same degree of complexity as is captured by the Voce-type expressions. Contributions from precipitation strengthening or grain size are also introduced as constant values in the generalized back-stress term. In some computer models, the work hardening rates associated with different slip modes are allowed to diverge considerably from one another. An example of this [30] has been given for $\langle c+a \rangle$ slip, which is considered to work harden more rapidly than $\langle a \rangle$ slip modes. It is certainly the case that strongly textured polycrystals oriented for $\langle c+a \rangle$ slip demonstrate a very high work hardening rate – at least for conditions of compression along the c-axis (pyramidal $\langle c+a \rangle$ slip shows asymmetric behavior). TEM observations [36] show that the high work hardening rate can be understood on the basis that double cross-slip occurs profusely, leaving large amounts of dislocation debris. Of course, such debris interferes not only with the $\langle c+a \rangle$ dislocations but with all other deformation modes as well. Our present approach will not include differential hardening for the sake of simplicity and because there are not as yet physical principles available on which to base it.

Our reasoning leads us to the view that when modeling the plasticity in polycrystalline hexagonal materials it should be sufficient to include only two or three free parameters in a single hardening expression, together with independent values of the fundamental CRSSs taken from literature sources. Contributions from grain size or second phases should also be included in the common back-stress term and often this can be done using independent data that are not evaluated within the model calculations. This would represent a major simplification over the 12 or 16 factors that are commonly invoked in current models. A fundamentally different view is adopted by Raeisinia *et al.* [37], who propose modifying the CRSS functions in the model to include not only work hardening but also the metallurgical parameters such as grain size and, in particular, solute. All these quantities are then obtained as solutions in the computer. In terms of the inner working of the VPSC model, the two approaches need not vary greatly but we believe that the present viewpoint is more elegant in its simplicity and more readily permits physical insights into the underlying mechanisms.

We take as an example of this approach a model to describe the tensile deformation in a magnesium alloy, including both the stress–strain relationship and the plastic strain ratio (r-value) to make the test as critical as possible.

The material chosen was the magnesium alloy AZ31 containing 2.94%Al, 0.90%Zn, and 0.50%Mn, which had been processed by hot rolling to a thickness of 1.7 mm and subsequently annealed for 17 hours at 500 °C to produce a well-recrystallized microstructure with mean linear intercept grain size of 40 μm. The texture of this material is shown in the $\{0001\}$, $\{10\bar{1}1\}$, and $\{10\bar{1}0\}$ pole figures in Figure 5.15 determined from EBSD measurements. It demonstrates a moderately strong $\langle 0001 \rangle //$ND fiber texture containing peaks at $\{0001\}\langle 10\bar{1}2 \rangle$, which is quite typical for this alloy in the rolled and annealed condition.

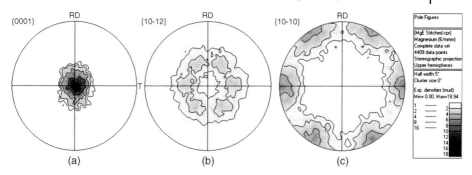

Figure 5.15 (a) {0001}, (b) $\{10\bar{1}1\}$, and (c) $\{10\bar{1}0\}$ pole figures for the present AZ31 alloy sheet.

Specimens for tensile testing were prepared in the transverse direction of the sheet and had a milled gauge length of 25 mm and a width of 5 mm prior to the annealing treatment. Tensile tests were carried out using an MTS universal machine at a nominal strain rate of 10^{-3} s^{-1} up to an elongation of 10%. Width and length strains were monitored continuously using an Instron noncontacting optical extensometer. True stress–true strain behavior was evaluated in the standard way. Determination of the plastic strain ratio requires some more explanation. The r-value is defined as the ratio of plastic strains in the width (w) and thickness (t) directions. However, it is a standard practice to measure strains in the width and length (l) directions and deduce the thickness strain assuming constancy of volume. Thus

$$r = \varepsilon_w/\varepsilon_t = -\varepsilon_w/(\varepsilon_l + \varepsilon_w) \tag{5.2}$$

The present experiments were carried out during continuous loading because it was known from previous reports [30] that the r-value is not constant during straining and this variation provides a sensitive test of the model's correctness. When measurements are made under load there will necessarily be elastic strains as well as plastic ones and these must be first eliminated since they are not relevant to the modes of plastic deformation. Elastic contributions to ε_l and ε_w were first subtracted using values of 45 GPa for Young's modulus and 0.29 for Poisson's ratio and the r-values were calculated based both on total strains (elastic + plastic, r_{tot}) and on plastic strains only (r_{plast}).

Figure 5.16 shows the variation of true stress with true plastic length strain and also the variation of the strain ratios (r) with and without the elastic components. It is clear that the r-values vary markedly with strain, being very low initially and rising sharply to approximate plateau values followed by a slight reduction. No values are included for length strains less than 0.004 since the errors become large in the ratios of very small experimental numbers. The r-value is closely related to the particular crystal deformation processes and to the slope of the yield locus, so it is a rather fundamental parameter in connection with plastic deformation. Its relevance is less evident, however, when its value changes during the test as is the case here. Since r is calculated from integral strains, it represents a historic average of events up to the strain where it is measured, which does not necessarily correspond to the actual

5 Plastic Anisotropy in Magnesium Alloys – Phenomena and Modeling

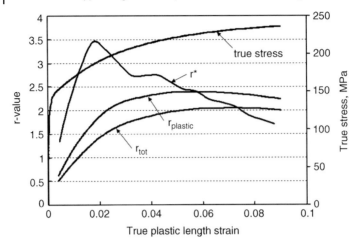

Figure 5.16 True stress and r-values plotted against the true plastic length strain for a tensile test of the sheet alloy AZ31. The r-values are evaluated from elastic + plastic strains (r_{tot}), plastic strains only ($r_{plastic}$), and from incremental plastic strains (r^*).

behavior at any point. The only physically meaningful value of the plastic strain ratio in this situation is the current actual value that is determined over small strain increments or, more strictly, from differentials of the width and thickness strains. Let us call this the r^*-value defined as

$$r^* = d\varepsilon_w / d\varepsilon_t \tag{5.3}$$

The present results were processed using cubic spline functions to smooth the data and provide differentiable expressions from which values of r^* could be evaluated at different stages through the test. The variation of r^* with length strain is also shown in Figure 5.16. (Some minor undulations appear in this curve, which are mathematical artifacts resulting from the different segments of the splines). The variation of r^* differs considerably from that of the conventional r-value. Its initial value is estimated to be about 0.5 (dashed line) since r^* and r must converge here. There is a very rapid rise in r^* up to 3.5 after only 2% strain, followed by an approximately linear decrease down to about 1.7 at 10% elongation.

Qualitatively, the r^* results in Figure 5.16 imply major changes in the ongoing deformation processes. The low initial value may be associated with basal $\langle a \rangle$ slip in grains that are not perfectly oriented and which results in a predominant thinning of the sheet. The very small fundamental CRSS for basal slip gives this deformation mode a relative advantage but an advantage that is rapidly lost with progressive straining, as shown in Figure 5.14. Accordingly, prismatic $\langle a \rangle$ slip that has a high Schmid factor may then become the most active mechanism and this produces mainly narrowing of the specimen and an associated high r^*-value. With further strain hardening, yet another mode becomes apparent that counteracts the effect of prismatic slip. Possible modes that would markedly reduce r^* are

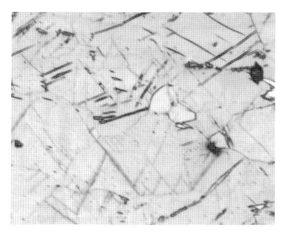

Figure 5.17 Optical micrograph of AZ31 after 10% elongation in tension along TD showing presence of narrow compression twins. Image width: 250 μm.

pyramidal $\langle c+a \rangle$ slip [16,17] or alternatively compression twinning, which has been observed by optical microscopy after completion of the tests, as shown in Figure 5.17. These processes should affect the anisotropy in similar manners, contributing to a large thinning of the sheet and so leading to a prominent reduction in r^*.

A quantitative analysis is now presented using the VPSC method [28]. Table 5.1 summarizes the various parameters that permit reasonably close agreement between the experimental values of flow stress and r-values and the corresponding values from the computer model. These parameters are of two types. One category is for independent input data regarding CRSS values for the different modes in accordance with the published values obtained from single crystal tests [13]. These CRSS values are taken to be the τ_0 terms in the VPSC model. Parameters of the other type include the microstructural state, corresponding mainly to the grain size via the Hall–Petch contribution, which is also constant throughout the test, as well as the work hardening factors τ_1 and θ_0 that are obtained by optimization in the model, where all the deformation modes are subjected to the same deformation substructure described by a single relationship, Eq. (5.4), where $\Sigma \gamma$ is the sum of shears

Table 5.1 Parameters used in modeling flow stress and r-values.

Independent parameters					Adjustable parameters		
CRSS basal $\langle a \rangle$	CRSS prism $\langle a \rangle$	CRSS pyramidal $\langle c+a \rangle$	Compression twinning		Microstructure constant	θ_0	τ_1
			CRSS	Shear			
1 MPa	46 MPa	65 MPa	85 MPa	0.14	35 MPa	1500 MPa	70 MPa

on all active systems. In the computer program this is achieved by assigning the same values of θ_0 and τ_1 to all deformation modes.

$$\Delta\tau = \tau_1\left\{1 - \exp\left(\sum \gamma \cdot \theta_0/\tau_1\right)\right\} \tag{5.4}$$

The deformation modes included in the model were basal $\langle a \rangle$ slip, prismatic $\langle a \rangle$ slip, pyramidal $\langle c+a \rangle$ slip, and compression twinning $\{10\bar{1}1\}\langle \bar{1}012\rangle$. Extension twinning was not included since it is largely precluded by the texture in this material, although a few such twins were seen in the microstructure. In terms of the model, their participation cannot readily be distinguished from that of basal slip. The microstructure constant is adjusted as a fitting parameter but is dominated by the Hall–Petch grain size contribution, ($\Delta\sigma = k_y \Delta d^{-1/2}$). The VPSC calculations were carried out using the affine grain interaction model, which was determined by Wang et al. [38] to be the most suitable for a similar magnesium alloy.

Figures 5.18 and 5.19 show the stress versus strain as well as r- and r^*-values versus strain relationships, respectively, comparing the experimental and calculated results. Agreement is not perfect but quite close in both respects and it appears that the model has indeed captured the salient features of the deformation process. In particular, the unusual pattern demonstrating both a rise and a fall in r^* with strain evolves naturally without any *ad hoc* assumptions. Yielding is somewhat more gradual in the model than in reality and the rise in r^*-value in the early stages is more rapid. These effects may be related to one another although the connection is not evident. Despite there being only three adjustable parameters in the present treatment, the degree of fit is barely inferior to corresponding published results (e.g., [15,37,39]) where 12 or more free parameters were applied in the optimization. We consider that this justifies the reasoning adopted here, notably that "fundamental" CRSS values are applicable to modeling exercises and that all hardening terms can be treated as being additive to these.

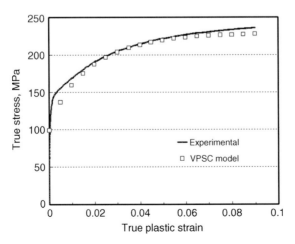

Figure 5.18 Comparison of experimentally measured and calculated stress–strain curves for AZ31.

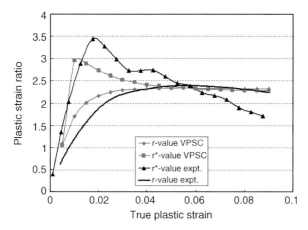

Figure 5.19 Comparison of experimentally measured and calculated r- and r^*-values as functions of strain.

Contributions from the different deformation modes during the test as determined by the VPSC model are plotted in Figure 5.20. Basal slip is important in the very beginning, benefitting from its low CRSS, and causing a very low initial value for r^*. This advantage relative to other modes is rapidly lost as work hardening commences, with the assertion of prismatic glide that makes for a rapid rise in the r^*-value. Pyramidal slip and twinning then come increasingly into play as the overall stress level rises further when their high CRSSs become relatively less restricting. It is the participation of these latter modes that causes the r^*-value to fall progressively after the peak. All these behaviors are explicable in terms of the earlier discussion as exemplified in Figure 5.14. The thin compression twins are not easy to measure quantitatively in optical micrographs such as Figure 5.16 but their volume (area) fraction appears compatible with the model prediction of 6% at the end of the test.

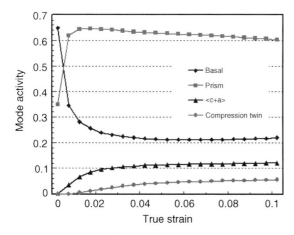

Figure 5.20 Fraction of strain accommodated by the different deformation modes during tensile testing as predicted by the VPSC model.

The present approach to using the VPSC model also opens up applications that were previously not self-evident. Strengthening mechanisms such as grain refinement and precipitation hardening make their contribution through the additive constant term and can be included as known quantitative parameters based on established physical metallurgical principles. It was not at first sight obvious that these will impact on anisotropic characteristics such as r-values but they are now expected to do so. In simple terms, these metallurgical strengthening mechanisms reduce the relative advantage of soft modes and permit the harder modes such as pyramidal slip and compression twinning to take up a larger share of the plastic activity. The results are quantitatively predictable with the aid of the VPSC model. Jain et al. [39] have examined the question of varying grain sizes and found a weak relationship in one case where a reduction in the r-value occurred for smaller grain sizes, although the situation was complicated since the texture was not entirely constant. The influence of twinning fraction was discussed in that work but the possibility that slip mode activity alone could be influential was not recognized. A similar tendency could be discerned in the work of Atwell et al. [6].

As an example, let us consider a case where the grain size of the AZ31 alloy is refined while the texture remains completely unchanged. The increase in strength due to grain refinement is calculable from the known Hall–Petch behavior such as in Figure 5.4 ($\Delta\sigma = k_y d^{-1/2}$). All other terms in the calculation such as the CRSS values and work hardening parameters are left unchanged except that the possibility of twinning has been removed for the purpose of demonstration. That is not, in fact, unreasonable for the texture in question and for finer grained structures. Yield stresses and r-values are then calculated and plotted together as shown in Figure 5.21. In this situation, the r-value is predicted to decrease as strength rises due to the grain size effect since this benefits the activity of the hard pyramidal $\langle c + a \rangle$ slip

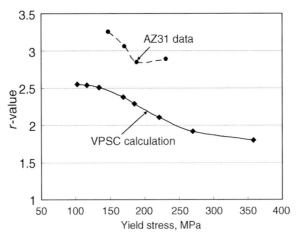

Figure 5.21 The effect of strengthening by grain refinement on the r-value after 10% elongation predicted by the VPSC model (solid line). Experimental data [6] on AZ31 (dashed line) are from a similar but not identical texture to that in the calculations.

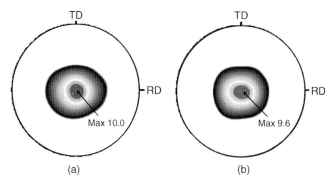

Figure 5.22 {0001} pole figures for (a) initial texture and (b) calculated texture after 10% strain.

systems. A direct experimental confirmation of this is hard to achieve as it is notoriously difficult to change grain sizes without simultaneously modifying the texture. However, it is encouraging that the trend is qualitatively in agreement with some of the observations reported by Jain *et al.* and is also compatible with results for room temperature data in Figure 5.10, shown by the dashed line.

It is important to recognize that the changes in anisotropy predicted here are not connected principally with any variations in the texture. As shown in Figure 5.22, the texture is virtually identical before and after an elongation of 10%. The influence of microstructural hardening is caused by the change in relative values of the apparent CRSSs. When there is little strengthening from microstructure sources, soft modes such as basal slip are favored but their advantage is progressively lost as the effects of grain refinement or precipitation increase. Then, the harder modes, first prismatic slip and subsequently $\langle c+a \rangle$ slip, take up larger fractions of the plastic activity (averaged after 10% strain) as can be seen in the calculations in Figure 5.23. In all cases, the fundamental CRSS values (τ_0) in the models remain the same.

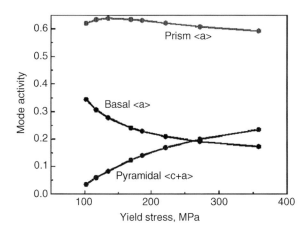

Figure 5.23 Calculated mode activity after 10% strain as the yield stress changes due to microstructural hardening.

5.4
Concluding Remarks

Examples of phenomena in Section 5.2 together with the modeling approach in Section 5.3 demonstrate the remarkable complexity of magnesium alloys in respect of their mechanical properties. The plastic deformation modes, of which there are at least five types, show wide variation in their intrinsic hardness, which manifests itself in many ways. Basal $\langle a \rangle$ slip and extension twinning can take place at very low intrinsic stresses so they tend to dominate whenever the texture is weak. These modes are most active during the break-down of coarse initial grain structures in as-cast ingots and account principally for the textures arising in wrought products. Important examples are the "basal" texture in hot-rolled plates and the $\{0001\}\langle 10\bar{1}0\rangle$ cylindrical textures in round section extrusions, which are largely a result of early basal slip. Although additional and harder deformation mechanisms are also necessitated, especially as texture becomes sharper during processing, their influence on the preferred orientation is only minor. Temperature affects the relative hardness of the various modes, in particular, prismatic slip becomes more favorable above approximately 200 °C, but even this has only a limited influence on the resulting texture. Further processing by annealing is commonly practiced in industry to generate a stable recrystallized condition. In pure magnesium and in the traditional alloys, recrystallization tends to recreate textures similar to those remaining after deformation although usually rotated around the c-axes, so it produces little change in the basal pole figures or in the anisotropy of properties. It should be noted, however, that newer alloy types containing rare earth metal additions may differ significantly from this pattern of behavior.

The strength of wrought magnesium alloys depends on the usual factors such as grain size, dislocation density, and precipitation and is also greatly influenced by the sharp textures that are typically present. This last effect depends on the relationship between the stress state during loading and the c-axis orientation in the predominant texture. Yield stresses are very low when shear components of the applied stress align with basal slip or extension twinning conditions. However, high strength may be achieved in other situations where the texture results in low Schmid factors for the soft modes and so where yielding necessitates slip on harder systems such as prismatic and pyramidal planes. Important examples of this effect are seen in the high tensile strength of extrusions along their axial direction and of rolled plates as conventionally measured along directions in the rolling plane.

A significant insight arising from the present modeling approach is that textural and microstructural hardening contributions are not independent but interact markedly with one another. This derives from the recognition that dislocation glide on any given slip mode always occurs at the same "fundamental" CRSS value. All forms of microstructure hardening raise the "apparent" CRSSs by constant amounts so that their differences relative to one another become reduced. Deformation modes that have low intrinsic CRSS levels such as basal slip are always preferred during the early stages of loading, but this advantage relative to other harder modes becomes much less as work hardening takes place. Equivalently, microstructural hardening due to grain refinement or precipitation also acts to disadvantage the soft deformation modes, permitting a greater share of the overall strain to be accommodated by the intrinsically harder process of prismatic and pyramidal slip or compression twinning. Change in the microstructural hardening contribution therefore modifies the shape of the yield locus, even when the texture is constant, which can be revealed in a sensitive manner by studying the large variations that occur in plastic strain ratios (*r*-values).

Acknowledgments

The authors thank colleagues (Dale Atwell, Filip Siska and Jain Jayant) from the Institute for Frontier Materials at Deakin University for assistance with measurements and analysis of the results. WBH wishes to thank Deakin University and the ARC Centre of Excellence for Design in Light Metals for providing facilities and financial support during this work. They also express their gratitude to Dr. Carlos Tomé for making available the VPSC program software.

References

1 Roberts, C.S. (1960) *Magnesium and its Alloys*, John Wiley & Sons, Inc., New York.
2 Hosford, W.F. (1993) *The Mechanics of Crystals and Textured Polycrystals*, Oxford University Press, Oxford.
3 Reed-Hill, R.E. (1964) *Deformation Twinning*, Gordon and Beach, New York.
4 Barnett, M.R. (2007) Twinning and the ductility of magnesium alloys. Part I tension twins. *Materials Science and Engineering*, **A464**, 1–7.
5 Barnett, M.R., Keshevarz, Z., Beer, A.G., and Atwell, D. (2004) Influence of grain size on the compressive deformation of wrought Mg 3Al 1Zn. *Acta Materialia*, **52**, 5093–5103.
6 Atwell, D., Barnett, M.R., and Hutchinson, W.B. (2012) The Effect of initial grain size and temperature on the tensile properties of magnesium alloy AZ31 sheet. *Materials Science and Engineering*, **A549**, 1–6.
7 Li, B., Joshi, S.P., Almagri, O., Ma, Q., Ramesh, K.T., and Mukai, T. (2012) Rate-dependent hardening due to twinning in an ultrafine-grained magnesium alloy. *Acta Materialia*, **60**, 1818–1826.
8 Mukai, T., Yamanoi, M., Watanabe, H., and Higashi, K. (2001) Ductility enhancement in AZ31 magnesium alloy by controlling its grain structure. *Scripta Materialia*, **45**, 89–94.
9 Li, B., Joshi, S.P., Azevedo, K., Ma, E., Ramesh, K.T., and Figueiredo, R.B. (2009) Dynamic testing at high strain rates of an ultrafine-grained magnesium alloy processed by ECAP. *Materials Science and Engineering*, **A517**, 24–29.
10 Safi Naqvi, S.F., Hutchinson, W.B., and Barnett, M.R. (2008) Texture and mechanical anisotropy in three extruded magnesium alloys. *Materials Science and Technology*, **24**, 1283–1292.
11 Lee, D. and Backofen, W.A. (1966) An experimental determination of the yield locus for titanium and titanium-alloy sheet. *Transactions of the Metallurgical Society of AIME*, **236**, 1077–1084.
12 Robson, J.D., Stanford, N., and Barnett, M.R. (2011) Effect of precipitate shape on slip and twinning in magnesium alloys. *Acta Materialia*, **59**, 1945–1956.
13 Chapuis, A. and Driver, J.H. (2011) Temperature dependency of slip and twinning in plane strain compressed magnesium single crystals. *Acta Materialia*, **59**, 1986–1994.

14 Chun, Y.B. and Davies, C.H.J. (2011) Negative lateral strain ratio induced by deformation twinning in magnesium alloy AZ31. *Materials Science and Engineering*, **A528**, 4941–4946.

15 Jain, A. and Agnew, S.R. (2007) Modelling the temperature dependent effect of twinning on the behaviour of magnesium alloy AZ31B sheet. *Materials Science and Engineering*, **A462**, 29–36.

16 Agnew, S.R. and Duygulu, O. (2005) Plastic anisotropy and the role of non-basal slip in magnesium alloy AZ31B. *International Journal of Plasticity*, **21**, 161–1193.

17 Hutchinson, B., Barnett, M.R., Ghaderi, A., Cizek, P., and Sabirov, I. (2009) Deformation modes and anisotropy in magnesium alloy AZ31 *International Journal of Materials Research*, **100**, 556–563.

18 Koike, J., Ohyama, R., Kobayashi, T., Suzuki, M., and Maruyama, K. (2003) Grain boundary sliding in AZ31 alloys at room temperature to 523K. *Materials Transactions*, **44**, 445–451.

19 Stanford, N., Sotoudeh, K., and Bate, P.S. (2011) Deformation mechanisms and plastic anisotropy in magnesium alloy AZ31. *Acta Materialia*, **59**, 4866–4874.

20 Ball, E.A. and Prangnell, P.B. (1994) Tensile-compressive yield asymmetries in high strength wrought magnesium alloys. *Scripta Metallurgica et Materialia*, **31**, 111–116.

21 Al-Samman, T. and Li, X. (2011) Sheet texture modification in magnesium-based alloys by selective rare earth alloying. *Materials Science and Engineering*, **A528**, 3809–3822.

22 Stanford, N. and Barnett, M.R. (2008) The origin of "rare earth" texture development in extruded Mg-based alloys and its effect on tensile ductility. *Materials Science and Engineering*, **A496**, 399–408.

23 Eckelmeyer, K.H. and Hertzberg, R.W. (1970) Deformation in wrought Mg 9 wt% Y. *Metall Trans*, **1**, 3411–3414.

24 Safi Naqvi, S.F. (1978) *Plastic anisotropy of some age hardening magnesium alloys*. Ph.D. Thesis, University of Birmingham, UK.

25 Stanford, N. (2010) Microalloying magnesium with Y, Ce Gd and La for texture modification – A comparative study. *Materials Science and Engineering*, **A527**, 2669–2677.

26 van Houtte, P. and Peeters, B. (2003) Contribution of deformation-induced intragranular microstructure to the deformation of metals. *Journal de Physique*, **105**, 207–214.

27 Dillamore, I.L., Haddon, P., and Stratford, D.J. (1972) Texture control and the yield anisotropy of plane strain magnesium extrusions. *Texture*, **1**, 17–29.

28 Lebensohn, R.A. and Tomé, C.N. (1993) A self consistent anisotropic approach for the simulation of plastic deformation and texture development of polycrystals: Application to zirconium. *Acta Metallurgica et Materialia*, **41**, 2611–2624.

29 Turner, P.A., Christodoulou, N., and Tomé, C.N. (1995) Modelling the mechanical response of rolled Zircaloy-2. *International Journal of Plasticity*, **11**, 251–265.

30 Agnew, S.R. (2002) Plastic Anisotropy of Magnesium Alloy AZ31B Sheet. *Magnesium Technology* (ed. H.I. Kaplan), TMS, Warrendale, PA, pp. 169–174.

31 Jain, J., Poole, W.J., and Sinclair, C.W. (2012) The deformation behaviour of the magnesium alloy AZ80 at 77 and 293K *Material Science Engineering*, **A547**, 128–137.

32 Hutchinson, W.B., Barnett, M.R., and Jain, J. (2012) A minimum parameter approach to crystal plasticity modelling. *Acta Materialia*, **60**, 5391–5398.

33 Schmid, E. and Boas, W. (1935) *Kristallplastizitaet*, English translation F. A. Hughes and Co. Ltd., London, p. 1950.

34 Hutchinson, W.B. and Barnett, M.R. (2010) Effective values of critical resolved shear stress for slip in polycrystalline magnesium and other hcp metals. *Scripta Materialia*, **63**, 737–740.

35 Kocks, U.F. (1970) The relationship between polycrystal deformation and single-crystal deformation. *Metallurgical Transactions*, **1**, 1121–1143.

36 Jones, I.P. and Hutchinson, W.B. (1981) Stress-state dependence of slip in, Titanium-6A1–4V and other H.C.P. metals. *Acta Metallurgica*, **29**, 951–968.

37 Raeisinia, B. and Agnew, S.R. (2010) Using polycrystal plasticity modeling to

determine the effects of grain size and solid solution additions on individual deformation mechanisms in cast Mg alloys. *Scripta Materialia*, **63**, 731–736.

38 Wang, H., Raeisinia, B., Wua, P.D., Agnew, S.R., and Tomé, C.N. (2010) Evaluation of self-consistent polycrystal plasticity models for magnesium alloy AZ31B sheet. *The International Journal of Solids and Structures*, **47**, 2905–2917.

39 Jain, A., Duygulu, O., Brown, D.W., Tomé, C.N., and Agnew, S.R. (2008) Grain size effects on the tensile properties and deformation mechanisms of a magnesium alloy AZ31B sheet. *Materials Science and Engineering*, **A486**, 545–555.

6
Application of Stochastic Geometry to Nucleation and Growth Transformations

Paulo R. Rios and Elena Villa

6.1
Introduction

The early theory developed by Johnson-Mehl [1], Avrami [2–4], and Kolmogorov [5] has constituted the foundation of formal kinetics theories applied today. Their theory, often referred to as the "JMAK" theory, and its developments and extensions find widespread application to transformations in both metallic and nonmetallic materials. Those transformations extend beyond strictly "nucleation and growth" first-order transitions to situations in which nucleation and growth have a purely "operational" meaning. Therefore, examples of formal kinetics modeling can be found in a diversity of transformations such as austenite to perlite transformation [1], recrystallization [6], abnormal grain growth in $BaTiO_3$ [7,8], martensite "spread" [9,10], and polymer crystallization [11,12].

Johnson-Mehl, Avrami, and Kolmogorov considered the transformation of a parent phase, α, into a new phase, β, by nucleation and growth and obtained a global kinetic measure, the volume fraction transformed, $V_V(t)$, as a function of the reaction time, t. Perhaps, their main contribution was to show how to deal with the problem of impingement, that is, the interference among growing regions. Johnson-Mehl and Avrami did this by defining another global quantity, $V_E(t)$, the extended volume fraction, as the sum of the volume of the grains, assuming that they could grow without impingement, divided by the total volume. For nuclei uniformly randomly located in space, they showed that $V_V(t)$ could be related to $V_E(t)$ by

$$V_V(t) = 1 - \exp(-V_E(t)) \tag{6.1}$$

Kolmogorov obtained the same result by means of a probabilistic approach. He determined the probability that an untransformed point x would transform in the time interval between t and $t + dt$. It is worthy of note that what is normally understood by "uniformly randomly" in this context can be more precisely defined as "homogeneous Poisson point process." More rigorous treatment and vocabulary are introduced later in this chapter.

Two exact analytical expressions are normally associated with this early work. Both were obtained supposing that the new phase grows with a spherical shape and constant velocity, G, but with different nucleation conditions. In the first of these, the nucleation is supposed to be site-saturated, which means that all possible nucleation sites are exhausted at the very beginning of the reaction. One often describes this by saying that all nuclei are already present at $t = 0$. If the MEAN number of nuclei per unit of volume is N_V, then

$$V_V(t) = 1 - \exp\left(-\frac{4\pi}{3} N_V G^3 t^3\right) \qquad (6.2)$$

whereas if one has a constant nucleation rate per unit of volume I_V, then

$$V_V(t) = 1 - \exp\left(-\frac{\pi}{3} I_V G^3 t^4\right) \qquad (6.3)$$

DeHoff and Gokhale [13,14] introduced the concept of microstructural path. They proposed the use of an additional microstructural measure, namely, the interfacial area per unit of volume between transformed and untransformed regions, S_V, and its corresponding extended interfacial area per unit of volume, S_E

$$S_V = (1 - V_V) S_E$$

The microstructural path of a transformation would then correspond to a curve on the (V_V, S_V) plane. Vandermeer and coworkers [15] further developed this concept into a theoretical treatment covering variable nucleation and growth rates, as well as inclusion of nonspherical regions. They called their extended analysis the microstructural path method (MPM). MPM has been extensively applied by Vandermeer and coworkers to analyze recrystallization in a variety of metallic materials. The advantage of their methodology is that the extended quantities are derived from experimental measurements of just V_V and S_V, both of which may be determined from planar sections using standard stereological techniques. Although theories that primarily rely on measures on a planar section continue to be useful, with the advent of 3D computer simulation of microstructures, and with recent developments in 3D characterization techniques, this situation has changed substantially. It is now possible, and desirable, to develop 3D theories that require 3D confirmatory measurements. This permits the construction of analytical models that relate directly to 3D microstructural evolution.

Most of the analytical developments building upon the JMAK theory have retained the core of their original assumptions regarding nucleation and growth. Specifically, they assumed nuclei to be uniformly randomly distributed in space and that the velocity of the moving boundary was the same for every growing grain. Recently, the present authors have introduced a new mathematical methodology involving stochastic geometry and geometric measure theory to deal with nucleation and growth problems. In a series of papers, new analytical expressions for nucleation and growth transformations have been derived. Those papers focused mainly on generalizing nuclei location in space by modeling it with the help of the mathematical concept of point process. The usage of this new mathematical methodology allowed the derivation of analytical expressions for several arrangements of nuclei in space, for example, nuclei located in space according to a homogeneous Poisson point process (i.e., uniformly randomly distributed nuclei, JMAK's assumption) and to a inhomogeneous Poisson point process. With respect to the moving boundary velocity, the assumption that all grains grew with the same velocity could also be generalized. Jensen and Godiksen recently reviewed 3D experimental results on recrystallization kinetics and concluded that there is compelling evidence that every grain has its own distinct growth rate even when the specimen has no macroscopic deformation gradients. Motivated by these practical recrystallization situations, "JMAK-like" analytical expressions for random velocities were also derived. Previously known results follow here as particular cases. Finally, it is possible that two or more transformations may take place simultaneously or sequentially; for example, the evolution of two or more texture components during recrystallization. A new methodology is presented to account for this.

This chapter presents these recent advances in JMAK theory brought about by this new mathematical methodology. Now one has analytical expressions that go well beyond JMAK original assumptions for nucleation and growth. Although the motivation for this chapter was recrystallization, the expressions derived here may be applied to nucleation and growth reactions in general.

6.2
Mathematical Background and Basic Notation

Detailed mathematical background may be found in previous work by the authors [16–18]; in particular, we refer to our recent paper [16] for basic definitions. Here only some essential definitions and some useful relationships are presented to make this chapter more self-contained and easier to read. For a brief description of homogeneous and inhomogeneous Poisson point process, the reader is referred to [16] and for a more detailed presentation to specific texts on Stochastic Geometry, to [19].

6.2.1
Modeling Birth-and-Growth Processes

A *birth-and-growth (stochastic) process* is a dynamic germ-grain model [19], used to model situations in which *nuclei* (germs) are born in time and are located in space randomly, and each nucleus generates a *grain* evolving in time according to a given growth law. Since, in general, nucleation and growth are random in time and space, the transformed region Θ^t at any time $t > 0$ is a random set [19] in $\mathbb{R}^d$, that is a measurable map from a probability space $(\Omega, \mathcal{F}, \mathbb{R})$ to the space of closed subsets in $\mathbb{R}^d$. The family $\{\Theta^t\}_t$ is called birth-and-growth process. Birth-and-growth and nucleation and growth are used as synonyms in this chapter.

Of course, different kinds of nucleation and growth models give rise to different kinds of processes $\{\Theta^t\}_t$.

As mentioned in the Introduction, we shall consider two kinds of nucleation of interest in recrystallization; the first one is that of *site-saturation*, in which all nucleation sites are exhausted at the beginning of the reaction, the second one is that of *time-dependent* nucleation, in which nuclei can be born randomly both in space and time. Site-saturated nucleation processes and space-time-dependent nucleation processes can be modeled by *point processes* and *marked point processes*, respectively (e.g., see [7,8]). Some basic concepts and definitions useful for the sequel are given here. (See also, for instance, [16], Section 6.2.3.)

Here $\mathcal{B}_{\mathbb{R}^d}$ denotes the Borel σ-algebra of $\mathbb{R}^d$ (e.g., see [20]). Note that a *point process* in $\mathbb{R}^d$ is, almost surely, a locally finite sequence of points $N = \{X_j\}_i$ randomly located in $\mathbb{R}^d$, in accordance with a given probability law. It can be equivalently described by the *counting process* associated to the sequence $\{X_j\}$ defined as

$$N(A) := \text{number of the } X_i\text{s that belong to } A,$$

for any $A \in \mathcal{B}_{\mathbb{R}^d}$.

By denoting X_j the spatial location of the jth nucleus of a site-saturated nucleation process, and $\Theta^t(X_j)$ the grain obtained as the evolution of the nucleus X_j up to time $t > 0$, the transformed region Θ^t at time $t > 0$ is

$$\Theta^t = \bigcup_j \Theta^t(X_j)$$

Given a complete separable metric space $\mathbf{K}$, such as $\mathbb{R}^d$, a *marked point process* in $\mathbb{R}_+ \times \mathbf{K}$ is a sequence $N := \{(T_i, K_i)\}_i$ of points such that the sequence $\{T_i\}_i$ is a point process in $\mathbb{R}_+$, while each $K_i \in \mathbf{K}$ is said to be the *mark* associated to the point T_i. $\mathbf{K}$ is called *mark space*. Then, a space-time nucleation process can be modeled by a marked point process, identifying T_i as the time of birth of the ith nucleus and X_i as its spatial location in $\mathbb{R}^d$. Analogously to the site-saturation case, a counting process N on $\mathbb{R}_+ \times \mathbb{R}^d$ can be defined as

$$N([s,t] \times A') := \text{number of nuclei, which are born in } A' \\ \text{during the time interval } [s,t]$$

By denoting $\Theta^t_{T_j}(X_j)$ the grain obtained as the evolution up to time $t \geq T_j$ of the nucleus born at time T_j in X_j, the transformed region Θ^t at time $t > 0$ is given by

$$\Theta^t = \bigcup_{T_j \leq t} \Theta^t_{T_j}(X_j)$$

Of course, a site-saturated process may be seen as a particular case of the time-dependent one by assuming $T_j \equiv 0$ for any j.

Given a point process N on $\mathbb{R}^d$, the measure

$$\Lambda(A) := \mathbb{E}[N(A)], \quad \forall A \in \mathcal{B}_{\mathbb{R}^d}$$

where $\mathbb{E}$ is "expectation", is called *intensity measure* of N; in other words, if N is a site-saturated nucleation process, then $\Lambda(A)$ represents the mean number of nuclei born in $A \subset \mathbb{R}^d$. Note that a measure μ on $\mathbb{R}^d$ admits density (i.e., its Radon–Nikodym derivative), if there exists a locally integrable function $f : \mathbb{R}^d \to \mathbb{R}$ such that $\mu(A) = \int_A f(x)\mathrm{d}x$ for any $A \in \mathcal{B}_{\mathbb{R}^d}$; the density of Λ, provided it exists, is called *intensity (of the process)*, and it is denoted here by λ, that is,

$$\Lambda(A) = \int_A \lambda(x)\mathrm{d}x, \quad \forall A \in \mathcal{B}_{\mathbb{R}^d} \tag{6.4}$$

Whereas if N is a marked point process in $\mathbb{R}_+ \times \mathbf{K}$, then its intensity measure Λ is a measure on $\mathbb{R}_+ \times \mathbf{K}$ defined as $\Lambda(A) := \mathbb{E}[N(A)]$ for all $A \in \mathcal{B}_{\mathbb{R}_+} \times \mathcal{B}_{\mathbf{K}}$; therefore, if $N = (T_j, X_j)_j$ is a time-dependent nucleation process, then $\Lambda([s, t] \times A')$ is the mean number of nuclei born in A' during a time interval $[s, t]$. If the marks are independent and identically distributed, and independent of the unmarked point process $\{T_i\}_i$, then the common probability distribution of the marks, say Q, is called *mark distribution*; in this case, the process N is said to be an independent marking of $\{T_i\}_i$, and Λ is of the type $\Lambda(\mathrm{d}(s, y)) = \lambda(s)\mathrm{d}s Q(\mathrm{d}y)$, where λ is the intensity of the process.

It should be mentioned that the notion of marked point process will be useful in the sequel for modeling birth-and-growth processes whose growth velocity for each grain is a random quantity, which can be described as a mark associated to the corresponding nucleus.

In order to define a birth-and-growth process we need to introduce also a growth model. Models of volume growth have been studied extensively, since the pioneering work by Kolmogorov [5]. We consider here a simple case of the so-called *normal growth model* (see also, e.g., [18] and reference therein); namely, we shall consider the case in which all the grains grow with velocity G constant in time or time dependent, so that for any time t all the grains have spherical shape (because G is not space-dependent). In Section 6.7, we treat random velocity.

6.2.2
Mean Densities Associated to a Birth-and-Growth Process

Since Θ^t is a random set, it gives rise to a random measure $\nu^d(\Theta^t \cap \cdot)$ in $\mathbb{R}^d$ for all $t > 0$, having denoted by ν^d the d-dimensional Lebesgue measure in $\mathbb{R}^d$. In particular, it is of interest to consider the *expected volume measure* $\mathbb{E}[\nu^d(\Theta^t \cap \cdot)]$ and its density, called *mean volume density of* Θ^t and denoted by V_V, provided it exists:

$$\mathbb{E}[\nu^d(\Theta^t \cap A)] = \int_A V_V(t, x)\mathrm{d}x \quad \forall A \in \mathcal{B}_{\mathbb{R}^d}$$

It is well known and easy to prove that

$$V_V(t, x) = \mathbb{P}(x \in \Theta^t) \quad \text{for } \nu^d - \text{a.e. } x \in \mathbb{R}^d \tag{6.5}$$

where $\mathbb{P}(x \in \Theta^t)$ is the probability that x belongs to Θ^t.

Whenever A is the region of the physical sample under observation, the ratio

$$\mathbf{V_V}(t, A) := \frac{\int_A V_V(t, x)\mathrm{d}x}{\nu^d(A)} \tag{6.6}$$

is called *volume fraction associated to the region A*. Let us note that whenever V_V is independent of x (e.g., under assumptions of homogeneous nucleation and growth), then $\mathbf{V}_V$ is independent of A and $\mathbf{V}_V(t) = V_V(t)$, representing in this case the mean volume of Θ^t per unit of volume, also called *volume fraction*. (See also p. 342 refers only to [19]). We also mention that other quantities of interest in real applications are the so-called *mean extended volume density* at time t, denoted by $V_E(t, \cdot)$, defined as the density of the *mean extended volume measure* at time t, $\mathbb{E}[\mu^{ex}_{\Theta^t}](\cdot) := \mathbb{E}\left[\sum_{j:T_j \leq t} \nu^d(\Theta^t_{T_j}(X_j) \cap \cdot)\right]$ on $\mathbb{R}^d$, that is

$$\mathbb{E}[\mu^{ex}_{\Theta^t}](A) = \int_A V_E(t, x)dx, \quad \forall A \in \mathcal{B}_{\mathbb{R}^d}$$

and the *mean surface density* $S_V(t, \cdot)$ and the *mean extended surface density* $S_E(t, \cdot)$ at time t, defined as the density of the *mean surface measure* at time t, $\mathbb{E}[\mu_{\partial\Theta^t}](\cdot) := \mathbb{E}[H^{d-1}(\partial\Theta \cap \cdot)]$ and of the *mean extended surface measure* at time t, $\mathbb{E}[\mu^{ex}_{\partial\Theta^t}](\cdot) := \mathbb{E}\left[\sum_{j:T_j \leq t} H^{d-1}(\partial\Theta^t_{T_j}(X_j) \cap \cdot)\right]$, respectively, where H^{d-1} is the $(d-1)$-dimensional Hausdorff measure, that is, the surface measure (for the mathematical definitions and properties, see, e.g., [21]). In other words, the mean extended volume and surface measures represent the mean of the sum of the volume measures and the surface measures of the grains which are born and grown until time t, supposed as *free to grow*, ignoring overlapping. (See also [16,22].)

It is clear that finding out formulas for the mean volume density V_V and for other quantities we mentioned above is of particular interest in real applications.

6.2.3
Causal Cone

Throughout this chapter $B_r(x)$ denotes the ball centered in x with radius r, and $1_A(a)$ is the indicator function of A, that is 1 if $a \in A$ and 0 otherwise.

The *causal cone* notion plays, together with Eq. (6.5), a fundamental role in evaluating the mean volume density $V_V(t, x)$ of Θ^t at a point x. It is defined as the space-time region in which at least one nucleation event has to take place in order to cover the point x at time t [16,17,23]. Explicit expressions for the causal cone will be provided throughout the text in relation with different kinds of birth-and-growth processes we are going to consider. In order to give a first more intuitive notion of the causal cone, let us consider the very particular case of site saturation with constant growth velocity of the grains G; in this case, a point x will be transformed at time t if at least one nucleus is born in the ball with radius $R = Gt$ centered in x; therefore, such a ball is the causal cone $C(t, x)$ of the point x at time t in this case.

Namely, $\mathscr{C}(t, x)$ is a subset of $\mathbb{R}^d$ in the site-saturated case, or a subset of $\mathbb{R}_+ \times \mathbb{R}^d$ in the time-dependent case, so defined

$$\mathscr{C}(t, x) := \begin{cases} \{y \in \mathbb{R}^d : x \in \Theta_t(y)\}, \\ \{(s, y) \in [0, t] \times \mathbb{R}^d : x \in \Theta^t_s(y)\} \end{cases}$$

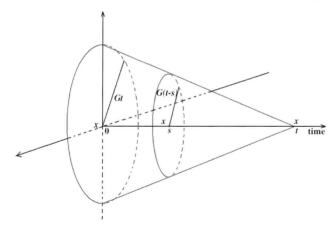

Figure 6.1 A geometric interpretation of the causal cone $\mathcal{C}(t,x)$ of a point $x \in \mathbb{R}^d$ at time t, in the time-dependent nucleation case, assuming G constant. The vertical and horizontal axes represent the space $\mathbb{R}^d$ and the time, respectively. Note that for any $s \in [0, t]$, the section of the causal cone is the set of points in $\mathbb{R}^d$ where a nucleation event has to take place in order to cover the point x at time t; so, in this case, it is the ball centered in x with radius $G(t-s)$.

in the site-saturated and the time-dependent cases, respectively (e.g., see also [16]). Let us observe that considering a constant G, it follows that

$$\mathcal{C}(t,x) = \begin{cases} B_{Gt}(x) \\ \{(s,y) \in [0,t] \times \mathbb{R}^d : y \in B_{G(t-s)}(x)\} \end{cases} \quad (6.7)$$

in the site-saturated and the time-dependent cases, respectively (see also Figure 6.1); whereas if $G = G(t)$ is time dependent, then for any time t, each grain has spherical shape, namely

$$\Theta^t(X_j) = B_{R(t)}(X_j), \quad R(t) = \int_0^t G(\tau)d\tau$$

in the site-saturated nucleation case, and

$$\Theta^t_{T_j}(X_j) = B_{R(T_j,t)}(X_j), \quad R(T_j,t) = \int_{T_j}^t G(\tau)d\tau$$

in the time-dependent nucleation case; as a consequence,

$$\mathcal{C}(t,x) = \begin{cases} B_{R(t)}(x) \\ \{(s,y) \in [0,t] \times \mathbb{R}^d : y \in B_{R(s,t)}(x)\} \end{cases} \quad (6.8)$$

in the site-saturated case and in the time-dependent case, respectively. For the general case of space-time dependent growth rate, see, for example, [12,16,22] and reference therein.

General results on $V_V(t,x)$ in terms of the causal cone are proved in [22]. In particular, we recall here that

$$V_V(t,x) = \mathbb{P}(N(C(t,x)) > 0)$$

as a consequence of (6.5) and the definition of the causal cone, and that

$$V_E(t,x) = \Lambda(\mathscr{C}(t,x)) \tag{6.9}$$

and

$$G(t) = \frac{1}{S_V(t,x)}\frac{\partial V_V(t,x)}{\partial t} = \frac{1}{S_E(t,x)}\frac{\partial V_E(t,x)}{\partial t} \tag{6.10}$$

These relationships are very important in practical applications. For instance, if the velocity G is known, Eq. (6.10) can be used to find out the mean interfacial area density or the interfacial area per unit of the volume from the corresponding mean volume density or volume fraction.

Explicit expressions for the causal cone will be provided throughout the chapter in relation with the different kinds of birth-and-growth processes we are going to consider.

6.3
Revisiting JMAK

The mathematical concepts outlined in the previous sections may be employed to revisit and generalize the original JMAK approach [16,24]. First, nucleation may be generalized by modeling it with an inhomogeneous Poisson point process, and then to more general point processes. Then, from the mean number of nuclei inside the causal cone one obtains the extended volume, with the help of (6.9). Finally, with the help of (6.13) to be derived below, which holds under Poissonian nucleation process, one can find the generalized expression for V_V.

The fundamental assumption of a Poisson point process is that nuclei are born independently of each other. If N is a *Poisson point process* in $\mathbb{R}^3$ with intensity measure Λ, then the probability $\mathbb{P}(N(A) = k)$ that there are exactly k points in A is given by

$$\mathbb{P}(N(A) = k) = \frac{\Lambda(A)^k}{k!}e^{-\Lambda(A)}, \quad \text{for any compact } A \subset \mathbb{R}^3 \tag{6.11}$$

If the intensity λ, see Eq. (6.4), of N is constant, then $\lambda =$ mean number of nuclei per unit of volume, then N is called *homogeneous* Poisson point process. If $\lambda \neq$ const, then $\lambda(x)dx =$ mean number of nuclei in the infinitesimal spatial region dx, then N is called *inhomogeneous* Poisson point process. Two-dimensional computer simulation of a homogeneous and inhomogeneous Poisson point processes using the software package "R" is illustrated in Figure 6.2 [26,27]. A unit square area with horizontal axis $x_1 = 0$ to $x_1 = 1$ and vertical axis $x_2 = 0$ to $x_2 = 1$ was used.

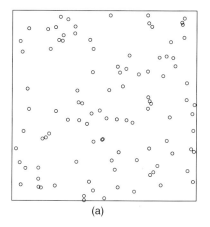
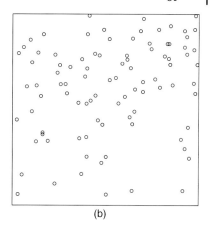

(a) (b)

Figure 6.2 Computer simulation of point processes in a square area with horizontal axis $x_1 = 0$ to $x_1 = 1$ and vertical axis $x_2 = 0$ to $x_2 = 1$. Figures show a realization of homogeneous Poisson point process (a) with intensity equal to 100 and inhomogeneous Poisson point process (b) with intensity varying linearly along the vertical axis (x_2) according to $20 + 160x_2$. Note that for $x_2 = 0.5$, the intensity of the inhomogeneous process is the same (100) as that of the homogeneous process.

Poisson point processes model birth processes in which nuclei are assumed to be born independently, in space in the site-saturated case, and also in time in time-dependent nucleation. This means that if one would model a nucleation process whose nuclei are dependent on each other, or the birth of a new nucleus depends on the history of the process, then the assumption of Poisson nucleation is not appropriate.

In case of time-dependent nucleation, when the nucleation is homogeneous in space only, then the intensity function λ is independent of x, whereas when the nucleation is homogeneous in time, but inhomogeneous in space, then λ is a function of the space variable x only. Moreover, for a Poisson point process, the probability that there is at least one nucleus inside the causal cone is

$$\mathbb{P}(N(\mathcal{C}(t,x)) > 0)) \stackrel{(11)}{=} 1 - e^{-\Lambda(\mathcal{C}(t,x))} \tag{6.12}$$

and therefore, using Eqs. (6.5) and (6.9) into (6.12)

$$V_V(t,x) = 1 - e^{-V_E(t,x)} \tag{6.13}$$

Thereby, Eq. (6.13) is a generalization of (6.1).

We observed in (6.8) that if G is constant or time dependent, then $\mathcal{C}(t,x)$ is a ball centered in x; therefore, using Eq. (6.9), if $\lambda = \lambda(y)$ is *harmonic* in y, that is, $\sum_{i=1}^{3} \partial^2 \lambda(y)/\partial y_i^2 = 0$, then, in $\mathbb{R}^3$

$$V_E(t,x) = \int_{B_{R(t)}(x)} \lambda(y) \mathrm{d}y = \lambda(x) \nu^3(B_{R(t)}(x)) = \frac{4}{3}\pi\lambda(x)\left(\int_0^t G(\tau)\mathrm{d}\tau\right)^3$$

and

$$V_V(t,x) = 1 - \exp\left\{-\frac{4\pi}{3}\lambda(x)\left(\int_0^t G(\tau)\mathrm{d}\tau\right)^3\right\} \tag{6.14}$$

Following a similar reasoning, in the time-dependent nucleation case, if the nucleation process is modeled by a marked Poisson point process homogeneous in time but inhomogeneous in space, such that the intensity $\lambda = \lambda(x)$ is harmonic, then

$$V_V(t, x) = 1 - \exp\left\{-\frac{4}{3}\pi\lambda(x) \int_0^t \left(\int_s^t G(\tau)d\tau\right)^3\right\} \quad (6.15)$$

note that if G is constant, then the above equation simplifies as

$$V_V(t, x) = 1 - \exp\left\{-\frac{\pi}{3}\lambda(x)G^3 t^4\right\}$$

Therefore, JMAK's expressions, Eqs. (6.2) and (6.3), may be seen as particular cases of Eqs. (6.14) and (6.15).

We also have

$$S_V(t, x) = (1 - V_V(t, x))S_E(t, x) \quad (6.16)$$

In order to exemplify the effect of a linear gradient on the transformation kinetics, it is interesting to show a numerical example. This is done here for the site-saturation case. We choose $m = 10^7$ per cubic unit of length, $n = 10^5$ per cubic unit of length, and $G = 0.001$ units of length per unit time. If the unit of length were set at 1 mm, $m = 10^7$ mm^{-3} and $n = 10^5$ mm^{-3} would yield final grain sizes roughly 0.005 and 0.02 mm, respectively, for specimens containing m and n homogeneously distributed nuclei.

For comparison, values of V_V and S_V at $x_1 = 0.05, 0.5$, and 0.95 were chosen. Note that the plane $x_1 = 0.5$ is the plane that has the average nuclei intensity, namely, $m/2 + n$.

In the comparison, we have also included the global quantities $\mathbf{V_V}$, defined in Eq. (6.6), and $\mathbf{S_V}$, defined analogously

$$\left(\mathbf{S_V}(t, A) := \frac{\int_A S_V(t, x)dx}{v^d(A)}\right).$$

By their definition with $A = [0, 1]^3$, we have

$$\mathbf{V_V}\left(t, [0, 1]^3\right) = \frac{\int_{[0,1]^3} V_V(x, t)dx}{v^3([0, 1]^3)} = \int_0^1 \left(1 - \exp\left(-\frac{4\pi(mx_1 + n)G^3 t^3}{3}\right)\right)dx_1$$

$$= 1 - \frac{-\exp(-(4\pi/3)(m + n)G^3 t^3) + \exp(-(4\pi/3)nG^3\pi t^3)}{(4\pi/3)mG^3 t^3},$$

$$\mathbf{S_V}\left(t, [0, 1]^3\right) = \frac{\int_{[0,1]^3} S_V(x, t)dx}{v^3([0, 1]^3)} = \int_0^1 4\pi(mx_1 + n)G^2 t^2 \exp\left(-\frac{4\pi(mx_1 + n)G^3 t^3}{3}\right)dx_1$$

$$= 3\left(-1 - \frac{4\pi}{3}(m + n)G^3 t^3 + \left(1 + \frac{4\pi}{3}nG^3 t^3\right)\exp\left(\frac{4\pi}{3}mG^3 t^3\right)\right)$$

$$\times \frac{\exp(-(4\pi/3)(m + n)G^3 t^3)}{(4\pi/3)mG^4 t^4}$$

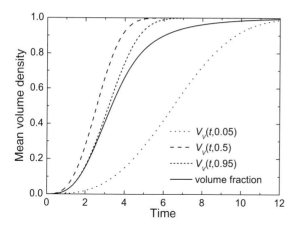

Figure 6.3 Mean volume density, $V_V(t,x)$ as a function of time at planes $x_1 = 0.05, 0.5, 0.95$. The volume fraction (solid line), $\mathbf{V_V}(t, [0,1]^3)$, is also shown.

It can be seen that both $\mathbf{V_V}$ and $\mathbf{S_V}$ go to 0 as t tends to 0^+, whereas $\mathbf{V_V} \to 1$ and $\mathbf{S_V} \to 0$ as t goes to ∞.

Figure 6.3 shows $V_V(t,x)$ as a function of time for the planes $x_1 = 0.05, 0.5$, and 0.95 as well as $\mathbf{V_V}$. A clear delay can be observed between the planes with a high and low nuclei intensity. Although the $\mathbf{V_V}(t, [0,1]^3)$ and $V_V(t, 0.5)$ curves are fairly close, their shape is very different. The microstructural path shown in Figure 6.4 also shows a significant difference between the $\mathbf{S_V}(\mathbf{V_V})$ and $S_V(V_V)$ curves. Therefore, this relatively simple case shows some important characteristics of inhomogeneous nuclei distributions.

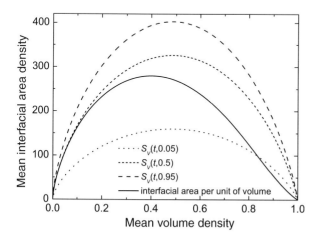

Figure 6.4 Microstructural path of the reaction is shown at different x_1 planes. The microstructural path for the interfacial area per unit of volume, $\mathbf{S_V}(t, [0,1]^3)$, as a function of the volume fraction, $\mathbf{V_V}(t, [0,1]^3)$, is the solid line and is significantly different.

6.4
Nucleation in Clusters

In order to model the situation in which nuclei are located in clusters, it is convenient to model the nucleation process by the so-called Matérn cluster process. Only site-saturated nucleation is discussed here.

6.4.1
The Matérn Cluster Process

Clustering is a fundamental operation on point processes well-known in Stochastic Geometry, which permits to construct new point processes. A brief description of such operation is given in the following text; refer to [19,26] for a more exhaustive treatment.

The clustering operation consists of replacing each point x of a given point process, Φ_p, called *parent point process*, by a cluster of points, N_x, called *daughter points*. Each cluster N_x is itself a point process and is assumed to have only a finite mean number of points. The point process

$$\Phi := \bigcup_{x \in \Phi_p} N_x$$

given by the union of all the clusters N_x is said to be a *cluster point process*.

An important class of cluster point processes, often used to model real phenomena, is the so-called *Neyman–Scott processes*. A Neyman–Scott process is a cluster point process whose parent points form a homogeneous Poisson point process, and whose clusters N_x are of the form $N_{x_i} = N_i + x_i$ for each $x_i \in \Phi_p$, where $N_1, N_2, \ldots$ are independent of Φ_p and identically distributed. ($N_i + x_i$ is the translation of N_i by vector x_i.) Let us denote the *representative (or typical) cluster*, centered in 0, by N_0. If the number of points of N_0 is a random variable with Poisson distribution of parameter n_c (and so the mean number of points of each cluster is n_c), and such points are independently and uniformly distributed in the ball $B_R(0)$ with center 0 and radius R, which is a further parameter of the model, then the resulting cluster point process Φ is called *Matérn cluster process*. Therefore, a Matérn cluster process Φ with parameters λ_p (the intensity of the parent process, Poisson homogeneous), n_c (the mean number of nuclei per cluster), and R (the radius of each cluster) has a constant intensity λ_Φ, that is given by

$$\lambda_\Phi = \lambda_p n_c$$

Figure 6.5 shows a 2-d computer simulation of Matérn cluster process within a square of unit area.

The Matérn cluster process is stationary, so that Θ^t is also stationary, which implies that V_V is independent of x. The Matérn cluster process Φ can be seen

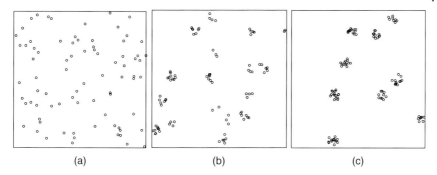

Figure 6.5 Computer simulation of Matérn cluster process in a square of unit area. Figures show homogeneous Poisson point process (a) with intensity equal to 100. Matérn cluster processes with cluster radius, R, equal to 0.04 and $\lambda_p = 20$ and $n_c = 5$ (b) and $\lambda_p = 10$ and $n_c = 10$ (c). Simulations were performed using the "R" software [26,27].

as a "Boolean' model" [19] with primary grain N_0; as a consequence, it follows that [25]

$$V_V(t) = 1 - \exp\left\{-\lambda_p \mathbb{E}\left[\nu^3\left(\bigcup_{x \in N_0} B_{Gt}(x)\right)\right]\right\}$$

The typical cluster N_0, centered in 0, is itself an inhomogeneous Poisson point process with intensity

$$\lambda_{N_0}(x) = \lambda_c 1_{B_R(0)}(x)$$

where $\lambda_c := n_c/(b_d R^d)$; therefore, $\mathbb{E}\left[\nu^3\left(\bigcup_{x \in N_0} B_{Gt}(x)\right)\right]$ is the mean volume of a birth-and-growth process with a site-saturated Poisson nucleation process and constant growth velocity G. Thus, a more useful expression may be derived (see [25] for a detailed proof):

$$V_V(t) = 1 - \exp\left\{-\lambda_p \int_{\mathbb{R}^3} \left(1 - e^{-\lambda_c \nu^3(B_R(0) \cap B_{Gt}(x))}\right) dx\right\} \quad (6.17)$$

In this form, this result is still not very useful to the materials scientist. The derivation of explicit expressions from these equations [25] is shown below.

6.4.2
Evaluation of the Integral in Eq. (6.17)

Now, in order to evaluate $\nu^3(B_{Gt}(x) \cap B_R(0))$ at varying of $x \in \mathbb{R}^3$, let us note that it depends on the distance $a := \text{dist}(x, 0)$ between the centers x and 0 of the two balls; therefore, the function can be defined as

$$\nu(a) := \nu_3(B_R(0) \cap B_{Gt}(x)), \quad a \in [0, \infty)$$

We must distinguish two cases: $Gt \geq R$ and $Gt < R$. It is convenient to define an auxiliary function $g(a)$ recalling that a is the distance of the point x to the origin

$$g(a) := \frac{\pi R^2 G^2 t^2}{2a} - \frac{\pi R^4}{4a} - \frac{1}{2}\pi R^2 a + \frac{2}{3}\pi R^3 - \frac{\pi G^4 t^4}{4a} - \frac{1}{2}\pi G^2 t^2 a + \frac{1}{12}\pi a^3 + \frac{2}{3}\pi G^3 t^3$$

It is not difficult to check that if $Gt \geq R$, then

$$v(a) = \begin{cases} \frac{4}{3}\pi R^3 & \text{if } 0 \leq a < Gt - R \\ g(a) & \text{if } Gt - R \leq a < R + Gt \\ 0 & \text{if } a \geq R + Gt \end{cases} \tag{6.18}$$

whereas if $Gt < R$, then

$$v(a) = \begin{cases} \frac{4}{3}\pi G^3 t^3 & \text{if } 0 \leq a < R - Gt \\ g(a) & \text{if } R - Gt \leq a < R + Gt \\ 0 & \text{if } a \geq R + Gt \end{cases} \tag{6.19}$$

The function $v(a)$ is continuous, and for $Gt = R$, it is given by

$$v(a) = \begin{cases} \frac{1}{12}\pi a^3 - \pi G^2 t^2 a + \frac{4}{3}\pi G^3 t^3 & \text{if } 0 \leq a < 2Gt \\ 0 & \text{if } a \geq 2Gt \end{cases}$$

Note that the integral in Eq. (6.17) is simplified as follows by using spherical coordinates:

$$\int_{\mathbb{R}^d} \lambda_p (1 - e^{-\lambda_c v^3 (B_{Gt}(x) \cap B_R(0))}) dx = \lambda_p 4\pi \int_0^\infty (1 - e^{-\lambda_c v(a)}) a^2 da \tag{6.20}$$

Then, combining Eqs. (6.18) and (6.19) with Eq. (6.20), we finally obtain

$$V_V(t) = 1 - \exp\left\{-\lambda_p \frac{4}{3}\pi (Gt - R)^3 \left(1 - e^{-\lambda_c \frac{4}{3}\pi R^3}\right) - \lambda_p 4\pi \int_{Gt-R}^{Gt+R} (1 - e^{-\lambda_c g(a)}) a^2 da\right\} \tag{6.21}$$

if $t \geq R/G$, and

$$V_V(t) = 1 - \exp\left\{-\lambda_p \frac{4}{3}\pi (R - Gt)^3 \left(1 - e^{-\lambda_c \frac{4}{3}\pi G^3 t^3}\right) - \lambda_p 4\pi \int_{R-Gt}^{R+Gt} (1 - e^{-\lambda_c g(a)}) a^2 da\right\} \tag{6.22}$$

if $t < R/G$.

Similar expressions in the case of time-dependent velocity $G = G(t)$ are provided in [25].

It is clear that for the birth-and-growth process with nucleation sites following a Matérn cluster process

$$V_V(t) \neq 1 - e^{-V_E(t)} \tag{6.23}$$

that is, the well-known relationship between V_V and V_E in the case of Poissonian nucleation, Eq. (6.1), does not hold. Indeed, as shown in [22], Eqs. (6.2) and (6.3) are true only if the nucleation process is given by a Poisson point process. Equation (6.23) deviates from Eq. (6.1) owing to the contribution from each individual nucleus of Matérn Cluster to the extended volume. The general form of Eqs. (6.21) and (6.22) is still $V_V(t) = 1 - e^{-f(t)}$ but $f(t) \neq V_E(t)$.

Noteworthy, it is straightforward to show that for $t \to 0$

$$V_V(t) \sim 1 - e^{-\lambda_p \frac{4}{3}\pi R^3 \left(\lambda_c \frac{4}{3}\pi G^3 t^3\right)} = 1 - e^{-V_E(t)} \sim V_E(t)$$

as expected, because for t close to 0, the grains do not overlap.

6.4.3
Numerical Examples

In this section, numerical examples of the equations derived in the previous section are presented. The numerical examples highlight the influence of the magnitude of relevant parameters of the Matérn cluster process upon subsequent transformation kinetics. First, the influence of a change in the cluster radius, R, is investigated, keeping both λ_p and n_c constant. Then we examine the influence of the mean number of nuclei per cluster, n_c, keeping both R and the total number of nuclei per unit of volume constant, and so changing λ_p. In what follows a cubic observation window with side measuring a unit length is used, $[0, 1]^3$. All quantities are referred to this unit volume. The boundary velocity was set to $G = 0.001$ units of length per unit of time. Time was taken in arbitrary units and ranged from 0 to about 300 arbitrary units in the calculations.

Just to establish the terminology, it is useful to distinguish between the intracluster impingement that takes place among grains of the same cluster and intercluster impingement that takes place between regions nucleated within distinct clusters.

6.4.3.1 Influence of Cluster Radius
Figure 6.6 shows the influence of increasing the cluster radius on transformation kinetics. Cluster radius changed from 0.05 to 0.200 units of length. Both $\lambda_p = 125$ per cubic unit of length and $n_c = \lambda_c\{(4\pi R^3)/3\} = 10$ were kept constant in Figures 6.6 and 6.7. It is intuitively clear that if each cluster has a very large radius, they will overlap and in the limit $R \to \infty$, one will have nuclei uniformly randomly located with intensity $\lambda_p \lambda_c\{(4\pi R^3)/3\}$, and so like a homogeneous Poisson nucleation process with intensity $\lambda_p n_c$. The leftmost solid line, that is, faster kinetics, in Figure 6.6 represents this situation. By contrast, when the radius R of the clusters tends to 0, the clusters tend to a point (the center of the cluster), and hence again one has nuclei uniformly randomly located, but with intensity λ_p, because at this limit, the nucleation process coincides with the Poisson parent point process. This situation is represented by the rightmost curve in Figure 6.6; for intermediate values of R, the transformation curves lie in between these two extremes.

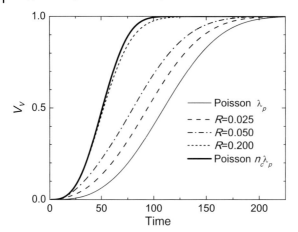

Figure 6.6 Mean volume density transformed as a function of time for cluster radii ranging from 0.025 to 0.200 units of length. Curves obtained for uniformly randomly located nuclei, Eq. (6.2), with intensities (N_V) equal to λ_p and $\lambda_p \cdot n_c$ are also shown.

The microstructural paths, S_V versus V_V plots, of the reactions depicted in Figure 6.6 are shown in Figure 6.7. For large radii, the microstructural path approaches that of $R \to \infty$, as mentioned above, and one will have nuclei randomly uniformly located with intensity $\lambda_p \lambda_c \{(4\pi R^3)/3\}$. Conversely, for small radii, the microstructural path will

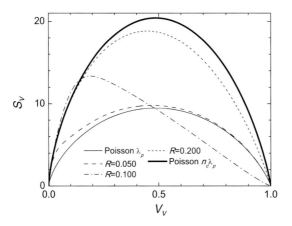

Figure 6.7 Microstructural path of the reactions for different cluster radii. Curves obtained for uniformly randomly located nuclei, Eq. (6.2), with intensities (N_V) equal to λ_p and $\lambda_p \cdot n_c$ are also plotted and appear as symmetrical curves with a maximum at $V_V = 0.5$ in this plot. For cluster radii ranging from 0.050 to 0.200 units of length, the curves exhibit an asymmetrical shape. It is interesting to point out that for the smaller cluster radius, 0.050, the curve is close to that of λ_p. For an even smaller radius, 0.025 (not shown in this plot), the cluster microstructural path practically coincides with that of λ_p. On the other extreme for a very large cluster radius, 0.200, both the V_V against time and the microstructural path curves approach the corresponding curves for $\lambda_p \cdot n_c$.

approach that of $R \to 0$ as expected from what was said above. More interesting is the microstructural path when cluster radii have intermediate values. In those cases, as shown in Figure 6.7, the microstructural path curve shape is significantly skewed to the left, substantially different from the symmetric shape of the curves for the two extreme cases.

In summary, for a very small radius (in this particular case less than 0.025) or very large cluster radius (in this particular case around 0.200), the transformation kinetics of clustered nuclei tends to one of the extreme cases, $R \to 0$ or $R \to \infty$, respectively.

6.4.3.2 Influence of Number of Nuclei per Cluster

In this section, the cluster radius, $R = 0.050$ and the total number of nuclei per unit of volume, $\lambda_p n_c = 16,000$ were maintained constant. Although λ_p increased from 125 to 1000 per unit of volume, n_c correspondingly decreased from 128 to 16, so that their product remained constant and equal to 16,000 nuclei per unit of volume. Figure 6.8 shows the plots of volume fraction transformed against time. It is clear that clustering retards the kinetics taking as reference a matrix with the same number of nuclei but uniformly randomly located. The microstructural path curves are depicted in Figure 6.9. For smaller values of n_c, these curves exhibits a skew to the right. However, as n_c increases, the curves develop a secondary peak for low volume fractions that is caused by intracluster impingement. Similar behavior was observed in computer simulation of clusters using cellular automata [28,29]. Therefore, clustering transformation kinetics may significantly depart from the usual JMAK kinetics. Both the volume fraction transformed and the microstructural

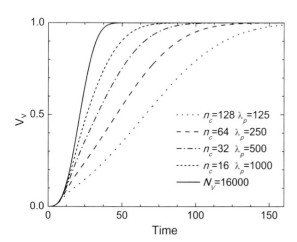

Figure 6.8 Mean volume density transformed as a function of time for a number of nuclei per cluster, n_c, ranging from 16 to 128. The total number of nuclei per unit of volume of all curves was kept constant equal to 16 000 nuclei per unit of volume. Curve obtained for uniformly randomly located 16 000 nuclei per unit of volume is represented by a solid line. It is clear that clustering significantly delays transformation kinetics.

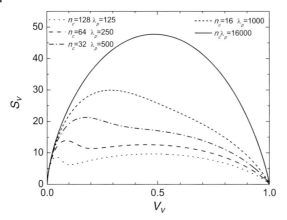

Figure 6.9 Microstructural path of the reactions for a number of nuclei per cluster, n_c, ranging from 16 to 128. The total number of nuclei per unit of volume was kept constant equal to 16,000 nuclei per unit of volume. Curve obtained for uniformly randomly located nuclei with intensity (N_V) equal to $\lambda_p \cdot n_c$ appears as a symmetrical curve with a maximum at $V_V = 0.5$. One distinct feature of this plot is that for the largest values of n_c, the curve develops a clear peak in the low $V_V \approx 0.05$–0.20 range. Thus, an increase in number of nuclei per cluster significantly changes the shape of the microstructural path curves.

path show well-defined differences from the corresponding kinetics with the same number of uniformly randomly located nuclei per unit of volume.

6.5
Nucleation on Lower Dimensional Surfaces

In practice, sometimes the transformation takes place in small specimens. Here the word "small" means that the external surface must be taken into account in the analytical solution. Examples of such specimens are (a) thin films, (b) wires, and (c) powders.

6.5.1
Derivation of General Expressions for Surface and Bulk Nucleation

6.5.1.1 Surface Nucleation
Let us consider a Poissonian nucleation on the surface $\partial \mathcal{D}$, first in the site-saturated case, then in the time-dependent case.

First, we consider site-saturated nucleation with λ_S nuclei per unit of area on $\partial \mathcal{D}$. Then the intensity λ of the nucleation process is given by

$$\lambda(y) = \lambda_S \delta_{\partial \mathcal{D}}(y)$$

where $\delta_{\partial D}$ is the Dirac delta function on $\partial \mathcal{D}$, so that

$$\Lambda(A) = \lambda_S \nu^2(A \cap \partial \mathcal{D}) \quad \forall A \in \mathcal{B}_{\mathbb{R}^3}$$

The causal cone $\mathcal{C}(t,x)$ at time t at point $x \in \mathbb{R}^3$ is given by the ball $B_{Gt}(x)$ centered in x with radius Gt; hence, by (6.9)

$$V_E(t,x) = \lambda_S \nu^2(B_{Gt}(x) \cap \partial \mathcal{D})$$

and, by (6.6), the volume fraction $\mathbf{V}_V(t,A)$ for $A \subseteq \mathcal{D}$ is given by

$$\mathbf{V}_V(t,A) = \frac{1}{\nu^3(A)} \int_A (1 - e^{-\lambda_S \nu^2(B_{Gt}(x) \cap \partial \mathcal{D})}) dx$$

We now consider the case in which there is a constant nucleation rate with I_S nuclei per unit of area on $\partial \mathcal{D}$ per unit of time. Then the intensity of the nucleation process is given by

$$\lambda(s,y) = I_S \delta_{\partial \mathcal{D}}(y)$$

hence,

$$V_E(t,x) = \Lambda(\mathcal{C}(t,x)) = \int_0^t \int_{B_{G(t-s)}(x)} \lambda(s,y) dy ds = \int_0^t I_S \nu^2(\partial \mathcal{D} \cap B_{G(t-s)}(x)) ds$$

and, by (6.6), the volume fraction $\mathbf{V}_V(t,A)$ for $A \subseteq \mathcal{D}$ is given by

$$\mathbf{V}_V(t,A) = \frac{1}{\nu^3(A)} \int_A \left(1 - \exp\left\{-\int_0^t I_S \nu^2(\partial \mathcal{D} \cap B_{G(t-s)}(x)) ds\right\}\right) dx$$

6.5.1.2 Bulk Nucleation

Let us consider a Poisson nucleation in the interior of $\mathcal{D}$, first in the site-saturated case, then in the time-dependent case. For the sake of simplicity, we still assume here that the growth rate of the grains is constant in every direction.

First, we deal with site-saturated nucleation with λ_B nuclei per unit of volume in $\mathcal{D}$. Then the intensity λ of the nucleation process is given by

$$\lambda(y) = \lambda_B 1_{\mathcal{D}}(y)$$

so that

$$\Lambda(A) = \lambda_B \nu^3(\mathcal{D} \cap A) \quad \forall A \in \mathcal{B}_{\mathbb{R}^3}$$

therefore, by (6.9),

$$V_E(t,x) = \lambda_B \nu^3(B_{Gt}(x) \cap \mathcal{D})$$

and, by (6.6), the volume fraction $\mathbf{V}_V(t,A)$ for $A \subseteq \mathcal{D}$ is given by

$$\mathbf{V}_V(t,A) = \frac{1}{\nu^3(A)} \int_A (1 - e^{-\lambda_B \nu^3(B_{Gt}(x) \cap \mathcal{D})}) dx$$

Next the case of a constant nucleation rate with I_B nuclei per unit of volume in $\mathcal{D}$ per unit of time is considered. Then the intensity of the nucleation process is given by

$$\lambda(s,y) = I_B 1_{\mathcal{D}}(y)$$

and so

$$V_E(t,x) = \Lambda(\mathcal{C}(t,x)) = \int_0^t \int_{B_{G(t-s)}(x)} \lambda(s,y)dy ds = \int_0^t I_B \nu^3(\mathcal{D} \cap B_{G(t-s)}(x))ds$$

and, by (6.6), for $A \subseteq \mathcal{D}$

$$V_V(t,A) = \frac{1}{\nu^3(A)} \int_A \left(1 - \exp\left\{-\int_0^t I_B \nu^3(\mathcal{D} \cap \mathcal{B}_{G(t-s)}(x))ds\right\}\right) dx$$

The derivation of the above equations was quite straightforward but obtaining an explicit expression for a particular shape may require lengthy calculations (see supplementary material to [30]). Explicit expressions for bulk and surface nucleation in the case of two parallel planes (thin films), cylinder (wires) and ball (powders), can be found in Villa and Rios [30].

6.5.2
Numerical Examples

In this section, we present some numerical examples of the equations presented in Sections 6.5.1.1 and 6.5.1.2. We restrict these examples to bulk and surface site-saturated nucleation in a spherical sample $B_R(0)$. The values of λ_S and λ_B were chosen to illustrate typical behavior of the analytical expressions. In all cases, the radius of the sphere, R, was taken to be equal to 1 unit of length and the velocity, G, was taken to be equal to 0.001 units of length per unit of time.

6.5.2.1 Surface Nucleation

Figure 6.10 shows the mean volume density, $V_V(t,r) := V_V(t, B_r(0))$, for the transformation nucleating on the surface of the sphere with intensity $\lambda_S = 10$, in the site-saturated case. The solid black line corresponds to the volume fraction, $\mathbf{V}_V(t)$. Starting from the left-hand side of the plot each curve corresponds to: $r = 1$ (on the extreme left), $r = 0.75$, $r = 0.5$, $r = 0.25$, and $r = 0.0$ (on the extreme right). The reason why these curves have the shape displayed in Figure 6.10 can be understood in terms of the causal cone. For instance, a certain point (on a spherical shell), say $r = 0.75$, remains untransformed until time $t = (1-r)/G = 250$, since its causal cone has empty intersection with the boundary of the sphere until that time. The curve $V_V(t, r = 0.75)$ shows indeed that the growing regions nucleated on the surface may transform a point on the spherical shell at $r = 0.75$ when $t > 250$, and this is accomplished very quickly as soon as $t = 250$. We may also note in Figure 6.10 that $V_V(t,0) = 0$ for any $t \leq 1000$ and $V_V(t,0) = 1 - e^{-10\pi} \sim 1$ for any $t > 1000$.

6.5.2.2 Bulk Nucleation

Figure 6.11 shows the mean volume density, $V_V(t,r)$ for the transformation nucleating within the bulk of a sphere with intensity $\lambda_B = 5$. The solid black line corresponds to the volume fraction, $\mathbf{V}_V(t)$. Starting from the left-hand side of the plot, each curve corresponds to the mean volume density, $V_V(t,r)$: $r = 0$ (on the extreme left), $r = 0.8$,

6.5 Nucleation on Lower Dimensional Surfaces

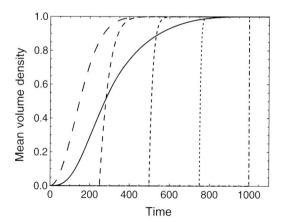

Figure 6.10 Surface nucleation, $r = 1$: mean volume density, $V_V(t,r)$ for the transformation nucleating on the surface of a sphere of unit radius with intensity $\lambda_S = 10$. The solid black line corresponds to the volume fraction, $\mathbf{V_V}(t)$. Starting from the left-hand side of the plot, each curve corresponds to, from left to right: $r = 1$ (long-dash line on the extreme left), $r = 0.75$ (medium-dash line), $r = 0.5$ (short-dash line), $r = 0.25$ (dot line), and the sphere center, $r = 0.0$ (dash-dot line on the extreme right).

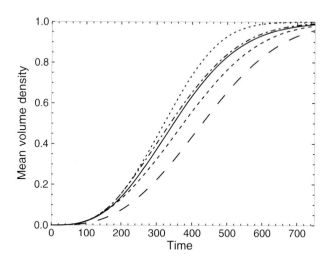

Figure 6.11 Bulk nucleation: mean volume density, $V_V(t,r)$, for the transformation nucleating within the bulk of a sphere with intensity $\lambda_B = 5$. The solid black line corresponds to the volume fraction, $\mathbf{V_V}(t)$. Starting from the left-hand side of the plot, each curve corresponds to the mean volume density, $V_V(t,r)$, from left to right: $r = 0$ (dot line on the extreme left), $r = 0.8$ (dash-dot line close to the solid line), $r = 0.9$ (short-dash line), $r = 1$ (long-dash line on the extreme right).

$r = 0.9$, and $r = 1$ (on the extreme right). It can be seen that the reaction close to the sphere surface, $r = 0.9$ and $r = 1$, proceeds slower than the overall reaction represented by the solid black line, $\mathbf{V_V}(t)$. Nevertheless, already for $r = 0.8$, the reaction proceeds faster, with the fastest rate taking place at $r = 0$. At $r = 0$ for $\lambda_B = 5$ the mean density curve, $V_V(t, 0)$ coincides with the curve calculated for an infinitely large specimen, which is not shown in the plot in order not to overload the figure.

6.5.2.3 Simultaneous Bulk and Surface Nucleation

The results of this section were obtained using *the superposition principle* that is presented below in Section 6.8. Figure 6.12 shows the mean volume density $V_V(t, r)$ at any point at distance r from the center of the sphere, for the transformation nucleating simultaneously on the surface with $\lambda_S = 10$, and within the bulk of a sphere with $\lambda_B = 5$. The solid black line corresponds to the volume fraction, $\mathbf{V_V}(t)$. Starting from the left-hand side of the plot, each curve corresponds to the mean volume density, $V_V(t, r)$: $r = 1$ (on the extreme left), $r = 0.9$, $r = 0.75$, and $r = 0$ (on the extreme right). The contribution of each nucleation mode is clear from Figure 6.12. Compare Figure 6.12 with Figures 6.10 and 6.11. Consider, for instance, the curve corresponding $V_V(t, r = 0)$ in Figure 6.12. The transformation is dominated by bulk nucleation. Therefore, for $\lambda_S = 10$ and $\lambda_B = 5$, the center of the specimen will be transformed before it "feels" any influence from the grains nucleated at the surface. In contrast, consider the curve corresponding to $V_V(t, r = 0.75)$ in Figure 6.12. Initially, up to a time equal to 250 time units, bulk nucleation dominates. However, at $t = 250$, the growing regions nucleated at surface reach $r = 0.75$. It is straightforward to see why

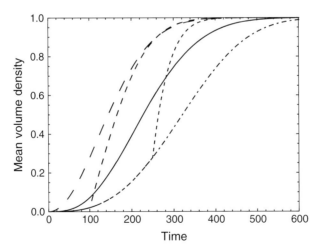

Figure 6.12 Simultaneous surface and bulk nucleation: mean volume density, $V_V(t, r)$ for the transformation nucleating simultaneously on the surface with $\lambda_S = 10$, and within the bulk of a sphere with $\lambda_B = 5$. The solid line corresponds to the volume fraction, $\mathbf{V_V}(t)$. Starting from the left-hand side of the plot, each curve corresponds to the mean volume density, $V_V(t, r)$, from left to right: $r = 1$ (long-dash on the extreme left), $r = 0.9$ (medium-dash line), $r = 0.75$ (short-dash line), $r = 0$ (dash-dot line on the extreme right).

this happens: $r = 0.75$ means a distance of 0.25 from the surface, that is covered in a time $t = 0.25/G = 250$. Alternatively, one can say that $t = 250$ is the time that the causal cone of $r = 0.75$ takes to reach the surface. From $t = 250$, $V_V(t, r = 0.75)$ changes its shape, and the transformation proceeds very quickly in the same manner as the $V_V(t, r = 0.75)$ in Figure 6.10.

6.6
Analytical Expressions for Transformations Nucleated on Random Planes

6.6.1
General Results for Nucleation on Random Planes

Let us briefly recall some general results on nucleation on planes randomly located in $\mathbb{R}^3$.

First of all, a plane B in $\mathbb{R}^3$ is uniquely determined by its distance from the origin, say u, and by its normal outer vector, say $w \in \mathbf{S}^2$ ($\mathbf{S}^2$ is the unit sphere in $\mathbb{R}^3$). The equation of $B = B(w, u)$ is then given by

$$B(w, u) := \{x \in \mathbb{R}^3 : \langle w, x \rangle = u\}$$

and it is well known that $\mathrm{dist}(x, B) = |\langle w, x \rangle - u|$ for any $x \in \mathbb{R}^3$ ($\langle w, x \rangle$ is the scalar product of w and x).

For the sake of simplicity, we assume that nuclei are homogeneously randomly located on B in the saturated case, and homogenously in time and space in the time-dependent nucleation case [31]. We denote the mean number of nuclei on the planes per unit of area of the plane by λ_S, and the constant nucleation rate per unit of area of the plane by I_S. Then, it can be proved [31] that the volume density $V_V^{u,w}(t, x)$ and the mean extended volume density $V_E^{u,w}(t, x)$ associated to the transformed region $\Theta^t = \Theta^t(w, u)$ at time t due to the nucleation on the plane $B(w, u)$ are given by

$$V_V^{u,w}(t, x) = 1 - e^{-V_E^{u,w}(t,x)}$$

and

$$V_E^{u,w}(t, x) = \begin{cases} \lambda_S \pi \left(G^2 t^2 - |\langle w, x \rangle - u|^2 \right) 1_{[0, Gt]}(|\langle w, x \rangle - u|) \\ I_S \pi \left(\dfrac{2|\langle w, x \rangle - u|^3}{3G} - |\langle w, x \rangle - u|^2 t + \dfrac{G^2 t^3}{3} \right) 1_{[0, Gt]}(|\langle w, x \rangle| - u|) \end{cases}$$

in the site-saturated and in the time-dependent cases, respectively.

If now we want to consider the same kind of nucleation on a plane $B = B(W, D)$ randomly located in $\mathbb{R}^3$ with random orientation W, and random distance from the origin D, then it can be shown (see [31]) that

$$V_V(t, x) = \int_{\mathbb{R}_+} \int_{S^2} V_V^{u,w}(t, x) P_{D,W}(d(u, w))$$

and

$$V_E(t,x) = \int_{\mathbb{R}_+}\int_{S^2} V_E^{u,w}(t,x) P_{D,W}(d(u,w))$$

where $P_{D,W}$ is the joint probability law of D and W in $\mathbb{R}_+ \times S^2$.

Note that $V_V(t,x) \neq 1 - e^{-V_E(t,x)}$; this is a remarkable difference between the case in which B is a fixed plane and the case in which B is a random plane.

Let us consider now the case in which nucleation takes place on the union of N random planes, where N is itself a random quantity. Namely, let N be a Poisson random variable with expected value n (i.e., n is the mean number of the planes where the nucleation process takes place), and let $B_1 = B_1(W_1, D_1), \ldots, B_N = B_N(W_N, D_N)$ be independent random planes, distributed as $B = B(W, D)$; we assume that $D_1, D_2, \ldots$ and $W_1, W_2, \ldots$ are independent and identically distributed as D and W, respectively, and independent of N, and that D is a continuous random variable (this implies that the probability that two or more planes B_i coincide is zero). It can be proved that (see [31])

$$V_V(t,x) = 1 - \exp\left\{-n \int_{\mathbb{R}_+}\int_{S^2}\left(1 - e^{-V_E^{u,w}(t,x)}\right) P_{D,W}(d(u,w))\right\} \qquad (6.24)$$

The above equations are quite general because they allow to obtain explicit expressions that depend on the chosen distribution $P_{D,W}$ of D and W. Equation (6.24) relies on two main assumptions:

1) nuclei are located in the planes homogeneously;
2) the *number of random planes* that contributes to the transformation taking place in a certain bounded region is a Poisson random variable.

6.6.2
Behavior at the Origin as a Model for the Behavior in an "Unbounded" Specimen

Sometimes the behavior of V_V at the origin might be of particular interest. At the origin $x = 0$, $\text{dist}(0, B) = |\langle w, 0 \rangle - u| = |u|$, and so $V_E^{u,w}(t) = V_E^u(t)$. Since the distance from the plane to the origin depends solely on $P_D(du)$, Eq. (6.24) simplifies

$$V_V(t) = 1 - \exp\left\{-n\int_{\mathbb{R}_+}\left(1 - e^{-V_E^u(t)}\right) P_D(d(u))\right\}$$

Clearly, if the specimens were very large relatively to the interplanar distance, all points located within the specimen that are sufficiently far from its boundary would be "equivalent" in the sense that they are surrounded by the same environment. More rigorously, if the plane process is stationary but not necessarily isotropic, that is, invariant under translation not under rotation, then any internal point that does not suffer influence from the boundaries may be treated as if it was an "origin point." A point cannot "see" the boundary if its distance to the boundary is less than the radius of the causal cone.

6.6.3
Nucleation on Random Parallel Planes Located Within a Specimen of Finite Thickness

A particular case of the above model is obtained by assuming that the planes are parallel. In such a case, the outer normal vector of each plane is fixed, say $\bar{w} = (0, 0, 1)$ (i.e., the planes are parallel to the xy plane); as a consequence, we obtain

$$V_V(t, x) = 1 - \exp\left\{-n \int_{\mathbb{R}_+} \left(1 - e^{-V_E^{u,\bar{w}}(t,x)}\right) P_D(du)\right\}$$

with $x = (x_1, x_2, x_3)$

$$V_E^{u,\bar{w}}(t, x) = \begin{cases} \lambda_S \pi (G^2 t^2 - (x_3 - u)^2) 1_{[x_3 - Gt, x_3 + Gt]}(u) \\ I_S \pi \left(\dfrac{2|x_3 - u|^3}{3G} - |x_3 - u|^2 t + \dfrac{G^2 t^3}{3}\right) 1_{[x_3 - Gt, x_3 + Gt]}(u) \end{cases} \quad (6.25)$$

in the site-saturated and the time-dependent case, respectively.

Explicit expressions for the case in which the parallel planes are *uniformly distributed* within $[0, M]$ (a specimen of finite thickness) can be now obtained easily by (6.25)

$$V_V(t, x) = 1 - \exp\left\{-S_V^{\text{planes}} \int_0^M \left(1 - e^{-\lambda_S \pi (G^2 t^2 - (x_3 - u)^2)}\right) 1_{[x_3 - Gt, x_3 + Gt]}(u) du\right\} \quad (6.26)$$

for the site-saturated case and

$$V_V(t, x) = 1 - \exp\left\{-S_V^{\text{planes}} \int_0^M \left(1 - e^{-I_S \pi\{(2|x_3 - u|^3)/(3G)\} - |x_3 - u|^2 t + (G^2 t^3)/3\}}\right)_{[x_3 - Gt, x_3 + Gt]}(u) du\right\} \quad (6.27)$$

for the constant nucleation rate case. S_V^{planes} is the mean area per unit of volume of the planes in $[0, M] \times [0, 1]^2$, equal to n/M.

6.6.4
Nucleation on Random Parallel Planes Located Within an "Unbounded Specimens"

Equations (6.26) and (6.27) may be used to obtain an expression for parallel planes in an "unbounded" specimen. In order to derive such an expression one must consider a specimen from $[-M, M]$, so that the plane $x_3 = 0$ is a symmetry plane. Second, one observes that a spherical region growing from a nucleus located at the plane $x_3 = 0$ would "see" an unbounded specimen, while its radius $R = Gt$ is less than or equal to the specimen with half-thickness M, in other words, if $Gt \leq M$. One can then make a transformation of variable $z = u/Gt$ and consider that the maximum (or minimum) value $|z| = 1$ is achieved when $|Gt| = M$. The final result for the mean volume density evaluated at any point x with $x_3 = 0$ for any $0 \leq t \leq M/G$ is, for site-saturation

$$V_V(t) = 1 - \exp\left(-2S_V^{\text{planes}} Gt \int_0^1 \left\{1 - e^{-\pi \lambda_S G^2 t^2 [1-z^2]}\right\} dz\right) \quad (6.28)$$

and

$$V_V(t) = 1 - \exp\left(-2S_V^{\text{planes}} Gt \int_0^1 \left\{1 - e^{-\frac{2}{3}\pi I_S G^2 t^3 \left[z^3 - \frac{3}{2}z^2 + \frac{1}{2}\right]}\right\} dz\right) \quad (6.29)$$

for constant nucleation rate. Equation (6.29) is identical to that obtained by Cahn assuming that the planes are "randomly" distributed in space.

When nucleation is so intense that the growth is essentially planar, Cahn obtained

$$V_V(t) = 1 - \exp\left(-2S_V^{\text{planes}} Gt\right)$$

It is not intuitive that such an expression should hold when the planes are parallel but indeed this is precisely the case.

Summarizing the above results, it can be concluded that Cahn's expression for transformations nucleated on random planes does not depend on the *orientation* of these planes but rather on the fact than the *number* of planes contained within a random sample of unit volume is a Poisson random variable. As a result, Cahn's expression is valid for random parallel planes, a somewhat surprising result. Computer simulation shown in the following section illustrates this.

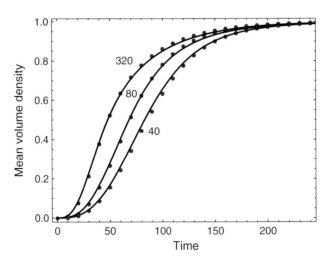

Figure 6.13 Mean volume density as a function of time for transformations nucleated at random parallel planes growing with a constant velocity equal to 0.001 mm/s. The mean distance between planes was equal to 0.1 mm and $S_V^{\text{planes}} = 10$ mm^{-1}. The number of nuclei per unit of area varied from 40 to 320 nuclei per mm^2, as indicated by the labels of each curve. The solid curve corresponds to the theoretical solution and the full circles to the computer simulation.

6.6.5
Computer Simulation Results

Figure 6.13 shows the influence of the number of nuclei per unit of area on the kinetics of the transformation nucleated on random parallel planes. The nucleation was site-saturated, that is, all nucleation sites were exhausted at $t = 0$. The solid curve corresponds to the theoretical solution, Eq. (6.28), and the full circles to the computer simulation. The mean distance between planes was equal to 0.1 mm or, equivalently, the area per unit of volume of the planes was S_V^{planes} equal to $10\,\text{mm}^{-1}$. The number of nuclei per unit of area varied from 40 up to 320 nuclei per mm^2, as indicated by the labels of each curve. The growth velocity was kept constant and equal to 0.001 mm/s.

6.7
Random Velocity

In a recent review of experimental 3D studies, Jensen and Godiksen concluded that in recrystallization there is compelling evidence to support that "every single grain has its own kinetics different from the other grains" [32]. Jensen and Godiksen then proposed that there is a continuous distribution of growth velocities. General expressions to deal with the situation in which the velocity is randomly distributed throughout the individual growing grains are presented in this section [33].

6.7.1
Time-Dependent, Random Velocity

Let us consider the case in which the velocity is time dependent and random, of the type

$$G(t) = G_0 g(t, \alpha)$$

where G_0 is a nonnegative random variable, and g is a nonnegative function depending on time and on a random vector parameter α in $\mathbb{R}^n$. Let us assume that G_0 and α are independent of the spatial location of the associated nuclei, and let us denote the joint probability distribution of G_0 and α by $Q(d(\xi, a))$. Then, we can model the nucleation process by a marked Poisson point process $\Phi = \{(X_i, (G_i, \alpha_i))\}$ in $\mathbb{R}^3$ with marks in $\mathbb{R}_+ \times \mathbb{R}^n$, with intensity measure

$$\Lambda(d(y, \xi, a)) = \lambda(y) dy Q(d(\xi, a))$$

It follows that the growing region Θ^t at time t is given by

$$\Theta^t = \bigcup_{(X_i,(G_i,\alpha_i))\in\Phi} B_{R_i(t)}(X_i)$$

with

$$R_i(t) := G_i \int_0^t g(\tau, \alpha_i) d\tau$$

and the causal cone $\mathcal{C}(t,x)$ of the point x at time t is now given by the subset of $\mathbb{R}^3 \times \mathbb{R}_+ \times \mathbb{R}^n$ so defined:

$$\begin{aligned}\mathcal{C}(t,x) &:= \{(y,\xi,a) \in \mathbb{R}^3 \times \mathbb{R}_+ \times \mathbb{R}^n : x \in B_{R(t)}(y)\} \\ &= \{(y,\xi,a) \in \mathbb{R}^3 \times \mathbb{R}_+ \times \mathbb{R}^n : y \in B_{R(t)}(x)\}\end{aligned}$$

where

$$R(t) := \xi \int_0^t g(\tau,a)d\tau$$

Taking into account Eq. (6.9), a general equation for the site-saturated case may be derived

$$V_E(t,x) = \int_{\mathbb{R}_+ \times \mathbb{R}^n} \int_{B_{R(t)}(x)} \lambda(y)dy Q(d(\xi,a))$$

If $\lambda(x)$ is harmonic, and if G_0 and α are independent of probability distribution Q_1 and Q_2, respectively, then $Q(d(\xi,a)) = Q_1(d\xi)Q_2(da)$, and so

$$V_E(t,x) = \lambda(x)\frac{4}{3}\pi \int_{\mathbb{R}_+} \xi^3 Q_1(d\xi) \int_{\mathbb{R}^n} \left(\int_0^t g(\tau,a)d\tau\right)^3 Q_2(da)$$

$$= \frac{4}{3}\pi\lambda(x)\mathbb{E}[G_0^3]\mathbb{E}\left[\left(\int_0^t g(\tau,\alpha)d\tau\right)^3\right]$$

since the nucleation process is assumed to be Poissonian, then

$$V_V(t,x) = 1 - \exp\left\{-\frac{4}{3}\pi\lambda(x)\mathbb{E}[G_0^3]\mathbb{E}\left[\left(\int_0^t g(\tau,\alpha)d\tau\right)^3\right]\right\} \tag{6.30}$$

$$S_V(t,x) = (1 - V_V(t,x))4\pi\lambda(x)\mathbb{E}[G_0^2]\mathbb{E}\left[\left(\int_0^t g(\tau,\alpha)d\tau\right)^2\right] \tag{6.31}$$

where the expected value of a random variable X with probability distribution $Q(dx) = q(x)dx$ is defined as $\mathbb{E}[X] = \int xQ(dx)$ or, more generally, $\mathbb{E}[f(X)] = \int f(x)Q(dx)$.

In the time-dependent nucleation case, if $\lambda(t,x)$ is bounded and continuous, and harmonic with respect to x in the spatial region where the nucleation takes place, then for $s \geq 0$

$$V_V(t,x) = 1 - \exp\left\{-\frac{4}{3}\pi\mathbb{E}[G_0^3]\int_0^t \lambda(s,x)\mathbb{E}\left[\left(\int_s^t g(\tau,\alpha)d\tau\right)^3\right]ds\right\} \tag{6.32}$$

and

$$S_V(t,x) = (1 - V_V(t,x))4\pi\mathbb{E}[G_0^2]\int_0^t \lambda(s,x)\mathbb{E}\left[\left(\int_s^t g(\tau,\alpha)d\tau\right)^2\right]ds \tag{6.33}$$

The overall velocity, $\mathcal{G}(t,x)$, may be obtained quite generally by

$$\mathcal{G}(t,x) = \frac{1}{S_V(t,x)} \frac{\partial V_V(t,x)}{\partial t} \tag{6.34}$$

6.7.2
Particular Cases

Particular cases of the general equations have been discussed above. For the site-saturated situation, it is supposed that: (a) nucleation sites are located according to a homogeneous Poisson point process with intensity N_V, the mean number of nuclei per unit of volume; (b) $G(t) = G_0$ with G_0 the random variable, that is, the velocity of each grain is constant during the transformation, but the value of such a constant is random. Under these assumptions, the mean volume density, $V_V(t,x)$, is equal to the volume fraction, $V_V(t)$, and therefore, Eq. (6.30) becomes

$$V_V(t) = 1 - \exp\left\{-\frac{4\pi}{3} N_V \mathbb{E}[G^3] t^3\right\} \tag{6.35}$$

Equation (6.35) is formally identical to the classical JMAK expression, Eq. (6.2), [1,2,5] for site-saturation, excepting that $\mathbb{E}[G^3]$ substitutes for a constant non-random G^3.

Equation (6.31) gives the mean interfacial area density (sometimes called the interfacial area per unit of volume), $S_V(t)$,

$$S_V(t) = 4\pi N_V t^2 \mathbb{E}[G^2] \exp\left\{-\frac{4}{3}\pi N_V \mathbb{E}[G^3] t^3\right\}$$

and Eq. (6.34)

$$\mathcal{G}(t) = \frac{1}{S_V(t)} \frac{\partial V_V(t)}{\partial t} = \frac{\mathbb{E}[G^3]}{\mathbb{E}[G^2]}$$

The velocity obtained from Eq. (6.34), $\mathcal{G}$, cannot be directly identified with the velocity of a growing grain, not even with a mean velocity.

If the nucleation is homogeneous in time, that is, λ is independent of time, Eqs. (6.32) and (6.33) simplify as follows:

$$V_V(t,x) = 1 - \exp\left\{-\frac{4}{3}\pi\lambda(x)\mathbb{E}[G_0^3]\int_0^t \mathbb{E}\left[\left(\int_s^t g(\tau,\alpha)d\tau\right)^3\right]ds\right\}$$

$$S_V(t,x) = (1 - V_V(t,x))4\pi\lambda(x)\mathbb{E}[G_0^2]\int_0^t \mathbb{E}\left[\left(\int_s^t g(\tau,\alpha)d\tau\right)^2\right]ds$$

A particular case of interest is that in which: (a) G is constant, that is, time and position independent, but random, that is, $g(\tau,\alpha) \equiv 1$ and $G = G_0$; (b) the nucleation rate is constant and position-independent so that $\lambda(t, x) = I_V$, where I_V is a constant nucleation rate per unit of volume; therefore,

$$V_V(t) = 1 - \exp\left\{-\frac{\pi}{3}I_V \mathbb{E}[G^3] t^4\right\}$$

Again a fundamental result from JMAK, Eq. (6.3), is recovered excepting that $\mathbb{E}[G^3]$ substitutes for a constant nonrandom G^3.

6.7.3
Computer Simulation

The results above may be compared with Godiksen et al. [35] simulations. Jensen et al. [34] in their analytical work and Godiksen et al. [35] in their computer simulation supposed that the velocity of each grain was time-dependent and random of the type or $G_{0g}(t, \alpha) = A(1 - \alpha)t^{-\alpha}$ or, equivalently, that the radius of a growing spherical region, R, was given by $R = At^{1-\alpha}$. Equation (6.30) becomes

$$V_V(t) = 1 - \exp\left\{-\frac{4}{3}\pi N_V \mathbb{E}[A^3] \mathbb{E}\left[t^{3(1-\alpha)}\right]\right\} \tag{6.36}$$

Godiksen et al. considered two cases. In the first case, A was random and $\alpha = 0$, $G_0 = A$, which reduces to Eq. (6.35) with $G = A$. Equation (6.35) shows that the form of the equation is independent of the particular probability distribution of A, which only influences the value of the expectation $\mathbb{E}[A^3]$. This is in agreement with previous work by Jensen et al. [34] and Godiksen et al. [35].

In the second case simulated by Godiksen et al., $A = 1$ and α was random. They used two probability distributions p for α: uniform distribution in $(0, 1)$ ($U(0, 1)$), $p(z) = 1$ with $z \in [0, 1]$, and beta distribution. Returning to Eq. (6.36), $A = 1$ and $p(z) = 1$ gives

$$V_V(t) = 1 - \exp\left\{-\frac{4}{3}\pi N_V \int_0^1 t^{3(1-z)} dz\right\} \tag{6.37}$$

$$S_V(t) = (1 - V_V(t)) 4\pi N_V \int_0^1 t^{2(1-z)} dz$$

and the overall velocity from Eq. (6.34)

$$\mathcal{G}(t) = \frac{\int_0^1 t^{-3z}(1-z) dz}{\int_0^1 t^{-2z} dz} \tag{6.38}$$

When α is $beta(a, b)$ distributed with parameters $a > 0$ and $b > 0$, its probability density function p is given by

$$p(z) = \frac{\Gamma(a+b)}{\Gamma(a)\Gamma(b)} z^{a-1}(1-z)^{b-1} 1_{[0,1]}(z) \tag{6.39}$$

Note that $beta(1, 1) = U(0, 1)$. Combining Eqs. (6.36) and (6.39)

$$V(t) = 1 - \exp\left\{-\frac{4}{3}\pi N_V \int_0^1 t^{3(1-z)} \frac{\Gamma(a+b)}{\Gamma(a)\Gamma(b)} z^{a-1}(1-z)^{b-1} dz\right\} \tag{6.40}$$

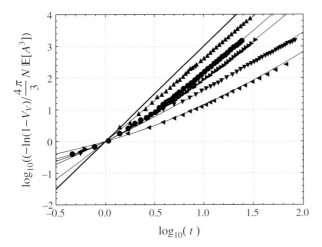

Figure 6.14 The straight line without any symbols on it corresponds to the JMAK kinetics for a constant, nonrandom velocity. The symbols correspond to simulated data of Godiksen et al. and the solid lines were calculated from uniform distribution, Eq. (6.37): (•) and beta distribution, Eq. (6.40): $a = 6, b = 2.5$ (◄); $a = 11, b = 11$ (▼); $a = 2, b = 2$ (►); $a = 2, b = 11$ (▲).

This equation was obtained by Jensen et al. [34] by a different method. They directly calculated the extended volume instead of using the causal cone method as was done here. The overall velocity from Eq. (6.34) is

$$\mathcal{G}(t) = \frac{\int_0^1 t^{-3z} z^{a-1} (1-z)^b dz}{\int_0^1 t^{-2z} z^{a-1} (1-z)^{b-1} dz} \tag{6.41}$$

Godiksen et al. simulated the case of uniform distribution and also simulated several cases corresponding to different values of the parameters a and b in the beta distribution. Table 1 of Godiksen et al. [35] contains all the information about their simulations. Using Eqs. (6.37) and (6.40), the same cases simulated by Godiksen et al. for $A = 1$ and α random can be calculated analytically. The result is plotted in Figure 6.14 together with the results from their simulations. The agreement is excellent showing the correctness of both analytical solution and the computer simulation.

Figure 6.15 shows the influence of the distribution of growth velocities on the overall velocity (Eq. (6.34)). The overall velocity is evidently the quantity more strongly affected by the distribution of growth velocities. This is apparent in Figure 6.15.

In summary, expressions were developed for the situation in which the growth velocity is not the same for all growing grains but is randomly distributed. The mean volume density, $V_V(t, x)$, and the mean interfacial area density, $S_V(t, x)$, were obtained for site saturation: Eqs. (6.30) and (6.31), and for time-dependent nucleation rate: Eqs. (6.32) and (6.33).

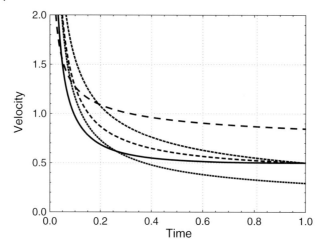

Figure 6.15 Overall velocities calculated from Eqs. (6.38) and (6.41). The solid line represents Eq. (6.38). The long-dashed, dashed, short-dashed and dotted lines were calculated using $a = 2, b = 11$; $a = 11, b = 11$; $a = 2, b = 2$; and $a = 6, b = 2.5$, respectively.

In addition to experimental methods [32] and computer simulation [35], the analytical approach presented here may be useful to analyze and understand the effect of the distribution of velocities on the transformation kinetics in general and on the recrystallization kinetics in particular.

6.8
Simultaneous and Sequential Transformations

Formal kinetics methodology has its origin in the early work of Kolmogorov [5], Johnson-Mehl [1], and Avrami [2] for modeling a single transformation. Nonetheless, two or more reactions may take place simultaneously independent of each other. Vandermeer and Jensen [36] and Jones and Bhadeshia [37] proposed a methodology to model simultaneous reactions. More generally, the reactions might not start at the same time. For example, reaction 1 starts at $t = 0$, whereas reaction 2 starts at $t = \tau > 0$. These may be called sequential transformations. A recent example is the potential overlap of recrystallization and austenite formation during heating of cold worked steels [41].

Modeling simultaneous or sequential reactions normally involve two distinct but closely related objectives:

1) Predict the overall kinetics quantities of the combined reactions from analytical models developed for each individual reaction.
2) Extract the kinetics for an individual reaction from experimental measurements.

Recently [38], Rios and Villa derived exact analytical expressions to model simultaneous and sequential reactions. In this section, we present a simplified version of the main results of their paper. Therefore, we will not present expressions containing interfacial area densities that can be found in Rios and Villa [38]. Moreover, Rios and Villa employed position-dependent quantities, such as the mean volume density, $V_V(t, x)$. In this section, for simplicity, only position-independent densities are used, for example, $V_V(t)$, which coincides with the volume fraction. We then use these results to model the kinetics of recrystallization of an IF steel [39] studied by Magnussen et al. [40].

6.8.1
Simultaneous Transformations

First, it is necessary to make a distinction between experimental and theoretical quantities. The theoretical volume fraction of reaction i is denoted by $V_{Vi}(t)$. The theoretical volume fraction associated with reaction i, V_{Vi}, is the value of this quantity if reaction i were the only reaction taking place. Note that in this case, depicted in Figure 6.16a for two reactions, there is no impingement between 1 and 2 transformations, but there is impingement (not shown in Figure 6.16a) between 1 and 1, and also between 2 and 2 transformations. The experimental volume fraction is directly measured from the transformed microstructure. The experimental volume fraction of a certain reaction i will be denoted $V^*_{Vi}(t)$. Note that in this case, as shown in Figure 6.16b, there is impingement between 1 and 2 and (not shown in Figure 6.16b) between 1 and 1 and between reactions 2 and 2.

The fundamental equations are Eqs. (6.42)–(6.44). Equation (6.42) is the superposition principle

$$V_V(t) = 1 - \prod_{i=1}^{n}(1 - V_{Vi}(t)) \tag{6.42}$$

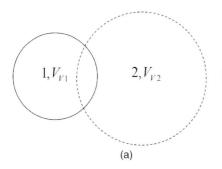
(a)

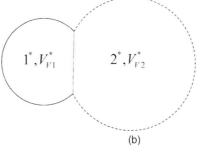
(b)

Figure 6.16 Illustration of the definitions of theoretical (a) and experimental quantities (b) for two transformations. Left: The transformed regions of each reaction are labeled as 1 and 2. The volume fractions of 1 and 2 are V_{V1} and V_{V2}, respectively. V_{V1} and V_{V2} are called "theoretical volume fractions." Right: The "visible" or "experimental" transformed region of each reaction is indicated by 1^* and 2^*. The volume fraction of 1^* and 2^* are V^*_{V1} and V^*_{V2}, respectively. V^*_{V1} and V^*_{V2} are called "experimental volume fractions." The dotted line on the right-hand side figure is the boundary between 1^* and 2^*.

where V_{Vi} is the theoretical volume fraction associated with transformation i. Equation (6.43) permits extracting theoretical quantities from experimental quantities:

$$V_{Vi}(t) = 1 - \exp\left(-\int_0^t \frac{1}{1 - V_V(s)} \frac{\partial V^*_{Vi}(s)}{\partial s} ds\right) \quad (6.43)$$

By contrast, Eq. (6.44) permits obtaining experimental quantities from theoretical quantities

$$V^*_{Vi}(t) = \int_0^t \frac{1 - V_V(s)}{1 - V_{Vi}(s)} \frac{\partial V_{Vi}(s)}{\partial s} ds \quad (6.44)$$

Note that $V_V(t)$ may be either obtained experimentally or from the theoretical quantities by means of Eq. (6.42). The fundamental assumption underlying Eqs. (6.42)–(6.44) is that the transformations are *independent* among themselves; see Rios and Villa [38] for more details.

6.8.2
Sequential Transformations

In many cases, transformations might not start at the same time but may start at different times so that the reactions overlap [41], known as sequential reactions. Simultaneous reactions may be seen as a particular case of sequential reactions when all transformations start at the same point in time. We present here a general model for sequential transformations based on the equations derived in the previous section. In order to modify previous equations to encompass these cases, we need to define the incubation time, τ_i, for reaction i to start. $\tilde{V}_{Vi}$ is the volume fraction associated with the same reaction i when $\tau_i = 0$. For sequential reactions, one has

$$V_{Vi}(t) = \tilde{V}_{Vi}(t - \tau_i) 1_{[\tau_i, \infty)}(t)$$

This is equivalent to say that $\tilde{V}_{Vi}$ is equal to zero for $t < \tau_i$. Moreover, using Eq. (6.42)

$$V_V(t) = 1 - \prod_{i=1}^n (1 - \tilde{V}_{Vi}(t - \tau_i) 1_{[\tau_i, \infty)}(t))$$

Also, for $t > \tau_i$

$$\tilde{V}_{Vi}(t - \tau_i) = 1 - \exp\left(-\int_{\tau_i}^t \frac{1}{1 - V_V(s)} \frac{\partial V^*_{Vi}(s)}{\partial s} ds\right)$$

and

$$V^*_{Vi}(t) = \int_{\tau_i}^t \frac{(1 - V_V(s))}{(1 - \tilde{V}_{Vi}(s - \tau_i))} \frac{\partial \tilde{V}_{Vi}(s - \tau_i)}{\partial s} ds$$

These equations may be used when one has sequential transformations.

6.8.3
Application to Recrystallization of an IF Steel

Magnusson et al. [40] studied the growth rates of the different texture components of an IF steel. They measured the growth rate of the "γ-fiber" or "γ" components and the growth rate of what they called "other" or "o" components. Magnusson et al. [40] did not report the experimental errors, but judging from the techniques they used, it is reasonable to expect the relative experimental errors to be about $\pm 10\%$. For details of their IF steel, experimental techniques, and results, the reader is referred to their work [40]. Here we apply the methodology described in previous sections to the results of Magnusson et al.

The analysis was carried out in four steps. First, the experimental volume fractions against time data were fitted with conveniently chosen functions, so that integrations and derivatives can be carried out more easily. Figure 6.17 shows that a good description of the experimental data could be achieved. The point here is that the fittings must be consistent: the sum of the volume fractions of the γ-fiber and other orientations must give a good fit to the total volume fraction. The corresponding fitting equations are

$$V_{V_o}^{*\text{fit}}(t) = 0.41(1 - \exp(-0.00049t)) \tag{6.45}$$

and

$$V_{V_\gamma}^{*\text{fit}}(t) = 0.59(1 - \exp(-0.00047t)) \tag{6.46}$$

where t is in seconds. The total volume fraction is simply $V_V^{\text{fit}}(t) = V_{V_\gamma}^{*\text{fit}}(t) + V_{V_o}^{*\text{fit}}(t)$. Least square method was used to fit all three equations simultaneously to the data. The coefficient of determination, R^2 was greater than 0.99 in all three cases.

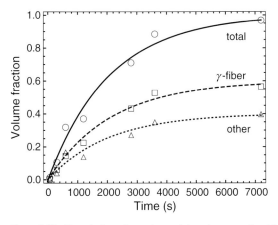

Figure 6.17 Description of experimental data (open markers) by conveniently chosen functions. Dotted line represents the curve fitted to the experimental data of other orientations, dashed line of γ-fiber orientations, and solid line of the sum of $\gamma + o$.

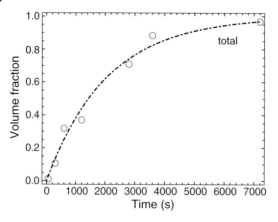

Figure 6.18 Comparison between total theoretical volume fraction, $V_V(t)$ (dash-dotted line) and experimental data (open circles). For explanation, see text.

Second, Eq. (6.43) was used to obtain the theoretical quantities from the experimental quantities represented by Eqs. (6.45) and (6.46),

$$V_{V\gamma}(t) = 1 - \exp\left(-\int_0^t \frac{0.0002773 e^{-0.00047s}}{1 - 0.41(1 - e^{-0.00049s}) - 0.59(1 - e^{-0.00047s})} ds\right) \quad (6.47)$$

$$V_{Vo}(t) = 1 - \exp\left(-\int_0^t \frac{0.0002009 e^{-0.00049s}}{1 - 0.41(1 - e^{-0.00049s}) - 0.59(1 - e^{-0.00047s})} ds\right) \quad (6.48)$$

These are the theoretical volume fraction of γ-fiber (Eq. (6.47)) and other (Eq. (6.48)) texture components. Note that one can calculate the total volume fraction from $V_{V\gamma}(t)$ and $V_{Vo}(t)$ with the help of Eq. (6.42):

$$V_V(t) = 1 - (1 - V_{V\gamma}(t))(1 - V_{Vo}(t)) = V^*_{V\gamma}(t) + V^*_{Vo}(t) \quad (6.49)$$

As shown in Eq. (6.49), the total volume fraction obtained from $V^*_{V\gamma}$ and V^*_{Vo} and $V_{V\gamma}$ and V_{Vo} must be the same. A comparison of both is depicted in Figure 6.18 showing the consistency of the methodology. In fact, the dash-dotted curve in Figure 6.18 is identical to the solid line in Figure 6.17.

Third, the theoretical volume fraction may be analyzed with the help of analytical models. The simplest assumption is that the nucleation of both γ and o components is site-saturated. Moreover, γ and o components grow with the velocities experimentally measured by Magnusson et al.: $G_\gamma = 0.14 t^{-0.58}$ µm/s and $G_o = 0.08 t^{-0.49}$ µm/s. For site-saturation, Eq. (6.2) may be used but with a time dependent velocity instead of a constant velocity. As a consequence, following equations are obtained:

$$V_{V\gamma}(t) = 1 - \exp\left(-\frac{4\pi N_{V\gamma}}{3}\left(\int_{s=0}^{s=t} G_\gamma(s) ds\right)^3\right) = 1 - \exp(-3.92 \times 10^{-5} t^{1.26}) \quad (6.50)$$

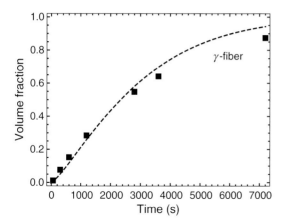

Figure 6.19 Comparison between model (dashed line), Eq. (6.50), and theoretical volume fraction (filled squares) obtained from experimental data of γ-fiber orientations by Eq. (6.43). The agreement is good.

$$V_{V_o}(t) = 1 - \exp\left(-\frac{4\pi N_{V_o}}{3}\left(\int_{s=0}^{s=t} G_o(s)ds\right)^3\right) = 1 - \exp(-2.57 \times 10^{-6} t^{1.53})$$
(6.51)

$$V_V(t) = 1 - \left(1 - V_{V\gamma}(t)\right)\left(1 - V_{V_o}(t)\right) = 1 - \exp\left(-10^{-5}(0.257 t^{1.53} + 3.92 t^{1.26})\right)$$
(6.52)

$N_{V\gamma} = 0.000253$ nuclei/µm³ and $N_{V_o} = 0.000159$ nuclei/µm³ were obtained by least-squares fit. $R^2 = 0.997$ for Eq. (6.50), $R^2 = 0.979$ for Eq. (6.51,) and $R^2 = 0.997$ for Eq. (6.52). The fitting of other orientations was worse than that for γ, Eq. (6.50), and than that for the total, Eq. (6.52). The models are compared with the theoretical quantities in Figures 6.19–6.21. The markers in Figures 6.19–6.21 represent the

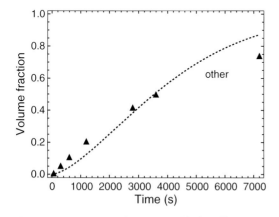

Figure 6.20 Comparison between model (dotted line), Eq. (6.51), and theoretical volume fraction (filled triangles) obtained from experimental data of other orientations by Eq. (6.43). The agreement is fair.

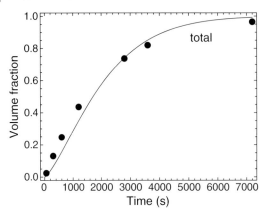

Figure 6.21 Comparison between model (solid line), Eq. (6.52), and the total theoretical volume fraction (filled circles). The agreement is good.

theoretical volume fractions calculated by means of Eqs. (6.47)–(6.49) at the same transformation times of the corresponding experimental volume fractions. The result suggests that the site-saturated model provides a reasonable description of the behavior of the kinetics of the γ-fiber and of other texture components in the IF steel studied by Magnusson et al. The higher interface velocity of the other components is compensated by a higher number of nuclei of the γ-fiber components. Moreover, the analysis provides quantitative estimates for the number of nuclei per unit of volume.

Finally, the analytical models above may be inserted in Eq. (6.44) to calculate the curves corresponding to the experimental quantities. The result is depicted in Figure 6.22. We started from experimental volume fraction data and obtained theoretical volume fractions. Theoretical volume fractions led to the analytical models. Finally, the analytical models permitted determining curves that satisfactorily

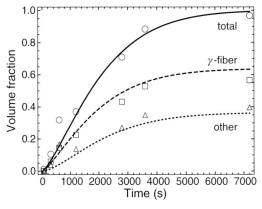

Figure 6.22 Comparison between the curves generated by the models (lines) with the help of Eq. (6.44) and the experimental data (open markers). Compare with Figure 6.17.

described the experimental data. The good agreement shown in Figure 6.22 is an indication of the consistency of the methodology.

In conclusion, by employing this new methodology, quantities suitable for theoretical analysis may be obtained from quantities measured experimentally. Conversely, experimentally measurable quantities may be predicted from theoretical ones. This methodology was applied here to analyze data from Magnusson *et al.* [40] on the kinetics of individual texture components during recrystallization of an IF steel. The analysis of IF steel data demonstrated that this methodology may be successfully applied to a practical case.

6.9
Final Remarks

In recent papers, the present authors have introduced modern stochastic geometry methods in order to obtain new solutions to the nucleation and growth ("birth-and-growth") problem. This is a rigorous approach to treat nucleation and growth transformations in Materials Science. If one could make a parallel with experimental work, the procedure is similar to employing a new technique to study a problem already studied by an early technique. This is done in order to gain new insights, new results, and also to understand better what was obtained with the old technique. In this regard, the present approach has been successful in several ways. First, it is methodologically new as our rigorous mathematical approach is more modern and rigorous than the mathematical approach used so far in the established literature to deal with kinetics of nucleation and growth transformations. Second, this methodology allows us to obtain new results, considerably expanding the range of exact analytical solutions available. Finally, in some specific cases, results already available were rederived (See Section 6.6), this lead to a better understanding of the assumptions and limitations of the original derivations.

We believe the modern complexity of Materials Science demands that problems be faced with a combination of experimental methods, computer simulation, and analytical results.

References

1 Johnson, W.A. and Mehl, R.F. (1939) Reaction kinetics in process of nucleation and growth. *Transactions AIME*, **135**, 416–442.
2 Avrami, M.J. (1939) Kinetics of phase change I. General theory. *Journal of Chemical Physics*, **7** (12), 1103–1112.
3 Avrami, M.J. (1940) Kinetics of phase change. II. Transformation-time relations for random distribution of nuclei. *Journal of Chemical Physics*, **8** (2), 212–224.
4 Avrami, M.J. (1941) Kinetics of phase change III. Granulation, phase change, and microstructure kinetics of phase change. *Journal of Chemical Physics*, **9** (2), 177–184.
5 Kolmogorov, A.N. (1937) On the statistical theory of the crystallization of metals. *Isvestiia Academii Nauk SSSR – Seriia Matematicheskaia*, **1**, 333–359.
6 Vandermeer, R.A. and Jensen, D.J. (2001) Microstructural path and temperature dependence of recrystallization in

commercial aluminum. *Acta Materialia*, **49** (11), 2083–2094.

7 Rios, P.R., Yamamoto, T., Kondo, T., and Sakuma, T. (1998) Abnormal grain growth kinetics in BaTiO$_3$ with an excess TiO$_2$. *Acta Materialia*, **46** (5), 1617–1623.

8 Yamamoto, T., Sakuma, T., and Rios, P.R. (1998) Application of microstructural path analysis to abnormal grain growth of BaTiO$_3$ with an excess TiO$_2$. *Scripta Materialia*, **39** (12), 1713–1717.

9 Rios, P.R. and Guimarães, J.R.C. (2007) Microstructural path analysis of athermal martensite. *Scripta Materialia*, **57** (12), 1105–1108.

10 Rios, P.R. and Guimarães, J.R.C. (2008) Formal analysis of isothermal martensite spread. *Materials Research*, **11** (1), 103–108.

11 Burger, M., Capasso, V., Micheletti, C., and Sala, C. (2002) Mathematical models for polymer crystallization processes, in *Mathematical Modelling for Polymer Industry*, vol. 2, Math in Industry Series (ed. V. Capasso), Springer, Berlin, pp. 167–242.

12 Burger, M., Capasso, V., and Salani, C.V. (2002) Modelling multi-dimensional crystallization of polymers in interaction with heat transfer. *Nonlinear Analysis – Series B*, **3** (1), 139–160.

13 Gokhale, A.M., Iswaran, C.V., and DeHoff, R.T. (1979) Use of the stereological counting measurements in testing theories of growth-rates. *Metallurgical Transactions A*, **10A** (9), 1239–1245.

14 DeHoff, R.T. (1986) in Annealing processes recovery, recrystallization and grain growth, in *Proceedings of the Seventh Conference of Risø National Laboratory, September, 8–12, 1986* (eds N. Hansen, D.J. Jensen, T. Leffers, and B. Ralph), RISØ National Laboratory, Roskilde, Denmark, pp. 35–52.

15 Vandermeer, R.A. and Rath, B.B. (1989) Modeling recrystallization kinetics in a deformed iron single crystal. *Metallurgical Transactions A*, **20A** (3), 391–491.

16 Rios, P.R. and Villa, E. (2009) Transformation kinetics for inhomogeneous nucleation. *Acta Materialia*, **57** (11), 1199–1208.

17 Capasso, V. and Villa, E. (2007) On the evolution equations of mean geometric densities for a class of space and time inhomogeneous stochastic birth-and-growth processes, in *Stochastic Geometry, Lecture Notes in Mathematics* (ed. W. Weil), Springer, Heidelberg, pp. 267–281.

18 Capasso, V. and Villa, E. (2008) On the geometric densities of random closed sets. *Stochastic Analysis and Applications*, **26** (4), 784–808.

19 Stoyan, D., Kendall, W.S., and Mecke, J. (1995) *Stochastic Geometry and Its Application*, 2nd edn, John Wiley & Sons, Ltd, Chichester, UK.

20 Billingsley, P. (1995) *Probability and Measure*, John Wiley & Sons, Ltd, Chichester, UK.

21 Evans, L.C. and Gariepy, R.F. (1992) *Measure Theory and Fine Properties of Functions*, CRC Press, Boca Raton, FL.

22 Villa, E. (2008) A note on mean volume and surface densities for a class of birth-and-growth stochastic processes. *International Journal Contemporary Mathematical Science*, **3** (23), 1141–1155.

23 Cahn, J.W. (1996) The time cone method for nucleation and growth kinetics on a finite domain. *MRS Proceedings*, **398**, 425–437.

24 Rios, P.R., Jardim, D., Assis, W.L.S., Salazar, T.C., and Villa, E. (2009) Inhomogeneous Poisson point process nucleation: Comparison of analytical solution with cellular automata simulation. *Materials Research*, **22** (2), 219–224.

25 Villa, E. and Rios, P.R. (2009) Transformation kinetics for nucleus clusters. *Acta Materialia*, **57** (13), 3714–3724.

26 Baddeley, A. (2007) Spatial point processes and their applications, in *Stochastic Geometry, Lecture Notes in Mathematics* (ed. W. Weil), Springer, Heidelberg, pp. 1–75.

27 The Comprehensive R Archive Network (October 2012) http://www.cran.r-project.org

28 Rios, P.R., Pereira, L.O., Oliveira, F.F., Assis, W.L.S., and Castro, J.A. (2007) Impingement function for nucleation on non-random sites. *Acta Materialia*, **55** (13), 4339–4348.

29 Rios, P.R., Pereira, L.O., Assis, W.L.S., Oliveira, F.F., and Oliveira, V.T. (2007) Analysis of transformations nucleated on non-random sites simulated by cellular

automata in three dimensions. *Materials Research*, **10** (2), 141–146.

30 Villa, E. and Rios, P.R. (2010) Transformation kinetics for surface and bulk nucleation. *Acta Materialia*, **58** (7), 2752–2768.

31 Villa, E. and Rios, P.R. (2011) Transformation kinetics for nucleation on random planes and lines. *Image Analysis Stereology*, **30** (3), 153–165.

32 Jensen, D.J. and Godiksen, R. (2008) Neutron and X-ray studies of recrystallization kinetics. *Metallurgical and Materials Transactions A*, **39** (13), 3065–3069.

33 Rios, P.R. and Villa, E. (2011) An analytical approach to the effect of a distribution of growth velocities on recrystallization kinetics. *Scripta Materialia*, **65** (11), 938–941.

34 Jensen, D.J., Lauridsen, E.M., and Vandermeer, R.A. (2002) In-situ determination of grain boundary migration during recrystallization, in *Science and Technology of Interfaces* (ed. A. Ankem), TMS, Warrendale, PA, pp. 361–374.

35 Godiksen, R.B., Schmidt, S., and Jensen, D.J. (2007) Effects of distributions of growth rates on recrystallization kinetics and microstructure. *Scripta Materialia*, **57** (4), 345–348.

36 Vandermeer, R.A. and Jensen, D.J. (1994) Modeling microstructural evolution of multiple texture components during recrystallization. *Acta Metallurgica Materialia*, **42** (7), 2427–2436.

37 Jones, S.J. and Bhadeshia, H.K.D.H. (1997) Kinetics of the simultaneous decomposition of austenite into several transformation products. *Acta Materialia*, **45** (7), 2911–2920.

38 Rios, P.R. and Villa, E. (2011) Simultaneous and sequential transformations. *Acta Materialia*, **59** (4), 1632–1643.

39 Rios, P.R., Villa, E., and Oliveira, S.C. (2012) New methodology to model simultaneous and sequential reactions: Main results and applications. *Materials Science Forum*, **706–709**, 149–156.

40 Magnusson, H., Jensen, D.J., and Hutchinson, B. (2001) Growth rates for different texture components during recrystallization of IF steel. *Scripta Materialia*, **44** (3), 435–441.

41 Huang, J., Poole, W.J., and Militzer, M. (2004) Austenite formation during intercritical annealing. *Metallurgical and Materials Transactions A*, **35A** (11), 3363–3375.

7
Implementation of Anisotropic Grain Boundary Properties in Mesoscopic Simulations
Anthony D. Rollett

7.1
Introduction

The objective of this chapter is to review how anisotropic grain boundary properties can be modeled in the various methods for simulating microstructural evolution. We limit the discussion to grain growth and recrystallization, although there is extensive related work on the simulation of solidification, particle coarsening, and sintering. Solidification in particular has received a considerable amount of attention because the phase-field method has proven to be very useful and the earliest references to this method are, in fact, motivated by solidification. The methods are grouped together so that the first set comprises image-based algorithms and the second set covers sharp interface models. Since in-depth reviews are available for each method in nearly all cases, the description of each method is brief and the main focus is on how (anisotropic) grain boundary properties are implemented in the context of each algorithm.

7.2
Overview of Simulation Methods

Simulation of microstructural evolution is accessible through a wide range of techniques, including the Potts model, cellular automata, level set, phase field, vertex model, and finite element. All the techniques permit anisotropic grain boundary properties to be imposed but to varying degrees. The Potts and cellular automata methods are image based and therefore use implicit definitions of the boundaries. Accordingly, the change in lattice orientation across a boundary is easily determined but the boundary normal is not readily available, at least not without substantial extra computational effort. Thus, one can include the anisotropy of grain boundary properties such as energy and mobility for only three out of the five macroscopic degrees of freedom. Level set and phase-field methods are also image based but use functions to model phases or grains (i.e., fields) that vary smoothly with position, resulting in diffuse boundaries. The gradients in these fields are an

Microstructural Design of Advanced Engineering Materials, First Edition. Edited by Dmitri A. Molodov.
© 2013 Wiley-VCH Verlag GmbH & Co. KGaA. Published 2013 by Wiley-VCH Verlag GmbH & Co. KGaA.

integral part of both methods and automatically provide information on boundary normal. Thus, the anisotropy can be made a function of all five parameters. The sharp interface models such as the vertex and finite element models contain explicit descriptions of the models via the meshes used to discretize the microstructures.

7.3
Anisotropy of Grain Boundaries

7.3.1
Energy

Our understanding of the anisotropy of grain boundary energy is reasonably good. For small enough misorientations, the Read–Shockley model has repeatedly been shown to be correct [1,2]. For high-angle boundaries, Rohrer recently reviewed the available literature [3] and suggested that a reasonable approximation is to consider grain boundaries as a pair of surfaces such that the boundary energy is approximately the sum of the two surface energies minus a binding energy. If a constant binding energy is assumed as a first approximation then the expected result is that the grain boundary energy should be the average of the two surface energies. Moreover, low energies should be observed when low energy surfaces form one or both sides of the boundary. This general result is observed for a wide variety of materials as shown in Figure 13 of [3]. A simulation to be discussed in more detail in the following [4] showed that in texture-free materials, the grain boundary character distribution (GBCD), that is, the relative populations of different boundary types, evolves during grain growth to be the inverse of the energy anisotropy, that is, small populations are observed for high-energy boundaries and *vice versa*. Thus, comparisons of GBCD against energy should show an inverse correlation in general. Rohrer *et al.* and Holm *et al.* provided such comparisons [5,6] between theoretical grain boundary energies computed with atomistic methods [7] and experimentally measured GBCD in nickel and aluminum. They found that for boundary types that occur with high relative frequency, the expected inverse relationship was indeed present; for infrequent boundary types and high energies, however, the relationship was not apparent, presumably because of poor (experimental) statistics.

The role of coincident site lattice (CSL) relationships [8–10] has been much discussed in the literature for the straightforward reason that the relative simplicity of the geometry involved makes the appealing suggestion that boundaries close to misorientations with a large fraction of coincident points should have low energy, at least in fcc metals. It has been apparent for some time, however, that this is too simple a view, as shown by the work of Hasson and Goux, for example [11,12]. Although the $\Sigma 3$ boundary type consistently shows low energy and high population in all fcc metals [13], even aluminum [14], other low sigma types such as $\Sigma 5$ and $\Sigma 7$, for example, do not show unusual character. It is clear that fcc metals such as nickel exhibit large populations of certain CSL types [15] and the types that occur with high frequencies are the so-called $\Sigma 3^n$ series [16] that are most likely generated via repeated twinning during grain growth. The current state of knowledge suggests that CSL types are only of interest in fcc metals and that grain boundaries in other crystal structures follow the average surface energy model [3].

7.3.2
Mobility

Grain boundary mobility concept is based on treating grain boundary motion as a thermally activated process that involves atom hopping from one side of the boundary to the other [17]. The most comprehensive review of this property is the book by Gottstein and Shvindlerman, *Grain Boundary Migration in Metals* [18]. The basic equation for curvature-driven motion is as follows, where v is the grain boundary migration rate (velocity), M is mobility, γ is the grain boundary energy, γ_{11} is the second derivative of the grain boundary energy in the direction of the first principal curvature, and κ_1 and similarly for the second principal curvature [19]. There are more elegant formulations of the boundary based on the divergence of the capillarity vector [20–22], which is the equivalent of the interface stiffness, but this is the more accessible form for most readers.

$$v = M((\gamma + \gamma_{\theta_1\theta_1})\kappa_1 + (\gamma + \gamma_{\theta_2\theta_2})\kappa_2) \tag{7.1}$$

Gottstein and Shvindlerman review in considerable detail the evidence for the validity of this equation by demonstrating that the migration rate depends linearly on the curvature of the boundary, for example. The presence of the derivative of the grain boundary energy indicates that the variation in energy with boundary normal (inclination) can affect migration rates. Even if the curvature driving force is replaced by a stored energy driving force as in recrystallization, the mobility is known to depend on boundary normal. Since boundary motion is affected by thermal activation, we write $M = M_0 \exp\{-Q/RT\}$ where the activation enthalpy, Q, is of the same order as for self-diffusion. Most mobility measurements have been conducted on aluminum because it is fcc and has a conveniently low melting point. Driving boundaries with either curvature or stored energy shows roughly similar results with $\langle 1\,1\,1 \rangle$ tilt boundaries exhibiting the highest mobilities [18,23,24]. Within the $\langle 1\,1\,1 \rangle$ tilt series, misorientations close to 40° typically exhibit the highest mobilities although the position of the peak appears to vary with measurement temperature. Although there have been a number of reports of theoretical estimates of mobility based on molecular dynamics [25–27], the most complete is by Olmsted *et al.* who calculated a large set of grain boundary mobilities in nickel for the same set of 388 different types that were used in previous computations of energy [28]. Although some broad trends were similar such as twist boundaries showing lower mobilities in general than corresponding tilt series, we lack detailed comparisons of these results with experimental data. Impurities play a major role in boundary migration. For example, the activation energies observed experimentally are much higher than found by molecular dynamics but, of course, the latter are calculated for exactly pure substances. Several studies show how solute levels and solute type both affect boundary mobility for example, [29]. Gottstein and Shvindlerman suggest that solute segregation as a function of boundary type is largely responsible for the variations in mobility by misorientation [18]. Recent work in ceramic systems suggests that distinct interface types or complexions can exist, of which some exhibit substantially higher mobilities than their pure, un-doped counterparts [30]. In conclusion, although there is a significant body of literature, we lack a consensus view on how to predict mobility and the mechanisms that control it. We do know, for example, the broad outlines of how mobility varies such as the transition from low to high mobility with the transition from low-angle to high-angle misorientation [31], which is frequently used in simulation of grain growth and recrystallization.

7.4
Simulation Approaches

7.4.1
Potts Model

Historically speaking, the Q-state Ising model or Potts model was the first to be developed as a general method for simulating microstructural evolution during annealing, including such processes as grain growth, recrystallization, and (late stage) sintering [32–45]. The method has been described in many papers so only a brief recap is given here. Starting in the early 1980s, the Potts model was applied to first grain growth, then grain growth in the presence of particles, followed by recrystallization and eventually sintering. Limited computer power and memory meant that simulation was limited to of order 10^5 points in two dimensions (2D) and expanded to roughly 10^7 points and three dimensions (3D) by the early 1990s, with parallel codes being written in the late 1990s [46]. Variable (anisotropic) grain boundary properties were included in simulations at an early stage but representation of anisotropy was mostly limited to one-parameter relationships based on solely the misorientation angle.

As mentioned earlier, one attractive feature of the Potts model is that it is straightforward to impose anisotropic properties, provided that they are limited to functions of the misorientation. The system energy E in the Potts model is calculated using

$$E = \frac{1}{2}\sum_i^N \sum_j^n \{\gamma(s_i, s_j)(1 - \delta_{s_i,s_j})\} + \sum_i^N H(s_i) \qquad (7.2)$$

where $\gamma(s_i,s_j)$ is the interaction energy and $H(s_i)$ is the contribution to the energy due to an external field. The interaction term (between unlike orientations) refers to grain boundary energy and the second term refers in general to external field which in this context means the volume averaged stored energy of deformation. The first summation, on j, is over the nearest neighbors (this is the distance over which the interactions between various lattice spins are considered to be significant). In the simple cubic three-dimensional lattice the count extends to the 26, 1st, 2nd, and 3rd nearest neighbors. The second summation is over all the lattice sites in the system. The transition probability P is given by

$$P = \begin{cases} \dfrac{\gamma(s_i, s_j)}{\gamma_{max}} \dfrac{\mu(s_i, s_j)}{\mu_{max}} & \Delta E \leq 0 \\ \dfrac{\gamma(s_i, s_j)}{\gamma_{max}} \dfrac{\mu(s_i, s_j)}{\mu_{max}} \exp\left(-\dfrac{\Delta E}{(\gamma_{max})kT}\right) & \Delta E > 0 \end{cases} \quad (7.3)$$

where $\mu(s_i,s_j)$ is the mobility of the boundary between grains s_i and s_j and ΔE is the change in energy associated with the transition. These two equations govern the evolution of the system. Input to the model in the form of microstructure and initial properties, namely stored energy, nucleation conditions, grain boundary mobility, and energy were determined from experimental results. The exact methodology to obtain the input values is described in detail in the following. $H(s_i)$ represents the volume averaged stored energy of deformation and is a function of grain orientation. $\gamma(s_i,s_j)$ represents the grain boundary energy between two neighboring grains and $\mu(s_i,s_j)$ represents the grain boundary mobility. Scaling the energy and mobility values by the maximum in the system is convenient for working with transition probabilities.

The form of these equations makes clear that, for the same driving force, the boundary velocity will be a linear function of the mobility. This can be readily tested with simple verification simulations such as allowing a single, isolated grain with area A to shrink under the influence of surface tension. In either 2D or 3D, the rate of change of area is constant and proportional to both the mobility and, in fact, the energy of the boundary, provided that the lattice temperature is corrected for the energy as noted by the γ_{max} factor in the denominator of the Boltzmann factor. In three dimensions, the inward velocity of an isolated grain is given as follows

$$\vec{v} = M\gamma \frac{2}{R} \quad (7.4)$$

Combining this with $dA/dt = 2\pi R \, dR/dt$:

$$\frac{dA}{dt} = -2\pi R \frac{dR}{dt} = -4\pi M\gamma \quad (7.5)$$

Verification of the correct operation of Potts algorithms should include demonstration that constant shrinkage rates are obtained for isolated grains in 2D and 3D that are in proportion to the assigned values of energy and/or mobility.

Kandel and Domany analyzed the Potts model and demonstrated that its kinetics conforms to motion by curvature [47]. While the response of boundaries to curvature driving forces is linear, the same is not true for stored energy (i.e., the term H in the Hamiltonian). If the stored energy difference that enters the ΔE term in the equation describing the reorientation probability is large enough (and the change decreases the total energy), then magnitude of ΔE does not affect the outcome. This means that, above a certain stored energy level in a recrystallization simulation, the boundary velocity is insensitive to that level [48]. Note, however, that the presence of the term in mobility as a pre-exponential in the switching probability means that the mobility anisotropy applies at all levels. If the magnitude of H is varied in a range that is comparable to the grain boundary energy, γ, however, then the response is linear, as discussed by Rollett and Raabe [49] and more recently by Zhang et al. [50].

Much has been made of the difficulty of associating a physical time with a simulation time in the Monte Carlo model [48], and, by implication, other models to varying degrees. This is an issue for any simulation of evolution but the problem is more apparent than real. For curvature-driven grain growth, at least, there are basic verification tests, such as the single shrinking grain simulation mentioned earlier, that allow rates to be compared between experiment and simulation. Such comparisons represent validation in some sense. A more mesoscopic approach is to compare coarsening rates between simulation and experiment. Such comparisons are approximate at best, however, unless the anisotropy of the boundaries is known and included in the simulation. In the following equation that describes the evolution kinetics of a polycrystal, self-similarity dictates that the exponent, n, should be unity [51]. Any algorithm for grain growth in which the boundary properties are isotropic (or invariant with time) should obey this relationship. The proportionality factor, k, can be shown (with varying degrees of sophistication) to be $k = \alpha M \gamma$, where $\alpha \sim 0.2$ [51,52].

$$R^2 - R_0^2 = kt^n \tag{7.6}$$

Brahme et al. applied the Potts model to simulation of texture development during recrystallization of warm rolled aluminum 1050 [53]. The texture characteristic of plane strain deformation with components on the beta fiber is replaced by an annealing texture dominated by the cube component [54]. They found that there are many factors that influence the texture development in addition to just the anisotropy of the grain boundary properties. One important aspect was that the initial, as-deformed microstructure had highly elongated grain shapes; these were instantiated using a tool called Microstructure Builder [55]. The final microstructure was in form of a cubical lattice of size $500 \times 200 \times 100$. Each of the lattice points was labeled by the grain number, which serves as the spin number, s, in the Monte Carlo simulations. All the Monte Carlo simulations used a lattice temperature, kT, equal to 0.9, which was judged to be high enough to minimize the effects of lattice anisotropy for these simulations. It was also found that the recrystallization nuclei had orientations that differed from the as-deformed texture, which is a phenomenon known as oriented nucleation, and their locations relative to the various orientations present in the deformed microstructure were nonrandom [54]; this meant that the placement of nuclei had to be controlled in the Potts model [53]. Finally, it was also found that the stored energy, H, varied by texture component; thus, this value was made to depend on the orientation just as described earlier.

Focusing for the purposes of this paper on the anisotropy of grain boundary properties, the anisotropy concerns energy and mobility. In the Potts model, only the sites adjacent to an interface of two grains are allowed to change state (flip spin). Hence, only the grain boundary energy and mobility are important.

Energy anisotropy. The energy anisotropy is characterized by the misorientation between the two neighboring grains. We divide the boundaries into low-angle boundaries (disorientation $\theta \leq 15°$) and high-angle boundaries ($\theta > 15°$). The energy of grain boundaries in the low-angle regime is expressed thus

$$\gamma_0 \theta^*(1 - \ln \theta^*) \tag{7.7}$$

where θ^* is $\theta/15°$ and γ_0 is the normalizing factor. For the high-angle boundaries with $\theta > 15°$ the energy associated is constant at γ_0. The only exceptions are certain "special" boundaries. In this study, the three-parameter misorientation was used although this is a more complete parameterization than the more typical (one-parameter) misorientation angle used in much of the literature. The energy and mobility for each possible boundary in the system is computed separately and stored in a look-up table at the beginning of the simulation. Figure 7.1a shows the variation of grain boundary energy as a function of the misorientation angle (θ) [53]. The value of energy increases as the misorientation angle increases up to the transition angle value ($\sim 15°$). The cusps in the energy value shown in the figure are associated with only special boundaries and not all boundaries having those misorientation angles. The special boundaries considered here are $\Sigma 3$, $\Sigma 7$, $\Sigma 13b$, and $\Sigma 19b$. These CSL boundaries have been reported in both experimental work [56] and MD simulations [57,58] to have lower energies.

Upmanyu et al. [57] addressed the influence of boundary anisotropy on microstructural evolution. They confined the anisotropy in boundary properties to variations with misorientation angle only but used both a phase field and a Potts model to compare the two algorithms. When only the mobility was made anisotropic and the initial structure was random, no effect on boundary population was observed. When the energy was made anisotropic, with sharp cusps corresponding to special boundary types (i.e., mimicking the effect of low energy CSL boundaries), the populations of those types increased with time. This was ascribed to the force balance at triple junctions allowing high energy boundaries to shorten and low energy boundaries to lengthen. This reciprocal relationship between boundary frequency (defined in terms of either area or number fractions) has been consistently observed for normal grain growth, for example, [59].

Rollett also used the Potts model to investigate the effect of energy and mobility anisotropy [60] during grain growth. By starting with nonrandom crystallographic textures typical of rolled fcc metals, mobility anisotropy was found to lead to measurable texture changes. This demonstrated by the relative importance of anisotropy in energy and mobility depends on the initial conditions.

Mobility anisotropy. As for the energy function, the mobility anisotropy is also characterized by the grain boundary misorientation. Unlike the grain boundary energy which has cusps, or valleys for the special boundary configurations, the mobility function has a low mobility for low-angle grain boundaries (LAGBs) with a sharp transition [61] to high-angle grain boundaries (HAGBs). Rising above the general level of HAGB mobility, there are peaks corresponding to the highly mobile boundary types at specific misorientations. Figure 7.1b shows the plot of mobility as a function of the misorientation angle.

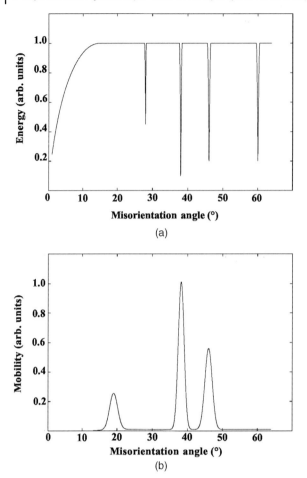

Figure 7.1 Plots of idealized functions to represent anisotropic grain boundary properties. (a) Grain boundary energy and (b) mobility as a function of misorientation angle. The cusps and peaks in the energy and mobility functions, respectively, are functions of the misorientation axis (not shown) in addition to the angle (horizontal axis) [53].

The peaks in plot correspond to Σ37c, Σ7, and Σ19b boundaries. These CSL values have been noted to have higher mobilities in both experimental and atomistic simulation results [23,24,26,61–64]. Note all these are in the ⟨1 1 1⟩ misorientation axis series. Thus, the peaks shown in Figure 7.1b are unique points in the 3D misorientation space.

A general high-angle boundary has high mobility and also high energy other than the special CSL boundaries. The energy anisotropy has cusps for the special boundaries while the mobility anisotropy has peaks. The cube texture component ({0 0 1}⟨1 0 0⟩) is related to S-orientation ({1 2 3}⟨6 3 4⟩) by a misorientation of 40° about the ⟨1 1 1⟩ axis, which happens to be close to the Σ7 boundary type. The

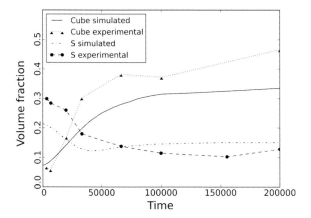

Figure 7.2 Comparison of the evolution of the two dominant texture components between experiment and simulation, namely the cube component (annealing texture) and S component (deformation texture). This shows results from including all the factors found to be important in influencing texture development, namely boundary anisotropy, stored energy, and placement and orientation of nuclei [53].

mobility curve has a peak at the $\Sigma 7$ position, Figure 7.1b, corresponding to common experimental observation that this boundary is highly mobile. Thus, the anisotropy in mobility can help the growth of the cube component when recrystallizing grains belonging to the cube component impinge on unrecrystallized S component. On the other hand the $\Sigma 7$ position in the energy curve has a cusp implying that the boundary has lower energy. Hence, if the effect of energy anisotropy is included and if the cube and S nuclei impinge, then the boundary will be sessile and the cube component can not grow at the expense of S. Hence, it appears that overall growth of S component is aided by the energy anisotropy.

The net result of incorporating this anisotropy is summarized in the texture evolution, Figure 7.2. This demonstrates that good agreement can be obtained provided that a complete range of factors is included in the simulation beyond simply the grain boundary anisotropy that is the focus of this review [53].

7.4.2
Cellular Automata

A very general approach to microstructural evolution is to use cellular automata (CA). Obviously this (discrete) mathematical model is extremely general and particular rule sets must be used in order to apply the method. The basic approach is to instantiate a microstructure in similar fashion to the Potts model and then advance in time with a suitable rule. The original application was to solidification and then recrystallization [65], which is an equivalent problem from a numerical perspective. Hesselbarth and Göbel used a 2D CA model to explore different numerical aspects such as the neighborhood considered to evaluate the change of a given cell, which affects the growth morphology of growing grains. They also explored how varying the way in which impingement occurs between growing grains affects a Johnson–Mehl–Avrami–Kolmogorov (JMAK) analysis, finding that any deviation from strictly random nucleation and uniform growth results in JMAK exponents that differ from the theoretical values.

Raabe described a detailed 3D CA model in which the anisotropy of grain boundary mobility was accounted for with a probabilistic switching algorithm [66–68]. The maximum probability is scaled to the maximum mobility of any boundary in the system. The driving force depends on the change in stored energy available for changing the orientation from the current value to that of a near neighbor cell or voxel. For any given transition or switching event, the probability is computed and an actual switch is decided by a standard Monte Carlo algorithm. A linear relationship between boundary velocity and mobility and/or driving force was demonstrated. For site-saturated nucleation of recrystallization, the JMAK exponents were within a few percent of the nominal value of 3. Variable mobility was demonstrated to affect growth patterns, that is, recrystallizing grains with a high mobility between them and the deformed matrix grew more rapidly. Detailed descriptions of mobility anisotropy based on experimental or theoretical data were not shown.

7.4.3
Phase Field

The phase-field method started with Cahn and Allen's [69] modeling of coarsening of ordered materials via the motion of antiphase boundaries. Papers on the application of the method to solidification problems and especially dendrite tip shape modeling followed [70]. An important advance was the development of asymptotics, which is mathematical analysis that shows that behaviors such as interface velocity reduce to the sharp interface values under certain limiting conditions [71,72]. The method was soon extended to grain growth by Fan and Chen [73,74]. Lusk extended phase field to recrystallization shortly thereafter [75] and a number of authors have used the method to model this phenomenon.

The word "field" refers to the use of fields to represent phases and grains in materials. In the case of solid state evolution, it is most convenient to use one field per orientation, which implicitly restricts representation of a polycrystalline solid to grains with no internal structure. Microstructure is represented by discretization on a regular grid for convenience in computing field gradients, that is, each point can, in principle, have a finite value for each field, φ_i. Some algorithms enforce the constraint that the fields sum to unity at every lattice point, $\Sigma_i \varphi_i = 1$.

$$\{\varphi\} = \{\varphi_1, \varphi_2, \ldots, \varphi_i, \ldots, \varphi_N\} \tag{7.8}$$

Note that the presence of stored energy can be represented by a volumetric energy. Each field varies between unity inside the grain and zero in the far field; it varies smoothly across each boundary with neighboring grains. Each field has a free-energy function associated with it that typically has a minimum at each extreme value. There is a second energy term associated with gradients in the fields that effectively penalizes the presence of any field gradient. The original approach to phase-field modeling was directed towards solidification, for which only two fields are required and then adapted to grain growth as noted above. This approach involves energy functions of the fields that ensure a local minimum in energy in each grain (within which the value of the associated field ~1) together with a term that provides an energy penalty for gradients in the fields [76].

Notwithstanding the earlier work on asymptotic analysis, the diffuseness of the representation of grain boundaries calls into question what happens at small grain sizes when the field gradients can interact across a grain. McKenna et al. [77] analyzed this effect for 2D phase simulations in terms of standard grain growth theory as tested by shrinking isolated grains (see above discussion). They found that approximately six grid points across each grain boundary are required in order for the results to be verified against von Neumann–Mullins theory [78,79]; in addition, the interface thickness needs to be a factor of 2 smaller than the smallest radius of curvature of a boundary.

This classical approach has been further developed for anisotropic grain boundary properties by Moelans, for example [80–82]. The field energy function contains both differences in neighboring fields at a boundary and the product of the two fields. By introducing terms that modify both the field energy function and the field gradient function, it was possible to assign specific grain boundary energy to each boundary defined by pair of fields. The algorithm was used to simulate test case microstructures in 2D and measure dihedral angles, which suggested that anisotropy corresponding to a range of angles 102°–138° was feasible. Polycrystal simulations yielded a similar inverse relationship between grain boundary energy and populations as previously described. Rather than describing this approach in detail, however, we describe an alternative approach to phase field modeling of grain growth with anisotropy that is based on interface fields and builds on the work of Steinbach et al. [83]. Use of the interface fields is considered less desirable from the perspective of obtaining asymptotic solutions that correspond to sharp interface theory; nonetheless, it has some practical advantages especially for implementation of grain boundary anisotropy. Full details of the approach are provided elsewhere [84].

The total free energy, F, of the system is the sum of the potential energy, f_{pot}, and gradient energy, f_{gr}, terms over the simulation domain. Where fields must change values across a boundary, the gradient penalty forces the fields to change at a finite rate, leading to diffuse boundaries [85]. The terms ε and w determine the energy of boundaries in the system, which we discuss in more detail in the following:

$$F(\{\phi\}) = \int_{\Omega} (f_{pot} + f_{gr}) dV \tag{7.9}$$

$$f_{gr} = \sum_{\gamma=1}^{N} \sum_{\delta=\gamma+1}^{N} \frac{\varepsilon_{\gamma\delta}}{2} \nabla\phi_{\gamma} \nabla\phi_{\delta} \tag{7.10}$$

$$f_{pot} = \sum_{\gamma=1}^{N} \sum_{\delta=\gamma+1}^{N} \omega_{\gamma\delta} |\phi_{\gamma}||\phi_{\delta}| \tag{7.11}$$

The system is allowed to evolve via the time-dependent Ginzburg–Landau equation:

$$\frac{\partial \phi_i}{\partial t} = -L \frac{\partial F}{\partial \phi_i} \quad (7.12)$$

We now introduce an interface field, $\psi_{\alpha\beta}$, which is simply the difference between the values of the two fields with finite values in a grain boundary, $\psi_{\alpha\beta} = \phi_\alpha - \phi_\beta$. This leads to the following expression for the time evolution of the interface fields.

$$\psi_{\alpha\beta}(t+\Delta t) - \psi_{\alpha\beta}(t)$$
$$= m_{\alpha\beta}\left[\sum_{\gamma \neq \alpha}\left(-\frac{\varepsilon_{\alpha\gamma}}{2}\nabla^2\phi_\gamma - \omega_{\alpha\gamma}\phi_\gamma\right) - \sum_{\gamma \neq \beta}\left(-\frac{\varepsilon_{\beta\gamma}}{2}\nabla^2\phi_\gamma - \omega_{\beta\gamma}\phi_\gamma\right)\right]\Delta t$$

(7.13)

The coefficients ε and w are related to the interface width, λ, and the grain boundary energy, γ, as $\lambda = \sqrt{(\varepsilon/w)}$ and $\gamma = \sqrt{(\varepsilon w)}$. Making the grain boundary energy anisotropic but maintaining a constant boundary width therefore prescribes the choice of the functions for these two coefficients. Expressed as a finite difference and including the coefficients from the two energy functionals we obtain this form for the evolution of the fields themselves, based on the interface fields:

$$\phi_\alpha(t+\Delta t) - \phi_\alpha(t) = \frac{\sum_{\beta \neq \alpha}(\psi_{\alpha\beta}(t+\Delta t) - \psi_{\alpha\beta}(t))}{N} \quad (7.14)$$

An important aspect of this approach to interface phase-field scheme is that the computation of boundary normal must be accurate even where they are normally poorly defined or exhibit jumps such as at triple lines. Therefore, it is helpful to average gradients in multiple directions using methods such as the weighted essentially nonoscillatory (WENO) approach [86]. To illustrate the functioning of this model, 2D phase field simulations were run on a 1500^2 grid with a particular form of energy anisotropy. In three dimensions, the boundary energy γ can be expressed as a cubic-symmetric function dependent on the inclination angles (θ, φ) relative to crystal axes of the boundary normal [n_x, n_y, n_z]:

$$\gamma(\theta, \varphi) = \gamma_0[1 + \delta(\cos^1\theta + \sin^1\theta\{1 - 2\cos^2\varphi\sin^2\varphi\})] \quad (7.15)$$

Figure 7.3 shows the contrast between a 2D simulation with isotropic boundary properties versus a similar simulation in which anisotropic properties were used and the force balance was enforced at triple points [84]. The constants in Eq. (7.12) were chosen such that the boundary with was eight grid points and the initial grain size was four times the boundary width, that is, 32 grid points. Uniform (isotropic) mobility was used and the initial microstructure was evolved for $N = 30\,000$ time steps. The result for isotropic case, Figure 7.4a, shows the expected 120° dihedral angles. For the anisotropic case, the boundary energy γ had the form shown in the Eq. (7.14) above, with $\delta = 0.3$ and rescaled so that $0.5 \leq \gamma \leq 1.5$. In this case, Figure 7.3b, the dihedral angles deviate from 120° and the boundary inclinations are clearly not random.

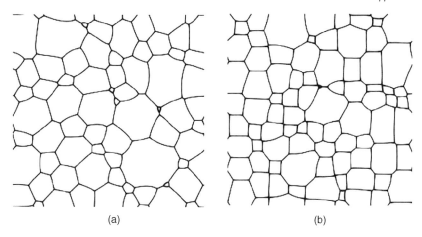

Figure 7.3 (a) Snapshot for 2D simulation using the interface phase-field method and isotropic boundary properties. (b) Snapshot for 2D simulation using the interface phase-field method and anisotropic boundary properties with enforcement of equilibrium at triple points.

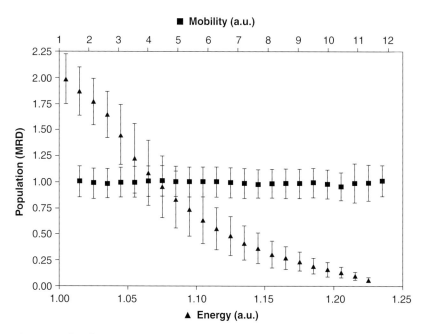

Figure 7.4 Plot of the population (area) of grain boundaries, measured in multiples of a random distribution (MRD) against the grain boundary energy (triangles), lower scale or against grain boundary mobility, and upper scale (squares). Where anisotropic energy was used, a reciprocal relationship between population and energy developed and stabilized after about one-half of the initial population of grains had been eliminated. Where anisotropic mobility was imposed on the model, no significant variation in population was observed. From Gruber et al. [4].

Chen et al. modeled grain growth in 3D [87] assuming uniform boundary energies and mobilities using a standard phase-field approach. Sekerka et al. considered anisotropy in the kinetic prefactor in the evolution equations [88], and observed that the morphology of the steady-state kinetic shapes depended both on the width and depth of the cusp of the particular anisotropy function used. Instead, when an inclination-based anisotropy in the gradient energy coefficient is considered, the variation of the free-energy functional yields additional terms, such as higher derivatives of the gradient energy prefactor with inclination [89]. The interface thickness in the aforementioned case scales with interfacial energy, hence, to maintain a constant interface width irrespective of the boundary inclination, both the energy coefficient and the depth of the potential function should have a similar anisotropic variation with inclination [90].

7.4.4
Cusps in Grain Boundary Energy

Since higher derivatives of the anisotropic function with inclination (such as γ_θ) can be discontinuous in the vicinity of a cusp, the method can become ill-posed and unstable, and additional treatment of the energy functional is essential for these inclinations. Karma et al. used a modified version of the following inclination-based energy anisotropy $\gamma(\theta) = \gamma_0(1 + \delta(|\sin\theta| + |\cos\theta|))$ where they approximated the sharp cusps with rounded ones to simulate growth of dendritic tips [91]. Amberg used a similar regularization [92] to model evolution of Widmanstatten morphology. Torabi et al. used the Willmore regularization method, where the square of the mean curvature is added to the energy term to avoid instability near cusps and to smooth corners [93]. Voorhees used the energy convexification scheme [94] which ensured that the interfacial stiffness $\gamma + \gamma_{\theta\theta}$ is positive for all inclinations, to model the shape evolution of anisotropic crystals using a Cahn–Hilliard formulation (and conserved phase fields). Wollants et al. simulated polycrystalline grain growth with arbitrary misorientation and inclination dependence in two dimensions, while maintaining a constant interface width [95]. Regularization of the energy function was necessary for the nonconvex regions of the interfacial stiffness (where $\gamma + \gamma_{\theta\theta} < 0$) when the grain boundary energy is strongly anisotropic with inclination.

7.4.5
Level Set

Smereka's group has developed level set methods for simulation of grain growth [96]. Bernacki has pioneered the use of the level set approach to model recrystallization [97]. This approach has not developed enough, however, to include anisotropic properties of boundaries to any significant degree. In many respects, anisotropy should be more straightforward than for the phase field because the availability of functions to describe interface position means that normal interface is readily available. Also the properties are directly imposed on the constitutive descriptions, for example, for boundary motion. However, ensuring local equilibrium at triple points or lines likely requires the same sort of effort to extend fields past the triple line positions are remarked on in connection with phase-field models.

7.4.6
Vertex

One of the earliest works to describe a "vertex model" was by Frost who outlined the procedure for instantiating a grain boundary network as a set of edges (in 2D), each one connecting two nodes [98]. Most nodes are twofold, that is, connected to two edges, and a minority are threefold, that is, connected to three edges at a triple junction. Twofold nodes are moved by an implicit computation of curvature due to Sullivan et al. [99] in their discrete modeling of solidification. Triple junction nodes were adjusted after each time step to achieve dihedral angles of 120° using parabolic arcs fitted to each boundary out to the second node away from the triple node. Switching events in the network allowed for grain disappearance and neighbor exchange but the details were not given. Computations of grain coarsening with this isotropic and periodic model gave an exponent very close to 0.5, as expected theoretically [51]. Different starting configurations all converged on similar grain size distributions.

In the same period, Kawasaki, Nagai, and Nakashima developed similar 2D vertex models [100–103]. As with the Frost approach, the approach assumed isotropic properties and the emphasis was on analyzing for topology and kinetics of coarsening. In this case, the switching events were explained as involving either exchange of nearest neighbors for groups of four grains, or the disappearance of small (generally three sided) grains. Fuchizaki and co-workers later extended the vertex models to 3D [104–106]. The model is based on two basic equations, the first of which describes the total energy, V, of the system integrating over the grain boundary area, a, and grain boundary energy, γ, in the polycrystal system:

$$V(\{\vec{r}\}) = \int_{GB} \gamma(\vec{n}_{GB}(a)) da \tag{7.16}$$

The second equation describes the dissipation, R, in the system associated with the motion (velocity), v, of the grain boundaries, where **r** describes the position of the boundaries and m the mobility.

$$R(\{\vec{r}\}, \{\vec{v}\}) = \frac{1}{2} \int_{GB} \frac{v(a)^2}{m_{GB}(a)} da \tag{7.17}$$

Using a Lagrange approach, the evolution of the system of nodes (vertices) and area elements (generally triangles in 3D surface meshes) can be described by the following:

$$\frac{\partial R(\{\vec{r}^A\}, \{\vec{v}^A\})}{\partial \vec{v}^B} + \frac{\partial V(\{\vec{r}^A\}, \{\vec{v}^A\})}{\partial \vec{r}^B} = 0, \quad A, B = 1 \ldots N \tag{7.18}$$

Further detail is given in the references about the numerical implementation of the algorithm with discrete representations of the grain boundary network using, in general, triangular surface meshes. The main point to note here is that the formulation permits grain boundary properties to be specified over all five macroscopic degrees of freedom, that is, including the two parameters associated with the boundary normal.

Shortly after this, Weygand developed 2D and 3D versions of the vertex model that allowed for the incorporation of inert pinning particles [107–112]. Maurice also used Fuchizaki's model as the starting point but incorporated anisotropy based on misorientation angle, that is, a one-parameter variation with the Read–Shockley model for energy and a sharp increase in mobility between low- and high-angle grain boundaries [113,114]. This was used to simulate the coarsening behavior of subgrain networks where strongly abnormal growth was identified as expected for systems in which a minority of grains have high angle and therefore mobile perimeters [115–117].

One reason to use such a model, these days typed as a sharp-interface model, is that anisotropic properties can be imposed on boundaries directly because all five macroscopic parameters, that is, lattice misorientation and normal, are known unambiguously. The theoretical dihedral angles can always be computed and triple junction adjustments made accordingly. However, if significant anisotropy is imposed then it is possible for one boundary to cross over another if sufficient care is not given to the numerical implementation. Also, switching events are far more complex in 3D than in 2D and many different possible rearrangements must be allowed for [107,114,118,119]. The most complete description of a 3D vertex model with anisotropy was provided by Syha and Weygand [120]. They provided detailed descriptions of the various mesh re-ordering steps required to deal with switching events and grain disappearance. They found excellent agreement between the simulated rate of change of individual grain volumes and the Macpherson–Srolovitz theory [121]. They also analyzed dihedral angles at triple lines for simulations performed with and without anisotropy. Since local equilibrium was not enforced along triple lines, the FWHM was of order $10°$ in the isotropic case and $\sim 35°$ in the anisotropic case. The anisotropy in grain boundary energy was mild with only a 10% variation as a function of deviation of either surface (adjacent to the boundary) from $\langle 1\,0\,0\rangle$; the latter surface was found to be a (minimum energy) cusp by Sano et al. [122].

7.4.7
Moving Finite Element

In the gradient-weighted moving finite element (GWMFE) method, which is similar to the vertex method, interfaces are represented as parameterized piecewise linear surfaces [123]:

$$x(s_1, s_2) = \sum_{\text{nodes} j} \alpha_j(s_1, s_2) x_j \qquad (7.19)$$

Here, (s_1, s_2) is the surface parameterization, the sum is over the N interface nodes, $\alpha_j(s_1,s_2)$ is the piecewise linear basis function ("hat function") that is unity at node j and zero at all other interface nodes, and $\mathbf{x}_j = (x_j^1, x_j^2, x_j^3)$ is the vector position of node j.

This leads to the following, where

$$\dot{x}(s_1, s_2) = \sum_j a_j(s_1, s_2) \dot{x}_j \tag{7.20}$$

is the velocity of the surface at the point $\mathbf{x}(s_1, s_2)$ (based upon linear interpolation of node velocities) and

$$v_n = \dot{x}(s_1, s_2) \cdot \hat{n} \; (\hat{n} \text{ is local surface normal}) \tag{7.21}$$

So

$$v_n = \sum_j (\hat{n} a_j) \cdot \dot{x}_j \tag{7.22}$$

In effect, there are $3N$ basis functions for v_n that are equal to $n_k a_j$, where $\hat{n} = (n_1, n_2, n_3)$. These basis functions are discontinuous piecewise linear, since the n_k are piecewise constant.

The GWMFE method seeks to minimize

$$\int (v_n - \mu\sigma K)^2 dS \tag{7.23}$$

over all possible values for the derivatives $\dot{x}_i$ (where the integral is over the surface area of the interfaces). Thus, one obtains

$$\begin{aligned} 0 &= \frac{1}{2} \frac{\partial}{\partial \dot{x}_i^k} \int (v_n - \mu\sigma K)^2 dS \quad 1 \leq k \leq 3 \quad 1 \leq i \leq N \\ &= \int (v_n - \mu\sigma K) n_k a_i dS \end{aligned} \tag{7.24}$$

Using Eq. (7.22), a system of $3N$ ODEs is obtained

$$\left[\int \hat{n}\hat{n}^T a_i a_j dS \right] \dot{x}_j = \int \mu\sigma K \hat{n} a_i dS \tag{7.25}$$

or

$$C(x)\dot{x} = g(x) \tag{7.26}$$

where $\mathbf{x} = (x_1^1, x_1^2, x_1^3, x_2^1, \ldots x_N^3)^T = (\mathbf{x}_1, \mathbf{x}_2, \ldots \mathbf{x}_N)^T$ is the 3N-vector containing the x, y, and z coordinates of all $\mathbf{N}$ interface nodes, $\mathbf{C}(x)$ is the matrix of inner products of basis functions, and $g(x)$ is the right-hand side of inner products involving surface curvature. Since $\hat{n}\hat{n}^T$ is a 3×3 matrix, it is clear that $C(x)$ has a 3×3 block structure. As explained in [123], the inner products involving curvature can equivalently be viewed as arising from a surface tension of magnitude μ on each of the planar triangular cells of our discretization of the interfaces. This enables the Herring condition to be satisfied at triple lines.

The system of ODEs, Eq. (7.25), is solved with an implicit second-order backward difference variable time-step ODE solver [124]. A generalized minimal residual (GMRES) iteration [125] is used with a block-diagonal preconditioner to solve the linear equations arising from the Newton's method. More details appear in [123].

An early application of the method was to run simulations of grain growth in 2D with comparisons against experimental data for thin foils of aluminum that had been processed to give a microstructure with essentially all low-angle boundaries (in a strong $\langle 1\,0\,0 \rangle$ fiber texture) [126]. The main result was that good agreement between the experiments and simulations was only obtained by including the strong anisotropy of boundary mobility associated with the transition from low to high-angle misorientation.

This method was also used to examine the relative effects of anisotropic interfacial energy and mobility on the grain boundary character distribution in materials undergoing normal grain growth in 3D [4]. As described earlier, nodal velocities are calculated by minimizing a functional that depends on the local geometry of the mesh and the (anisotropic) properties of the grain boundaries. Grain boundary properties are calculated based on the grain boundary type. The grain boundary energy $\gamma(\Delta g, n)$ and mobility $M(\Delta g, n)$ functions we use are defined by the two surfaces adjacent to each GB, that is, the two surfaces bounding the grains on either side of the interface [127]. Taking n_1 to be the interface normal pointing into grain one and indexed in the crystal reference frame of that grain, and n_2 to be the interface normal pointing into grain two and indexed in the crystal reference frame of that grain, the energy and mobility are assigned thus:

$$\gamma = (E(n_1) + E(n_2))/2 \tag{7.27}$$

$$M = (\mu(n_1) + \mu(n_2))/2 \tag{7.28}$$

where the functions $E(n)$ and $\mu(n)$ are chosen as

$$E = \alpha \left[\sum_{i=1,3} (|n_i| - 1/\sqrt{3})^2 \right] + 1 \tag{7.29}$$

$$\mu = \beta \left[\sum_{i=1,3} \left(|n_i| - 1/\sqrt{3}\right)^2 \right] + 1 \tag{7.30}$$

where α and β are positive constants. Minima for either functional then occur with normal vectors of $\langle 1\,1\,1 \rangle$ type and maxima with normal vectors of $\langle 1\,0\,0 \rangle$ type. Note that Eqs. (7.27) and (7.28) imply cubic crystal symmetry. Using cubic symmetry minimizes the number of grain boundaries necessary to produce a statistically significant data set and simplifies the effort required. The particular functional form was inspired by investigations of GB properties in aluminum [14] and was also used in recent use of the vertex model [120]. Note that although the symbol for misorientation, Δg, was used earlier, the misorientation enters only indirectly since in this work the GB properties were controlled entirely by the two normals. One might write under this circumstance that functions of the form $\gamma(n_1, n_2)$ and mobility $M(n_1, n_2)$ were used. Nevertheless it is useful to examine particular misorientation types for comparison with previous work [14] and the same discretization scheme was used with an approximately $10°$ bin size in all five dimensions (3 in misorientation and 2 in boundary normal).

The parameter α controls the anisotropy of the GB energy such that isotropy corresponds to α = 0 and for α = 0.2957, the ratio of the maximum to minimum GB energies is 1.25. Similarly for mobility β = 0 and for β = 13·6, the ratio of the maximum to minimum GB energies is 12.5; full details are given in [4]. The GWMFE method requires a meshed representation of the microstructure and in this case a tetrahedral mesh was used. To simplify the mesh generation procedure, a unit cube was filled with tetrahedral, grains assigned via random choice of 5000 grain centers followed by grouping of tetrahedra with their nearest grain center. Normal, isotropic grain growth was used to coarsen the microstructure down to about 2500 grains to allow the grain boundaries to adopt smooth shapes according to local equilibrium along triple lines, following which anisotropic grain growth simulations were performed. For each combination of grain boundary properties, 20 simulations were run, which provided data from about 20 000 grains or 120 000 grain boundaries. The distributions of grain boundary normals were inversely related to the energy function used. Figure 7.4 from [4] shows a plot of population against either variable energy (lower scale, triangles) or mobility (upper scale, squares). Since full three-parameter orientations were used, various different boundary types could give rise to different energies or mobilities; the standard deviation in population is shown as an error bar. The lack of dependence of populations on mobility variations is evident. The dependence observed for anisotropic energy is also evident and remains unchanged if mobility anisotropy is added to the model. As noted earlier, it is clear that the development of nonuniform (anisotropic) grain boundary character distributions depends on grain boundary energy and the equilibrium at triple lines.

7.4.8
Particle Pinning of Boundaries

An important application of grain growth simulation is to the problem of particle pinning. Pinning occurs via the subtraction of grain boundary area as it moves into contact with a particle and, more importantly, the re-creation of area as a boundary pulls away from a particle. Harare *et al.* [128] investigated and compared different methods for simulating the interaction between particles and moving boundaries such as the Potts model, phase-field and finite element. Fortunately for the field, the various methods yielded largely similar results. Couturier *et al.* [129] performed detailed of interactions between a single moving boundary and an array of particles and determined the prefactor in the Smith–Zener equation [130]. Unfortunately there has been very little attention paid to the effect of variations in grain boundary properties and so we leave this topic at this point. The more important effect is in any case that of coherency of particles with their matrix; if a high-angle boundary passes over such a particle, the particle-matrix interface must lose its coherency, which should enhance the drag effect because a low energy interface is replaced by a high energy one. Nevertheless, Ringer *et al.* found that small coherent particles in a steel remained coherent after grain growth, suggesting that the particles rotated back into a coherent relationship during (or after) passage of a grain boundary [131].

7.5
Summary

The role of anisotropy in grain boundary properties has been reviewed with a particular emphasis on implementation in methods for simulation of microstructural evolution. Image-based models with implicit descriptions of interfaces such as the Potts model can accommodate anisotropy that is a function of the misorientation parameters only. Models with explicit descriptions of interfaces such as finite element or vertex models, along with diffuse interface models such as level set or phase field, can accommodate anisotropy that is a function of all five macroscopic degrees of freedom. For grain growth in the absence of preferred orientation, energy anisotropy dominates via its control of triple junction equilibrium and gives rise to boundary populations that are inversely related to the energy function whereas mobility plays no role. In recrystallization, however, mobility tends to play a more dominant role because boundary motion is dominated by the stored energy driving force.

References

1 Read, W.T. and Shockley, W. (1950) Dislocation models of crystal grain boundaries. *Physical Review,* **78,** 275–289.

2 Gjostein, N.A. and Rhines, F.N. (1959) Absolute interfacial energies of [001] tilt and twist grain boundaries in copper. *Acta Metallurgica,* **7,** 319–330.

3 Rohrer, G.S. (2011) Grain boundary energy anisotropy: A review. *Journal of Materials Science,* **46,** 5881–5895.

4 Gruber, J., George, D.C., Kuprat, A.P., Rohrer, G.S., and Rollett, A.D. (2005) Effect of anisotropic grain boundary properties on grain boundary plane distributions during grain growth. *Scripta Materialia,* **53,** 351–355.

5 Holm, E.A., Rohrer, G.S., Foiles, S.M., Rollett, A.D., Miller, H.M., and Olmsted, D.L. (2011) Validating computed grain boundary energies in fcc metals using the grain boundary character distribution. *Acta Materialia,* **59,** 5250–5256.

6 Rohrer, G.S., Holm, E.A., Rollett, A.D., Foiles, S.M., Li, J., and Olmsted, D.L. (2010) Comparing calculated and measured grain boundary energies in nickel. *Acta Materialia,* **58,** 5063–5069.

7 Olmsted, D.L., Foiles, S.M., and Holm, E. A. (2009) Survey of computed grain boundary properties in face-centered cubic metals: I. Grain boundary energy. *Acta Materialia,* **57,** 3694–3703.

8 Kronberg, M.L. and Wilson, F.H. (1949) Secondary recrystallization in copper. *Transactions of the Metallurgical Society of AIME,* **185,** 501–514.

9 Brandon, D.G. (1966) The structure of high-angle grain boundaries. *Acta Metallurgica,* **14,** 1479–1484.

10 Hardouin Duparc, O.B.M. (2011) A review of some elements in the history of grain boundaries, centered on Georges Friedel, the coincident site lattice and the twin index. *Journal of Materials Science,* **46,** 4116–4134.

11 Hasson, G.C. and Goux, C. (1971) Interfacial energies of tilt boundaries in aluminum. Experimental and theoretical determination. *Scripta Metallurgica,* **5,** 889–894.

12 Hasson, G.C., Guillot, J.B., Baroux, B., and Goux, C. (1970) Structure and energy of grain boundaries: Application to symmetrical tilt boundaries around [1 0 0] in aluminum and copper. *Physica Status Solidi,* **2,** 551–558.

13 Li, J., Dillon, S.J., and Rohrer, G.S. (2009) Relative grain boundary area and energy

distributions in nickel. *Acta Materialia*, **57**, 4304–4311.

14 Saylor, D.M., El Dasher, B.S., Rollett, A.D., and Rohrer, G.S. (2004) Distribution of grain boundaries in aluminum as a function of five macroscopic parameters. *Acta Materialia*, **52**, 3649–3655.

15 Hefferan, C.M., Li, S.F., Lind, J., Lienert, U., Rollett, A.D., Wynblatt, P., and Suter, R.M. (2009) Statistics of high purity nickel microstructure from high energy X-ray diffraction microscopy. *Computers Materials and Continua*, **14**, 209–219.

16 Hefferan, C.M. (2012) Measurement of annealing phenomena in high purity metals with near-field high energy X-ray diffraction microscopy, *PhD thesis*, Carnegie Mellon University, Pittsburgh.

17 Burke, J.E. and Turnbull, D. (1952) Recrystallization and grain growth. *Progress in Metal Physics*, **3**, 220–292.

18 Gottstein, G. and Shvindlerman, L.S. (1999) *Grain Boundary Migration in Metals*, CRC Press, Boca Raton, FL.

19 Herring, C. (1951) *The Physics of Powder Metallurgy* (ed. W.E. Kingston), McGraw-Hill Book Co., New York, pp. 143–179.

20 Cahn, J.W. (1985) Thermodynamic driving forces for interface migration. *American Ceramic Society Bulletin*, **64**, 1344.

21 Cahn, J.W. and Hoffman, D.W. (1974) A vector thermodynamics for anisotropic surfaces: II. Curved and faceted surfaces. *Acta Metallurgica*, **22**, 1205–1214.

22 Hoffman, D.W. and Cahn, J.W. (1972) Vector thermodynamics for anisotropic surfaces: I. Fundamentals and application to plane surface junctions. *Surface Science*, **31**, 368–388.

23 Taheri, M.L., Molodov, D., Gottstein, G., and Rollett, A.D. (2005) Grain boundary mobility under a stored-energy driving force: a comparison to curvature-driven boundary migration. *Zeitschrift für Metalkunde*, **96**, 1166–1170.

24 Huang, Y. and Humphreys, F.J. (1999) Measurements of grain boundary mobility during recrystallization of a single-phase aluminium alloy. *Acta Materialia*, **47**, 2259–2268.

25 Upmanyu, M., Srolovitz, D., and Smith, R. (1998) Atomistic simulation of curvature driven grain boundary migration. *Interface Science*, **6**, 41–58.

26 Upmanyu, M., Srolovitz, D.J., Shvindlerman, L.S., and Gottstein, G. (1999) Misorientation dependence of intrinsic grain boundary mobility: Simulation and experiment. *Acta Materialia*, **47**, 3901–3914.

27 Gottstein, G., Molodov, D.A., Shvindlerman, L.S., Srolovitz, D.J., and Winning, M. (2001) Grain boundary migration: Misorientation dependence. *Current Opinion in Solid State and Materials Science*, **5**, 9–14.

28 Olmsted, D.L., Holm, E.A., and Foiles, S.M. (2009) Survey of computed grain boundary properties in face-centered cubic metals. II: Grain boundary mobility. *Acta Materialia*, **57**, 3704–3713.

29 Boutin, F.R. (1975) Experimental demonstration of the influence of impurities on the grain boundary migration rate in aluminum recrystallization. *Journal de Physique*, **C4**, C43.55–C4.65.

30 Dillon, S.J., Tang, M., Carter, W.C., and Harmer, M.P. (2007) Complexion: A new concept for kinetic engineering in materials science. *Acta Materialia*, **55**, 6208–6218.

31 Winning, M., Rollett, A.D., Gottstein, G., Srolovitz, D.J., Lim, A., and Shvindlerman, L.S. (2010) Mobility of low-angle grain boundaries in pure metals. *Philosophical Magazine*, **90**, 3107–3128.

32 Hassold, G.N., Chen, I.W., Srolovitz, D.J., and Visscher, W. (1988) Monte Carlo simulation of sintering. *Journal of Metals*, **40**, A44–A50.

33 Grest, G.S., Anderson, M.P., and Srolovitz, D.J. (1988) Domain-growth kinetics for the Q-state Potts-model in 2-dimension and 3-dimension. *Physical Review B*, **38**, 4752–4760.

34 Srolovitz, D.J. and Hassold, G.N. (1987) Effects of diffusing impurities on domain growth in the Ising-model. *Physical Review B*, **35**, 6902–6910.

35 Doherty, R.D., Srolovitz, D.J., Rollett, A.D., and Anderson, M.P. (1987) On the

volume fraction dependence of particle limited grain-growth. *Scripta Metallurgica*, **21**, 675–679.

36 Srolovitz, D.J., Grest, G.S., and Anderson, M.P. (1986) Computer simulation of recrystallization: I. Homogeneous nucleation and growth. *Acta Metallurgica*, **34**, 1833–1845.

37 Srolovitz, D.J. (1986) Grain-growth phenomena in films: A Monte Carlo approach. *Journal of Vacuum Science and Technology A – Vacuum Surfaces and Films*, **4**, 2925–2931.

38 Srolovitz, D.J., Grest, G.S., and Anderson, M.P. (1985) Computer-simulation of grain growth: V. Abnormal grain growth. *Acta Metallurgica*, **33**, 2233–2247.

39 Srolovitz, D.J. and Grest, G.S. (1985) Impurity effects on domain-growth kinetics: II. Potts-model. *Physical Review B*, **32**, 3021–3025.

40 Grest, G.S., Srolovitz, D.J., and Anderson, M.P. (1985) Computer simulation of grain growth: IV. Anisotropic grain-boundary energies. *Acta Metallurgica*, **33**, 509–520.

41 Grest, G.S. and Srolovitz, D.J. (1985) Impurity effects on domain-growth kinetics: I. Ising model. *Physical Review B*, **32**, 3014–3020.

42 Anderson, M.P., Grest, G.S., and Srolovitz, D.J. (1985) Grain growth in three dimensions: A lattice model. *Scripta Metallurgica*, **19**, 225–230.

43 Srolovitz, D.J., Anderson, M.P., Sahni, P.S., and Grest, G.S. (1984) Computer simulation of grain growth: II. Grain size distribution, topology and local dynamics. *Acta Metallurgica*, **32**, 793–802.

44 Srolovitz, D.J., Anderson, M.P., Grest, G.S., and Sahni, P.S. (1984) Computer simulation of grain growth: III. Influence of a particle dispersion. *Acta Metallurgica*, **32**, 1429–1438.

45 Anderson, M.P., Srolovitz, D.J., Grest, G.S., and Sahni, P.S. (1984) Computer simulation of grain growth: I. Kinetics. *Acta Metallurgica*, **32**, 783–791.

46 Wright, S.A., Plimpton, S.J., Swiler, T.P., Fye, R.M., Young, M.F., and Holm, E.A. (1997) *Potts-model Grain Growth Simulations: Parallel Algorithms and Applications*, Sandia National Laboratories, Albuquerque, NM, USA.

47 Kandel, D. and Domany, E. (1990) Rigorous derivation of domain growth kinetics without conservation laws. *Journal of Statistical Physics*, **58**, 685–706.

48 Raabe, D. (2000) Scaling Monte Carlo kinetics of the Potts model using rate theory. *Acta materialia*, **48**, 1617–1628.

49 Rollett, A.D. and Raabe, D. (2001) A hybrid model for mesoscopic simulation of recrystallization. *Computational Materials Science*, **21**, 69–78.

50 Zhang, L., Rollett, A.D., Bartel, T., Wu, D., and Lusk, M.T. (2012) A calibrated Monte Carlo approach to quantify the impacts of misorientation and different driving forces on texture development. *Acta Materialia*, **60**, 1201–1210.

51 Mullins, W.W. (1998) Grain growth of uniform boundaries with scaling. *Acta materialia*, **46**, 6219–6226.

52 Hillert, M. (1965) On the theory of normal and abnormal grain growth. *Acta Metallurgica*, **13**, 227.

53 Brahme, A., Fridy, J., Weiland, H., and Rollett, A.D. (2009) Modeling texture evolution during recrystallization in aluminum. *Modeling and Simulation in Materials Science and Engineering*, **17**, 015005.

54 Alvi, M.H., Cheong, S.W., Suni, J.P., Weiland, H., and Rollett, A.D. (2008) Cube texture in hot-rolled aluminum alloy 1050 (AA1050): Nucleation and growth behavior. *Acta materialia*, **56**, 3098–3108.

55 Brahme, A., Alvi, M.H., Saylor, D., Fridy, J., and Rollett, A.D. (2006) 3D reconstruction of microstructure in a commercial purity aluminum. *Scripta Materialia*, **55**, 75–80.

56 Barmak, K., Kim, J., Kim, C.S., Archibald, W.E., Rohrer, G.S., Rollett, A.D., Kinderlehrer, D., Ta'asan, S., Zhang, H., and Srolovitz, D.J. (2006) Grain boundary energy and grain growth in Al films: Comparison of experiments and simulations. *Scripta Materialia*, **54**, 1059–1063.

57 Upmanyu, M., Hassold, G.N., Kazaryan, A., Holm, E.A., Wang, Y., Patton, B., and Srolovitz, D.J. (2002) Boundary mobility and energy anisotropy effects on microstructural evolution during grain growth. *Interface Science*, **10**, 201–216.

58 Zhang, H., Upmanyu, N., and Srolovitz, D.J. (2005) Curvature driven grain boundary migration in aluminum: Molecular dynamics simulations. *Acta materialia*, **53**, 79–86.

59 Saylor, D.M., Morawiec, A., and Rohrer, G.S. (2002) Distribution and energies of grain boundaries in magnesia as a function of five degrees of freedom. *Journal of The American Ceramic Society*, **85**, 3081–3083.

60 Rollett, A.D. (2004) Crystallographic texture change during grain growth. *JOM*, **56**, 63–68.

61 Winning, M. (2003) Motion of ⟨1 0 0⟩-tilt grain boundaries. *Acta materialia*, **51**, 6465–6475.

62 Upmanyu, M., Srolovitz, D.J., Shvindlerman, L.S., and Gottstein, G. (2002) Molecular dynamics simulation of triple junction migration. *Acta materialia*, **50**, 1405–1420.

63 Aristov, V.Y., Kopetskii, C., and Shvindlerman, L.S. (1976) Mobility of (1 1 1) intergrain tilt boundary in aluminium. *Soviet Physics Solid State*, **18**, 137–142.

64 Taheri, M.L., Rollett, A.D., and Weiland, H. (2004) In-situ quantification of solute effects on grain boundary mobility and character in aluminum alloys during recrystallization. *Materials Science Forum*, **467–470**, 997–1002.

65 Hesselbarth, H.W. and Göbel, I.R. (1991) Simulation of recrystallization by cellular automata. *Acta Metallurgica*, **39**, 2135.

66 Raabe, D. (1999) Introduction of a scalable three-dimensional cellular automaton with a probabilistic switching rule for the discrete mesoscale simulation of recrystallization phenomena. *Philosophical Magazine A*, **79**, 2339–2358.

67 Raabe, D. (2002) Cellular automata in materials science with particular reference to recrystallization simulation. *Annual Review of Materials Research*, **32**, 53–76.

68 Mukhopadhyay, P., Loeck, M., and Gottstein, G. (2007) A cellular operator model for the simulation of static recrystallization. *Acta Materialia*, **55**, 551–564.

69 Allen, S.M. and Cahn, J.W. (1979) Microscopic theory for antiphase boundary motion and its application to antiphase domain coarsening. *Acta Metallurgica*, **27**, 1085–1095.

70 Wheeler, A.A., McFadden, G.B., and Boettinger, W.J. (1996) Phase-field model for solidification of a eutectic alloy. *Proceedings of The Royal Society of London Series A – Mathematical Physical and Engineering Sciences*, **452**, 495–525.

71 Wheeler, A.A. and McFadden, G.B. (1996) A xi-vector formulation of anisotropic phase-field models: 3D asymptotics. *European Journal of Applied Mathematics*, **7**, 367–381.

72 Sandhage, K.H., Schmutzler, H.J., Wheeler, R., and Fraser, H.L. (1996) Mullite joining by the oxidation of malleable, alkaline-earth-metal-bearing bonding agents. *Journal of The American Ceramic Society*, **79**, 1839–1850.

73 Chen, L.Q. (1995) A novel computer simulation technique for modeling grain growth. *Scripta Metallurgica*, **32**, 115–120.

74 Fan, D. and Chen, L.-Q. (1997) Computer simulation of grain growth using a continuum field model. *Acta materialia*, **45**, 611–622.

75 Lusk, M.T. (1999) A phase-field paradigm for grain growth and recrystallization. *Proceedings: Mathematical, Physical and Engineering Sciences*, **455**, 677–700.

76 Fan, D. and Chen, L.Q. (1997) Computer simulation of grain growth using a continuum field model. *Acta Materiala*, **45**, 611–622.

77 McKenna, I.M., Gururajan, M.P., and Voorhees, P.W. (2009) Phase field modeling of grain growth: Effect of boundary thickness, triple junctions, misorientation, and anisotropy. *Journal of Materials Science*, **44**, 2206–2217.

78 von Neumann, J. (1952) discussion of article by In: *Metal Interfaces* (ed. C.S., Smith), American Society for Testing of Materials, Cleveland, OH, p. 108.

79 Mullins, W.W. (1956) Two-dimensional motion of idealized grain boundaries. *Journal of Applied Physics*, **27**, 900–904.

80 Vanherpe, L., Moelans, N., Blanpain, B., and Vandewalle, S. (2011) Bounding box framework for efficient phase field

simulation of grain growth in anisotropic systems. *Computational Materials Science*, **50**, 2221–2231.

81 Moelans, N., Blanpain, B., and Wollants, P. (2008) Quantitative phase-field approach for simulating grain growth in anisotropic systems with arbitrary inclination and misorientation dependence. *Physical Review Letters*, **101**, 025502.

82 Moelans, N., Spaepen, F., and Wollants, P. (2010) Grain growth in thin films with a fibre texture studied by phase-field simulations and mean field modelling. *Philosophical Magazine*, **90**, 501–523.

83 Steinbach, I. and Pezzolla, F. (1999) A generalized field method for multiphase transformations using interface fields. *Physica D*, **134**, 385–393.

84 Kar, D., Wilson, S., Gruber, J., Rohrer, G. S., and Rollett, A.D. (2012) Effect of triple junction equilibria on the development of anisotropic boundary character in polycyrstalline systems: An interface field study in three dimensions, submitted.

85 Fan, D.N., Chen, L.Q., and Chen, S.P. (1997) Effect of grain boundary width on grain growth in a diffuse-interface field model. *Materials Science and Engineering A – Structural Materials Properties Microstructure and Processing*, **238**, 78–84.

86 Shu, C.W. (2003) High-order finite difference and finite volume WENO schemes and discontinuous Galerkin methods for CFD. *International Journal of Computational Fluid Dynamics*, **17**, 107–118.

87 Krill, C.E. and Chen, L.Q. (2002) Computer simulation of 3-D grain growth using a phase-field model. *Acta Materialia*, **50**, 3057–3073.

88 Uehara, T. and Sekerka, R.F. (2003) Phase field simulations of faceted growth for strong anisotropy of kinetic coefficient. *Journal of Crystal Growth*, **254**, 251–261.

89 McFadden, G.B., Wheeler, A.A., Braun, R.J., Coriell, S.R., and Sekerka, R.F. (1993) Phase-field models for anisotropic interfaces. *Physical Review E*, **48**, 2016–2024.

90 Ma, N., Chen, Q., and Wang, Y.Z. (2006) Implementation of high interfacial energy anisotropy in phase field simulations. *Scripta Materialia*, **54**, 1919–1924.

91 Debierre, J.M., Karma, A., Celestini, F., and Guerin, R. (2003) Phase-field approach for faceted solidification. *Physical Review E*, **68**, 041604.

92 Loginova, I., Amberg, G., and Agren, J. (2001) Phase-field simulations of non-isothermal binary alloy solidification. *Acta materialia*, **49**, 573–581.

93 Torabi, S., Lowengrub, J., Voigt, A., and Wise, S. (2009) A new phase-field model for strongly anisotropic systems. *Proceedings of the Royal Society A – Mathematical Physical and Engineering Sciences*, **465**, 1337–1359.

94 Eggleston, J.J., McFadden, G.B., and Voorhees, P.W. (2001) A phase-field model for highly anisotropic interfacial energy. *Physica D*, **150**, 91–103.

95 Moelans, N., Blanpain, B., and Wollants, P. (2008) Quantitative analysis of grain boundary properties in a generalized phase field model for grain growth in anisotropic systems. *Physical Review B*, **78**, 024113.

96 Elsey, M., Esedoglu, S., and Smereka, P. (2009) Diffusion generated motion for grain growth in two and three dimensions. *Journal of Computational Physics*, **228**, 8015–8033.

97 Bernacki, M., Chastel, Y., Coupez, T., and Loge, R.E. (2008) Level set framework for the numerical modelling of primary recrystallization in polycrystalline materials. *Scripta Materialia*, **58**, 1129–1132.

98 Frost, H.J., Thompson, C.V., Howe, C.L., and Whang, J.H. (1988) A 2-dimensional computer-simulation of capillarity-driven grain-growth preliminary results. *Scripta Metallurgica*, **22**, 65–70.

99 Sullivan, J.M., Lynch, D.R., and Oneill, K. (1987) Finite-element simulation of planar instabilities during solidification of an undercooled melt. *Journal of Computational Physics*, **69**, 81–111.

100 Nagai, T., Ohta, S., Kawasaki, K., and Okuzono, T. (1990) Computer simulation of cellular pattern growth in two and three dimensions. *Phase Transitions*, **28**, 177–211.

101 Nakashima, K., Nagai, T., and Kawasaki, K. (1989) Scaling behavior of two-dimensional domain growth: Computer

simulation of vertex models. *Journal of Statistical Physics*, **57**, 759–787.

102 Kawasaki, K., Nagai, T., and Nakashima, K. (1989) Vertex models for two-dimensional grain growth. *Philosophical Magazine Part B*, **60**, 399–421.

103 Nagai, T., Kawasaki, K., and Nakamura, K. (1988) Vertex dynamics of two-dimensional cellular-patterns. *Journal of the Physical Society of Japan*, **57**, 2221–2224.

104 Fuchizaki, K. and Kawasaki, K. (1996) Three-dimensional computer modeling of grain growth: A Vertex model Approach, in *Grain Growth in Polycrystalline Materials II, Pts 1 and 2, Materials Science Forum*, **204–206**, 267–278.

105 Fuchizaki, K., Kusaba, T., and Kawasaki, K. (1995) Computer modeling of 3-dimensional cellular-pattern growth. *Philosophical Magazine B*, **71**, 333–357.

106 Fuchizaki, K. and Kawasaki, K. (1995) Time evolution of 3-dimensional cellular-systems: Computer modeling based on vertex-type models. *Physica A*, **221**, 202–215.

107 Weygand, D., Brechet, Y., Lepinoux, J., and Gust, W. (1999) Three-dimensional grain growth: A vertex dynamics simulation. *Philosophical Magazine B*, **79**, 703–716.

108 Weygand, D., Brechet, Y., and Lepinoux, J. (1998) A vertex dynamics simulation of grain growth in two dimensions. *Philosophical Magazine B*, **78**, 329–352.

109 Weygand, D., Brechet, Y., and Lepinoux, J. (2000) Inhibition of grain growth by particle distribution: Effect of spatial heterogeneities and of particle strength dispersion. *Materials Science and Engineering A – Structural Materials Properties Microstructure and Processing*, **292**, 34–39.

110 Weygand, D., Brechet, Y., and Lepinoux, J. (1999) Zener pinning and grain growth: A two-dimensional vertex computer simulation. *Acta Materialia*, **47**, 961–970.

111 Weygand, D., Brechet, Y., and Leepinoux, J. (1999) Reduced mobility of triple nodes and lines on grain growth in two and three dimensions. *Interface Science*, **7**, 285–295.

112 Weygand, D., Brechet, Y., and Lepinoux, J. (1998) Influence of a reduced mobility of triple points on grain growth in two dimensions. *Acta Materiala*, **46**, 6559–6564.

113 Maurice, C. (2000) 3D-Network simulation of deformation structure recovery, in *Recrystallization – Fundamental Aspects* (ed. N. Hansen), Risoe National Laboratory, Risoe, Denmark, pp. 431–438.

114 Maurice, C. and Humphreys, F.J. (1998) 2- and 3-D curvature driven simulations of grain growth, in *Grain Growth in Polycrystalline Materials III* (ed. H. Weiland et al.), TMS, Pittsburgh, pp. 81–90.

115 Rollett, A.D. and Holm, E.A. (1996) Abnormal grain growth – the origin of recrystallization nuclei? in *ReX-96* (ed. T.R. McNelley), TMS, Warrendale, PA Monterey, CA, pp. 31–42.

116 Humphreys, F.J. (1997) Recovery, recrystallization and grain growth - stability and instability of cellular microstructures, in *Recrystallization-96* (ed. T.R. McNelley), ReX'96, Monterey, CA, pp. 1–14.

117 Rollett, A.D., Srolovitz, D.J., and Anderson, M.P. (1989) Simulation and theory of abnormal grain-growth anisotropic grain-boundary energies and mobilities. *Acta Metallurgica*, **37**, 1227–1240.

118 Kuprat, A. (2000) *Inspecting and Repairing Physical Topology in a Moving Grid Grain Growth Simulation*, Los Alamos National Laboratory, Los Alamos, NM.

119 Weygand, D., Brechet, Y., and Lepinoux, J. (2001) A vertex simulation of grain growth in 2D and 3D. *Advanced Engineering Materials*, **3**, 67–71.

120 Syha, M. and Weygand, D. (2010) A generalized vertex dynamics model for grain growth in three dimensions. *Modelling and Simulations in Materials Science and Engineering*, **18**, 015010.

121 MacPherson, R.D. and Srolovitz, D.J. (2007) The von Neumann relation generalized to coarsening of three-dimensional microstructures. *Nature*, **446**, 1053–1055.

122 Sano, T., Saylor, D.M., and Rohrer, G.S. (2003) Surface energy anisotropy of

SrTiO$_3$ at 1400 degrees C in air. *Journal of the American Ceramic Society*, **86**, 1933–1939.

123 Kuprat, A. (2000) Modeling microstructure evolution using gradient-weighted moving finite elements. *SIAM Journal of Scientific Computing*, **22**, 535–560.

124 Carlson, N.N. and Miller, K. (1998) Design and application of a gradient-weighted moving finite element code: I. In one dimension. *SIAM Journal of Scientific Computing*, **19**, 728–765.

125 Saad, Y. and Schultz, M.H. (1986) GMRES: a generalized minimal residual algorithm for solving nonsymmetric linear systems. *SIAM Journal on Scientific and Statistical Computing*, **7**, 856–869.

126 Demirel, M.C., Kuprat, A.P., George, D.C., Straub, G.K., and Rollett, A.D. (2002) Linking experimental characterization and computational modeling of grain growth in Al-foil. *Interface Science*, **10**, 137–141.

127 Rohrer, G.S., Saylor, D.M., El Dasher, B., Adams, B.L., Rollett, A.D., and Wynblatt, P. (2004) The distribution of internal interfaces in polycrystals. *Zeitschrift für Metallkunde*, **95**, 197–214.

128 Harun, A., Holm, E.A., Clode, M.P., and Miodownik, M.A. (2006) On computer simulation methods to model Zener pinning. *Acta Materiala*, **54**, 3261–3273.

129 Couturier, G., Maurice, C., and Fortunier, R. (2003) Three-dimensional finite-element simulation of Zener pinning dynamics. *Philosophical Magazine*, **83**, 3387–3405.

130 Zener, C. (1948) communication to C.S. Smith. In: *Trans. Met. Soc. AIME*, p. 15.

131 Ringer, S.P., Li, W.B., and Easterling, K.E. (1992) On the rotation of precipitate particles. *Acta Metallurgica et Materialia*, **40**, 275–283.

Part II
Interfacial Phenomena and their Role in Microstructure Control

8
Grain Boundary Junctions: Their Effect on Interfacial Phenomena
Lasar S. Shvindlerman and Günter. Gottstein

8.1
Introduction

The major microstructural elements of a polycrystal are grain boundaries. Traditionally, grain growth in polycrystals is solely attributed to the motion of grain boundaries. In polycrystalline materials, their energy determines the driving force for grain growth. Also with growing interest in fine grained and even nanocrystalline materials, the need arose to consider the role of other structural elements of grain boundaries: facets, grain boundary, and interfacial triple and quadruple junctions as well as grain boundary ridges [1–4]. In particular, it was shown that the kinetics of grain boundary triple and quadruple junctions may be different from the kinetics of the adjoining grain boundaries. We would like to stress that even though facets and grain boundary triple junctions have rather long been known, their effect on grain boundary migration is still poorly understood. This owes mainly to the fact that their study entails the observation of complex dynamic phenomena, like recrystallization or grain growth, where the motion of grain boundaries is influenced by a multitude of factors that complicate the association of its fundamental dependencies with microstructural features. Moreover, it was shown that these interfacial elements affect the kinetics of grain boundary motion and grain growth. They impact the mechanisms of interaction of grain boundaries with second phase particles, the equilibrium concentration of vacancies in the vicinity of voids at a grain boundary, and the process of sintering of nanopowders.

The influence of grain boundary triple junctions on grain boundary motion and grain growth will not be addressed since it has been discussed already rather comprehensively [1–7]. In this chapter, we discuss the results of some most recent studies on the effect of the energy of grain boundary and interfacial triple junctions on grain boundary controlled phenomena in solids.

Microstructural Design of Advanced Engineering Materials, First Edition. Edited by Dmitri A. Molodov.
© 2013 Wiley-VCH Verlag GmbH & Co. KGaA. Published 2013 by Wiley-VCH Verlag GmbH & Co. KGaA.

8.2
Experimental Measurement of Grain Boundary Triple Line Energy

The problem of the triple line energy has been addressed by several authors in the past. Gibbs [8] speculated that the excess free energy of a triple line between fluid phases might be positive or negative. Computer simulations by Srinivasan et al. [9] came to the conclusion that a negative triple line energy is possible, whereas Caro and van Swygenhoven [10] found in simulation studies that the obtained triple line tension was always positive. Typically, the experimental measurements were based on very simplified assumptions, the majority of them related to the shape of the triple junction etch pit. Fortier et al. [11] and Nishimura [12] approximated the measured crater at the intersection of a triple line with the crystal surface by a tetrahedron and estimated the energy of a triple line as 5×10^{-7} J/m, that is, two orders of magnitude higher than the line tension of a dislocation.

A thermodynamically correct approach was put forward only recently [13,14]. The basic idea of the approach is as follows: the equilibrium of four line tensions is considered, that is, grain boundary triple line and three triple lines at the bottom of the thermal grooves (Figure 8.1) attached to the triple line. From the equilibrium of the four line tensions, it follows for the triple line tension γ_{tj}^{l}

$$\gamma_{tj}^{l} = \gamma_{1-2}^{lS} \sin \zeta_{1-2} + \gamma_{1-3}^{lS} \sin \zeta_{1-3} + \gamma_{2-3}^{lS} \sin \zeta_{2-3} \qquad (8.1)$$

where γ_{i-j}^{lS} are the line tensions of triple lines at the bottom of the thermal grooves, respectively, ζ_{i-j} are the angles of inclination of each groove root with respect to the plane perpendicular to the triple line. If u_{i-j} denotes the shape function of the groove–root $i-j$ (Figure 8.1a) then

$$\sin \zeta_{i-j} = \sin\left(\arctan \frac{\partial u_{i-j}}{\partial r}\bigg|_{r=0}\right) = \frac{\dfrac{\partial u_{i-j}}{\partial r}}{\sqrt{1 + \left(\dfrac{\partial u_{i-j}}{\partial r}\right)^{2}}}\Bigg|_{r=0} \qquad (8.2)$$

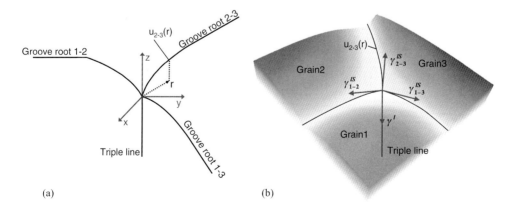

Figure 8.1 (a) Schematic three-dimensional (3D) view of the line geometry at a triple junction; (b) Atomic force microscopy (AFM) 3D view of the line tension equilibrium at a triple junction [14].

The line tension of the groove root triple lines γ_{i-j}^{ls} can be determined by comparing the angle at the root of a grain boundary groove formed at a straight root line and at a curved root line, in particular close to the crater of a grain boundary triple junction. In case that the grain boundary remains flat but its rim at the root of the groove is curved [8], one has to take into account an additional line tension term that is akin to the Laplace pressure for 3D-curved grain boundary surfaces. In equilibrium,

$$\gamma_B - \gamma^{ls} \frac{\frac{\partial^2 u}{\partial r^2}}{\left[1+\left(\frac{\partial u}{\partial r}\right)^2\right]^{3/2}} = 2\gamma_S \cos\frac{\xi}{2} \tag{8.3}$$

The angle ξ denotes the dihedral groove angle for a curved groove root, $u(r)$ is the shape function of the respective groove root (Figure 8.1a), γ_B and γ_S are the grain boundary and the surface energy, respectively. Equations (8.1)–(8.3) can be used to calculate the triple line energy from the geometry of the merging boundaries at the triple junction.

A copper tricrystal was investigated to determine the triple line energy.

From the AFM measurements (Figure 8.2), all necessary parameters can be derived to extract the triple line tension, such as the grain boundary groove angles, the groove root angles at the curved part of the grain boundary rim, and the curvature of the groove roots. To measure the dihedral angles of the grain boundary grooves and therewith the grain boundary energy, we investigated the groove profiles perpendicular to the grain boundary in bicrystals (Figures 8.2 and 8.3).

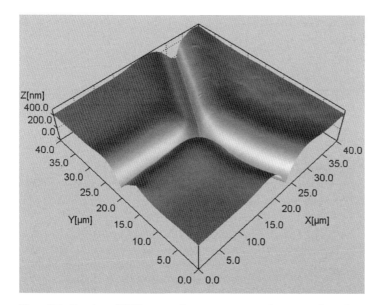

Figure 8.2 Top view of AFM topography measurement in the vicinity of a triple junction after annealing at 980 °C [14]. Note the different scale in the z-direction.

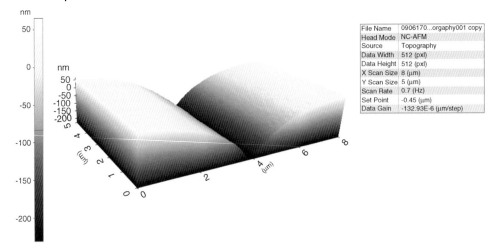

Figure 8.3 Top view of AFM topography measurement on the grain boundary groove of a Cu-bicrystal after annealing at 980 °C for 2 h [14].

The measurements were performed with high-aspect-ratio tips (OLYMPUS AC11160BN-A2) in an atomic force microscope in noncontact mode.

With the determined grain boundary-free surface line tension, the grain boundary triple line tension can be derived from the equilibrium of four line tensions at their point of intersection (Figure 8.1). For the respectively measured grain boundary-free surface line tension and grain boundary triple line tension in copper, we obtained the values: $\gamma^{lS} = (2.5 \pm 1.1) \times 10^{-8}$ J/m and $\gamma^{l}_{TP} = (6.0 \pm 3.0) \times 10^{-9}$ J/m [15].

The measured absolute values of the grain boundary-free surface and grain boundary triple line can then be utilized to estimate the impact of triple line tension on the thermodynamics and kinetics of interfaces in polycrystals.

8.3
Impact of Triple Line Tension on the Thermodynamics and Kinetics in Solids

8.3.1
Grain Boundary Triple Line Contribution to the Driving Force for Grain Growth

As shown in [15,16], the total driving force from both grain boundaries and triple junctions reads

$$P = \frac{2\gamma_B}{\langle D \rangle} + \frac{36\gamma^{\ell}}{\pi \langle D \rangle^2} \qquad (8.4)$$

where $\langle D \rangle$ is the mean grain size, γ_B, γ^{ℓ} are grain boundary surface tension and grain boundary triple junction line tension, respectively, and we tacitly assume that the grain boundary radius of curvature is equal to the average grain diameter.

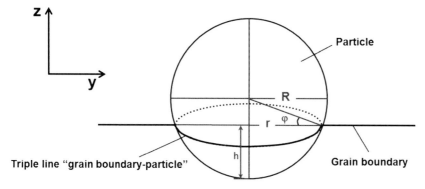

Figure 8.4 Schematic view of a particle intersecting a grain boundary.

The ratio $\frac{p^l}{p_B} = \frac{18\gamma^l}{\pi\gamma_B} = D_c$ defines the grain size D_c for which grain boundary and triple junction contribution to the driving force are equal. Using the measured values for Cu grain boundary surface tension and triple line tension (with a grain boundary surface tension in Cu of $\gamma_B = 0.6\,\mathrm{J/m^2}$), we arrive at $D_c \approx 55$ nm. In other words, up to a mean grain size of about 55 nm, the driving force stemming from triple junctions is larger than that of the boundaries. As a consequence, a correct examination of grain growth in nanocrystalline materials at least up to a mean grain size of D_c also requires to account for the energy of triple junctions [16,17].

8.3.2
Effect of the Triple Junction Line Tension on the Zener Force

In 1948, Zener put forward a concept how the interaction between a grain boundary and particles can be estimated quantitatively [15].

When a spherical solid particle with radius R (Figure 8.4) intersects a (nonflexible, i.e., flat) grain boundary, it replaces part of the grain boundary area, and thus, reduces the total boundary energy:

$$\Delta G = -\pi r^2 \gamma_B \tag{8.5}$$

However, this area has to be regenerated when the grain boundary detaches from the particle. The respective retarding "Zener force" is given by

$$f = -2\pi r \gamma_B \tag{8.6}$$

During the past 60 years, all considerations of the grain boundary motion and the grain growth in materials with both immobile and mobile particles were based on this concept.

Since the grain boundary triple junction energy and grain boundary-free surface line tension are now known by experiment, this permits to take into account the triple line tension "grain boundary-particle" γ^{BP} [15]. In a first approximation, we assume that $\gamma^l = \gamma^{BP}$, that is, the line tension of a groove root which would be the exact value if the particles were a void. Then Eq. (8.5) needs to be rewritten as

$$\Delta G_I^0 = -\pi r^2 \gamma_B + 2\pi r \gamma^{BP} \tag{8.7}$$

where γ^{BP} is the line tension of the triple line between boundary and particle. From the derivative of Eq. (8.7)

$$\frac{d\Delta G_I^0}{d\varphi} = 2\pi R \gamma^{BP} \sin\varphi (\eta \cos\varphi - 1) \tag{8.8}$$
$$\varphi = \arccos(r/R)$$

where $\eta = \frac{\gamma_B R}{\gamma^{BP}}$, we can derive the interaction between boundary and particle.

When $\eta < 1$, there is a maximum value of ΔG_I^0 at $\varphi = 0$. For $\eta > 1$, value of ΔG_I^0 appears at $\cos\varphi = \frac{1}{\eta}$, and $\varphi = 0$ corresponds to a minimum value of ΔG_I^0. The sign of the energy ΔG_I^0 is a function of the particle radius, as shown in Figure 8.5.

The critical particle size is $R^* = \frac{\gamma^{BP}}{\gamma_B}$ for $\eta = 1$. When the particle is smaller than the critical size, it will not attach to the grain boundary. In case, the particle radius is larger than R^*, only as $\varphi < \arccos\frac{1}{\eta}$, the grain boundary would intersect the particle as a circle, and $f^* = \frac{d\Delta G_I^0}{dh}(R^*)$ is the retarding force for grain boundary motion.

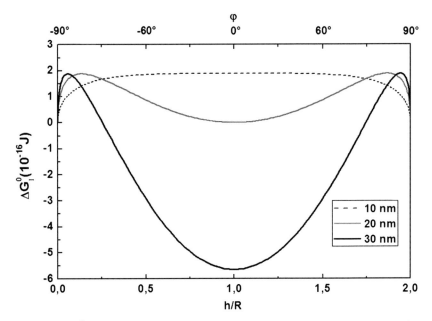

Figure 8.5 $\Delta G_I^0(\varphi)$ for different particle radii in copper, $\eta_{R=10\,nm} = 0.95$, $\eta_{R=20\,nm} = 1.90$, and $\eta_{R=30\,nm} = 2.85$.

With the given values for $\gamma^{BP} = \gamma^l$ and γ_B, the critical size R^* is about 40 nm. If we take the value $\gamma^l_{TP} = (6.0 \pm 3.0) \times 10^{-9}$ J/m for the grain boundary–particle line tension γ^{BP}, the critical size R^* is about 10 nm. From Figure 8.5, it is evident that ΔG^0_I is positive and increases when the particle is smaller than R^*, that is, $R < R^*$. The larger the area where the particle intersects the grain boundary, the larger the energy ΔG^0_I. The minimum value of ΔG^0_I is attained at $\varphi = \pm 90°$. At that point, the particle will not attach to the grain boundary at all. When $R > R^*$, another minimum of the curve appears at $\varphi = 0°$.

In other words, the impact of triple junctions brings about the formation of an energetic barrier for the particle–grain boundary interaction. This barrier cannot be overcome without external work PV, P is the driving force on the boundary, V is the volume swept by the moving grain boundary:

$$\Delta G_I = -\pi r^2 \gamma_B + 2\pi r \gamma^{BP} - PV \tag{8.9}$$

The condition $\Delta G_I = 0$ defines the critical driving force P^* that needs to be applied to overcome the barrier that prevents the boundary to touch the particle.

8.3.3
Effect of Triple Junction Line Tension on the Gibbs–Thompson Relation

The shape and behavior of a small inclusion (a particle or a void) on a grain boundary has been considered previously [16]. Such issue is especially important for small inclusions and voids in nanocrystalline materials since in such a systems the majority of inclusions is located at grain boundaries.

For the contact angle θ (Figure 8.6), we obtain

$$\cos\theta = \frac{\gamma_B}{2\gamma_S} - \frac{\gamma^{lS}}{2x_0 \gamma_S} \tag{8.10}$$

where γ_B, γ_S, and γ^{lS} are the boundary energy, the surface tension, and the line tension for the triple line inclusion–boundary, respectively. From Eq. (8.10) it follows that the triple line tension reduces the wetting of the inclusion at the grain boundary. The Gibbs–Thompson relation determines the difference between the pressure

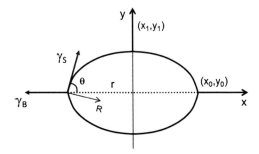

Figure 8.6 Schematic view of a void intersecting a grain boundary.

inside and outside of an inclusion. The triple junction between the inclusion and the grain boundary changes the classical Gibbs–Thompson relation to

$$\Delta p = \gamma_s \frac{dA_S}{dV} - \gamma_b \frac{dA_b}{dV} + \gamma^{IS} \frac{dL}{dV} \tag{8.11}$$

where A_s and A_b are the area of the surface (interface) and grain boundary respectively, and L is the length of the triple line. The first term on the right-hand side is equal to $\gamma_s\left(\frac{1}{R_1} + \frac{1}{R_2}\right)$, where R_1 and R_2 are the principal radii of curvature; for a lens-shaped void that can be considered as a body of rotation. For a spherical void this term is equal to $\frac{2\gamma_s}{R}$. The second term is equal to $\frac{\gamma_b}{2R\sin\theta}$ (Figure 8.6), where R is the radius of curvature at the vertex of the lens (Figure 8.6). The third term on the right-hand side is equal to $\frac{\gamma^{IS}\sin\theta}{2R^2}$. As shown in [16] up to $R \approx 200$ nm, $\sin\theta$ is larger than 0.99. In other words, relation (8.11) can be expressed in good approximation as

$$\Delta p = \frac{2\gamma_s}{R} - \frac{\gamma_b}{2R} + \frac{\gamma^{IS}}{2R^2} \tag{8.12}$$

With Eq. (8.12), the equilibrium concentration of impurities in the vicinity of the inclusion surface reads

$$c_R = c_\infty \exp\left[\frac{\Omega}{kT}\left(\frac{2\gamma_s}{R} - \frac{\gamma_b}{2R} + \frac{\gamma^{IS}}{2R^2}\right)\right] \tag{8.13}$$

where c_∞ is the impurity concentration far away from the inclusion, Ω is the atomic volume, which, for the simplicity is assumed to be the same for matrix and solute atom, k is the Boltzmann constant, and T is the absolute temperature. It follows from Eq. (8.13) that the solubility of foreign atoms increases to account for the triple junction line tension inclusion-grain boundary. For nanocrystalline systems where the grain size is extremely small and the majority of the particles is located at grain boundaries, this phenomenon is especially important. Along the same consideration relation (8.14) defines the equilibrium concentration c_v of the vacancies in the vicinity of a void at a grain boundary:

$$c_V = c_v^0 \exp\left[\frac{\Omega}{kT}\left(\frac{2\gamma_s}{R} - \frac{\gamma_b}{2R} + \frac{\gamma^{IS}}{2R^2}\right)\right] \tag{8.14}$$

where c_V^0 is the vacancy concentration far away from the void. For copper, in accordance with measurements, the terms $\frac{2\gamma_s}{R} - \frac{\gamma_b}{2R}$ and $\frac{\gamma^{IS}}{2R^2}$ are of the same magnitude at $R \approx 4$ nm. We would like to stress that relation (8.14) also controls the kinetics of sintering or, generally speaking, how fast an arrangement of separate particles will transform to one piece of material.

8.4
Why do Crystalline Nanoparticles Agglomerate with Low Misorientations?

Hertz demonstrated in 1886 that the size and the shape of the zone of contact of two spherical bodies under load are determined by their elastic deformation [18]. Only in the second half of the twentieth century, it was understood that apart from the elastic deformation the surface forces play an important role [19], that is, by adhesion caused by the change of the surface energy of the system. The formation of a strong bond between two nanoparticles leads to the formation of a strong agglomerate. It proceeds in two stages: first, an adhesive contact between the nanoparticles is formed. In the approximation of two ideal elastic spheres, the energy loss associated with the elastic deformation of the particles is compensated by elimination of two free surfaces and the formation of a grain boundary or a perfect crystal region in the circular adhesive contact spot between two particles. At this stage, the deformation of the particles is fully elastic. Second, the size of the contact spot increases by diffusion-controlled mechanisms and an interparticle neck is formed. At this stage, the deformation of the particles becomes irreversible and the strength of the agglomerate increases because the size of the contact spot (grain boundary) increases. The description of this process was given by Johnson, Kendall, and Roberts (JKR) [19]. However, they did not take into account the triple junction "grain boundary-free surface." In [20], this problem was considered in the framework of a model, where the elastic stress state was described in accordance with the JKR approach, whereas the presence of a "grain boundary-free surface" triple line was introduced by additional forces along the perimeter of the adhesion spot. The total energy of the two particles U_T reads [20]

$$U_T = \frac{1}{K^{2/3}R^{1/3}}\left[\frac{1}{15}P_1^{5/3} - \frac{1}{3}P_0^2 P_1^{2/3}\right] - \Delta\gamma\pi\frac{R^{2/3}P_1^{2/3}}{K^{2/3}} + 2\gamma^l\pi\frac{R^{1/3}P_1^{1/3}}{K^{1/3}} \quad (8.15)$$

where P_1 is the Hertz load–contact attraction load,

$$\Delta\gamma = 2\gamma_s - \gamma_b$$

$$K = \frac{4}{3\pi(k_1+k_2)}; \quad k_{1,2} = \frac{1-\nu_{1,2}^2}{\pi E_{1,2}}$$

$$R = \frac{R_1 R_2}{R_1 + R_2}$$

As shown in [20] in equilibrium, the natural condition $dU_T/dP_1 = 0$ is satisfied

$$(P_1 - P_0)^2 - 6\Delta\gamma\pi R P_1 + 6\gamma^l\pi R^{2/3}P_1^{2/3}K^{1/3} = 0 \quad (8.16)$$

Here, the indices 1 and 2 refer to the two spherical particles of radius R_1 and R_2, E_i and ν_i are Young's modulus and the Poisson ratio of the ith particle, respectively (assumed to be elastically isotropic); P_0 and P_1 are the externally applied load and apparent Hertzian load acting between the particles, respectively; γ^l denotes the energy of the "grain boundary-free surface" triple line.

At zero applied load ($P_0 = 0$), Eq. (8.16) can be re-written in the following dimensionless form

$$p^{4/3} - p^{1/3} + \frac{\gamma^l}{\gamma^{l*}} = 0 \quad (8.17)$$

where

$$p = \frac{P_1}{P_{JKR}}$$

$$\gamma^{\ell*} = \frac{(6\pi)^{1/3} \Delta\gamma^{4/3} R^{2/3}}{K^{1/3}}$$

Here, $P_{JKR} = 6\Delta\gamma\pi R$ is the Hertzian attraction load in the absence of external forces in the framework of the JKR model [19]. The loads P_1 and P_{JKR} are directly related to the radius of the contact spot. The attraction load decreases with increasing triple line energy but only up to a critical value

$$\gamma^{\ell}_{crit} = \frac{3}{4^{4/3}} \gamma^*_T = \frac{3}{4}\left(\frac{3\pi}{2}\right)^{1/3} \frac{\Delta\gamma^{4/3} R^{2/3}}{K^{1/3}} \tag{8.18}$$

Equations (8.17) and (8.18) represent the central result of [20]. They show that for high triple line energies, $\gamma^{\ell} > \gamma^{\ell}_{crit}$, an adhesive contact between two spherical particles does not form at all, since the energetic cost of the newly created triple line cannot be compensated for by the energy gain from the elimination of two free surfaces in the contact spot.

It is obvious that the triple line energy depends on the geometric degrees of freedom of the grain boundary formed in the contact zone between the particles. Evidently, in case of a full crystallographic alignment of the particles, no grain boundary is formed in the contact zone and $\gamma^{\ell} = 0$. Therefore, it is reasonable to assume that γ^{ℓ} is small for low-angle grain boundaries and will increase with increasing misorientation angle. Based on this assumption, we envision the following scenario of particle agglomeration: during compaction or intermixing, the nanoparticles randomly collide with each other, but in the case of high crystallographic misorientation between the contacting particles the adhesive contact is not formed, because the corresponding γ^{ℓ} value is too large. The particles in the compact collide, slide and rotate with respect to each other, until a sufficiently low misorientation between two neighboring particles is achieved by chance. In this case, the condition $\gamma^{\ell} < \gamma^{\ell}_{crit}$ is fulfilled, and an adhesive contact between the particles will be swiftly formed. This couple of adhering particles will serve as a nucleus of a larger agglomerate in which the individual nanoparticles are separated by low-angle grain boundaries, or are perfectly aligned.

8.5
Concluding Remarks

We have shown how the energy (line tension) of grain boundary triple lines can be experimentally determined. With the obtained results on copper, the effect of the energy of triple lines on interfacial phenomena like Zener drag, Gibbs–Thompson relation, and particle agglomeration has been investigated. In particular for small particle and grain size, the energy of triple lines may substantially influence microstructural evolution during materials processing and cannot be neglected for accurate predictions.

References

1 Gottstein, G. and Shvindlerman, L.S. (2010) *Grain Boundary Migration in Metals: Thermodynamics, Kinetics, Applications*, 2nd edn, CRC Press, Boca Raton, FL.

2 Gottstein, G., Shvindlerman, L.S., and Sursaeva, V.G. (2008) Migration of faceted high-angle grain boundaries in Zn. *International Journal of Materials Research*, 52, 491–495.

3 Sursaeva, V.G., Straumal, B.B., Gornakova, A.S., Shvindlerman, L.S., and Gottstein, G. (2008) Effect of faceting on grain boundary motion in Zn. *Acta Materialia*, 56, 2728–2734.

4 Sursaeva, V.G., Gottstein, G., and Shvindlerman, L.S. (2010) Effect of a first-order ridge on grain boundary motion in Zn. *Acta Materialia*, 59, 623–629.

5 Barrales-Mora, L.A., Shvindlerman, L.S., Mohles, V., and Gottstein, G. (2008) Effect of a finite quadruple junction mobility on grain microstructure evolution: Theory and simulation. *Acta Materialia*, 56, 1151–1164.

6 Barrales-Mora, L.A., Gottstein, G., and Shvindlerman, L.S. (2008) Three-dimensional grain growth: Analytical approaches and computer simulations. *Acta Materialia*, 56, 5915–5926.

7 Molodov, D.A. and Shvindlerman, L.S. (2009) Interface migration in metals (IMM): "Vingt Ans Après" (twenty years later). *International Journal of Materials Research*, 100, 461–482.

8 Gibbs, J.W. (1874) On the equilibrium of heterogeneous substances. *Transactions Connecticut Academy of Arts and Sciences*, 3, 198–248.

9 Srinivasan, S.G., Cahn, J.W., Jónsson, H., and Kalonji, G. (1999) Excess energy of grain-boundary trijunctions: An atomistic simulation study. *Acta Materialia*, 47, 2821–2829.

10 Caro, A. and Van Swygenhoven, H. (2001) Grain boundary and triple junction enthalpies in nanocrystalline metals. *Physical Review B*, 63, 134101.

11 Fortier, P., Palumbo, G., Bruce, G.D., Miller, W.A., and Aust, W.A. (1991) Triple line energy determination by scanning tunneling microscopy. *Scripta Metallurgica*, 25, 177–182.

12 Nishimura, G.S. (1973) Thermal etching at triple junctions on copper and silver surfaces. M.A. Sc. Thesis, University of Toronto, Toronto.

13 Galina, A.V., Fradkov, V.E., and Shvindlerman, L.S. (1988) Crystal surface shape near a threefold grain boundary vertex. *Physics, chemistry and mechanics of surfaces*, 1, 100–110.

14 Zhao, B., Verhasselt, J., Shvindlerman, L.S., and Gottstein, G. (2010) Measurement of grain boundary triple line energy in copper. *Acta Materialia*, 58, 5646–5653.

15 Zhao, B., Gottstein, G., and Shvindlerman, L.S. (2011) Triple junction effects in solids. *Acta Materialia*, 59, 3510–3518.

16 Gottstein, G., Shvindlerman, L.S., and Zhao, B. (2010) Thermodynamics and kinetics of grain boundary triple junctions in metals – recent developments. *Scripta Materialia*, 62, 914–917.

17 Zhao, B., Ziemons, A., Shvindlerman, L.S., and Gottstein, G. (2012) Surface topography and energy of grain boundary triple junctions in copper tricrystals. *Acta Materialia*, 60, 811–818.

18 Hertz, H. (1886) *Miscellaneous Papers*, Macmillan, London, p. 146.

19 Johnson, K.L., Kendall, K., and Roberts, A.D. (1971) Surface energy and the contact of elastic solids. *Proceedings of the Royal Society of London, A*, 324, 301–313.

20 Rabkin, E., Gottstein, G., and Shvindlerman, L.S. (2011) Why do crystalline nanoparticles agglomerate with low misorientations? *Scripta Materialia*, 65, 1101–1104.

9
Plastic Deformation by Grain Boundary Motion: Experiments and Simulations

Dmitri A. Molodov and Yuri Mishin

"Things that exist move and nothing remains still."
Heraclitus (about 535 – about 475 BCE)

9.1
Introduction

Grain boundaries (GBs) are important elements of materials microstructure that impact mechanical behavior, thermal stability, phase transformations, and many other properties and processes in technological materials. Plastic deformation of many polycrystalline materials is strongly influenced by crystallographic characteristics, structure, and chemistry of GBs. The impact of GBs is especially profound in ultrafine-grained and nanostructured materials [1–5], in which the atoms residing at or near GBs constitute a significant fraction. The GB effect on mechanical behavior is manifested, for example, in the Hall–Petch [6–9] relation predicting an increase in material's strength with decreasing grain size D due to the formation of dislocation pileups near GBs [6–9]. At small grain sizes on the order of 20–30 nm, the strength ceases to increase and the material can even become softer with further decrease in D. In this so-called inverse Hall–Petch regime [4,10–12], the deformation is dominated by GB sliding, GB migration, and other GB-related processes, with interiors of the grains remaining nearly dislocation free. Individual dislocations nucleate at GBs, traverse the grain, and get absorbed by another boundary without accumulation inside the grains.

A deeper fundamental understanding of the GB processes and their impact on mechanical behavior remain the goal of current research efforts employing both experimental and modeling approaches. In this chapter, we review the role of GB migration in the plastic deformation of polycrystalline materials. Specifically, we will focus on the phenomenon of GB motion coupled to shear deformation [13–15] and its role in microstructure evolution and plastic deformation. Some of the questions to be discussed in this chapter include: What is the coupled GB motion and how can it contribute to the deformation of polycrystalline materials? What are the mechanisms and geometric rules of coupling? What can we learn about stress-driven GB

Microstructural Design of Advanced Engineering Materials, First Edition. Edited by Dmitri A. Molodov.
© 2013 Wiley-VCH Verlag GmbH & Co. KGaA. Published 2013 by Wiley-VCH Verlag GmbH & Co. KGaA.

motion by computer simulations? Has the coupled GB motion been confirmed by experiments and do the experimental results always agree with theoretical predictions? What further work is needed in this field in order to elucidate the nature and mechanisms of the stress-driven GB motion and its role in microstructure development and mechanical behavior of materials?

We will start by describing the coupled GB motion and its main characteristics as revealed by experiments and simulations. To provide the background for the subsequent discussion, we briefly review the simulation and experimental methodologies currently employed in the studies of coupled GB motion. The rest of the chapter is devoted to an overview of the most important directions of the recent experimental and simulation work in the field, including the geometric theory of coupling and multiplicity of coupling modes, GB dynamics, stress-driven migration of asymmetric and more general GBs, and the interplay between the coupling effect and grain rotation. We conclude the chapter by giving a quick bird's eye view of the field: its retrospect, current status, and anticipated future developments. The common thread of all discussions throughout this chapter is the recognition that the recent breakthroughs in the understanding of this interesting phenomenon were made possible due to the synergistic interaction between the experiments, theory, and simulations.

9.2
What is the Coupled Grain Boundary Motion?

Consider a plane GB subject to an applied shear load. Figure 9.1 illustrates some of the possible mechanical responses of the boundary to the stress. Such responses include dislocation emission into the grains, crack nucleation and propagation, GB sliding (defined as rigid grain translation without GB motion), and GB migration coupled to shear deformation.

The latter response is caused by the existence of the *coupling effect*, in which an applied shear stress couples to the GB and induces its normal motion. In turn, the normal GB motion induced by the stress, or by a capillary or any other driving force whatsoever, produces shear deformation of the lattice swept by the GB motion. Because this lattice deformation is a simple shear parallel to the GB plane, the GB motion is also coupled to relative translation of the grains. Furthermore, it has been suggested [13], and verified by computer simulations [16–18], that coupled motion of a curved GB induces grain rotation. As proposed by Cahn and Taylor [13], a number of seemingly disparate materials phenomena originate from essentially the same physical effect, the coupling between normal GB motion and grain translations.

The coupling effect is characterized by a factor β defined as the ratio of the tangential grain translation s to the associated normal GB displacement d. Equivalently, β is the ratio of the tangential and normal GB velocities, $\beta = v_{\parallel}/v_n$ (Figure 9.1). The coupling is called perfect, or pure, if β is a geometric constant that depends only on the GB bicrystallography and not on the GB velocity, driving force, or other physical parameters.

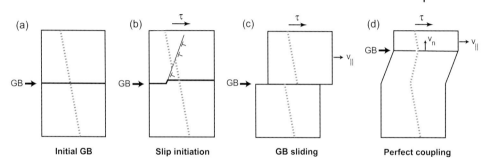

Figure 9.1 Schematic presentation of possible mechanical responses of a bicrystal with a plane GB to an applied shear stress τ [15]. (a) Initial bicrystal with the dotted line showing a set of inert markers. (b) The GB initiates slip by emitting a lattice dislocation. (c) Rigid GB sliding with a grain translation velocity $v_\parallel$; note the discontinuity of the marker line. (d) GB motion coupled to shear deformation, v_n being the normal GB velocity.

It should be emphasized that the coupled state of a GB represents its physical property that does not depend on the state of motion or presence or absence of applied driving forces. The coupled state exists due to atomically ordered structure of the GB core and its tendency to preserve this structure during displacements, regardless of whether such displacements are caused by a driving force or occur spontaneously. For example, it has been demonstrated [14,19,20] that spontaneous GB displacements induced by thermal fluctuations produce simultaneous grain translations that follow all geometric rules of coupling. In fact, the coupling factor can be extracted as the correlation coefficient between the GB displacements and concurrent grain translations during a simulated random walk of a stress-free boundary [19,20]. The stress-induced GB motion is one of many manifestations of the coupled state of the GB.

Stress-induced GB motion was observed by first-principles calculations [21–23], atomistic computer simulations [14,19,24–33], and early experiments on low-angle GBs [34,35]. More recently, the stress-driven motion has been found in experiments on high-angle GBs in both metals [36–52] and ceramic materials [42]. Over the recent years, significant progress has been achieved in understanding the atomic mechanisms, geometric rules, and dynamics of coupled GB motion. The coupling effect can be responsible for the stress-induced grain growth in nanocrystalline materials [45,53–56] and can play a role in the nucleation of new grains during recrystallization [57].

Coupled GB motion is a purely mechanical process similar to dislocation glide. It does not require atomic diffusion and is implemented by highly correlated atomic movements producing distortion and rotation of structural units forming the boundary structure. Diffusion may be needed, however, as an accommodation mechanism allowing the moving boundary to overcome obstacles and constraints imposed by the microstructure of a real material. Both simulations and experiments indicate that at high temperatures many GBs lose their coupling ability and respond to applied stresses by rigid sliding [14,17,32,48]. The nature of the coupling to sliding transition is not fully understood and remains the subject of current research [14,17,20,32].

9.3
Computer Simulation Methodology

Most of the simulations of GB motion conducted so far employed molecular dynamics (MD) with classical interatomic potentials [15]. While *ab initio* MD would guarantee more accurate interatomic forces, the large length and timescales required for the simulations of GB migration necessitate the use of computationally more efficient empirical models. Most of the MD studies employ the embedded-atom method [58], particularly for modeling aluminum and copper for which accurate interatomic potentials are now available [59,60].

A typical simulation geometry is illustrated in Figure 9.2a for the case of asymmetrical [001] tilt GBs in cubic materials [32]. The GB is created by constructing two separate crystals with appropriate lattice orientations and joining them along a chosen plane normal to the y-direction. Periodic boundary conditions are imposed in the x- and z-directions parallel to the GB plane, ensuring that the GB can freely migrate up and down without constraints. The tilt angle θ and the inclination angle φ are chosen so that to create a coincident site lattice (CSL) and align one of its planes parallel to the GB plane. The dimensions of the simulation block in the x- and z-directions comprise integral numbers of the CSL periods in order to avoid coherency strains. In the y-direction, the grains are terminated at free surfaces. Several atomic layers near each surface are exempt from the MD simulation and are used only to control the boundary conditions. The atoms in surface layer 2 are fixed in their perfect lattice positions, whereas atoms in surface layer 1 are fixed only in the

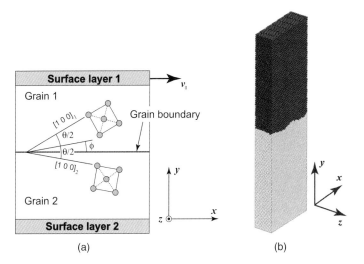

Figure 9.2 (a) Typical geometry of MD simulations of asymmetrical tilt GBs in cubic crystals [32]. The GB is characterized by a misorientation angle θ and an inclination angle φ. (b) Typical image from MD simulations of a Cu GB with $\theta = 16.26°$ and $\varphi = 14.04°$ at the temperature of 500 K. The bright and dark colors mark atoms assigned to different grains using the local orientation parameter [32].

y- and z-directions and can be moved with the same velocity $v_\parallel$ in the x-direction. Depending on the goal of the simulation, the system usually contains from 10^5 to 10^7 atoms.

The equilibrium GB structure is obtained by global energy minimization, which includes not only local atomic displacements but also rigid translations of the grains [61]. Often, the energy minimization requires removal or insertion of atoms, or heating the boundary to a high temperature and slowly cooling it down to allow a diffusional redistribution of atoms. The MD simulations are usually performed in the canonical (NVT) ensemble with temperature controlled by a Nose–Hoover or Langevin thermostat. After initial thermal equilibration of the system, a shear is applied by moving one surface layer relative to the other as described above. The stress arising in the system is constantly monitored using the virial expression for the stress tensor [62].

The MD simulations are usually run for time periods up to tens of nanoseconds. For typical system sizes, this time is close to the upper limit of MD timescales dictated by the speeds of today's computers. To collect detailed information on the GB motion, its displacement during the simulation run should be at least a few nanometers. Thus the lower bound of accessible GB velocities is on the order of 0.01–0.1 m/s. The MD timescale limitation creates a large gap between the simulated GB velocities and those implemented in typical experiments (0.1–10 μm/s). One way to narrow this gap is to apply the recently developed accelerated MD methodologies [63–67]. It has been demonstrated, for example, that by using the parallel replica dynamics method one can reach GB velocities as low as 1 mm/s [31].

A number of sophisticated visualization methods have been developed for examining the detailed GB structures and automatically tracking the GB position in the course of the simulation. One of such methods, based on a local orientation parameter, is illustrated in Figure 9.2b. For GBs composed of discrete dislocations, their slip traces contain important information about the dislocation motion and reactions, providing insights into GB migration mechanisms (Figure 9.3). Such traces can be visualized using the deformation fields method recently proposed by Tucker *et al.* [68].

Besides MD, the phase field crystal (PFC) methodology [18,32,69–76] has recently emerged as an effective tool for studying GB motion. The PFC methodology has been shown to reproduce the coupling effect for both two-dimensional [18,32,74] and three-dimensional GBs [76]. The MD and PFC methods are highly complementary to each other and work best in tandem as has been demonstrated in [32]. MD simulations are quantitative, usually conducted for a specific material, and are well suited for studying detailed atomic-level mechanisms of GB migration by examining multiple atomic trajectories. The already mentioned weakness of MD is that the timescale is limited to tens of nanoseconds, preventing access to diffusion-controlled processes. The latter limitation is overcome by the PFC methodology, which permits simulations on diffusive timescales and provides access to diffusion-controlled processes such as dislocation climb [69]. However, while certain geometric aspects of glide-mediated conservative GB motion are well reproduced by PFC on a quantitative level, the description of nonconservative climb-mediated motion

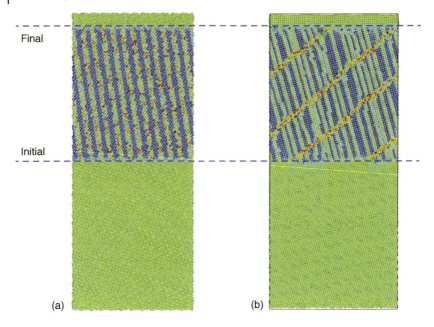

Figure 9.3 Traces of gliding dislocations in MD simulations of coupled GB motion in Cu at 500 K [32]. The dislocation traces are revealed using the microrotation vector method from [68]. The green and blue colors represent different localized lattice deformations. The initial and final GB positions are indicated. (a) Symmetrical tilt GB with $\theta = 16.26°$ and $\varphi = 0$ moves by collective glide of $\langle 100 \rangle$ dislocations. (b) Asymmetrical tilt GB with $\theta = 16.26°$ and $\varphi = 38.06°$ moves by collective glide and reactions of $\langle 100 \rangle$ (minority) and $1/2\langle 110 \rangle$ (majority) dislocations.

remains largely qualitative [32]. In addition, analysis of atomistic mechanisms presents certain challenges discussed in the recent literature [32].

9.4
Experimental Methodology

Stress-induced GB motion was investigated in dog-bone-shape specimens with a cross-section of about 4.9 mm × 4.7 mm (Figure 9.4a) fabricated from specially grown Al bicrystals. The specimens were subject to a constant tensile load ranging from 5 to 20 N at different elevated temperatures. The load created a shear stress parallel to the GB plane, which was aligned at 45° to the tensile axis. Details related to the crystal growth, bicrystal characterization, and sample preparation can be found elsewhere [43,44,49]. Both discontinuous and continuous methods were used for tracking the GB migration. Most experiments were conducted using a step-wise annealing technique in which the boundary positions were measured prior to loading and after unloading. The boundary displacement was revealed by optical

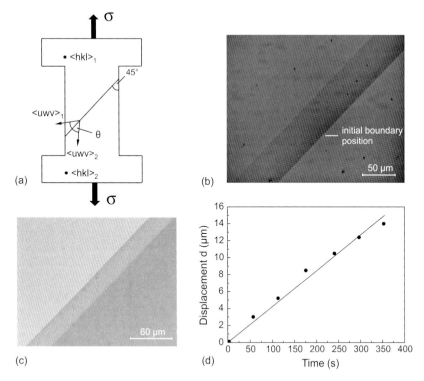

Figure 9.4 (a) Geometry of bicrystalline specimens for the tensile loading used in experiments. (b) Optical contrast due to altered reflectivity of the aluminum oxide layer in the sheared crystal region behind a moving 76.4°⟨001⟩ tilt boundary. The sample was annealed for 30 min at 380 °C under an applied tensile stress of 0.26 MPa. Note the traces of the initial and final GB positions on the specimen surface. (c) Orientation contrast image of the bicrystal surface in SEM; (d) Displacement versus time plot for the GB migration at 380 °C.

contrast on the surface of the bicrystal (Figure 9.4b) and was measured by optical microscopy [43,44,49]. Most recent measurements were performed by *in situ* observations in which the GB migration was recorded in a scanning electron microscope (SEM) using a commercial hot deformation (tension-compression) stage [1] integrated in a SEM JEOL 820. The stage was adapted for application of normal stresses to bicrystalline samples with the required geometry (Figure 9.4a) at elevated temperatures up to 850 °C [49]. GB motion in response to the applied shear stress was observed and measured *in situ* using the orientation contrast on the specimen surface revealed by an electron backscatter detector (Figure 9.4c and d) [49]. For geometric characterization of the shear-migration relation, the surface of the specimen was intentionally scratched parallel to the tensile axis prior to the

1) Kammrath & Weiss company.

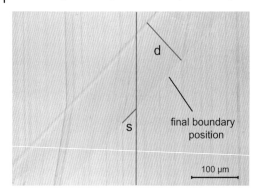

Figure 9.5 Coupling between GB migration and shear strain is characterized by the factor β defined as the ratio of the lateral grain translation s to the normal boundary displacement d: $\beta = s/d$ [43]. Note the zig-zag shape of the initially vertical marker lines in the region swapped by the boundary motion.

annealing in order to produce reference marks (Figure 9.5). Analysis of the recorded orientation image sequences with a special software reveals the normal boundary displacement d and the lateral grain translation s (Figure 9.5), permitting calculation of the migration velocity v and the coupling factor $\beta = s/d$.

The GB mobility m was determined from the constitutive relation

$$v = mp \tag{9.1}$$

where p is the driving force for the GB motion. It can be shown that for a tilt boundary subject to a shear stress τ,

$$p = 2\tau\cos(\psi)\tan(\psi/2) \tag{9.2}$$

Here $\psi/2$ is the angle between the GB normal and the effective Burgers vector **B** per unit GB length as determined from the Frank–Bilby equation [14,30]. For example, for $\langle 001 \rangle$ tilt boundaries $\psi = \theta$ for the $\langle 100 \rangle$ coupling mode and $\psi = \theta - 90°$ for the $\langle 110 \rangle$ coupling mode. For low-angle GBs, for which ψ is either small or close to 90°, Eq. (9.2) can be approximated by [44]

$$p \approx \tau\sin(\psi) \tag{9.3}$$

9.5
Multiplicity of Coupling Factors

One of the fascinating features of the coupling effect is that the coupling factor β is a *multivalued* function of crystallographic angles of the boundary. Due to this multiplicity, physically the same GB can behave as having different coupling factors and can demonstrate different mechanical responses to the same applied stress. These different *coupling modes* are characterized by not only different β values but also different atomic mechanisms of GB motion, critical stresses, temperature dependences of β, and other characteristics [14,29,30]. The active coupling mode not only depends on the geometric parameters of the particular boundary but also varies with the orientation of the applied shear stress and its magnitude relative to the critical resolved shear stresses of different coupling modes [14]. In turn, the critical stresses of the coupling modes can vary with temperature and their crossovers can produce mode switches.

The multiplicity of coupling modes originates from point symmetry of the crystal lattices and can be explained as follows. The shear deformation produced by the coupled GB motion can be thought of as resulting from two processes: (i) plastic deformation of the lattice by the passage of the dislocation content of the boundary, and (ii) rotation of the deformed region to align its lattice with the lattice of the advancing grain. The shear deformation produced at the first step depends on the dislocation content of the boundary, which for a general GB is defined by the Frank–Bilby equation [30,77]. The latter is known [30,77] to generate multiple solutions due to the crystal symmetry, leading to the multiplicity of the dislocation content of any GB. But different dislocation contents produce different shear deformations of the lattice leading to different coupling factors. Thus, the existence of the coupling modes can be traced back to different solutions of the Frank–Bilby equation.

The coupling modes can also be explained by focusing on the lattice rotation step. To ensure continuity of the lattice left behind the moving boundary, the receding lattice must be rotated around the tilt axis by the misorientation angle θ. If the lattice possesses n-fold rotational symmetry around the tilt axis, rotations by the angles $\theta + k(2\pi/n)$, $k = 1, \ldots, n-1$, produce physically identical states of the lattice. But these different rotation angles lead to different relative translations of the two grains and thus different coupling factors.

For the particular case of $\langle 001 \rangle$ tilt GBs in cubic materials, the fourfold symmetry around the tilt axis generates four possible coupling modes, with the coupling factors $\beta = 2\tan(\theta/2 + \pi k/4)$, $k = 0, 1, 2, 3$ [14,30]. Although all four modes are geometrically possible, only two of them, corresponding to the smallest magnitude of β, have been seen in MD simulations [14,29,30] and experiments [44,47–49,52]. These two coupling modes are referred to as $\langle 100 \rangle$ and $\langle 110 \rangle$ type and have the coupling factors [14,29,30]

$$\beta_{\langle 100 \rangle} = 2\tan\left(\frac{\theta}{2}\right) \tag{9.4}$$

and

$$\beta_{\langle 110 \rangle} = 2\tan\left(\frac{\theta}{2} - \frac{\pi}{4}\right) \tag{9.5}$$

respectively. In the limit of small θ, the symmetrical tilt GB is composed of dislocations with the Burgers vector $\langle 100 \rangle$ moving in the $\langle 100 \rangle$ mode. In the other limit, when θ is close to $90°$, we again have a low-angle GB but this time composed of $1/2\langle 110 \rangle$ dislocations moving in the $\langle 110 \rangle$ mode. In between, the boundary can, in principle, exhibit either of the two coupling modes. Note that the two coupling

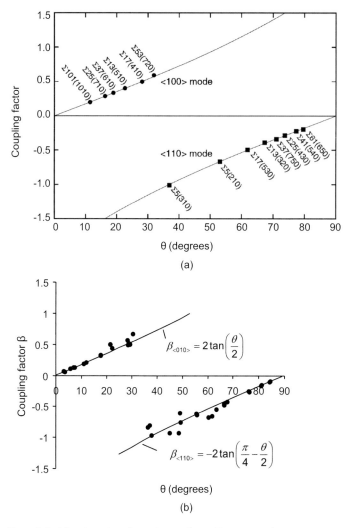

Figure 9.6 Misorientation dependence of the coupling factor β of $\langle 001 \rangle$ symmetrical tilt GBs in FCC metals. (a) MD simulations for copper at 800 K [14,29,30]. (b) Experimental measurements on Al bicrystals [48]. The lines indicate the geometric predictions by Eqs. (9.4) and (9.5).

factors have opposite signs, with $\beta_{\langle 100 \rangle} > 0$ and $\beta_{\langle 110 \rangle} < 0$. In other words, in response to the same shear, the boundary can move in one direction when coupled in the $\langle 100 \rangle$ mode and in the opposite direction when coupled in the $\langle 110 \rangle$ mode.

Figure 9.6a displays the misorientation dependence of the coupling factors computed by MD simulations for $\langle 001 \rangle$ symmetrical tilt GBs in Cu. Note the excellent agreement between the MD results and the predictions of Eqs. (9.4) and (9.5) for both low-angle and high-angle GBs. An important effect revealed by this plot is that the coupling factor is discontinuous: it abruptly switches from one coupling mode to the other at an angle θ_c of approximately 36°. This mode switching is accompanied by a change of sign of β. As a consequence of this discontinuity, all GBs with $\theta < 36°$ move in one direction, whereas the boundaries with $\theta > 36°$ move in the opposite direction in response to the same shear. A closer examination shows that near the critical angle of about 36°, the GBs can display the so-called dual behavior [14,30]. Namely, the GB starts moving in one coupling mode but later switches to the other and begins to move in the opposite direction, occasionally followed by a switch back to the initial mode, and so on.

For comparison, Figure 9.6b shows the results of recent experimental measurements on Al bicrystals [44,47–49,52]. The striking observation is that the experimental results closely follow the theoretical predictions. The coupling factors agree well with Eqs. (9.4) and (9.5) within the statistical scatter of the data points. Furthermore, the same discontinuity is observed between the angles of 30.5° and 36.5°, which is consistent with the simulation predictions. As noted in [15,32], the exact value of the critical angle θ_c is not dictated by crystal symmetry and may depend on the particular material, crystal structure, and temperature. The good agreement between the simulations for Cu and experiments on Al suggests that the material dependence of the critical angle is rather weak, as long as both metals are FCC. Furthermore, the fact that this agreement holds at several different temperatures indicates that the temperature dependence is likely to be also small, provided that the temperature is below the coupling to sliding transition. On the other hand, simulations suggest that a temperature independence of the switching angle may appear at low temperatures $< 0.4T_m$ as indicated on the coupling diagram in Figure 8 of Ref. 14 (T_m being the melting temperature).

An important observation is that the mode switching angle found in both experiments and simulations is smaller than 45° (Figure 9.6). This nontrivial fact has the following dislocation interpretation. The motion of dislocation boundaries in the two coupling modes is mediated by glide in the $\langle 100\rangle\{100\}$ and $\langle 110\rangle\{110\}$ slip systems, respectively. The competition between the two coupling modes is controlled by the respective Peierls–Nabarro stresses as well as the Schmidt factors on the two slip planes [14]. The magnitude of the Schmidt factor is symmetric with respect to the 45° tilt angle. Therefore, if the Peierls–Nabarro stresses of the two types of dislocations were equal, the coupling mode would switch at exactly 45°. That the switching angle is in fact substantially below 45° indicates that the $\langle 100\rangle\{100\}$ slip has a higher Peierls–Nabarro stress than the $\langle 110\rangle\{110\}$ slip.

This dislocation argument can be extended to high-angle GBs by analyzing gamma-surfaces corresponding to the coupling modes. For $\langle 001\rangle$ symmetrical tilt GBs in Cu, comparison of the $\langle 100\rangle\{100\}$ and $\langle 110\rangle\{110\}$ gamma surfaces (Figure 24 in Ref. 14) revealed that the slip responsible for the $\langle 100\rangle$ coupling mode is more difficult than the slip responsible for the $\langle 110\rangle$ mode, explaining why the switching angle is less than 45°. This link between the mode-switching angle and the Peierls–Nabarro stresses was additionally confirmed by parametric calculations in the recent PFC study, in which the model parameters were adjusted to increase or decrease the Peierls–Nabarro stress for the $\langle 100\rangle\{100\}$ slip system [32]. As expected, a decrease in the Peierls–Nabarro stress resulted in an increase of the switching angle to nearly 45°.

9.6
Dynamics of Coupled GB Motion

Significant research efforts were put in understanding the dynamics of stress-driven GB motion. The attention was focused on explaining the frequently seen stop-and-go character of the GB motion, the velocity-force relation, and the GB mobility.

MD simulations have shown that at relatively low temperatures and/or large velocities, the GB motion exhibits a remarkably regular stick–slip behavior [19,31,32]. At a fixed GB velocity, the stick–slip motion is characterized by a sawtooth time dependence of the shear stress as illustrated in Figure 9.7. The magnitude of the peak stress increases with the GB velocity v and decreases with temperature T. This can be explained by the thermally activated nature of the mechanisms responsible for the GB motion. The elementary steps of this process require overcoming thermal barriers after numerous attempts. For example, at low temperatures, the elementary step consists in nucleation of a disconnection loop of a critical size. The applied stress suppresses the nucleation barrier, which increases the probability of successful attempts. At a critical value of the stress, the barrier turns to zero and the GB motion becomes athermal. This athermal migration mechanism usually dominates at very low temperatures. At high temperatures, thermal fluctuations help the boundary overcome the barrier before it disappears, that is, at stresses below critical. The higher the temperature, the lower is the stress needed for sustaining the GB motion at a given speed. This qualitatively explains the reduction of the peak stress with temperature. On the other hand, at a fixed temperature, higher GB velocities give the thermal fluctuations less time to overcome the barrier and the peak stress increases.

The stick–slip GB dynamics can be described by a simple one-dimensional model proposed in [19,31]. The model conceptually represents the boundary by a particle attached to an elastic rod that drags the particle through a periodic potential with a constant velocity v. In the overdamped regime of friction, the model predicts stick–slip motion of the particle with $\log v$ proportional to $-\left(\tau_c^0 - \tau_c\right)^{3/2}/k_B T$, where τ_c is the critical resolved shear stress at the simulated temperature T and τ_c^0 is its value at 0 K. This relation correctly predicts that at a fixed temperature, the critical stress increases with the velocity in a nonlinear manner. When the velocity is fixed, the model predicts that τ_c must decrease with temperature as $T^{2/3}$. This behavior is qualitatively consistent with the MD simulations of an Al GB shown in Figure 9.7. The $T^{2/3}$ relation has been found to hold quantitatively for both symmetrical [19] and asymmetrical [33] GBs in copper over a wide temperature range. This agreement demonstrates that, despite the obvious simplicity, the proposed model captures the essential physics of the stick–slip process at GBs. This agreement also highlights the similarity between the coupled GB motion and many other natural phenomena

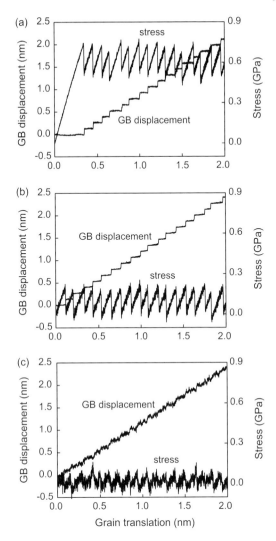

Figure 9.7 GB displacement and shear stress at (a) 100 K, (b) 500 K, and (c) 900 K at an imposed grain translation velocity of 1 m/s for the Σ21 GB in Al [19]. The average level of the peak stress decreases with temperature. Above 900 K, the stress behavior becomes noisy and the boundary makes occasional backward jumps. Through these changes in GB dynamics, the coupling factor β remains very close to its ideal (geometric) value.

involving stick–slip dynamics, ranging from squeaking door hinges to the sound of a violin to the tip movements in atomic friction microscopy [78–81].

In the stick–slip mode, the GB executes a series of stop-and-go displacements (jumps) in the direction of the applied force. At high temperatures and/or slow velocities, the saw-tooth stress behavior becomes increasingly noisier due to the growing thermal fluctuations. Eventually, the fluctuations begin to cause occasional backward jumps whose probability increases with temperature. At even higher temperatures, the stick–slip GB dynamics transform to driven random walk (Brownian regime) [19,31]. In this regime, the role of the applied stress is to bias the rates of the spontaneous forward and backward jumps, driving the GB position on average forward. The Brownian regime is characterized by a linear stress–velocity relation represented by Eq. (9.1). It should be emphasized that the dynamic transition from the stick–slip dynamics to the Brownian motion does not necessarily affect the coupling factor. In fact, the MD simulations [19] present examples of GBs that remain perfectly coupled through this dynamic transition and maintain an accurate geometrical value of β.

In experimental studies, the GB is usually moved by a fixed level of applied stress and the GB dynamics are characterized by the mobility coefficient m and its Arrhenius parameters. As one illustration, Figure 9.8 shows velocity plots for two $\langle 001 \rangle$ symmetrical tilt GBs in Al bicrystals [43]. Note the linear relation between the GB velocity and the applied shear stress over the range of stresses from 0.05 to

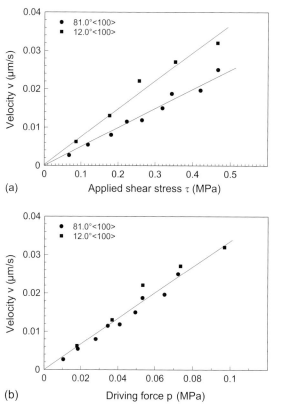

Figure 9.8 Experimental migration rate of two $\langle 001 \rangle$ symmetrical tilt GBs in Al at 320 °C [43]. (a) Velocity versus shear stress τ parallel to the boundary plane. (b) Velocity versus the driving force p.

0.5 MPa. This linearity verifies Eq. (9.1) and is well consistent with theoretical models of coupling [14,19,20] and the MD simulations of Al GBs [19]. As evident from Figure 9.8a, the 12.0° boundary moves with a larger velocity than the 81.0° boundary over the entire loading range. This does not mean, however, that the 12.0° boundary has a higher mobility m. As seen in Figure 9.8b, when replotted as velocity versus the actual driving force p, the results for the two GBs nearly coincide, giving statistically identical mobilities of about $m = 3.3 \times 10^{-13}$ m^4/Js.

Mobility measurements at different temperatures confirm that stress-driven GB migration is a thermally activated process whose rate at a fixed driving force increases with temperature according to the Arrhenius relation

$$v = v_0 \exp\left(-\frac{H}{k_B T}\right) \tag{9.6}$$

where H is the activation enthalpy of migration. Figure 9.9 displays misorientation dependences of the activation parameters H and $m_0 = v_0/p$ for $\langle 001 \rangle$ Al GBs

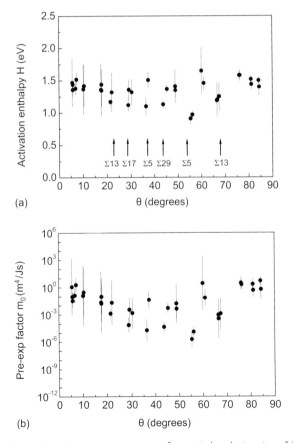

Figure 9.9 Activation parameters of stress-induced migration of $\langle 001 \rangle$ symmetrical tilt GBs in Al [48]. (a) The activation enthalpy. (b) The preexponential factor of mobility.

measured in the temperature range from 280 °C to 400 °C [48]. Note that in the low-angle regimes of $\theta < 18°$ and $\theta > 76°$, the activation parameters do not vary significantly, with the activation enthalpy remaining around 1.45 eV. In contrast, the high-angle region is characterized by a marked misorientation dependence of the activation parameters, with lower values of H and m_0 corresponding to GBs with misorientations close to low-Σ CSLs. This behavior suggests that the migration mechanism of low-angle GBs is independent of the tilt angle but is significantly influenced by the GB structure variations in the high-angle regime.

9.7
Coupled Motion of Asymmetrical Grain Boundaries

Most of the simulations and experiments have been focused on symmetrical tilt GBs due to the simplicity of their crystallographic description and atomistic structure. In real polycrystalline materials, however, most GBs have a mixed (tilt/twist) character, and those that are nearly tilt type are usually not symmetrical. It is, therefore, important to push the modeling and experimental efforts toward more complex boundaries. It is only since recently that stress-driven motion of *asymmetrical* tilt GBs has became the subject of modeling and experiment. Among many questions that need to be answered, the most basic ones are: Can such boundaries be moved by applied shear stresses? And if they can, what are the geometric rules and atomic-level mechanisms of their shear-coupled motion?

As already mentioned, both simulations and experiments confirm that low-angle symmetrical tilt GBs subject to shear stresses move by collective dislocation glide in parallel slip planes [34,35,43,44,48,82]. The relative simplicity of this mechanism derives from the fact that the dislocations forming such boundaries have the same Burgers vector and can glide in the same slip system. In contrast, low-angle asymmetrical GBs are composed of at least two different types of dislocations. For example, $\langle 001 \rangle$ asymmetrical tilt GBs contain a mixture of $\langle 100 \rangle$ and $1/2\langle 110 \rangle$ dislocations. It could be expected that the dislocations gliding in intersecting slip planes would block each other and prevent the coupled motion as suggested by Read and Shockley [82]. Nonetheless, recent MD simulations [32,83–85], bicrystal experiments [52,86], and the fact that coupled GB motion is observed in polycrystalline materials suggest that the ability to couple to applied stresses is not limited to symmetrical boundaries. To date, however, the exact geometric rules that would permit predictions of coupling factors from crystallographic data have not been established. An understanding of the mechanisms by which the dislocations gliding in intersecting slips plane can avoid blocking each other is only beginning to emerge [32]. Much less is known about the mechanisms of coupled motion of large-angle asymmetrical tilt GBs, let alone the mixed boundaries.

Geometric analysis [14] suggests that the coupling factor should be independent of the inclination angle φ as long as the coupling mode remains the same. According to this theory, departures of the GB plane from a symmetrical tilt inclination should preserve not only the coupled motion itself but also the coupling factor for both low-

and high-angle boundaries. Some of the MD simulations attempting to test this prediction produced inconclusive results, with coupling factors varying with the inclination angle in both magnitude and sign [83–85]. A more coherent picture has emerged from the recent study [32] combining crystal symmetry analysis with computer simulations by two complementary methodologies: MD and PFC.

The crystal symmetry analysis [32] suggests that for $\langle 001 \rangle$ tilt GBs, the plot of the coupling factor β as a function θ and φ can be represented by two surfaces separated by a cut as shown schematically in Figure 9.10a. The surfaces represent the $\langle 100 \rangle$ and $\langle 110 \rangle$ modes of coupling mentioned above and the cut corresponds to the discontinuous transition between the two modes accompanied by a reversal of sign. Note that symmetrical GBs occur at both $\varphi = 0$ and $\varphi = \pm 45°$. If for symmetrical GBs with $\varphi = 0$ the mode switching is found at particular tilt angle,

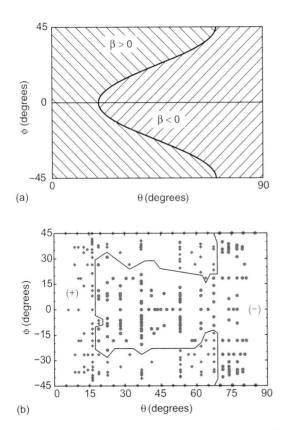

Figure 9.10 (a) Schematic dependence of the coupling factor β on the tilt angle θ and inclination angle φ, showing the areas of positive and negative β. The line represents the discontinuity of β and has a tentative shape reflecting only the crystal symmetry. (b) Summary of PFC calculations of the coupling factor for $\langle 001 \rangle$ tilt GBs [32]. The diamond and circle symbols indicate positive and negative β values, respectively. The line outlines the approximate boundary between the two coupling modes with different signs of β.

say $\theta_c = 36°$, then for symmetrical boundaries with $\varphi = \pm 45°$ it should occur at $\theta_c = 90° - 36° = 54°$. Accordingly, all boundaries with $\theta < 36°$ are predicted to have a positive coupling factor given by Eq. (9.4), whereas all boundaries with $\theta > 54°$ are predicted to have a negative coupling factor given by Eq. (9.5). For boundaries with $36° < \theta < 54°$, β is positive at and near $\varphi = 0$ but must switch to negative values when approaching $\varphi = 45°$ and $\varphi = -45°$.

These predictions were tested by MD and PFC simulations of a large set of $\langle 001 \rangle$ boundaries. Figure 9.10b summarizes the results of the PFC simulations, indicating the positive and negative values of β by different symbols. Despite the statistical scatter of the points, the shape of the line separating positive and negative coupling factors is in qualitative agreement with the theoretical predictions. This diagram demonstrates that the discontinuity between the two coupling modes exists not only for symmetrical GBs [14,29,30,44,47–49,52] but also extends over the entire range of inclination angles. This diagram also explains the sudden changes in the sign of β observed in some of the previous MD simulations [83–85]. More importantly, it demonstrates that the asymmetrical GBs couple to shear stresses for *all* possible tilt and inclination angles.

The same set of GBs was studied by MD simulations using Cu and Al as model materials [32]. The results were found to be qualitatively similar for both metals and in close agreement with the PFC findings, suggesting a generic character of the results. In particular, the overwhelming majority of the asymmetrical GBs studied by MD coupled to shear stresses and was moved by them. This study clearly demonstrates that the coupling effect is *not* an attribute of only symmetrical GBs. The existence and the shape of the discontinuity line shown in Figure 9.10 were also verified by the MD results. At the same time, both MD and PFC simulations indicate that the magnitude of β for a given coupling mode does depend on the inclination angle, a finding which does not confirm the geometric predictions [14]. Figure 9.11 shows that the deviations from the geometrically predicted β values are largest at the inclination angles of about $|\varphi| \approx 15°$ and can be quite significant in magnitude. Such deviations are predominantly positive, suggesting the existence of a partial sliding component along with coupling. Note that the plot in Figure 9.11 is not symmetrical about $\phi = 0$, which does not contradict the crystal symmetry of these boundaries. Changing the sign of the inclination angle can produce a different coupling factor. Equivalently, reversing the direction of the applied shear stress can change the magnitude of the measured β.

The MD simulations revealed several mechanisms of stress-driven GB migration that were presented in detail in [32]. In particular, this study provides an answer to the long-standing mystery [82] of how the dislocations gliding in intersecting slip planes avoid blocking each other. The dislocations alleviate the locks by several mechanisms, most notably by chains of dislocation dissociation–recombination reactions propagating along the boundary. When diffusion is allowed, the locks can also be overcome by dislocation climb as suggested by PFC observations [32]. The operation of complex mechanisms responsible for the locking-unlocking of the dislocations can explain the observed deviations of the coupling factor from the ideal

9.7 Coupled Motion of Asymmetrical Grain Boundaries

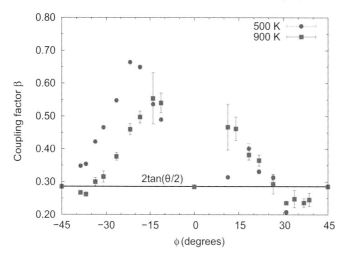

Figure 9.11 MD simulation results for the coupling factor of asymmetrical GBs in Al with $\theta = 16.26°$ as a function of the inclination angle φ [32]. The simulation temperatures are indicated in the legend. The horizontal line indicates the ideal coupling factor.

geometrical value (Figure 9.11). However, a more detailed mechanistic understanding of such deviations requires further studies.

As indicated above, the simulation finding that asymmetrical GBs can be moved by applied shear stresses is well consistent with many experimental observations, including the stress-driven grain growth in nanocrystalline materials [45,55,56,87]. A more detailed quantitative information supporting this conclusion comes from experiments on bicrystalline samples subject to tensile loads [52,86]. As an example, Figure 9.12 demonstrates stress-driven motion of an Al $\langle 001 \rangle$ tilt boundary with $\theta = 17.4°$ and $\varphi = 19.1°$ [52]. The coupling factor measured for this particular GB is $\beta = 0.39$, which is higher than the ideal geometric value of 0.31 computed from Eq. (9.4). The trend for higher than ideal coupling factors is well consistent with the MD and PFC simulations [32]. In fact, the experimentally measured coupling factor can be *directly* compared with the MD result shown in Figure 9.11. The boundary closest to the experimental conditions is the one with $\theta = 16.26°$ and $\varphi = 18.44°$. The computed coupling factor for this boundary is 0.40 at the temperature of 500 K and 0.38 at 900 K. Both numbers are in excellent agreement with the experiment value of 0.39 [52].

Attempts to extend the simulations and experiments to more complex GBs are at early stages. An example of work in this direction is offered by the recent study of the $\Sigma 7$ tilt boundary with different directions of the applied shear. The goal was to explore the multiplicity of possible crystallographic descriptions of the same GB using stress-driven motion as a test. Indeed, in a cubic crystal physically the same GB can have 24 different axis-angle descriptions, all of which are crystallographically legitimate. For example, the same $\Sigma 7\{123\}$ boundary can be described as $\theta = 38.2°$ tilt around $\langle 111 \rangle$, or $\theta = 73.4°$ tilt around $\langle 210 \rangle$, or $\theta = 135.6°$ tilt around $\langle 112 \rangle$, all three symmetrical tilt descriptions being equally legitimate. However, according to

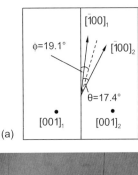

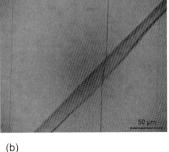

Figure 9.12 (a) Geometry of a bicrystal with a ⟨001⟩ asymmetrical tilt GB with $\theta = 17.4°$ and $\varphi = 19.1°$. (b) Optical image of the bicrystal surface demonstrating shear-coupled GB migration [52].

the geometric theory [14] each description predicts a different coupling factor. The experiment was intended to demonstrate that these different descriptions can be implementing as the respective coupling modes.

To this end, bicrystals were prepared that contained the same $\Sigma 7\{123\}$ boundary but with orientations of the crystallographic axes relative to the GB plane and the tensile load direction that would approximately correspond to each of the three alternate descriptions [51]. The results come out to be interesting but mixed. On one hand, all three GBs coupled to the applied stress and moved with markedly different velocities and activation energies. This pointed to the existence of three modes of coupling corresponding to the three crystallographic descriptions. On the other hand, in all three cases the moving boundary did not produce any shear deformation of the lattice, which corresponds to the coupling factor of *zero* (Figure 9.13). To highlight the contrast, a non-Σ tilt boundary with $\theta = 23.6°$ around ⟨111⟩ was also tested [51] and was found to exhibit shear-coupled motion with a β value matching the geometric prediction [14]. To make things even more complicated, MD simulations performed on the same set of GB orientations revealed perfect coupling with geometric β values in two cases and a mixture of coupling and sliding in the third case (V.A. Ivanov and Y. Mishin, 2008, unpublished research). The discrepancy between the simulations and experiment, and especially the observation of the extremely puzzling shear-driven GB motion with zero coupling factor call for continuing efforts combining theory, modeling, and experiment.

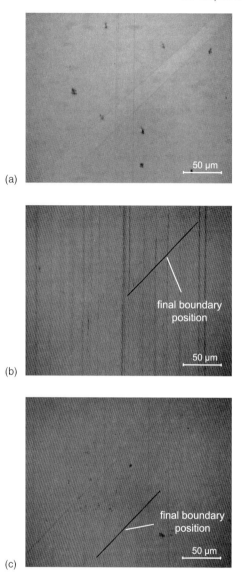

Figure 9.13 Stress-induced migration of (a) 38.1°⟨111⟩ GB, (b) 74.2°⟨210⟩ GB, and (c) 136.8°⟨112⟩ GB in Al [51]. Note that the GBs are moved by the applied shear stress without producing any shear deformation (no bending of the marker lines).

9.8
Coupled Grain Boundary Motion and Grain Rotation

An interesting consequence of the coupling effect is that coupled motion of a *curved* GB must induce grain rotation. Cahn and Taylor [13] proposed a grain rotation model for an isolated cylindrical grain with a circular cross-section shrinking by capillary forces. One of the remarkable predictions of their model is that a low-angle GB can induce grain rotation toward *higher* misorientation angles and thus larger GB free energy γ. This unusual rotation was indeed observed in two-dimensional MD and PFC simulations [16,18]. Many questions remain unresolved and more work is needed to better understand the driving forces and mechanisms of the grain rotation. The interest in this topic is largely motivated by the role of grain rotation in microstructure evolution, plastic deformation, phase transformations, recrystallization, and grain coarsening [88–96].

Recently, the grain rotation effect and its relation to coupling have been studied by MD simulations [17] employing the cylindrical grain geometry shown in Figure 9.14. The capillary-driven shrinkage of this grain was studied over a wide range of temperatures (from room temperature to the melting point), grain sizes, initial misorientation angles, and other conditions. The study included the effect of imposed constraints and an applied torque on the GB dynamics. Some of the findings of this work are summarized below.

At relatively low temperatures ($\lesssim 0.8 T_m$) when the GB was coupled, its capillary-driven shrinkage produced grain rotation in agreement with the theoretical predictions [13]. The grain rotation is illustrated in Figure 9.15 using a colored vertical stripe playing the role of a marker line. Note the rotation of the GB segment located inside the shrinking grain relative to the outer grain. The zig-zag shape of the marker line left after the grain has completely disappeared is a signature of the shear deformation produced by the GB migration. The grain rotated with an increasing or decreasing misorientation angle θ, depending on whether the initial angle θ_0 was below or above $\theta_c \approx 36°$ (Figure 9.16). Thus, at $\theta_0 < \theta_c$ the simulations confirm the rotation from low-angle to high-angle misorientations. The reversal of the rotation direction at θ_c can be explained by the existence of the discontinuous transition

Figure 9.14 Simulation block used in MD simulations of grain rotation [32]. The atomic interactions are described by an embedded-atom method potential for Cu [60]. The atoms are colored by the orientation parameter. The dark color reveals the circular cross-section of the embedded cylindrical grain.

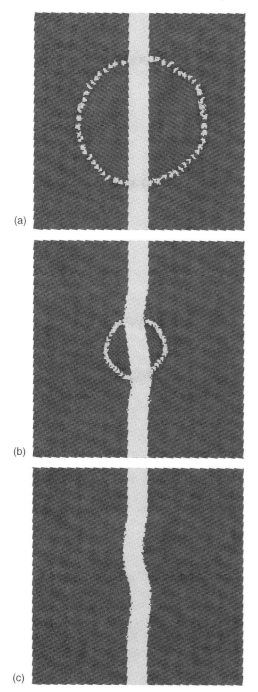

Figure 9.15 Selected snapshots of a shrinking grain with the initial misorientation $\theta_0 = 16°$ at 900 K [32]. To illustrate the grain rotation, the atoms initially located within a vertical stripe were colored in a bright color to track their motion. (a) Early stage of simulation; (b) The grain is close to disappearing; (c) The grain has disappeared.

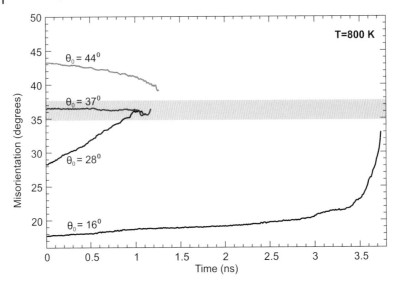

Figure 9.16 Time dependence of the lattice misorientation angle θ for different initial misorientations θ_0 of an inclosed cylindrical grain in MD simulations at the temperature of 800 K [17]. The horizontal stripe shows the narrow angular range in which the rotation direction changes sign.

between the two coupling modes with opposite signs of the coupling factor (cf. Figure 9.10). A boundary with exactly this angle is "frustrated" by the two conflicting coupling modes and is unable to rotate. Although the critical angle θ_c did not show any temperature dependence within the statistical errors [17], it is likely that it can depend on the crystal structure and other material properties.

Many results of the simulations [17] were found to be in quantitative agreement with predictions of the Cahn and Taylor model [13]. In particular, in several cases when the model predicts parabolic kinetics, such kinetics were indeed observed in the simulations. However, there is one important aspect in which the simulation results departed from the theoretical predictions. No matter what the initial angle, temperature, or other simulation conditions, the coupling was never perfect. This conclusion is in stark contrast to the simulations of plane GBs which, as discussed above, can exhibit perfect coupling with β values matching the geometric predictions with astonishingly high accuracy. During the shrinkage of a *curved* GB, its coupled motion was *always* accompanied by at least some amount of GB sliding.

The reason for the imperfect coupling during the GB shrinkage can be explained by the fact that the dislocation content of the GB was never conserved. Analysis of the GB structures revealed that the GB was gradually and irreversibly losing part of its dislocation content during its motion. This loss of dislocation content produced additional grain rotation toward smaller angles and thus GB sliding. Generally, sliding along a curved GB always alters its dislocation content in order to accommodate the change in θ. Thus, the sliding process requires either nucleation or annihilation of dislocations, depending on the direction of the sliding. It is only for a plane GB that sliding can preserve the misorientation and does not require nucleation or annihilation of dislocations. In the simulations reported in [17], the sliding component was always toward decreasing angles and thus required annihilation of dislocations. Analysis of MD configurations revealed the typical mechanisms of the dislocation annihilation. In most cases, the dislocation content rapidly propagated along the GB by a chain of dislocation reactions until the dislocations with opposite Burgers vectors were able to recombine. Furthermore, it was found that the same reaction-based mechanism is responsible for the motion of a curved GB without dislocation climb or formation of locked configurations. The analysis also suggests that the rate-controlling step in the motion of *curved* GBs is the annihilation or nucleation of GB dislocations rather than their motion.

In real polycrystalline materials, the GBs are subject to various constraints that may affect both their coupled motion and grain rotation. In some of the simulations [17], the central region of the grain was artificially fixed in order to prevent its rotation. The GB was nevertheless shrinking by the capillary forces but now without grain rotation. This constrained grain shrinkage occurred significantly slower than the shrinkage of the free grain. The "frozen" grain rotation resulted in additional GB sliding, which required higher stresses than the coupled motion and led to the retardation of the GB motion. In other words, free grain rotation assists the GB motion, whereas constraints on rotation can significantly reduce the GB mobility.

To explore the effect of applied stresses on the capillary-driven GB motion, simulations were performed with a torque applied to the shrinking grain [17]. When the direction of the torque coincided with the direction of grain rotation, the GB migration was significantly accelerated. When the torque was against the grain rotation, the GB migration could be stopped or even *reversed*. In the latter case, the GB began to move against its curvature resulting in growth of the grain. The motion against curvature is perhaps one of the most impressive manifestations of the coupling effect. It is also a demonstration that due to the existence of coupling, applied stresses can induce grain growth, a process that was indeed observed in experiments on nanocrystalline Al thin films subject to tensile loads [45,55,56].

The most direct and convincing evidence of grain rotation accompanying coupled GB motion was reported in recent experiments on Al bicrystals [50]. The boundary studied was mixed type, with the $\langle 001 \rangle$ misorientation of $\theta = 18.2°$ and a twist component of $\psi = 20°$. This boundary readily coupled to stress as was evidenced by its stress-driven motion producing shear deformation parallel to the GB plane (Figure 9.17). At the same time, examination of the polished surface of the sample revealed displacements of its ends in opposite directions suggesting mutual rotation of the two grains (Figure 9.18). This was confirmed by measuring the surface topography on opposite surfaces of the bicrystal by means of AFM (Figure 9.19). The angle of grain rotation increased with time simultaneously with the normal GB motion, providing evidence that the two processes were related.

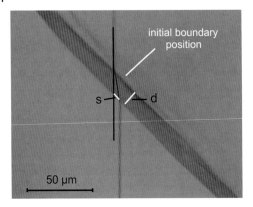

Figure 9.17 Stress-driven migration of an Al GB with the $\langle 001 \rangle$ rotation angle of $\theta = 18.2°$ and a twist component of $\psi = 20°$ [50]. The optical image of the surface of the bicrystal after annealing under a stress of 0.34 MPa at the temperature of 350 °C.

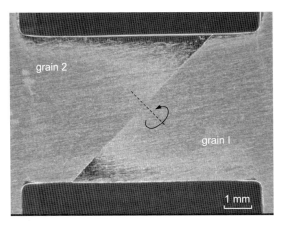

Figure 9.18 Mechanically polished surface of an Al bicrystal with the mixed GB ($\theta = 18.2°$, $\psi = 20°$) after 29-min annealing at 392 °C under applied tensile stress of 0.26 MPa [50].

The results of this experiment can be rationalized in terms of the dislocation structure of the mixed boundary. The latter is assumed to contain a set of edge dislocations accommodating the tilt component of the misorientation and a set of screw dislocations accommodating the twist component. As these dislocation arrays move in the direction normal to the GB plane, they produce shear deformation of the lattice and thus the coupled GB motion. At the same time, the applied shear stress imposes Peach–Kohler forces on the screw dislocations causing their glide in the GB plane. This process can induce grain rotation, the sign of which depends on the direction of the shear stress. It is noteworthy that this grain rotation changes the twist component of the grain misorientation (increases or decreases), demonstrating the interesting effect of altering the GB character by applied stresses.

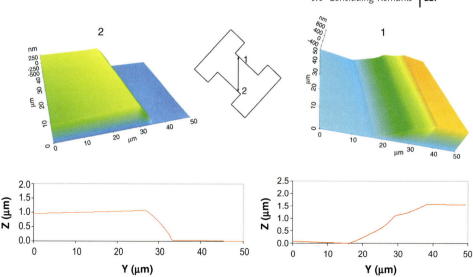

Figure 9.19 Surface topography of the Al bicrystal with the mixed GB ($\theta = 18.2°$, $\psi = 20°$) after annealing for 42 min at 392 °C under applied tensile stress of 0.23 MPa [50]. The different signs of the step on the opposite surfaces confirm the grain rotation.

9.9
Concluding Remarks

Stress-driven GB motion is an interesting, practically important, and rather complex materials phenomenon which we are only beginning to understand. Its relatively short history could make an excellent case study of how mutual interactions of theory, modeling, and experiment can lead to rapid advancement of knowledge. The early observation of stress-induced motion of low-angle symmetrical GBs [34,35] was more of a demonstration of dislocation glide and was not recognized as an intrinsic property of all GBs. This recognition emerged much later, within a two-year period in 2004–2005 [13,14,29,30] as a product of a theoretical insight and a computer simulation discovery. Without direct experimental evidence at the time, it was proposed, and confirmed by simulations, that the ability to couple to applied stresses is not limited to GBs composed of dislocation but must be a generic property of *all*, including high angle, boundaries. The coupling factor β was introduced, the Frank relation was replaced by the Frank–Bilby equation valid for arbitrary misorientation angles, the link between crystal symmetry and the existence of multiple coupling modes was established, and first glimpses into atomic mechanisms of coupled GB motion were obtained.

The simulation work was initially focused on symmetrical tilt GBs with little proof of the existence of coupling for asymmetrical boundaries. However, those initial simulations as well as the theory motivated experimental studies aimed at direct observation and quantification of the coupled motion of high-angle GBs.

Within a few years, most of the simulation predictions were confirmed by bicrystal experiments, culminating in the reproduction of the theoretically predicted coupling factors for the two coupling modes and even the correct angle of discontinuity between them (Figure 9.6). Before long, the experiments were leading the theory by exploring the coupled motion of asymmetrical and mix-type boundaries and providing direct information on GB dynamics. This, in turn, stimulated further simulation work toward more complex cases, such as asymmetrical tilt GBs [32], coupling-induced grain rotation [17], and more detailed investigations of the velocity-driving force relations for coupled GBs [19,31].

At the present stage, it appears that the experiments provide more information than theory and simulations can explain and more questions than they can answer. One of the unsolved mysteries is the stress-driven motion of the $\Sigma 7$ GBs in Al bicrystals [51], which move with different velocities depending on the stress direction but do not produce any shear. One possible explanation is that this boundary executes multiple switches back and forth between two coupling modes with opposite signs of β, which cancel the net shear of the lattice as discussed in the end of Ref. 14 (p. 4974). However, testing this hypothesis requires further simulations, which presents a challenging task since this regime of coupled motion has never been observed in the computer. A more detailed crystal symmetry analysis is needed for the identification of all possible coupling modes and prediction of symmetry-dictated zeros and infinities of β for a general GB. The geometric theory of coupling is also facing serious challenges. For example, none of the existing theoretical treatments provides an analytical expression for $\beta(\theta, \varphi)$ for even the relatively simple case of asymmetrical $\langle 001 \rangle$ tilt boundaries. If proposed, such an expression could be immediately tested by both atomistic simulations and experiment. Another interesting and practically important direction for simulations would be to study the effect of triple lines on coupled GB motion, starting with a simple tricrystal geometry and later building up the complexity. Finally, the effect of solute segregation on the coupled motion [97,98] remains a largely unexplored area, in which the simulations could benefit from incorporation of multiscale approaches to beat the slow diffusion problem.

At the same time, simulations continue to provide inspiration if not guidance to further experiments. As one example, the recent study of asymmetrical $\langle 001 \rangle$ tilt GBs [32] suggests the following set of experiments that could be done on Al bicrystals. Recall that the critical switching angle θ_c between the two coupling modes $\langle 100 \rangle$ and $\langle 110 \rangle$ has been found between 30.5° and 36.5° [48]. This bracket could probably be narrowed down to some extent, but suppose a reasonable estimate of θ_c is available. The following experiments could test the simulation predictions of the coupling regimes shown in Figure 9.10 [32]:

1) Measurements of β at a fixed tilt angle $\theta < \theta_c$ and varying inclination angle φ. The coupling factor is predicted to remain positive but may deviate from the geometric value given by Eq. (9.4) (most probably in the positive direction).

2) Measurements of β at a fixed tilt angle $\theta > (90° - \theta_c)$ and varying φ. The coupling factor is predicted to remain negative but may deviate from the geometric value (9.5) (most probably in the negative direction).
3) Measurements of β at a fixed tilt angle $\theta_c < \theta < (90° - \theta_c)$ and varying inclination angle φ. The coupling factor is predicted to switch from positive at small φ to negative as φ approaches $\pm 45°$. At angles close to the sign-change region, the boundary may exhibit a sliding-like behavior.

It is also believed that the recent grain rotation simulations [17] have provided interesting predictions that could be tested by specially designed experiments on bicrystalline samples containing an isolated cylindrical grain. This could be followed by studying coupled motion in tricrystalline samples to examine the role of the triple-line constraint, and then moving toward more complex grain geometries that mimic the microstructure of real materials.

Acknowledgments

DM expresses his gratitude to the Deutsche Forschungsgemeinschaft for financial support (Grants MO 848/10 and MO 848/14). YM was supported by the U.S. Department of Energy, the Physical Behavior of Materials Program, through Grant No. DE-FG02-01ER45871.

References

1 Gleiter, H. (2000) Nanostructured materials: Basic concepts and microstructure. *Acta Materialia*, **48**, 1–29.
2 Valiev, R.Z., Islamgaliev, R.K., and Alexandrov, I.V. (2000) Bulk nanostructured materials from severe plastic deformation. *Progress in Materials Science*, **45**, 103–189.
3 Kumar, K.S., Van Swygenhoven, H., and Suresh, S. (2003) Mechanical behavior of nanocrystalline metals and alloys. *Acta Materialia*, **51**, 5743–5774.
4 Meyers, M.A., Mishra, A., and Benson, D.J. (2006) Mechanical properties of nanocrystalline materials. *Progress in Materials Science*, **51**, 427–556.
5 Dao, M., Lu, L., Asaro, R.J., De Hosson, J.T.M., and Ma, E. (2007) Toward a quantitative understanding of mechanical behavior of nanocrystalline metals. *Acta Materialia*, **55**, 4041–4065.
6 Hall, E.O. (1951) The deformation and ageing of mild steel.3. Discussion of results. *Proceedings of the Physical Society of London B*, **64**, 747–753.
7 Petch, N.J. (1953) The cleavage strength of polycrystals. *Journal of the Iron and Steel Institute*, **174**, 25–28.
8 Eshelby, J.D., Frank, F.C., and Nabarro, F.R.N. (1951) The equilibrium of linear arrays of dislocations. *Philosophical Magazine*, **42**, 351–364.
9 Hirth, J.P. (2006) Dislocation pileups in the presence of stress gradients. *Philosophical Magazine*, **86**, 3959–3963.
10 Yip, S. (1998) Nanocrystals – the strongest size. *Nature*, **391**, 532–533.
11 Schiotz, J. and Jacobsen, K.W. (2003) A maximum in the strength of nanocrystalline copper. *Science*, **301**, 1357–1359.
12 Argon, A.S. and Yip, S. (2006) The strongest size. *Philosophical Magazine Letters*, **86**, 713–720.
13 Cahn, J.W. and Taylor, J.E. (2004) A unified approach to motion of grain boundaries,

relative tangential translation along grain boundaries, and grain rotation. *Acta Materialia*, **52**, 4887–4998.

14 Cahn, J.W., Mishin, Y., and Suzuki, A. (2006) Coupling grain boundary motion to shear deformation. *Acta Materialia*, **54**, 4953–4975.

15 Mishin, Y., Asta, M., and Li, J. (2010) Atomistic modeling of interfaces and their impact on microstructure and properties. *Acta Materialia*, **58**, 1117–1151.

16 Srinivasan, S.G. and Cahn, J.W. (2002) Challenging some free-energy reduction criteria for grain growth, in *Science and Technology of Interfaces* (eds S. Ankem, C.S. Pande, I. Ovidko, and R. Ranganathan), TMS, Seattle, pp. 3–14.

17 Trautt, Z.T. and Mishin, Y. (2012) Grain boundary migration and grain rotation studied by molecular dynamics. *Acta Materialia*, **60**, 2407–2424.

18 Wu, K.A. and Voorhees, P.W. (2012) Phase field crystal simulations of nanocrystalline grain growth in two dimensions. *Acta Materialia*, **60**, 407–419.

19 Ivanov, V.A. and Mishin, Y. (2008) Dynamics of grain boundary motion coupled to shear deformation: An analytical model and its verification by molecular dynamics. *Physical Review B – Condensed Matter*, **78**, 064. 106.

20 Karma, A., Trautt, Z.T., and Mishin, Y. (2012) Relationship between equilibrium fluctuations and shear-coupled motion of grain boundaries. *Physical Review Letters*, **109**, 095 501.

21 Molteni, C., Francis, G.P., Payne, M.C., and Heine, V. (1996) First-principles simulation of grain boundary sliding. *Physical Review Letters*, **76**, 1284–1287.

22 Molteni, C., Morzani, N., Payne, M.C., and Heine, V. (1997) Sliding mechanisms in aluminum grain boundaries. *Physical Review Letters*, **79**, 869–872.

23 Hamilton, J.C. and Foiles, S.M. (2002) First-principles calculations of grain boundary theoretical strength using transition state finding to determine generalized gamma surface cross sections. *Physical Review B – Condensed Matter*, **65**, 064 104.

24 Sansoz, F. and Molinari, J.F. (2005) Mechanical bahavior of Σ tilt grain boundaries in nanoscale Cu and Al: A quasicontinuum study. *Acta Materialia*, **53**, 1931–1944.

25 Chen, L.Q. and Kalonji, G. (1992) Finite temperature structure and properties of $\Sigma = 5$ (310) tilt grain boundaries in NaCl. *Philosophical Magazine A-Physics of Condensed Matter Structure Defects and Mechanical Properties*, **66**, 11–26.

26 Shiga, M. and Shinoda, W. (2004) Stress-induced grain boundary sliding and migration at finite temperatures: A molecular dynamics study. *Physical Review B – Condensed Matter*, **70**, 054102.

27 Chandra, N. and Dang, P. (1999) Atomistic simulation of grain boundary sliding and migration. *Journal of Materials Science*, **34**, 655–666.

28 Haslam, A.J., Moldovan, D., Yamakov, V., Wolf, D., Phillpot, S.R., and Gleiter, H. (2003) Stress-induced grain growth in a nanocrystalline material by molecular-dynamics simulation. *Acta Materialia*, **51**, 2097–2112.

29 Suzuki, A. and Mishin, Y. (2005) Atomic mechanisms of grain boundary motion. *Materials Science Forum*, **502**, 157–162.

30 Cahn, J.W., Mishin, Y., and Suzuki, A. (2006) Duality of dislocation content of grain boundaries. *Philosophical Magazine*, **86**, 3965–3980.

31 Mishin, Y., Suzuki, A., Uberuaga, B., and Voter, A.F. (2007) Stick-slip behavior of grain boundaries studied by accelerated molecular dynamics. *Physical Review B – Condensed Matter*, **75**, 224101.

32 Trautt, Z.T., Adland, A., Karma, A., and Mishin, Y. (2012) Coupled motion of asymmetrical tilt grain boundaries: Molecular dynamics and phase field crystal simulations. *Acta Materialia*, **60**, 6528–6546.

33 Fensin, S.J., Asta, M., and Hoagland, R.G. (2012) Temperature dependence of the structure and shear response of a $\Sigma 11$ asymmetric tilt grain boundary in copper from molecular-dynamics. *Philosophical Magazine*, **92**, 4320–4333.

34 Li, C.H., Edwards, E.H., Washburn, J., and Parker, J. (1953) Stress-induced movement of crystal boundaries. *Acta Metallurgica*, **1**, 223.

35 Bainbridge, D.W., Li, C.H., and Edwards, E.H. (1954) Recent observations on the

motion of small angle dislocation boundaries. *Acta Metallurgica*, **2**, 322–333.

36 Biscondi, M. and Goux, C. (1968) Fluage intergranulaire de bicristaux orientes d'aluminium. *Mémoires Scientifiques Revue de Métallurgie*, **65**, 167–179.

37 Fukutomi, H., Iseki, T., Endo, T., and Kamijo, T. (1991) Sliding behavior of coincident grain boundaries deviating from ideal symmetric tilt relationship. *Acta Materialia*, **39**, 1445–1448.

38 Winning, M., Gottstein, G., and Shvindlerman, L.S. (2001) Stress induced grain boundary motion. *Acta Materialia*, **49**, 211–219.

39 Winning, M., Gottstein, G., and Shvindlerman, L.S. (2002) On the mechanisms of grain boundary migration. *Acta Materialia*, **50**, 353–363.

40 Winning, M. and Rollett, A.D. (2005) Transition between low and high angle grain boundaries. *Acta Materialia*, **53**, 2901–2907.

41 Sheikh-Ali, A.D., Szpunar, J.A., and Garmestani, H. (2003) Stimulation and suppression of grain boundary sliding by intergranular slip in zinc bicrystals. *Interafce Science*, **11**, 439–450.

42 Yoshida, H., Yokoyama, K., Shibata, N., Ikuhara, Y., and Sakuma, T. (2004) High-temperature grain boundary sliding behavior and brain boundary energy in cubic zirconia bicrystals. *Acta Materialia*, **52**, 2349–2357.

43 Molodov, D., Ivanov, A., and Gottstein, G. (2007) Low angle tilt boundary migration coupled to shear deformation. *Acta Materialia*, **55**, 1843–1848.

44 Molodov, D., Gorkaya, T., and Gottstein, G. (2007) Mechanically driven migration of ⟨1 0 0⟩ tilt grain boundaries in Al-bicrystals. *Materials Science Forum*, **558–559**, 927–932.

45 Legros, M., Gianola, D.S., and Hemker, K.J. (2008) In situ tem observations of fast grain-boundary motion in stressed nanocrystalline aluminum films. *Acta Materialia*, **56**, 3380–3393.

46 Mompiou, F., Caillard, D., and Legros, M. (2009) Grain boundary shear-migration coupling. I. *In situ* tem straining experiments in Al polycrystals. *Acta Materialia*, **57**, 2198–2209.

47 Molodov, D.A. and Shvindlerman, L.S. (2009) Interface migration in metals (IMM): "Vingt ans apres" (twenty years later). *International Journal of Materials Research*, **100**, 461–482.

48 Gorkaya, T., Molodov, D.A., and Gottstein, G. (2009) Stress-driven migration of symmetrical ⟨1 0 0⟩ tilt grain boundaries in Al bicrystals. *Acta Materialia*, **57**, 5396–5405.

49 Gorkaya, T., Burlet, T., Molodov, D.A., and Gottstein, G. (2010) Experimental method for true in situ measurements of shear-coupled grain boundary migration. *Scripta Materialia*, **63**, 633–636.

50 Gorkaya, T., Molodov, K.D., Molodov, D.A., and Gottstein, G. (2011) Concurrent grain boundary motion and grain rotation under an applied stress. *Acta Materialia*, **59**, 5674–5680.

51 Molodov, D.A., Gorkaya, T., and Gottstein, G. (2011) Migration of the Σ7 tilt grain boundary in Al under an applied external stress. *Scripta Materialia*, **65**, 990–993.

52 Molodov, D.A., Gorkaya, T., and Gottstein, G. (2011) Dynamics of grain boundaries under applied mechanical stress. *Journal of Materials Science*, **46**, 4318–4326.

53 Monk, J., Hyde, B., and Farkas, D. (2006) The role of partial grain boundary dislocations in grain boundary sliding and coupled grain boundary motion. *Journal of Materials Science*, **41**, 7741–7746.

54 Zhu, T., Li, J., Samanta, A., Kim, H.G., and Suresh, S. (2007) Interfacial plasticity governs strain rate sensitivity and ductility in nanostructured materials. *Proceedings of the National Academy of Sciences of the United States of America*, **104**, 3031–3036.

55 Hemker, K.J. and Sharpe, W.N. (2007) Microscale characterization of mechanical properties. *Annual Review of Materials Research*, **37**, 93–126.

56 Gianola, D.S., Eberl, C., Cheng, X.M., and Hemker, K.J. (2008) Stress-driven surface topography evolution in nanocrystalline Al thin films. *Advanced Materials*, **20**, 303–308.

57 Cahn, J.W. and Mishin, Y. (2009) Recrystallization initiated by low-temperature grain boundary motion coupled to stress. *International Journal of Materials Research*, **100**, 510–515.

58 Daw, M.S. and Baskes, M.I. (1984) Embedded-atom method: Derivation and application to impurities, surfaces, and other defects in metals. *Physical Review B – Condensed Matter*, **29**, 6443–6453.

59 Mishin, Y., Farkas, D., Mehl, M.J., and Papaconstantopoulos, D.A. (1999) Interatomic potentials for monoatomic metals from experimental data and ab initio calculations. *Physical Review B – Condensed Matter*, **59**, 3393–3407.

60 Mishin, Y., Mehl, M.J., Papaconstantopoulos, D.A., Voter, A.F., and Kress, J.D. (2001) Structural stability and lattice defects in copper: Ab initio, tight-binding and embedded-atom calculations. *Physical Review B – Condensed Matter*, **63**, 224106.

61 Suzuki, A. and Mishin, Y. (2003) Atomistic modeling of point defects and diffusion in copper grain boundaries. *Interface Science*, **11**, 131–148.

62 Frenkel, D. and Smit, B. (2002) *Understanding Molecular Simulation: From Algorithms to Applications*, Academic, San Diego.

63 Voter, A.F. (1997) A method for accelerating the molecular dynamics simulation of infrequent events. *Journal of Chemical Physics*, **106**, 4665–4677.

64 Voter, A.F. (1998) Parallel replica method for dynamics of infrequent events. *Physical Review B – Condensed Matter*, **57**, R13985–R13988.

65 Sørensen, M.R. and Voter, A.F. (2000) Temperature-accelerated dynamics for simulation of infrequent events. *Journal of Chemical Physics*, **112**, 9599–95606.

66 Voter, A.F., Montalenti, F., and Germann, T.C. (2002) Extending the time scale in atomistic simulation of materials. *Annual Review of Materials Research*, **32**, 321–346.

67 Uberuaga, B.P., Stuart, S.J., and Voter, A.F. (2007) Parallel replica dynamics for driven systems: Derivation and application to strained nanotubes. *Physical Review B – Condensed Matter*, **75**, 014301.

68 Tucker, G.J., Zimmerman, J.A., and McDowell, D.L. (2010) Shear deformation kinematics of bicrystalline grain boundaries in atomistic simulations. *Modelling and Simulations in Materials Science and Engineering*, **18**, 015 002.

69 Berry, J., Grant, M., and Elder, K.R. (2006) Diffusive atomistic dynamics of edge dislocations in two dimensions. *Physical Review E*, **73**, 031 609.

70 Berry, J., Elder, K.R., and Grant, M. (2008) Melting at dislocations and grain boundaries: A phase field crystal study. *Physical Review B – Condensed Matter*, **77**, 224 114.

71 Mellenthin, J., Karma, A., and Plapp, M. (2008) Phase-field crystal study of grain-boundary premelting. *Physical Review B – Condensed Matter*, **78**, 184 110.

72 Stefanovic, P., Haataja, M., and Provatas, N. (2009) Phase field crystal study of deformation and plasticity in nanocrystalline materials. *Physical Review E*, **80**, 046 107.

73 Jaatinen, A., Achim, C.V., Elder, K.R., and Ala-Nissila., T. (2010) Phase field crystal study of symmetric tilt grain boundaries of iron. *Technische Mechanik*, **30**, 169–176.

74 Spatschek, R. and Karma, A. (2010) Amplitude equations for polycrystalline materials with interaction between composition and stress. *Physical Review B – Condensed Matter*, **81**, 214 201.

75 Olmsted, D.L., Buta, D., Adland, A., Foiles, S.M., Asta, M., and Karma, A. (2011) Dislocation-pairing transitions in hot grain boundaries. *Physical Review Letters*, **106**, 046101.

76 Adland, A., Spatschek, R., Buta, D., Asta, M., and Karma, A. (2013) Phase-field-crystal study of grain-boundary premelting and shearing for bcc iron. *Phys. Rev. B* **87**, 024110.

77 Sutton, A.P. and Balluffi, R.W. (1995) *Interfaces in Crystalline Materials*, Clarendon Press, Oxford.

78 Mate, C.M., McClelland, G.M., Erlandsson, R., and Chiang, S. (1987) Atomic-scale friction of a tungsten tip on a graphite surface. *Physical Review Letters*, **59** (17), 1942–1945. doi: 10.1103/PhysRevLett.59.1942

79 Gnecco, E., Bennewitz, R., Gyalog, T., Loppacher, C., Bammerlin, M., Meyer, E., and Güntherodt, H.J. (2000) Velocity dependence of atomic friction. *Physical Review Letters*, **84** (6), 1172–1175. doi: 10.1103/PhysRevLett.84.1172

80 Gnecco, E., Bennewitz, R., Gyalog, T., and Meyer, E. (2001) Friction experiments on the nanometre scale. *Journal of Physics: Condensed Matter*, **13**, R619–R642.

81 Socoliuc, A., Gnecco, E., Maier, S., Pfeiffer, O., Baratoff, A., Bennewitz, R., and Meyer, E. (2006) Atomic-scale control of friction by actuation of nanometer-sized contacts. *Science*, **313**, 207.

82 Read, W.T. and Shockley, W. (1950) Dislocation models of crystal grain boundaries. *Physical Review*, **78**, 275–289.

83 Taylor, J.E. and Cahn, J.W. (2007) Shape accommodation of a rotating embedded crystal via a new variational formulation. *Interfaces and Free Boundaries*, **9**, 493–512.

84 Zhang, H., Du, D., and Srolovitz, D.J. (2008) Effect of boundary inclination and boundary type on shear-driven grain boundary migration. *Philosophical Magazine*, **88**, 243–256.

85 Zhang, H. (2009) Atomistic simulation of sliding of [1010] tilt grain boundaries in Mg. *Journal of Materials Research*, **24**, 3446–3453.

86 Syed, B., Catoor, D., Mishra, R., and Kumar, K. (2012) Coupled motion of [10 − 10] tilt boundaries in magnesium bicrystals. *Philosophical Magazine*, **92** (12), 1499–1522.

87 Zhang, K., Weertman, J.R., and Eastman, J.A. (2005) Rapid stress-driven grain coarsening in nanocrystalline Cu at ambient and cryogenic temperatures. *Applied Physics Letters*, **87**, 061 921.

88 Rath, B.B., Winning, M., and Li, J.C.M. (2007) Coupling between grain growth and grain rotation. *Applied Physics Letters*, **90**, 161 915.

89 Liu, P., Mao, S.C., Wang, L.H., Han, X.D., and Zhang, Z. (2011) Direct dynamic atomic mechanisms of strain-induced grain rotation in nanocrystalline, textured, columnar-structured thin gold films. *Scripta Materialia*, **64**, 343–346.

90 Kim, B.N., Hiraga, K., and Morita, K. (2005) Viscous grain-boundary sliding and grain rotation accommodated by grain-boundary diffusion. *Acta Materialia*, **53**, 1791–1798.

91 Margulies, L., Winther, G., and Poulsen, H.F. (2001) In situ measurement of grain rotation during deformation of polycrystals. *Science*, **291**, 2392–2394.

92 Harris, K.E., Singh, V.V., and King, A.H. (1998) Grain rotation in thin films of gold. *Acta Materialia*, **46**, 2623–2633.

93 Doherty, R.D. and Szpunar, J.A. (1984) Kinetics of sub grain coalescence – a reconsideration of the theory. *Acta Materialia*, **32**, 1789–1798.

94 Li, J.C.M. (1969) Irreversible thermodynamics of a curved grain boundary. *Transactions of AIME*, **245**, 1587–1590.

95 Li, J.C.M. (1969) A dislocation mechanism for the shrinking of a cylindrical tilt boundary. *Transactions of AIME*, **245**, 1591–1593.

96 Li, J.C.M. (1962) Possibility of subgrain rotation during recrystallization. *Journal of Applied Physics*, **33**, 2958–2965.

97 Elsener, A., Politano, O., Derlet, P.M., and Van Swygenhoven, H. (2009) Variable-charge method applied to study coupled grain boundary migration in the presence of oxygen. *Acta Materialia*, **57**, 1988–2001.

98 Tang, T., Gianola, D.S., Moody, M.P., Hemker, K.J., and Cairney, J.M. (2012) Observations of grain boundary impurities in nanocrystalline Al and their influence on microstructural stability and mechanical behaviour. *Acta Materialia*, **60**, 1038–1047.

10
Grain Boundary Migration Induced by a Magnetic Field: Fundamentals and Implications for Microstructure Evolution
Dmitri A. Molodov

10.1
Introduction

The physical and mechanical properties of crystalline solids are to a great extent determined by their grain microstructure, that is, the grain size and the orientation distribution, which develops in the course of recrystallization and grain growth during a heat treatment following plastic deformation. The recrystallization and grain growth proceed by the motion of grain boundaries, which in turn is controlled by their mobility m and the applied driving force p: as was shown theoretically [1,2] and proved by experiment [3–7], the rate of boundary migration v is proportional to the acting driving force

$$v = mp \qquad (10.1)$$

A control of grain boundary migration in response to exerted forces means a control of grain microstructure and texture evolution in crystalline solids and, therefore, is crucial for the design of advanced materials for various engineering applications.

The grain microstructure evolution can be influenced by the application of an external energetic field [8–13]. In particular, grain boundary motion can be affected by a magnetic field, owing to a driving force that arises in the field due to a crystal magnetic anisotropy. There is a large body of evidence of a substantial magnetic field effect on recrystallization and texture development in ferromagnetic materials [9,11–21]. Much fewer data can be found in literature concerning a magnetically affected grain microstructure evolution in nonferromagnetic (i.e., paramagnetic and diamagnetic) materials. Most reported observations in nonferromagnetics relate to crystal alignment during solidification in the presence of a magnetic field [11–13,22,23].

The grain boundary migration induced by an applied magnetic field was first analyzed and experimentally observed during grain growth in polycrystalline Bi by Mullins *et al.* [24,25]. The occurrence of a driving force for the grain boundary migration in a magnetically anisotropic crystalline solid was definitely proved by experiments with individual boundaries in specially grown bismuth and zinc bicrystals, where planar boundaries moved under the action of a magnetic driving force only [6,26–28]. Investigations on polycrystalline cold-rolled zinc [29], titanium [30–34], and zirconium [35] revealed that owing to this additional driving force a magnetic field can cause significant changes in the evolution of the microstructure with respect to the grain size, orientation distribution, and grain topology.

The impact of a magnetic field on the microstructure development, however, is not only restricted by magnetically anisotropic metals, but can also be observed in materials with magnetically isotropic properties [36,37]. Moreover, the microstructure evolution can be influenced by a magnetic field not only at the stage of grain growth, but also at the earlier stages of recovery and primary recrystallization.

In this chapter, we will briefly review the results of the experimental and modeling efforts over the last 15 years to study a magnetically driven grain boundary migration and its impact on texture and grain microstructure evolution in nonferromagnetic metals.

10.2
Driving Forces for Grain Boundary Migration

The driving force p for the grain boundary migration occurs if the boundary displacement is accompanied by a decrease of the total free energy of the system G: $p = -dG/dV$, where V is the volume swept by the boundary during its displacement. The driving force has therefore the dimension of energy per unit volume that is equivalent to a force per unit area.

The conventional curvature (capillary) driving force p_c for boundary motion during grain growth originates from the free energy of the boundary itself. A curved boundary reduces its area and therefore its free energy as it moves toward its center of curvature. Each boundary segment with the energy γ and the local curvature κ experiences a force per unit area

$$p_c = \gamma \kappa \qquad (10.2)$$

A way to obtain an artificial driving force is to create a difference in free energy density across the boundary between two grains. For example, this can be accomplished by the application of a stress field to a polycrystal with an anisotropic Young's modulus or by exposing a polycrystal with an anisotropic dielectric susceptibility to an electrostatic field. The coupling of an appropriately directed external field with an anisotropic property of a material will generate a free energy difference between adjacent grains that create a driving force for boundary displacement. This driving force does not depend on boundary properties and moves a boundary from a grain with lower free energy toward grain with a higher free energy.

Also, such a condition can be obtained by the action of an external magnetic field on a crystalline material with an anisotropic magnetic susceptibility. As it has been initially shown by Mullins [24], if the volume density of the magnetic free energy ω in a crystal, induced by a uniform magnetic field, is independent of the crystal shape and size then the magnetic driving force p_m, acting on the boundary between two adjacent crystals with different susceptibilities χ_1 and χ_2 along the field direction and consequently different magnetic energy densities ω_1 and ω_2, is given by

$$p_m = \omega_1 - \omega_2 = \frac{\mu_0 H^2}{2}(\chi_1 - \chi_2) \tag{10.3}$$

where χ_1 and χ_2 are the magnetic susceptibilities of adjacent crystals 1 and 2, respectively, along the magnetic field direction. This driving force depends only on the strength of the magnetic field and its orientation with respect to the two neighboring crystals. The direction of p_m remains the same when the sense of the field is reversed. According to elementary crystallography, the magnetic susceptibility of a uniaxial crystal can be written as $\chi = \chi_\perp + \Delta\chi \cos^2\varphi$, where $\Delta\chi$ is the difference in susceptibility parallel $\chi_\parallel$ and perpendicular $\chi_\perp$ to the principal (or c) axis of the crystal. Here φ is the angle between the c-axis and the magnetic field. Substituting this expression into Eq. (10.3) leads to

$$p_m = \frac{1}{2}\mu_0 \Delta\chi H^2 (\cos^2\varphi_1 - \cos^2\varphi_2) \tag{10.4}$$

A maximum magnetic driving force

$$p_m^{max} = \frac{1}{2}\mu_0 \Delta\chi H^2 \tag{10.5}$$

arises when the angles between the field and the c-axis in both neighboring grains are $\varphi_1 = 0$ and $\varphi_2 = 90°$. For $p_m = 0$, the grain orientations need not to be identical but merely need to fulfill $\varphi_1 = \varphi_2$. The sign of p_m depends on the magnetic anisotropy of a material ($\Delta\chi$) and the asymmetry of the spatial orientation of both adjacent grains with respect to the magnetic field direction.

10.3
Magnetically Driven Grain Boundary Motion in Bicrystals

10.3.1
Specimens and Applied Methods to Measure Grain Boundary Migration

The magnetically induced motion of specific planar grain boundaries was first measured by applying a strong magnetic field to specially prepared high-purity bismuth (99.999%) bicrystals [6,26]. Bismuth is known to possess the largest magnetic anisotropy with different susceptibilities parallel and perpendicular to the trigonal axis (at 22 °C $\Delta\chi = 0.53 \times 10^{-4}$; at 250 °C $\Delta\chi = 0.23 \times 10^{-4}$) [38,39] and it is thus a most suitable material for a model experiment to measure a magnetically driven grain boundary migration. Symmetrical and asymmetrical pure tilt boundaries (Figure 10.1) with a 90°⟨112⟩ misorientation were examined.

Recent experiments to measure the grain boundary migration in a magnetic field were conducted on pure zinc (99.995%) bicrystals [27,28]. The volume susceptibility difference in pure zinc crystals was measured to be $\Delta\chi = 0.5 \times 10^{-5}$ independent of temperature (F. Hüning, and D.A. Molodov, 1999, Crystal magnetism of zinc single

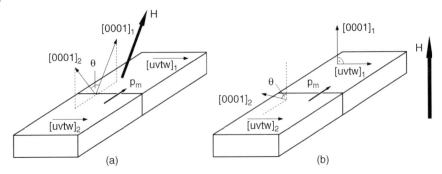

Figure 10.1 Geometry of zinc bicrystal specimens with magnetically driven (a) symmetrical and (b) asymmetrical $\langle 10\bar{1}0 \rangle$ and $\langle 11\bar{2}0 \rangle$ tilt grain boundaries and their orientation with respect to a magnetic field H [27,28]. Geometry of bismuth bicrystal specimens with $\langle 112 \rangle$ tilt boundaries [6,26] was similar with $\theta = 90°$ misorientation between the trigonal $[11\bar{1}]$ axes in the both adjacent grains. A magnetic susceptibility parallel and perpendicular to the trigonal axis in bismuth is different with $|\chi_{\parallel}| < |\chi_{\perp}|$ and the direction of the driving force p_m was opposite to that shown in the sketch, from grain 1 toward grain 2.

crystals, unpublished research). Planar symmetrical and asymmetrical $\langle 10\bar{1}0 \rangle$ and $\langle 11\bar{2}0 \rangle$ tilt (Figure 10.1) as well as nontilt grain boundaries with misorientation angles in the range between 60° and 90° were investigated. As also in bismuth bicrystals, the relatively large misorientation angles were chosen in order to gain sufficiently high magnetic driving forces (see Eq. (10.4)).

Both discontinuous and continuous methods of grain boundary motion measurements were used in the experiments. The migration of grain boundaries in bismuth and symmetrical tilt boundaries in zinc bicrystals was measured discontinuously, that is, by measuring the boundary location prior and after annealing for discrete time intervals. High temperature annealing was conducted stepwise under an atmosphere of high-purity argon using a cartridge heater. The main advantage of this method is its simplicity; the major shortcoming is that only an average migration rate can be obtained from recording the boundary position prior and subsequent to annealing. Besides, the boundary location is revealed by a groove, which is formed during specimen cooling (thermal "etching") or by chemical etching of the bicrystal surface. After each annealing, the specimen surface must be re-polished to remove the groove, otherwise, it can hinder boundary migration during subsequent annealing.

For the migration measurements of asymmetrical tilt boundaries in zinc (Figure 10.1b), an *in situ* technique on the basis of orientation contrast imaging was applied. In this technique, not only the boundary position but also its shape at any moment during annealing can be observed and recorded. Specifically, the boundary location is revealed by the contrast between differently oriented grains when illuminated by plane polarized light in an optical microscope (Figure 10.2). The reflection of polarized light was utilized in the special microscopy probe designed for use in high field magnets for measurements of magnetically induced boundary migration [40]. This probe can operate inside a magnet bore with a

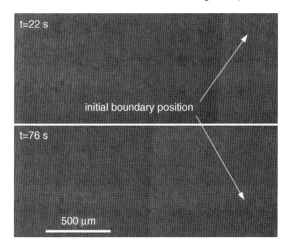

Figure 10.2 Images of a 71°⟨10$\bar{1}$0⟩ tilt grain boundary moving in a field of 12 T at 410 °C obtained by polarization microscopy [40].

diameter of 50 mm and consists of a remote controlled polarizing microscope equipped with a video camera and a heated specimen chamber for annealing at elevated temperatures in an inert gas atmosphere.

For creation of a magnetic driving force, the specimen was mounted onto a special holder or heater such that the trigonal [111] or hexagonal [0001] axis of crystal 1 in a bismuth or zinc bicrystal, respectively, was parallel to the direction of the magnetic field (Figure 10.1). In bismuth, due to the difference of a magnetic susceptibility parallel and perpendicular to the trigonal axis with $|\chi_\parallel| < |\chi_\perp|$ [38,39], the direction of the driving force p_m was from grain 1 toward grain 2 (Figure 10.1). By contrast, zinc has a different susceptibility parallel and perpendicular to its hexagonal axis such that $|\chi_\parallel| > |\chi_\perp|$ (F. Hüning, and D.A. Molodov, 1999, Crystal magnetism of zinc single crystals, unpublished research), [41]. Therefore, the investigated boundary moved during annealing in a field in the direction of crystal 1 (Figure 10.1) due to its higher magnetic free energy density.

For measurements of magnetically driven boundary migration, a magnetic field with strength between 0.6×10^7 A/m (7 T) and 2.2×10^7 A/m (28 T) was applied [6,26–28].

10.3.2
Measurements of Absolute Grain Boundary Mobility

Due to the fact that the magnetic driving force is generated by an external field and does not depend on boundary properties, that is, its energy and shape, the method provides an opportunity to investigate the motion of specific planar grain boundaries with well-defined structures and to determine their absolute mobility. By contrast,

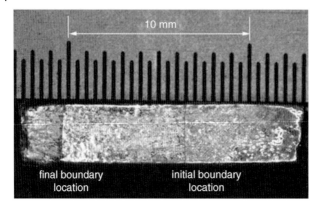

Figure 10.3 Migration of a 90°⟨112⟩ tilt grain boundary in a bismuth bicrystal during annealing for 180 s at 252 °C in a magnetic field of 20.45 T (1.63 × 10^7 A/m) [26].

from the measurements on bicrystals with curved boundaries driven by a capillary driving force, only the reduced boundary mobility $A = m\gamma$ can be obtained, where γ is the grain boundary energy, which is not exactly known [42]. Furthermore, the mobility obtained in such experiments cannot be related to a specific grain boundary structure, since the structure of a curved grain boundary, which is composed of differently inclined boundary sections, changes along the boundary.

First experiments performed on bismuth bicrystals [6,26] provided unambiguous evidence that the investigated grain boundaries actually moved due to the magnetic driving force only (Figure 10.3). It is remarkable that this driving force depends on the orientation of the neighboring grains with regard to the magnetic field (see Eq. (10.4)) and thus by rotation of the sample with respect to the field it can be reversed so that the grain boundary can be driven back and forth (Figure 10.4). In

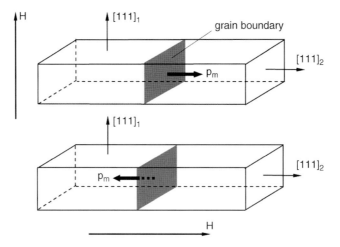

Figure 10.4 Magnetically driven boundary motion in opposite directions in the same specimen (Bi bicrystal) dependent on its position with regards to the magnetic field [26].

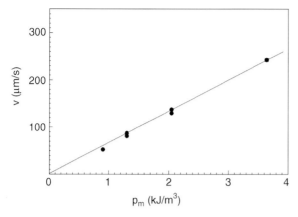

Figure 10.5 Dependence of the velocity of a symmetrical 90°⟨112⟩ tilt grain boundary in bismuth on the magnetic driving force at 250 °C [6].

fact, the boundary motion in opposite directions was observed in the same specimen dependent on its position with regards to the magnetic field [6,26].

The possibility to change the magnitude of the driving force for boundary migration by exposing the samples to magnetic fields of different strength yields the unique opportunity to change the driving force on a specific grain boundary and thus, to obtain the driving force dependence of grain boundary velocity $v = v(p)$. The measurements of boundary migration in different magnetic fields confirmed that the boundary velocity changes linearly with the driving force (Figure 10.5). Therefore, from these result the absolute value of the grain boundary mobility can be immediately extracted (Eq. (10.1)). By measuring boundary migration at various temperatures in bicrystals with structurally different boundaries, the dependence of grain boundary mobility on temperature and on the specific grain boundary character can be determined.

The experiments on zinc bicrystals [27,28] revealed that at a constant temperature in a field of a constant strength all investigated ⟨10$\bar{1}$0⟩ boundaries and symmetrical ⟨11$\bar{2}$0⟩ tilt boundaries moved at a constant rate (Figure 10.6), which proved the free character of their motion. For each investigated specimen with the known orientation of the grains with respect to the field direction, the magnetic driving force was determined according to Eq. (10.4). The absolute values of the grain boundary mobility were determined from the boundary migration rate and the acting driving force according to Eq. (10.1): $m = v/p_m$. In order to prove the proportionality between boundary velocity and driving force, the migration rate of some boundaries in zinc was measured at a constant temperature in fields of different strength. As seen in Figure 10.7, the boundary velocity rises linearly with the driving force, as was also obtained for a 90°⟨112⟩ boundary in bismuth (Figure 10.5).

Therefore, the measurements of the grain boundary migration in a magnetic field of different strength have confirmed that, if a boundary moves without any disturbing effects, for example, from solutes, particles or surface grooves, the

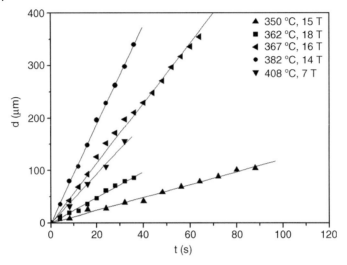

Figure 10.6 Displacement d versus time t for a $69.9°\langle10\bar{1}0\rangle$ tilt grain boundary during annealing at various temperatures in a field of different strength [28].

boundary velocity grows in proportion to the acting driving force according to Eq. (10.1). Such proportionality was also observed in experiments on aluminum bicrystals with curved boundaries [5] as well as planar boundaries driven by an applied shear stress [7]. Also, a proportional gain of the average grain boundary migration rate with increasing driving force during primary recrystallization was reported for a polycrystalline aluminum alloy [3] and cold-worked Cu [4]. On the other hand, a very large deviation from a linear relationship between boundary

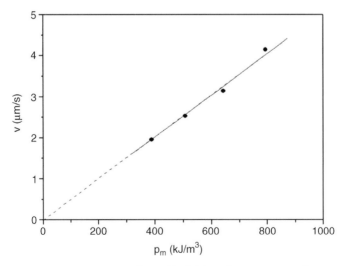

Figure 10.7 Migration rate of a $89.1°\langle10\bar{1}0\rangle$ boundary at 352°C as a function of the driving force.

migration rate and applied driving force was observed in experiments by Rath and Hu [43,44]. This discrepancy, however, was later explained by groove dragging [5,42].

It is worth noting that the motion of a specific boundary in specially grown zinc bicrystals in a magnetic field was first measured in experiments, where the boundary moved under both the magnetic and curvature driving forces [45,46]. The latter, however, was effectively zero for the section of the boundary in the middle of its length. Hence, that section moved under a magnetic force only, and its displacement could be used to evaluate the mobility of the investigated $89°\langle 10\bar{1}0\rangle$ tilt boundary, which at 390 °C amounted to about $m = 2.5 \times 10^{-8}$ m^4/J s [46]. This is close to the mobility value obtained in the recent experiments for the $89.5°\langle 10\bar{1}0\rangle$ tilt boundary with $m = 4.5 \times 10^{-8}$ m^4/J s at 390 °C [28].

The mobility of magnetically driven boundaries can also be compared with other previous data on the grain boundary migration in Zn related to experiments on bicrystals with curvature-driven $\langle 10\bar{1}0\rangle$ tilt boundaries [42,47], where the reduced mobility A was measured. For example, the reduced mobility of a curved $75°\langle 10\bar{1}0\rangle$ tilt boundary was $A^c_{75°} = 1.1 \times 10^{-8}$ m^2/s [47]. In recent experiments with zinc bicrystals, the grain boundary energy was estimated to be $\gamma \cong 0.35$ J/m^2 [48]. Accordingly, the absolute boundary mobility amounts to $m^c_{75°} \cong 3.1 \times 10^{-8}$ m^4/J s, which is in a good agreement with the result obtained for a magnetically driven $\langle 10\bar{1}0\rangle$ tilt boundary with similar misorientation (76.0°) at 400 °C, $m_{76°} = 5.8 \times 10^{-8}$ m^4/J s.

Apparently, the mobility values measured in zinc for planar boundaries moving under the magnetic driving force [28] are in reasonable agreement with results obtained by the other methods [46,47].

The temperature dependence of boundary migration in bicrystals measured in the temperature ranges between 205 °C and 260 °C for bismuth [6] and 330 °C and 415 °C for zinc [27,28] revealed that the grain boundary mobility follows an Arrhenius dependence $m = m_0 \exp(-Q/kT)$, where Q is the activation enthalpy of the grain boundary migration (Figure 10.8). The values of Q obtained from the Arrhenius plots for all investigated $\langle 10\bar{1}0\rangle$ tilt boundaries in zinc bicrystals are shown in Figure 10.9.

10.3.3
Misorientation Dependence of Grain Boundary Mobility

The results of the experiments with planar tilt grain boundaries in zinc bicrystals revealed that the grain boundary mobility in zinc and specifically its temperature dependence substantially changes with the misorientation angle between adjacent grains (Figures 10.8 and 10.9) [28]. Figure 10.9 depicts a pronounced misorientation dependence of the migration activation enthalpy in the investigated angular range for both asymmetrical and symmetrical $\langle 10\bar{1}0\rangle$ tilt boundaries. As seen, the activation enthalpy measured for migration of asymmetrical boundaries varied with misorientation within a wide range between 1.2 and 3.7 eV. For the two investigated symmetrical 64.2° and 85.2° $\langle 10\bar{1}0\rangle$ tilt boundaries, the migration activation enthalpy was also considerably different, 1.55 eV and 0.8 eV, respectively [28].

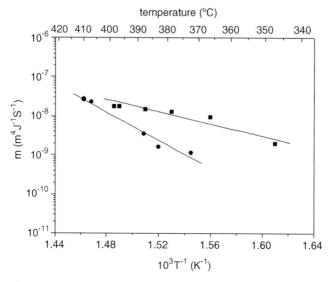

Figure 10.8 Temperature dependence of the mobility for investigated symmetrical 64.2° (square symbols) and asymmetrical 65.0° $\langle 10\bar{1}0 \rangle$ tilt boundaries.

Previously, the misorientation dependence of grain boundary mobility in Zn was investigated for curvature-driven $\langle 10\bar{1}0 \rangle$ and $\langle 11\bar{2}0 \rangle$ tilt boundaries [42,47]. The reduced boundary mobility was found to vary nonmonotonously with misorientation angle: minima of activation enthalpy and preexponential factor corresponded to special misorientations associated with low Σ CSL orientation relationships.

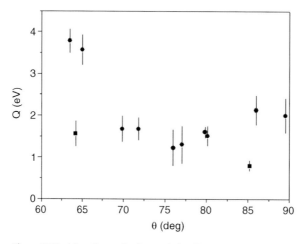

Figure 10.9 Migration activation enthalpy Q versus misorientation angle θ for the investigated symmetrical (square symbols) and asymmetrical $\langle 10\bar{1}0 \rangle$ tilt grain boundaries.

The pronounced misorientation dependence $Q(\theta)$ for investigated asymmetrical tilt boundaries in Figure 10.9, however, cannot be interpreted in terms of the low Σ CSL concept. In this concept, the "special" tilt boundaries with a high density of coincidence sites in the boundary plane are symmetrical boundaries between grains with low Σ CSL orientation relationships, whereas the dependence $Q(\theta)$ in Figure 10.9 was obtained for asymmetrical boundaries with boundary planes, which deviate from the symmetrical position by the angle $\psi = \theta/2$ (Figure 10.1b). On the other hand, for the symmetrical 85.2° $\langle 10\bar{1}0 \rangle$ tilt boundary, which can be associated with the $\Sigma 15$ CSL misorientation, the activation enthalpy of $Q = 0.8\,\text{eV}$ was substantially lower than that for another symmetrical 64.2° boundary (Figure 10.9) [28].

Anyway, the dependence of boundary mobility on the tilt angle, which was previously observed for zinc [47] as well as for fcc metals [42,49] in experiments with curvature-driven boundaries, is unambiguously confirmed by the measurements of planar boundary migration driven by the magnetic driving force. It is worth noting that the recent investigations of stress-induced migration of planar symmetrical tilt boundaries [50,51] also revealed the misorientation dependence of boundary mobility, where lower values on the activation enthalpy and preexponential factor corresponded to boundaries with low Σ misorientations.

10.3.4
Effect of Boundary Plane Inclination on Tilt Boundary Mobility

The results of experiments with planar magnetically driven tilt boundaries in zinc showed that the misorientation is not the only geometrical parameter, which defines the migration behavior of grain boundaries, but that the inclination of the tilt boundary, that is, its position in the bicrystal, has a very strong influence on boundary mobility as well. The effect of boundary plane inclination is apparent from a comparison of the migration behavior of the symmetrical and asymmetrical $\langle 10\bar{1}0 \rangle$ tilt boundaries with similar misorientation angles. The temperature dependence of the mobility of two such boundaries in Figure 10.8 showed that the migration activation enthalpy and preexponential mobility factor for the symmetrical boundary ($Q = 1.55\,\text{eV}$, $m_0 = 1.0 \times 10^4\,\text{m}^4/\text{J s}$) considerably differed from the corresponding parameters of the asymmetrical boundary ($Q = 3.57\,\text{eV}$, $m_0 = 5.4 \times 10^{18}\,\text{m}^4/\text{J s}$) of almost the same misorientation. The same holds for the migration parameters for symmetrical 85.2° and asymmetrical 86.0° $\langle 10\bar{1}0 \rangle$ tilt boundaries ($Q = 0.8\,\text{eV}$, $m_0 = 1.6 \times 10^{-2}\,\text{m}^4/\text{J s}$ and $Q = 2.13\,\text{eV}$, $m_0 = 2.04 \times 10^8\,\text{m}^4/\text{J s}$, respectively) [28].

Further evidence of the essential impact of the tilt boundary plane for the migration behavior was obtained in experiments with $\langle 11\bar{2}0 \rangle$ tilt boundaries with misorientation angles in the range between 88° and 91°: asymmetrical boundaries did not move at any temperature up to the melting point of zinc and a magnetic field up to $2.23 \times 10^7\,\text{A/m}$ (28 T), whereas a symmetrical 88.1° $\langle 11\bar{2}0 \rangle$ tilt boundary moved in a field of $1.6 \times 10^7\,\text{A/m}$ (20 T) and its migration was measured in the temperature regime between 350 °C and 405 °C [28].

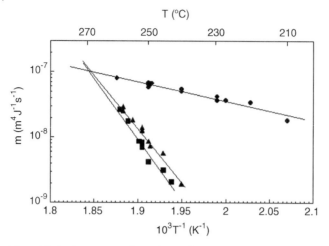

Figure 10.10 Temperature dependence of the mobility of a 90°⟨112⟩ symmetrical (●) and asymmetrical (▲,■) boundaries in Bi-bicrystals, moving in opposite directions. Trigonal axis in the growing grain parallel (▲) or perpendicular (■) to the growth direction [6].

The dependence of tilt grain boundary migration on inclination was also established by measuring the migration of 90°⟨112⟩ tilt grain boundaries in bismuth, where Q and m_0 values for the symmetrical boundary ($Q = 0.51\,\mathrm{eV}$, $m_0 = 0.67\,\mathrm{m^4/J\,s}$) substantially differed from these parameters for the asymmetrical boundary ($Q_\| = 3.38\,\mathrm{eV}$, $m_{0\|} = 2.04 \times 10^{24}\,\mathrm{m^4/J\,s}$ and $Q_\perp = 3.79\,\mathrm{eV}$, $m_{0\perp} = 1.10 \times 10^{28}\,\mathrm{m^4/J\,s}$, where the symbols $\|$ and $\perp$ refer to the orientation of the c-axis with regard to the boundary plane normal [6,26].

A consequence of the inclination dependence of tilt boundary mobility is that the symmetrical boundary has a much higher mobility than the asymmetrical boundary at all investigated temperatures up to the melting point of Zn and/or Bi but particularly at low temperatures (Figures 10.8 and 10.10).

It is important to note that the impact of tilt grain boundary orientation/inclination on its mobility can be reliably ascertained only in experiments on bicrystals with planar grain boundaries, which do not change their geometry during the migration process.

10.4
Selective Grain Growth in Locally Deformed Zn Single Crystals under a Magnetic Driving Force

In a series of experiments, an application of a high magnetic field during annealing of locally deformed zinc single crystals resulted in growth of new specifically oriented grains, that is, in the generation and migration of individual grain boundaries [32,52]. Zinc single crystal specimens of about 15 mm length, 4 mm width, and 2 mm thickness were locally deformed by a Vickers hardness indentation along either the $[10\bar{1}0]$ or $[11\bar{2}0]$ direction. A successive annealing at 400 °C in a high-purity inert gas atmosphere resulted in a local recrystallization of the deformed volume only. Typically, a few new grains appeared. Then concurrently to the heat treatment, a magnetic field of 1.99×10^7 A/m (25 T) was applied parallel to the hexagonal (c) axis of the single crystal. As a usual result, soon after applying the field the initial single crystals converted into bicrystals with a nearly flat grain boundary moving in the direction of the monocrystalline host matrix. The time of a magnetic annealing ranged between 3 and 8 min.

Obviously, during annealing the stored energy from cold work is dissipated by the appearance of new grains. The grains resulted from recrystallization could not grow outside the deformed volume, since there were no longer driving force for their growth. However, if recrystallization was followed by a magnetic annealing with the field directed along the hexagonal axis of the initial single crystal, an additional driving force occurred. This driving force acted on all boundaries between the host matrix and those new grains whose c-axes were not aligned along the field and thus, new grains could continue to grow at the expense of the host matrix and their less favorably oriented neighbors. Apparently, grains with the c-axis perpendicular to the field direction experienced the maximum magnetic driving force. As a result, usually only one grain with the best conditions regarding driving force and boundary mobility remained and grew at the expense of the host single crystals.

The selectively grown grains revealed a strong preferential orientation. This is illustrated by the frequency distribution of the angle between the c-axis in the selectively grown grains and the field direction (Figure 10.11). Whereas the principal c-axis of the host single crystal was aligned with the field, in most selectively grown grains this axis is tilted preferentially perpendicular to the field direction. The

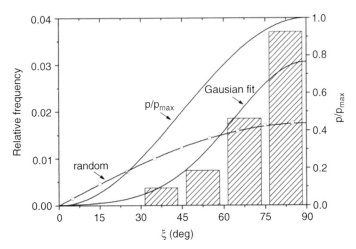

Figure 10.11 Relative frequency of angles ξ between the c-axis in new selectively grown grains and the field direction and normalized magnetic driving force p/p_{max} (right axis) for grain growth as a function of this angle [32].

experimental distribution is distinctly different from a random one. Large angles appeared more frequently than random and vice versa for small angles. Therefore, as seen in Figure 10.11, a true bias toward large inclination angles exists for grains that appeared from the selective growth process.

10.5
Impact of a Magnetic Driving Force on Texture and Grain Structure Development in Magnetically Anisotropic Polycrystals

10.5.1
Texture Evolution during Grain Growth

The observations of a magnetically driven grain boundary migration and magnetically induced selective grain growth convincingly evidenced that grain boundaries in magnetically anisotropic (nonferromagnetic) materials can be selectively affected by means of annealing in a magnetic field.

In a polycrystal exposed to a magnetic field boundaries of each grain experience an effective magnetic driving force that can be expressed as the difference between the magnetic free energy density of this grain ω and an average magnetic free energy density $\bar{\omega}$ of its neighboring grains $p_m = \omega - \bar{\omega}$. According to Eqs. (10.3) and (10.4) this can be also written as [53]

$$p_m = \frac{1}{2}\mu_0 \Delta \chi H^2 \left(\cos^2 \theta - \frac{\sum_j^n \cos^2 \theta_j}{n} \right) \tag{10.6}$$

where θ and θ_j are the angles between the field direction and principal axes of the considered grain and its n neighboring grains. As illustrated in Figure 10.12, for

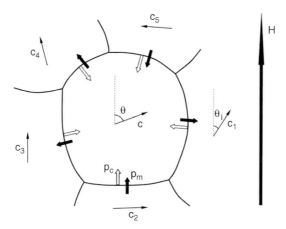

Figure 10.12 Curvature p_c and magnetic p_m driving forces acting on the boundaries of a grain in a polycrystal of a magnetically anisotropic material with $\Delta \chi = |\chi_\parallel| - |\chi_\perp| > 0$ exposed to an external magnetic field H.

10.5 Impact of a Magnetic Driving Force on Texture and Grain Structure Development

each grain in the polycrystal depending on particular grain orientation with regard to its nearest neighbors and field direction, the magnetic force can act on its boundaries both with and against the curvature forces. If, for instance, $\Delta\chi > 0$ and for the considered grain and its surrounding the condition $\cos^2\theta - 1/n \sum_j^n \cos^2\theta_j < 0$ is completed, the magnetic energy density in this grain is lower than the average energy density of the adjacent grains ($\omega < \bar{\omega}$) and the magnetic driving force ($p_m < 0$) favors the growth of this grain.

The orientation dependence of the magnetic driving force is its major property that provides an opportunity to affect the grain orientation distribution in a polycrystal by means of magnetic annealing and, therefore, to control the development of crystallographic texture during grain growth.

Experimentally, this was first demonstrated for grain growth in a zinc-1.1% aluminum alloy [29]. Annealing of sheet specimens after 99% reduction by rolling without field resulted in a typical symmetrical two-component texture with peaks in the (0002) pole figure, which can be described by the orientation $(0001)\langle 10\bar{1}0\rangle \pm 20°$ RD. If, however, annealing was occurred in a magnetic field and the sample was specifically oriented with respect to the field, the one or the other texture component could be additionally favored for growth, whereas the other component was disfavored and eventually disappeared, and thus a one component texture could develop (Figure 10.13) [29].

Similar results, that is, strengthening of one and weakening of another component in a sharp two-component texture, were observed in further experiments with cold-rolled paramagnetic titanium [30–34] and zirconium [35]. Similarly to the cold-rolling texture, the texture after annealing of Ti and Zr sheet at 750 °C and 700 °C, respectively, was characterized by the two symmetrical peaks in the (0002) pole figures (Figures 10.14 and 10.15). The application of a magnetic field substantially changed the texture evolution during grain growth. The texture peaks in the (0002) pole figures did not remain symmetrical anymore. After magnetic annealing of 75%

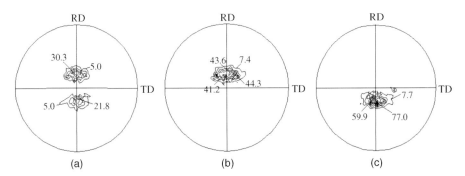

Figure 10.13 (0002) pole figures for the rolled (99%) zinc-1.1% aluminum alloy sheet samples after (a) annealing for 55 min at 390 °C without a magnetic field and annealing in a magnetic field of 32 T in the configuration rotated by (b) +19° and (c) −19° with regard to the field direction around the transverse direction (TD) [29]. (RD: rolling direction, TD: transverse direction).

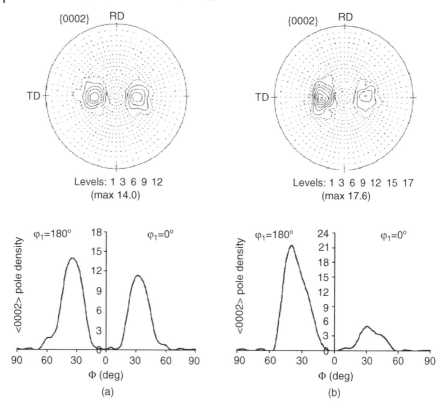

Figure 10.14 (0002) pole figures and (0002) pole density in the ND–TD plane for 75%-rolled Ti sheet samples after annealing at 750 °C for 15 min out of field and (a) 60 min and (b) 240 min in a magnetic field of 17 T [34].

cold-rolled Ti sheet at 750 °C for 60 min the intensity of the orientation distribution function (ODF) at the orientation (φ_1, Φ, φ_2) = (180°, 35°, 30°) exceeds the corresponding intensity after annealing at zero field by a factor of 1.34, whereas the intensity at (0, 35, 30) decreases to about 65% of the zero field intensity. With annealing time, the texture asymmetry continuously increased. After annealing in a field for 240 min, the intensity ratio of both major ODF peaks amounted to 3.5 [34]. Similarly, after magnetic annealing of 80%-rolled Zr specimens at 700 °C for 60 min the intensity ratio of both major ODF peaks at (180,35,30) and (0, 35, 30) was 1.7 and after annealing in the field for 90 min rose up to 2.2 [35]. Since a magnetic field affects the intensity of all orientations with $\varphi_1 = 0°/180°$ and $\Phi = 35°$ in the entire interval of φ_2, the magnetically induced texture anisotropy and its increase with annealing time is most appropriate revealed by a comparison of the (0002) pole density of both major peaks centered around (0,35, φ_2) and (180,35, φ_2) (Figures 10.14 and 10.15).

The development of an asymmetrical texture during annealing in a magnetic field is obviously caused by the magnetic driving force during grain growth. During

10.5 Impact of a Magnetic Driving Force on Texture and Grain Structure Development

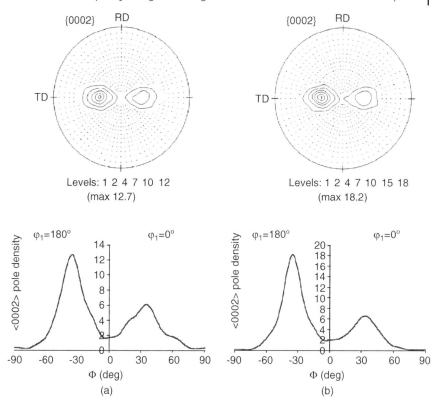

Figure 10.15 (0002) pole figures and (0002) pole density in the ND–TD plane for 80%-rolled Zr sheet samples after annealing at 700 °C (a) for 60 min and (b) 90 min in a magnetic field of 17 T [35].

magnetic annealing, the titanium and zirconium specimens were oriented in such a manner that the c-axes of the grains that compose a texture peak around the orientations (180, 35, φ_2) were aligned perpendicular to the field direction. With the difference in magnetic susceptibility parallel and perpendicular to the c-axis $\Delta\chi > 0$ for both titanium and zirconium, the magnetic free energy density of these grains attains a minimum, which results in an additional driving force for their growth ($p_m < 0$).

As apparent from (0002) pole density distributions along the transverse direction in Figures 10.14 and 10.15, an increase of the (180, 35, φ_2) peak intensity during annealing in a magnetic field corresponds to a decrease of the intensity of the (0, 35, φ_2) peak. This means that on average, grains close to the energetically favorable orientations (180, 35, φ_2) grow at the expense of grains oriented close to (0, 35, φ_2), whose free energy is increased in a magnetic field by the amount $p_m > 0$ (Eq. (10.6)). Accordingly, grains with the (180, 35, φ_2) orientations can be expected to grow faster and eventually be significantly bigger than grains, which compose the other peak.

The magnetic field effect on texture and microstructure evolution was also analyzed by computer simulations of 2D grain growth [34,53]. The grain

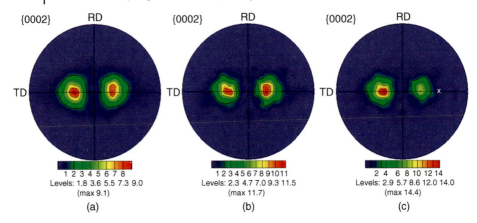

Figure 10.16 (a) (0002) pole figure for a 75%-rolled Ti sheet sample after annealing at 750 °C for 15 min; (b) and (c) – simulated pole figures for a 2D Ti polycrystal after additional 10-min annealing (b) without field and (c) in a magnetic field of 17 T (field direction is indicated by a white cross) [34].

structure in the applied simulation algorithm based on vertex and front-tracking models [54–57] is represented by differently oriented grains, separated by boundaries, which intersect at triple junctions (real vertices) and consist of points (virtual vertices) along the boundary length according to the boundary curvature [58]. During conventional (without field) grain growth, all virtual vertices move under the local curvature driving force p_c (Eq. (10.2)) perpendicular to the boundary. In the presence of an external magnetic field, the kinetic equation of the grain boundary migration (Eq. (10.1)) is solved for all virtual vertices with a driving force p equal to the sum of curvature and magnetic driving forces $p = p_c + p_m$, where p_m is calculated according to Eq. (10.4) [58].

Figure 10.16 depicts the texture in a 2D-polycrystal as obtained by computer simulations of 2D grain growth [34]. For these simulations, the actual experimental conditions were used: the initial texture and microstructure were reconstructed from individual orientation data measured by electron backscatter diffraction (EBSD) on a recrystallized Ti specimen annealed for 15 min at 750 °C. Grain boundary energy and mobility were assumed to be uniform. The magnetic energy density was computed for each grain and the condition that the magnetic field was applied perpendicular to the rolling direction of the specimen and inclined 32° from the transverse direction, as indicated by a cross in the pole figure in Figure 10.16c. Therefore, depending on their surroundings, grains with orientations close to (180, 35, φ_2) experience an additional force to grow, since their c-axes are aligned nearly normal to the field direction and thus, their magnetic energy density is the smallest. The results of simulations agree with experimental observations. After annealing out of field, the simulated texture remains symmetrical (Figure 10.16b), whereas after magnetic annealing the (180, 35, φ_2) texture peak becomes much stronger than the other peak (Figure 10.16c).

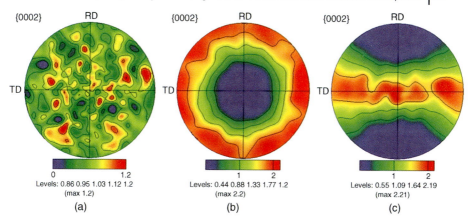

Figure 10.17 Simulated texture development during grain growth in a 2D Ti-polycrystal in a magnetic field. (0002) pole figures for the specimen (a) with an initially random texture and for the same specimen after 25-min annealing in a magnetic field of 20 T directed (b) parallel to the sheet normal (ND) and (c) parallel to the rolling direction (RD) [34].

The ability of magnetic annealing to produce preferred orientations in a polycrystal during grain growth is perfectly demonstrated by computer simulations, where a magnetic field was applied to a 2D polycrystal with initially random orientation distribution (Figure 10.17a). Simulations were performed for two different specimen orientations. First, the magnetic field was applied parallel to the sheet normal (ND). In this case, grains with c-axis parallel to the (φ_1, 90, φ_2) orientations (i.e., with c-axes perpendicular to ND) possess a minimum magnetic free energy and therefore experience an additional driving force for their growth ($p_m < 0$). This results in a texture with maximum (0002) pole intensity at (φ_1, 90, φ_2) and minimum at (φ_1, 0, φ_2) (Figure 10.17b). Second, a magnetic field was directed parallel to the rolling direction (RD). Correspondingly, a maximum magnetic energy was generated in grains with the c-axis aligned along (90, 90, φ_2) (i.e., RD), and the magnetic driving force promoted their shrinkage ($p_m > 0$), whereas in grains with the c-axis perpendicular to RD (i.e., aligned along (0/180, Φ, φ_2)) this force favored their expansion. As a consequence, a texture with a maximum intensity at (0/180, Φ, φ_2) developed (Figure 10.17c). Therefore, the simulations unambiguously proved that the texture evolution during grain growth in a magnetic field of magnetically anisotropic titanium is due to an orientation-dependent magnetic driving force.

10.5.2
Microstructure Evolution and Growth Kinetics

The magnitude of the texture peaks obtained by X-ray diffraction is determined by the total area of grains with a respective orientation. Therefore, global texture measurements do not render information on the grain microstructure with respect to grain size and number of grains, which compose different texture components. This information, however, was obtained by orientation imaging with EBSD [33–35].

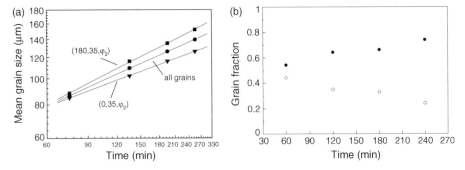

Figure 10.18 (a) Mean grain size versus annealing time and (b) grain fractions in the (180, 35, φ_2) (solid symbols) and (0, 35, φ_2) (open symbols) subsets in Ti specimens after annealing at 750 °C in the field of 17 T.

Since the observed textures in titanium and zirconium were characterized by a high ODF density at the orientations (0/180, 35, φ_2), the whole set of grains was subdivided into two subsets with orientations around (0, 35, φ_2) and (180, 35, φ_2). These two subsets represent the two texture peaks in the (0002) pole figures in Figures 10.14–10.16.

The analysis of individual orientation measurements on both titanium and zirconium sheet samples as well as the results of computer simulations [34,35,53] revealed that a magnetic field affects the grain growth kinetics in such a way that the growth of grains with energetically favored orientations is enhanced, whereas the growth of grains with disfavored orientations is retarded (Figure 10.18a). However, the relatively slight difference in the mean size of grains in different grain subsets cannot comprehensively explain the observed texture anisotropy after magnetic annealing. For instance, the ratio of the mean grain area $\bar{S}$ in both subsets after 240-min magnetic annealing of titanium sample is $\bar{S}_{(180,35,\varphi_2)}/\bar{S}_{(0,35,\varphi_2)} = 1.45$ [34], whereas the ratio of the pole intensity for both texture components in Figure 10.14b amounts to 4.3. As obtained by the analysis of EBSD data, the crucial reason for the developing texture anisotropy is that the number fraction of grains with energetically preferred orientations becomes larger than half of all grains and continuously rises during magnetic annealing, whereas the fraction of grains with disfavored orientations decreases (Figures 10.18b and 10.19).

This behavior can be understood by considering grain growth kinetics. According to Hillert [59], the rate of grain size change under a curvature driving force can be expressed as

$$\frac{dR}{dt} = -\alpha m \gamma \left(\frac{1}{\bar{R}} - \frac{1}{R} \right) \tag{10.7}$$

where the grain size is expressed by the radius R of a circle or sphere of the same area or volume, respectively, and α is a dimensionless constant. The grain boundary energy γ and mobility m are assumed to be equal for all grain boundaries. The mean grain size $\bar{R}$ denotes a critical size, such that at any time grains with $R > \bar{R}$ grow

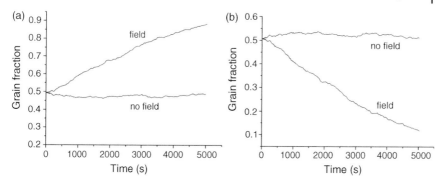

Figure 10.19 Computed grain fractions in different grain subsets versus annealing time at zero field and in a field of 17 T. (a) and (b) – grain subsets (180, 35, φ_2) and (0, 35, φ_2), respectively; initial grain structure – microstructure after 15-min annealing of the 75%-rolled Ti sheet at 750 °C; the magnetic field was directed as shown in Figure 10.16c.

($dR/dt > 0$), grains with $R < \bar{R}$ shrink ($dR/dt < 0$) and grains with $R = \bar{R}$ maintain their size ($(dR/dt)_{R=\bar{R}}$).

For the rate of grain size change under a magnetic driving force only, we can write

$$\frac{dR}{dt} = -mp_m \qquad (10.8)$$

According to Eq. (10.8) grains with a higher magnetic energy density ($p_m = \omega - \bar{\omega} > 0$) shrink, whereas grains with energetically favorable orientations ($\omega < \bar{\omega}$) grow ($dR/dt > 0$).

The forces acting on a grain boundary during grain growth can safely be assumed additive [60,61]. Therefore, for the simultaneous action of both driving forces we arrive at[1]

$$\frac{dR}{dt} = -am\gamma\left(\frac{1}{\bar{R}} - \frac{1}{R}\right) - mp_m \qquad (10.9)$$

Apparently, a magnetic driving force being superimposed to a curvature driving force during grain growth changes the size R_{th} of grains that neither grow nor shrink. This threshold size can be derived from

$$\frac{dR}{dt}\bigg|_{R=R_{th}} = am\gamma\left(\frac{1}{\bar{R}} - \frac{1}{R_{th}}\right) - mp_m = 0, \qquad (10.10)$$

which leads to the expression for R_{th}

$$R_{th} = \frac{\bar{R}}{1 - \dfrac{\bar{R}p_m}{a\gamma}} \qquad (10.11)$$

---

1) A similar equation for grain growth driven by both curvature driving force and differences in energy density was suggested by Deus et al. [10].

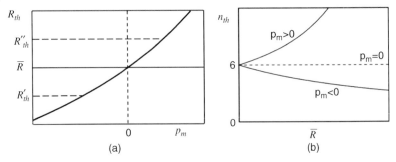

Figure 10.20 (a) Schematic dependence of a threshold grain size R_{th} on a magnetic driving force p_m, as given by Eq. (10.11); (b) Topological class of stable grains versus mean grain size, as given by Eq. (10.16).

According to Eq. (10.11), the threshold grain size for grains with a lower magnetic free energy density ($p_m = \omega - \bar{\omega} < 0$) is reduced to $R'_{th} < \bar{R}$ (Figure 10.20a). Hence in contrast to a purely curvature-driven grain growth, grains with a size $R'_{th} < R < \bar{R}$ will grow in the presence of a magnetic field. Opposite to this, the threshold grain size for grains with a higher magnetic free energy density ($\omega > \bar{\omega}$) is increased to $R''_{th} > \bar{R}$ and therefore, grains with $\bar{R} < R < R''_{th}$ will shrink, although in a purely curvature-driven process such grains would grow.

Based on this, the experimentally observed difference in grain numbers and grain size of the two texture components after magnetic annealing can be interpreted in terms of magnetically shifted grain growth kinetics. Due to their "high energy" orientations, grains composing the (0, 35, φ_2) texture peak (Figures 10.14–10.16) on average experience a positive magnetic driving force, $p_m > 0$, while for "low energy" grains of the (180, 35, φ_2) peak a magnetic driving force is negative, $p_m < 0$. Therefore, according to Eq. (10.11), the threshold grain size for shrinkage or growth of high energy grains increases, whereas it decreases for low energy grains. As a result, more grains of the high energy component shrink and disappear than it would be the case without a magnetic field. Vice versa, more grains of the low-energy component will grow in the presence of a magnetic field. As observed in experiments and simulations [34,53], this eventually leads to the difference between both grain subsets.

An application of a magnetic field during annealing results not only in the change of the growth rate of differently oriented grains, but also substantially alters the final distribution of the grains of different topological classes [34]. A topological approach for grain growth in 2D systems was developed by Mullins [62], who applied von Neumann's description of 2D soap froth [63] to curvature driven grain growth. All boundaries also were assumed to possess equal energy γ and mobility m, and considered to be in equilibrium at their triple junctions. According to this approach, there is a unique relation between the rate of area change dS/dt and the topological class n (number of immediate neighbors) of a grain [62]

$$\frac{dS}{dt} = \frac{\gamma m \pi}{3}(n - 6) \quad (10.12)$$

10.5 Impact of a Magnetic Driving Force on Texture and Grain Structure Development

Grains with $n > 6$ will grow, and those with $n < 6$ will disappear. Using an equivalent grain size (radius R of a circular grain of the same area), we can write for the grain size change rate

$$\frac{dR}{dt} = \frac{m\gamma}{R}\left(\frac{n}{6} - 1\right) \tag{10.13}$$

In the presence of a magnetic field, the topological class n_{th} of stable grains of size R_{th} can be derived from the condition

$$\frac{dR}{dt} = m\left[\frac{\gamma}{R_{th}}\left(\frac{n_{th}}{6} - 1\right) - p_m\right] = 0 \tag{10.14}$$

and reads

$$n_{th} = 6\left(1 + \frac{R_{th} p_m}{\gamma}\right) \tag{10.15}$$

Substituting the expression for R_{th} (Eq. (10.11)) into Eq. (10.15) and taking into account that for a 2D grain system $\alpha \cong 0.5$ [59], we find for the threshold topological class

$$n_{th} = 6\left(\frac{\gamma - \bar{R} p_m}{\gamma - 2\bar{R} p_m}\right) \tag{10.16}$$

According to Eq. (10.16), the topological class of stable grains in preferred energetic configuration ($p_m < 0$), is reduced to $n_{th} < 6$ and vice versa raised to $n_{th} > 6$ for grains in unfavored configuration (Figure 10.20b). This practically means that grains with topological class 6 do not remain stable in the magnetic field, but either grow or shrink that results in a significant decrease of the fraction of grains with $n = 6$ in a 2D-polycrystal after magnetic annealing [34].

Obviously, the texture and grain microstructure evolution can be affected by the magnetic force only if its magnitude is comparable with the conventional curvature forces acting on the grain boundaries during grain growth. In the experiments with cold-rolled titanium and zirconium, the c-axes of grains oriented close to the (180, 35, φ_2) – peaks on left side of (0002) pole figures in Figures 10.14–10.16 – are almost normal to the field direction and, therefore, the magnetic energy density in these grains is lower than in grains of any other orientations. The maximum magnetic driving force for grain boundary motion is determined by the difference of the magnetic energy density between these grains and grains with ideal orientations (0, 35, φ_2). According to Eq. (10.4), this force in titanium at 750° ($\Delta\chi_{Ti}^{750\,°C} = 1.18 \times 10^{-5}$ [64]) and zirconium at 700° ($\Delta\chi_{Zr}^{700\,°C} = 1.18 \times 10^{-5}$ [64]) in the field of 17 T (field strength H = 1.35×10^7 A/m) amounts to about $p_m \cong$ 1.5 kJ/m^3 and 2.0 kJ/m^3, respectively. The curvature driving force p_c for normal (continuous) grain growth can be expressed as $p_c = 2\gamma/r$, where r is the radius of curvature, which can be assumed to exceed the grain size by about an order of magnitude. In contrast to the magnetic force that does not depend on the grain size, the curvature driving force decreases with the increasing grain size. For a mean grain size of about 40 μm that corresponds to the grain structure in the investigated titanium before the application of a magnetic field [34], the curvature force attains the same magnitude as the magnetic driving force, that is, amounts to about $2\,kJ/m^3$ (assuming grain boundary energy γ of typically $0.4\,J/m^2$) and decreases further with a rising mean grain size during the annealing. Therefore, any concerns that the additional driving forces occurred in a magnetic field are insufficient to cause the observed changes in the orientation distribution of grains and growth kinetics of differently oriented grains in polycrystalline zinc, titanium, and zirconium can be completely eliminated.

10.6
Magnetic Field Influence on Texture and Microstructure Evolution in Polycrystals Due to Enhanced Grain Boundary Motion

Apparently, for fine-grained structures the magnetic driving force generated by a magnetic field is much smaller than the curvature force and, therefore, is unlikely to cause measurable texture anisotropy during the magnetic annealing. It is worth noting, however, that the impact of a magnetic field on grain growth is not only confined to a change of the net driving force for grain boundary motion.

As was observed in experiments with cold-rolled titanium [65] and zirconium [35], annealing in a magnetic field of 19 T at relatively low temperature (530 °C and 550 °C, respectively) distinctly modified the texture, although the peaks in the (0002) pole figure remained symmetrical (Figure 10.21). The density of the (0/180, 35, 0) orientations decreased, whereas the density of orientations close to (0/180, 35, 30) increased. As seen in Figure 10.22, where the orientation density along specific fibers containing major texture components is depicted, the peak maximum in the $\varphi_2 = 0°$ curves degrades after annealing in a magnetic field, whereas the intensity of the peaks in the $\varphi_2 = 30°$ curves rises.

Obviously, the observed texture changes cannot be caused by the magnetic driving force, since this behavior was observed for both components, that is, $\varphi_1 = 0°$ and $\varphi_1 = 180°$ (Figure 10.22), although the grains of the $\varphi_1 = 180°$ – component were favored and the grains of the $\varphi_1 = 0°$ – component were disfavored for growth by a difference in the magnetic energy density between differently oriented grains.

The preferred development of texture components (0/180, 35, 30) at the expense of (0/180, 35, 0) components in the annealing texture is typical for the progress of grain growth during annealing of cold-rolled titanium [66,67] and zirconium sheet [68]. Therefore, the magnetically induced texture changes (e.g., Figures 10.21 and 10.22) suggest that a magnetic field promotes grain growth in the investigated materials. This was also confirmed by orientation imaging measurements by EBSD, which revealed the markedly larger mean grain size in both the zirconium and titanium specimens after magnetic annealing compared with specimens annealed at zero field (Figure 10.23) [35,65]. The enhanced grain growth kinetics are obviously due to accelerated grain boundary motion, which in turn depends on driving force and boundary mobility (see Eq. (10.1)). For a mean grain size of about 5 μm (in Ti)

10.6 Magnetic Field Influence on Texture and Microstructure Evolution in Polycrystals

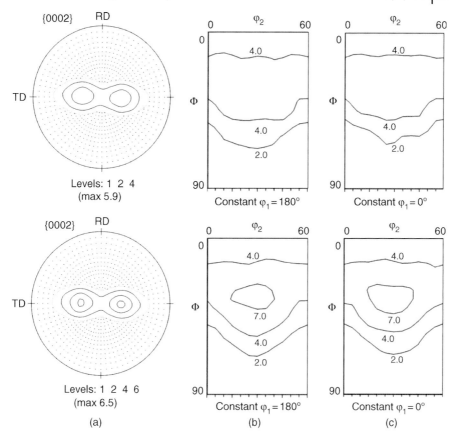

Figure 10.21 (a) (0002) pole figures and ODF sections at (b) $\varphi_1 = 180°$ and (c) $\varphi_1 = 0°$ for 80% cold-rolled Zr sheet specimens annealed at 550 °C for 45 min at zero field (top) and in a magnetic field of 19 T (bottom) [35].

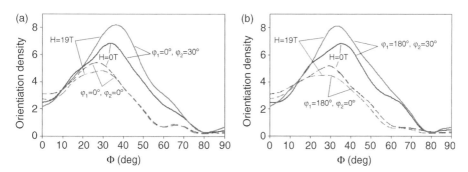

Figure 10.22 Orientation density along the fibers (a) $\varphi_1 = 0°$, $\varphi_2 = 0°$ as well as $\varphi_1 = 0°$, $\varphi_2 = 30°$, and (b) $\varphi_1 = 180°$, $\varphi_2 = 0°$ as well as $\varphi_1 = 180°$, $\varphi_2 = 30°$ after annealing of the 80% cold-rolled Zr sheet at 550 °C for 45 min at zero field and at 19 T [35].

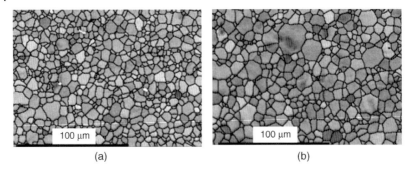

Figure 10.23 Grain microstructures after 45 min annealing at 550 °C (a) at zero field and (b) at 19 T obtained by orientation imaging with EBSD [35].

and 2 μm (in Zr) as obtained after annealing in the described experiments, the acting driving force remained practically unaffected by the magnetic field. Therefore, the observed microstructural changes during magnetic annealing of titanium and zirconium have to be attributed to magnetically increased grain boundary mobility.

Further evidence that the impact of an external magnetic field on texture and microstructure evolution must not be necessarily associated with a magnetic driving force for boundary migration, but can be caused by an enhanced capability of grain boundaries to move, was provided by the results of experiments with commercial aluminum alloys [36,37,69,70].

The magnetically affected annealing behavior of cold-rolled (71%) alloy AA3103 was investigated by measuring the crystallographic texture and the grain microstructure during heat treatment in a magnetic field [36,37]. The results revealed that the volume fraction of β – fiber orientations decreased markedly earlier and faster during annealing in a field than without field for all three investigated temperatures (Figure 10.24a). Since a degradation of the deformation texture components is a

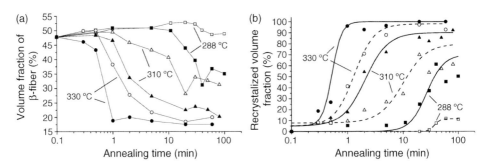

Figure 10.24 (a) Volume fraction of deformation texture components (β-fiber: {112} $\langle 111 \rangle$ (Cu−), {123} $\langle 634 \rangle$ (S−), and {011} $\langle 211 \rangle$ (Brass-component)) calculated from the ODFs versus annealing time; (b) the temporal change of the recrystallized volume fraction of 71% cold-rolled AA3103 during annealing out of field (open symbols) and in a magnetic field of 17 T (solid symbols) [37].

typical sign of the progress of recrystallization in deformed Al-alloy, this behavior indicates that recrystallization is promoted by a magnetic field. The incubation time of recrystallization during magnetic annealing is substantially decreased compared to conventional annealing (Figure 10.24b). This can be attributed to magnetically enhanced recovery in the deformed aluminum alloy as observed in [36]. Furthermore, an analysis has shown that the recrystallization kinetics are significantly accelerated by an applied magnetic field [37].

As discussed in [36,37], the raised kinetics of recovery and recrystallization can be attributed to a magnetically affected mobility of dislocations [71,72] and grain boundaries, which can be enhanced in the field due to the changing their interaction with defects in the crystal structure (precipitations, solute atoms, vacancies, etc.) acting as the pinning centers for dislocations and grain boundaries. This hypothesis was proved by measurements of the kinetics of the curvature driven grain growth in AA3103. The completely recrystallized specimens were annealed for 65 min at 610 °C in the field of 17 T and out of field. The EBSD analysis revealed that the mean grain size after magnetic annealing (33 μm) was distinctly bigger than after conventional annealing (27.5 μm) (Chr. Günster, S. Bhaumik, X. Molodova, and D.A. Molodov, 2009, unpublished research.).

10.7
Concluding Remarks

Under the impact of an external magnetic field grain boundaries in magnetically anisotropic (nonferromagnetic) materials can be driven by a driving force, which occurs due to a volume magnetic free energy difference across the boundary. This was proved by experiments with individual boundaries in specially grown bismuth and zinc bicrystals, where planar boundaries moved under the action of a magnetic driving force only. Since the magnetic driving force does not depend on boundary energy and shape, the measurements of magnetically driven boundary migration provides an opportunity to investigate the motion of specific planar grain boundaries with well-defined structures and to determine their absolute mobility. The experiments with various symmetrical and asymmetrical tilt boundaries revealed that grain boundary mobility essentially depends on the misorientation angle and the inclination of the boundary plane.

In polycrystals during annealing in a magnetic field the magnetic force acts supplementary to the conventional curvature driving force. It additionally impels each grain of a polycrystal to grow or to shrink, depending on its orientation and the orientation of its nearest neighbors with respect to the field direction. Just owing to the orientation dependency of the magnetic driving force, a magnetic field possesses a remarkable ability to produce preferred orientations in a polycrystal during grain growth. This was perfectly captured by computer simulations, where a magnetic field was applied to a 2D polycrystal with initially random orientation distribution. Experiments on polycrystalline cold-rolled zinc, titanium, and zirconium revealed that magnetic annealing can cause significant changes in the evolution of the microstructure with respect to grain size, orientation distribution, and grain topology. It could be also shown that recrystallization and grain growth can be affected by a magnetic field not only due to the altered net driving force for grain boundary motion, but due to enhanced grain boundary mobility by an applied magnetic field as well.

In summary, the experimental and modeling efforts overviewed in this Chapter have proved that a magnetic field can be effectively utilized as an additional degree of control of texture and microstructure development in polycrystals, which is indispensable for design and fabrication of materials with desired properties.

Acknowledgment

The author expresses his gratitude to the Deutsche Forschungsgemeinschaft for financial support (Grants MO 848/4, MO 848/6, and MO 848/12).

References

1 Smoluchowski, R. (1951) Theory of grain boundary motion. *Physical Review*, **83** (1), 69–70.
2 Turnbull, D. (1951) Theory of grain boundary migration rates. *Journal of Metals*, **3** (8), 661–665.
3 Vandermeer, R.A. (1965) Dependence of grain boundary migration rates on driving force. *Transactions AIME*, **233** (1), 266–271.
4 Vandermeer, R.A., Juul-Jensen, D., and Woldt, E. (1997) Grain boundary mobility during recrystallization of copper. *Metallurgical and Materials Transactions A*, **28A** (3), 749–754.
5 Aristov, V.Y., Fradkov, V.E., and Shvindlerman, L.S. (1979) Interaction of a moving grain boundary with a crystal surface. *Physics of Metals and Metallography*, **45** (5), 83–94.
6 Molodov, D.A., Gottstein, G., Heringhaus, F., and Shvindlerman, L.S. (1998) True absolute grain boundary mobility: Motion of specific planar boundaries in Bi-bicrystals under magnetic driving force. *Acta Materialia*, **46** (16), 5627–5632.
7 Molodov, D.A., Ivanov, V.A., and Gottstein, G. (2007) Low angle tilt boundary migration coupled to shear deformation. *Acta Materialia*, **55** (5), 1843–1848.
8 McLean, M. (1982) Microstructural changes in presence of externally applied potential gradients. *Metal Science*, **16**, 31–36.
9 Watanabe, T. (2001) External field applied grain boundary engineering for high performance materials, in *Recrystallization and Grain Growth* (eds G. Gottstein and D. A. Molodov), Springer, Berlin, pp. 11–20.
10 Deus, A.M., Fortes, M.A., Ferreira, P.J., and Vander Sande, J.B. (2002) A general approach to grain growth driven by energy density differences. *Acta Materialia*, **50**, 3317–3330.
11 Wada, H. and Schneider-Muntau, H.J. (eds) (2005) *Materials Processing in Magnetic Fields*, World Scientific Publishing, Singapore.
12 Han, Q.Ludtka, G. and Zhai, Q. (eds) (2007) *Materials Processing under the Influence of External Fields*, TMS, Warrendale, PA.
13 Yamaguchi, M. and Tanimoto, Y. (eds) (2006) *Magneto-Science. Magnetic Field Effect on Materials: Fundamentals and Applications*, Kodansha and Springer, Berlin.
14 Boothby, O.L., Wenny, D.H., and Thomas, E.E. (1958) Recrystallization of MnBi induced by a magnetic field. *Journal of Applied Physics*, **29**, 353.
15 Chen, T. and Stutius, W.E. (1974) Magnetic-field-induced recrystallization in

16. Marticainen, H.O. and Lindroos, V.K. (1981) Observations on the effect of magnetic field on the recrystallization in ferrite. *Scandinavian Journal of Metallurgy*, **10**, 3–8.
17. Watanabe, T., Suzuki, Y., Tanii, S., and Oikawa, H. (1990) The effects of magnetic annealing on recrystallization and grain-boundary character distribution (GBCD) in iron–cobalt alloy polycrystals. *Philosophical Magazine Letters*, **62**, 9–17.
18. Masahashi, N., Matsuo, M., and Watanabe, K. (1998) Development of preferred orientation in annealing of Fe–3.25%Si in a high magnetic field. *Journal of Materials Science*, **13**, 457–461.
19. Bacaltchuk, C.M.B., Castello-Blanco, G.A., Ebrahimi, M., Garmestani, H., and Rollett, A.D. (2003) Effect of magnetic field applied during secondary annealing on texture and grain size of silicon steel. *Scripta Materialia*, **48**, 1343–1347.
20. Tsurekawa, S., Harada, K., Sasaki, T., Matsuzaki, T., and Watanabe, T. (2000) Magnetic sintering of ferromagnetic metal powder compacts. *Materials Transactions JIM*, **41**, 991–999.
21. Harada, K., Tsurekawa, S., Watanabe, T., and Palumbo, G. (2003) Enhancement of homogeneity of grain boundary microstructure by magnetic annealing of electrodeposited nanocrystalline nickel. *Scripta Materialia*, **49**, 367–372.
22. Mikelson, A.E. and Karklin, Ya.Kh. (1981) Control of crystallization processes by means of magnetic fields. *Journal of Crystal Growth*, **52**, 524–529.
23. Sugiyma, T., Tahashi, M., Sassa, K., and Asai, S. (2003) The control of crystal orientation in non-magnetic metals by imposition of a high magnetic field. *ISIJ International*, **43**, 855–861.
24. Mullins, W.W. (1956) Magnetically induced grain boundary motion in bismuth. *Acta Metallurgica*, **4** (4), 421–432.
25. Fraser, M.J., Gold, R.E., and Mullins, W.W. (1961) Grain boundary mobility in bismuth. *Acta Metallurgica*, **9**, 960–961.
26. Molodov, D.A., Gottstein, G., Heringhaus, F., and Shvindlerman, L.S. (1997) Motion of planar grain boundaries in bismuth-bicrystals driven by a magnetic field. *Scripta Materialia*, **37** (8), 1207–1213.
27. Günster, Ch., Molodov, D.A., and Gottstein, G. (2010) Magnetically driven migration of specific planar grain boundaries in Zn bicrystals. *Scripta Materialia*, **63** (3), 300–303.
28. Günster, Chr., Molodov, D.A., and Gottstein, G. (2013) Migration of grain boundaries in Zn. *Acta Materialia*, **61** (7), 2363–2375.
29. Sheikh-Ali, A.D., Molodov, D.A., and Garmestani, H. (2002) Magnetically induced texture development in zinc alloy sheet. *Scripta Materialia*, **46** (12), 857–862.
30. Molodov, D.A. and Sheikh-Ali, A.D. (2004) Effect of magnetic field on texture evolution in titanium. *Acta Materialia*, **52** (14), 4377–4383.
31. Molodov, D.A., Konijnenberg, P.J., Bozzolo, N., and Sheikh-Ali, A.D. (2005) Magnetically affected texture and grain structure development in titanium. *Materials Letters*, **59** (26), 3209–3213.
32. Molodov, D.A. and Konijnenberg, P.J. (2005) Grain boundary dynamics and selective grain growth in non-ferromagnetic metals in high magnetic fields. *Zeitschrift für Metallkunde*, **96** (10), 1158–1165.
33. Molodov, D.A. and Konijnenberg, P.J. (2006) Grain boundary and grain structure control through application of a high magnetic field. *Scripta Materialia*, **54** (6), 977–981.
34. Molodov, D.A., Bollmann, Chr., Konijnenberg, P.J., Barrales-Mora, L.A., and Mohles, V. (2007) Annealing texture and microstructure evolution in titanium during grain growth in an external magnetic field. *Materials Transactions*, **48** (11), 2800–2808.
35. Molodov, D.A. and Bozzolo, N. (2010) Observations on the effect of a magnetic field on the annealing texture and microstructure evolution in zirconium. *Acta Materialia*, **58** (19), 3568–3581.
36. Molodov, D.A., Bhaumik, S., Molodova, X., and Gottstein, G. (2006) Annealing behaviour of cold rolled aluminum alloy in a high magnetic field. *Scripta Materialia*, **54** (12), 2161–2164.

37 Bhaumik, S., Molodova, X., Molodov, D.A., and Gottstein, G. (2006) Magnetically enhanced recrystallization in an aluminum alloy. *Scripta Materialia*, **55** (11), 995–998.

38 Goetz, A. and Focke, A. (1934) The crystal magnetism of bismuth crystals. *Physical Review*, **45** (2), 170–199.

39 Otake, S., Momouchi, M., and Matsuno, N. (1980) Temperature dependence of the magnetic susceptibility of bismuth. *Journal of the Physical Society of Japan*, **49** (5), 1824–1828.

40 Konijnenberg, P.J., Ziemons, A., Molodov, D.A., and Gottstein, G. (2008) Polarization microscopy probe for applications in a high magnetic field. *Review of Scientific Instruments*, **79**, 013701.

41 Mac Clure, J.W. and Marcus, J.A. (1951) The magnetic susceptibility of zinc at liquid nitrogen temperatures. *Physical Review*, **84**, 787–788.

42 Gottstein, G. and Shvindlerman, L.S. (2010) *Grain Boundary Migration in Metals: Thermodynamics, Kinetics, Applications*, 2nd edn, CRC Press, Boca Raton, FL.

43 Rath, B.B. and Hu, H. (1969) Effect of driving force on the migration of high-angle tilt grain boundaries in aluminum bicrystals. *Transactions AIME*, **245**, 1577–1585.

44 Rath, B.B. and Hu, H. (1972) Influence of solutes on the mobility of tilt boundaries, in *The Nature and Behavior of Grain Boundaries* (ed. H. Hu), Plenum Press, New York, pp. 405–435.

45 Sheikh-Ali, A.D., Molodov, D.A., and Garmestani, H. (2003) Boundary migration in Zn bicrystal induced by a high magnetic field. *Applied Physics Letters*, **82** (18), 3005–3007.

46 Sheikh-Ali, A.D., Molodov, D.A., and Garmestani, H. (2003) Migration and reorientation of grain boundaries in Zn bicrystals during annealing in a high magnetic Field. *Scripta Materialia*, **48** (5), 483–488.

47 Sursaeva, V.G., Andreeva, A.V., Kopezkii, Ch.V., and Shvindlerman, L.S. (1976) Mobility of a $\langle 10\bar{1}0 \rangle$ tilt boundary in zinc. *Physics of Metals and Metallography*, **41** (5), 98–103.

48 Molodov, D.A., Günster, Chr., Gottstein, G., and Shvindlerman, L.S. (2012) A novel experimental approach to determine the absolute grain boundary energy. *Philosophical Magazine*, **92** (36), 4588–4598.

49 Molodov, D.A., Czubayko, U., Gottstein, G., and Shvindlerman, L.S. (1998) On the effect of purity and orientation on grain boundary motion. *Acta Materialia*, **46** (2), 553–564.

50 Gorkaya, T., Molodov, D.A., and Gottstein, G. (2009) Mechanically driven migration of symmetrical $\langle 100 \rangle$ tilt grain boundaries in Al bicrystals. *Acta Materialia*, **57** (18), 5396–5405.

51 Molodov, D.A., Gorkaya, T., and Gottstein, G. (2011) Dynamics of grain boundaries under applied mechanical stress. *Journal of Materials Science*, **46** (12), 4318–4326.

52 Konijnenberg, P.J., Molodov, D.A., and Gottstein, G. (2004) Magnetically driven selective grain growth in locally deformed Zn single crystals. *Materials Science Forum*, **467–470**, 763–770.

53 Molodov, D.A., Konijnenberg, P.J., Barrales-Mora, L.A., and Mohles, V. (2006) Magnetically controlled microstructure evolution in non-ferromagnetic metals. *Journal of Materials Science*, **41** (23), 7853–7861.

54 Kawasaki, K., Nagai, T., and Nakashima, K. (1989) Vertex models for two-dimensional grain growth. *Philosophical Magazine B – Physics of Condensed Matter Statistical Mechanics Electronic Optical and Magnetic Properties*, **60**, 399–421.

55 Weigand, D., Brechet, Y., and Lepinoux, J. (1998) A vertex dynamics simulation of grain growth in two dimension. *Philosophical Magazine B – Physics of Condensed Matter Statistical Mechanics Electronic Optical and Magnetic Properties*, **78**, 329–352.

56 Frost, H.J., Thompson, C.V., Howe, C.L., and Whang, J. (1988) A two-dimensional computer simulation of capillarity-driven grain growth: Preliminary results. *Scripta Metallurgica*, **22**, 65–70.

57 Fayad, W., Thompson, C.V., and Frost, H.J. (1999) Steady-state grain-size distributions resulting from grain growth in two dimensions. *Scripta Materialia*, **40**, 1199–1204.

58 Barrales-Mora, L.A., Mohles, V., Konijnenberg, P.J., and Molodov, D.A. (2007) A novel implementation for the simulation of 2-D grain growth with consideration to external energetic fields. *Computational Materials Science*, **39** (1), 160–165.

59 Hillert, M. (1965) On the theory of normal and abnormal grain growth. *Acta Metallurgica*, **13**, 227–238.

60 Novikov, V.Yu. (1999) Simulation of grain growth in thin films with columnar microstructure. *Acta Materialia*, **47** (18), 4507–4514.

61 Gangulee, A. (1974) Structure of electroplated and vapor-deposited copper films. III. Recrystallization and grain growth. *Journal of Applied Physics*, **45**, 3749–3756.

62 Mullins, W.W. (1956) Two-dimensional motion of idealized grain boundaries. *Journal of Applied Physics*, **27**, 900–904.

63 von Neumann, J. (1952) *Metal Interfaces*, American Society of Metals, Cleveland, OH, pp. 108–110.

64 Volkenshtein, N.V., Galoshina, E.V., and Panikovskaya, T.N. (1975) Temperature dependence of magnetic susceptibility anisotropy of transition metal crystals with a hexagonal closely-packed structure. *Soviet Physics JETP*, **40**, 730–734.

65 Molodov, D.A., Bollmann, Chr., and Gottstein, G. (2007) Impact of a magnetic field on the annealing behavior of cold rolled titanium. *Materials Science and Engineering A*, **467** (1–2), 71–77.

66 Singh, A.K. and Schwarzer, R.A. (2000) Texture and anisotropy of mechanical properties in titanium and its alloys. *Zeitschrift für Metallkunde*, **91**, 702–716.

67 Bozzolo, N., Dewobroto, N., Grosdidier, T., and Wagner, F. (2005) Texture evolution during grain growth in recrystallized commercially pure titanium. *Materials Science and Engineering A*, **397**, 346–355.

68 Dewobroto, N., Bozzolo, N., Barberis, P., and Wagner, F. (2006) On the mechanisms governing the texture and microstructure evolution during static recrystallization and grain growth of low alloyed zirconium sheets (Zr 702). *International Journal of Materials Research*, **97**, 826–833.

69 Adedokun, S.T., Ojo, O., Fashanu, A., and Ogunmola, B. (2009) Pole figure characteristics of annealed aluminum alloy 6061 in different magnetic fields up to 30 Tesla, in *Aluminim Alloys: Fabrication, Characterization and Applications II* (eds W. Yin, S. K. Das, Z. Long), TMS (The Minerals, Metals and Materials Society), Warrendale, PA, pp. 115–119.

70 Adedokun, S.T., Ojo, O., and Aluko, O. (2009) Annealing behavior of an heavily deformed aluminum alloy in 20 Tesla magnetic field, in *Aluminim Alloys: Fabrication, Characterization and Applications II* (eds W. Yin, S. K. Das, Z. Long), TMS (The Minerals, Metals and Materials Society), Warrendale, PA, pp. 69–74.

71 Golovin, Yu.I. (2004) Magnetoplastic effects in solids. *Physics of the Solid State*, **46**, 789–824.

72 Morgunov, R.B. (2004) Spin micromechanics in the physics of plasticity. *Physics-Uspekhi*, **47**, 125–147.

11
Interface Segregation in Advanced Steels Studied at the Atomic Scale

Dierk Raabe, Dirk Ponge, Reiner Kirchheim, Hamid Assadi, Yujiao Li, Shoji Goto, Aleksander Kostka, Michael Herbig, Stefanie Sandlöbes, Margarita Kuzmina, Julio Millán, Lei Yuan, and Pyuck-Pa Choi

11.1
Motivation for Analyzing Grain and Phase Boundaries in High-Strength Steels

Developing strong, damage-tolerant, and functional steels forms the backbone for multiple industrial innovations in the fields of energy, transportation, health, safety, machinery, and industrial infrastructure. Examples are advanced Fe–Cr steels for high-temperature applications in emission-reduced turbines; weight-reduced and ultrahigh-strength Fe–Mn steels for lightweight and yet passenger-protecting mobility; soft magnetic Fe–Si steels, semiamorphous steels, and iron-based metallic glasses for low-loss electrical motors and wind energy generators; radiation-resistant steels for nuclear and fusion power plants; or Fe–Cr–Ni stainless steel tubes for direct solar thermal power plants.

The fact that new steels are often "hidden" behind the final product sometimes conceals the truth that without permanent innovation in this field many advanced industrial products would be impossible to realize. Awareness of the profound recent advances in steel research sometimes suffers a bit from its engineering success: Although steels are key components in many zero-damage-tolerance safety design parts (planes, trains, cars, bridges, wind mills, power plants, etc.) they very rarely fail, hence, awareness of the high relevance of steels is not as ubiquitous as the products that are made of it. This is surprising when considering that the annual steel production exceeds 1.4 billion tons and that most high-performance steels are complex nanostructured materials (Figure 11.1) [1–4].

The aforementioned applications require the development of improved high-strength and yet ductile and damage-tolerant steels. Most traditional hardening mechanisms, however, such as solute solution, dislocation multiplication, or precipitation, albeit leading to high strength, often may cause a decrease in ductility, that is, rendering the material brittle and susceptible for failure. This phenomenon is sometimes referred to as the inverse strength-ductility problem (Figure 11.2) [5].

In this context, the average grain size and the nature of the grain boundaries involved in terms of their thermodynamic, kinetic, and structural characteristics are

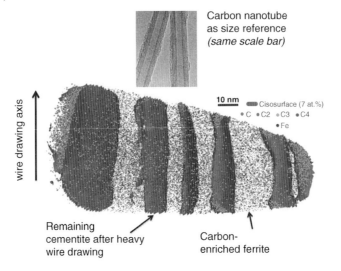

Figure 11.1 Most high-performance steels are nanostructured materials often consisting of a set of phases and complex arrangements of lattice defects. The image shows a wire drawn near-eutectoid pearlite steel, characterized by local electrode atom probe tomography (Imago Leap) [3]. The analysis (atomic envelopes placed at 7 at% carbon) reveals that some of the cementite lamellae that prevail after heavy wire deformation are thinner than a carbon nanotube (shown in the analysis at same size resolution). The symbols C, C2, C3, and C4 indicate the positions of corresponding evaporated carbon clusters.

important factors [6–23]. A better understanding of the role of grain boundaries plays particularly an essential role for designing and manufacturing nanostructured steels with higher ductility. Here we specifically address grain boundaries as defects in advanced high-strength steels and their susceptibility to chemical decoration by attracting solute atoms [14–33]. This phenomenon, also referred to as grain boundary segregation, is characterized by the inhomogeneous distribution of solute atoms between the grain boundary and the abutting grains. When interpreting the interface as a separate phase, this effect resembles a partitioning phenomenon as observed among abutting phases. The concentration of the solutes on the grain boundary exceeds their solubility in the grain interior, sometimes by a factor of 2–3 and sometimes even by up to several orders of magnitude [26,28,33]. A rough approximation for the segregation tendency of a solute is its bulk solubility, namely the smaller the bulk solubility, the higher is the enrichment factor of that element on the interface. Naturally, exceptions may apply from this simple rule.

The topic of grain boundary segregation is of high relevance in the context of alloy design, because the damage tolerance of structural materials, such as steels and Ti alloys, is often limited by segregation at grain boundaries. This phenomenon can hence profoundly affect the structure and the mechanical properties of advanced materials: The overall resistance of steels against damage and failure depends on two main material properties. First, during plastic straining, steels require a high, if possible, permanent capability to undergo additional deformation stimulated strain

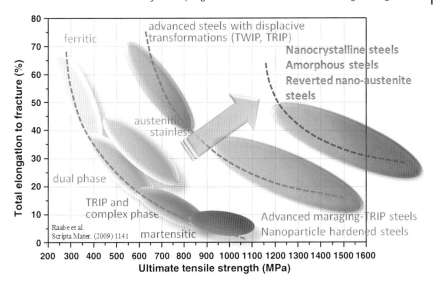

Figure 11.2 Inverse strength–ductility relationship for some steels. Activating special deformation mechanisms that provide additional hardening capacity at high strains, such as the TRIP or TWIP effects, can open pathways to developing novel alloys that deviate from this inverse relation as indicated by the broad arrow pointing upward (TRIP: transformation induced plasticity; TWIP: twinning induced plasticity). Also, UFG structuring and nanoparticle hardening via the maraging effect are promising strategies for designing advanced steels (UFG: ultrafine grained; maraging: martensite plus nanoprecipitate hardening) [5].

hardening. This property of the material, for example, such as that realized in transformation-induced plasticity (TRIP), twinning-induced plasticity (TWIP), and TRIPLEX (weight-reduced, austenitic or austenitic-ferritic with kappa carbides) steels, ensures that any tendency toward macroscopic and microstructural localization associated with geometrical softening during deformation is compensated for by strain hardening. This makes the material locally stronger and hence resistant to the further progress of plastic localization zones that might otherwise become fatal to the material. Second, steels require mechanisms that suppress the formation and specifically the growth of cracks. While tiny flaws, that is, initial crack nuclei, are always present in bulk materials, specific attention and understanding has to be placed on mechanisms that help avoiding, branching, deflecting, or even closing cracks that are capable of growing or already started to grow.

In the latter context, it has been often observed that in high-strength steels it is particularly the internal interfaces that are relatively weak against crack formation and growth. This is because elementary crack growth proceeds by debonding and dislocation emission at the crack tip, that is, by decohesion of the atomic bonds and the ability of the material to create a plastic relaxation zone around the tip of a proceeding crack tip. Owing to the reduced atomic stacking and hence smaller lattice coherence in the grain boundary zones than the bulk grain interiors, in high-strength steels cracks can often proceed more easily along grain boundaries than through the grain interior. This applies particularly to plastically strained steels, often associated with strain path changes, when the intrinsic strain hardening capacity of the material is gradually exhausted.

How does solute atomic segregation to internal interfaces in steels matter in such a scenario? When grain boundaries are affected by equilibrium segregation, either via inherited segregated solutes from a preceding high-temperature processing step (i.e., decoration of former austenite grain boundaries) or upon modest tempering after quenching (i.e., decoration of martensite grain boundaries), the structure and the atomic bonds change within the decorated grain boundary. Here, different scenarios are conceivable: First, segregation of certain atomic species might enhance the grain boundary structure, coherence, and bonding (grain boundary strengthening). Second, the reverse might be true, namely the further loss of coherence and electronic bonding (interface weakening) at grain boundaries due to segregation. Third, segregation could be strong enough to finally result in a phase transformation of the grain boundary region or support the formation of second phases (grain boundary precipitation, complexions, phase formation on grain boundaries, phase transformation of grain boundaries).

In addition to these possible direct mechanical and structural changes in the intrinsic grain boundary properties due to equilibrium segregation, the grain boundary energy is affected [21,23,25,27–36]. More precisely, the reduction in grain boundary energy provides the driving force behind many of these phenomena. As subsequently outlined in more detail the original formulation of the Gibbs adsorption isotherm and also the McLean variant of this concept, when being adapted to segregation at internal interfaces, describes the reduction in the total system free energy as the driving force for equilibrium segregation phenomena [14]. Since the bulk grain depletion and the associated bulk free energy change are negligible in this context, the main thermodynamic driving force for segregating solutes to grain boundaries lies in the reduction of the interface free energy. Kirchheim generalized this concept to the defectant theory, describing the general phenomenon of solute segregation to lattice defects with the result that their free energy is reduced [37,38]. In a thought experiment, he shows that in extreme cases the defect-free energy could even become smaller than that of the bulk phase surrounding it, meaning that the defects would be stabilized against the defect-free bulk phase [35,36]. This case will not be considered further here though.

The energy reduction due to solute segregation also plays an indirect role for the relationship between segregation and ductility because the lower grain boundary energy reduces the driving force for competitive capillary-driven grain coarsening (grain growth). Hence, materials with smaller grain boundary energy can, under the same thermomechanical treatment as applied to materials without grain boundary segregation, have a smaller and at the same time rather stable average grain size. This effect could also improve both the strength and the ductility of the steel via the Hall–Petch effect provided that the specific type of segregation does not lead to interface embrittlement.

When properly understood and quantitatively characterized, these various interconnected effects between grain boundaries and the associated equilibrium segregation effects in high-strength steels may open novel pathways to nanoscale engineering of damage-tolerant high-strength steels.

11.2
Theory of Equilibrium Grain Boundary Segregation

11.2.1
Gibbs Adsorption Isotherm Applied to Grain Boundaries

The thermodynamics of grain boundary segregation phenomena can be introduced in close analogy to monolayer gas adsorption at solid surfaces according to Gibbs [14–16,26]. More specifically, the solute decoration of interfaces can be formalized in terms of the Gibbs adsorption isotherm. This mechanism in principle also describes the equilibrium segregation of solutes to defects other than interfaces [38]. According to the Gibbs–Duhem equation, the amount and type of surface or, respectively, interface solute, which is also referred to as surfactant i, that is adsorbed at an interface (surface) per unit area influences the interfacial tension, more specific here, the grain boundary energy, that is

$$d\gamma = -\sum_i \Gamma_i d\mu_i$$

where $d\gamma$ is the change in grain boundary energy on segregation, Γ_i the grain boundary excess concentration of the solute element of type "i", and $d\mu_i$ the change in chemical potential of the solute type "i". In thermodynamic equilibrium, changes in the chemical potential depend on the chemical activity of the ingredients involved,

$$d\mu_i = RT \, d\ln a_i$$

where a_i is the activity of solute "i" in the bulk phase, R the gas constant, and T the absolute temperature. The activity can be derived by multiplying the molar fraction by the activity coefficient of the corresponding species in the system. Hence, we obtain

$$d\gamma = -RT \sum_i \Gamma_i \, d\ln a_i$$

For a dilute solution containing only a few types of nondissociating surfactant or, respectively, defectant element we assume that the activity can be replaced by the concentration rendering this equation into

$$d\gamma = -RT \sum_i \Gamma_i \, d\ln c_i$$

where c_i are the molar concentrations of the segregating element types i in the bulk. At constant temperature T and volume V the grain boundary excess concentration of the segregating solute is

$$\Gamma_i = -\frac{1}{RT} \left(\frac{d\gamma}{d\ln c_i} \right)_{T,V}$$

This expression is referred to as the Gibbs adsorption isotherm for dilute systems. It establishes a quantitative relationship between the interface energy and the segregating element of type i. The value for Γ_i can be obtained by measuring the change in interface energy as a function of concentration changes in logarithmic presentation.

11.2.2
Langmuir–McLean Isotherm Equations for Grain Boundary Segregation

Although the concept of the Gibbs adsorption isotherm outlined previously enables a quantitative analysis of grain boundary segregation, the actual measurement of the interface energy as a function of bulk composition and temperature is experimentally quite challenging in the case of grain boundaries [34]. Therefore, it is pertinent to pursue an alternative, more phenomenological approach to the analysis of interface segregation, which introduces a relationship between grain boundary segregation and the composition of the abutting bulk grains.

Such an alternative is formulated by Langmuir–McLean type approaches [14–19]. While the derivation of the Gibbs adsorption isotherm is based on changes of the interfacial energy with the bulk concentration (or activity) of the solute, the Langmuir–McLean isotherm describes the segregation equilibrium using a dynamic approach to the minimization of the total Gibbs free energy of a bulk system that contains an interface. These models assume that a state of dynamic equilibrium is established between the segregating solute element and the adsorbent interface and that adsorption is limited to one monolayer. In this approach, the adsorbent interface is thought to be composed of a regular array of energetically homogeneous adsorption atomic sites upon which an adsorbed monolayer is assumed to form. The rate of segregation is assumed to be equal to the rate of solute escape from the adsorbed monolayer at a given relative pressure and at constant temperature.

The dynamical equilibrium in a binary system is described in terms of the exchange rates of the components M and I between the grain boundary ϕ and the bulk grain B. The segregation of a solute I at an interface ϕ of a matrix (solvent majority element), M, in a binary M–I system can be quantified via the dynamic equilibrium relation

$$M^B + I^\phi \leftrightarrow M^\phi + I^B$$

In thermodynamic equilibrium the equality of the chemical potential holds for both element types. that is

$$\mu_M^B = \mu_M^\phi \text{ and } \mu_I^B = \mu_I^\phi$$

Using these qualities in the equation for the dynamic equilibrium means that the chemical potential of the left-hand side must equal that on the right-hand side, hence, we obtain

$$\mu_M^B + \mu_I^\phi - \left(\mu_M^\phi + \mu_I^B\right) = \Delta G = 0$$

11.2 Theory of Equilibrium Grain Boundary Segregation

where μ_M^B is the chemical potential of the majority component M in the grain interior; μ_M^φ is the chemical potential of the component M on the grain boundary; and μ_I^B and μ_I^φ are the corresponding terms for the solute component I. In dynamical equilibrium ΔG is zero.

Rewriting this equation in terms of the dependence of the chemical potential on the corresponding chemical activities, a_M^B, a_I^φ, a_M^φ, and a_I^B yields

$$\mu_M^{B0} + RT\ln(a_M^B) + \mu_I^{\varphi 0} + RT\ln(a_I^\varphi) - \left(\mu_M^{\varphi 0} + RT\ln(a_M^\varphi) + \mu_I^{B0} + RT\ln(a_I^B)\right) = 0$$

This can be reformulated according to

$$\left(\mu_M^{B0} - \mu_M^{\varphi 0}\right) + \left(\mu_I^{\varphi 0} - \mu_I^{B0}\right) = RT\ln(a_M^\varphi) + RT\ln(a_I^B) - RT\ln(a_M^B) - RT\ln(a_I^\varphi)$$

When defining the molar Gibbs free energy ΔG^0 from the activity terms according to

$$\Delta G^0 = \left(\mu_M^{\varphi 0} - \mu_M^{B0}\right) + \left(\mu_I^{B0} - \mu_I^{\varphi 0}\right)$$

we obtain

$$\Delta G^0 = RT\left(\ln(a_M^B) + \ln(a_I^\varphi) - \ln(a_M^\varphi) - \ln(a_I^B)\right)$$

This can be rewritten as

$$\frac{\left(a_M^B a_I^\varphi\right)}{\left(a_M^\varphi a_I^B\right)} = \exp\left(\frac{\Delta G^0}{RT}\right)$$

Expanding this expression by replacing the activity coefficients in terms of the molar concentrations and the activity coefficients yields

$$\frac{X_I^\varphi}{1 - X_I^\varphi} = \frac{X_I^B}{1 - X_I^B} \exp\left(\frac{\Delta G_I^0 + \Delta G_I^E}{RT}\right)$$

where ΔG^E is the excess molar Gibbs free energy associated with the interfacial segregation. The excess Gibbs energy of segregation is calculated from a combination of the activity coefficients

$$\Delta G_I^E = RT\ln\left(\frac{\gamma_M^B \gamma_I^\varphi}{\gamma_M^\varphi \gamma_I^B}\right)$$

When approximating the system in terms of a simplified dilute case, the two terms $1 - X_I^\varphi$ and $1 - X_I^B$ can be approximated as $1 - X_I^\varphi \approx 1$ and $1 - X_I^B \approx 1$. When further assuming that the impurities follow Henry's law, the effect of the excess Gibbs energy on the grain boundary segregation, ΔG_I^E, can be neglected, so that we obtain an expression for the grain boundary segregation coefficient

$$\beta_i = \frac{X_I^\varphi}{X_I^B} = \exp\left(\frac{\Delta G_I^0}{RT}\right) = \exp\left(-\frac{S_I^0}{R}\right)\exp\left(\frac{\Delta H_I^0}{RT}\right) = \beta_0 \exp\left(\frac{\Delta H_I^0}{RT}\right)$$

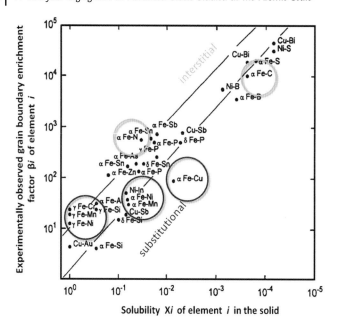

Figure 11.3 Grain boundary segregation data compiled from published values of Shea and Hondros [15–18], Lejček [25,26,28,33], and own measurements conducted by atom probe tomography [1,3–5]. It shows for a number of steels the grain boundary enrichment factor relative to the bulk solubility of the same solute element. The correlations are expressed in terms of the grain boundary enrichment ratio β, and solid solubility X for the specific solute element indicated at the data points. These data can help to identify elements of preferred (or reduced) segregation tendency to internal interfaces.

where ΔH_I^0 is the enthalpy of segregation. The segregation coefficient is often also referred to as grain boundary enrichment factor, Figure 11.3.

The Langmuir [19] and McLean models [14] for grain boundary segregation is, for solid state interfaces, usually more pertinent than the Gibbs treatment of segregation, since their approach does not require a detailed knowledge of the interfacial energy, and its variation with temperature or composition. However, the free energy of segregation in this model is usually an unknown property, and its theoretical prediction requires modeling at the electronic scale at the interface. It is noteworthy that the Langmuir–McLean model treats the grain boundary implicitly as an individual phase with specific thermodynamic properties, which are different from those of the matrix.

11.2.3
Phase-Field Modeling of Grain Boundary Segregation and Phase Transformation at Grain Boundaries

In contrast to statistically formulated thermodynamic sharp interface models presented earlier, phase-field modeling is an alternative method to treat

11.2 Theory of Equilibrium Grain Boundary Segregation

microstructural–kinetic problems pertaining to interfaces and the associated solute redistribution [39]. In this section we briefly outline how to apply phase-field modeling to the problem of grain-boundary segregation in steels [40–43].

An advantage of phase-field modeling is that it allows one to represent the grain boundary as a partially disordered region of finite thickness. In this way, thermodynamic properties of grain boundary can be linked to those of the bulk phases. In contrast to the existing segregation models, a grain boundary in the phase field approach does not have to be conceived as a phase-type area of unique thermodynamic properties, neither is there a sharp interface between this phase and the matrix. Instead, the grain boundary is represented as smooth changes in the structural order (phase-field) parameters, with related thermodynamic properties as well as temporal interaction with the adjacent grains.

In the approach outlined subsequently, we use a multiphase field approach, considering fcc structure, bcc structure, and the liquid phases in the Fe–C system (fcc: face centered cubic; bcc: body centered cubic). The main idea for the incorporation of the liquid phase – which is not expected to exist in bulk form at the temperatures of interest – is to represent the partial structural disorder at the grain boundaries. The structural disorder is represented by nonzero values of the liquid (amorphous) phase-field variable. In this way, thermodynamic properties of the grain boundary can be linked to those of the liquid (amorphous) phase, and, hence, segregation can be modeled quantitatively without a need to assume the grain boundary as a separate phase of adjustable thermodynamic properties. The multiphase field model used in this chapter is further combined with a probabilistic model of orientation evolution in polycrystalline materials.

The overall free energy of the system is considered to have the following form

$$F = \int \left[f(p_1, p_2, p_3, x_C, T) + \tfrac{1}{2}\varepsilon^2 \sum_{i=1}^{3} |\nabla p_i|^2 + \tfrac{1}{2}\varepsilon_c^2 |\nabla x_C|^2 + \sum_{i=2}^{3} g_i^{GB} \right] dV$$

where f is the local volumetric free energy density as a function of phase-field variables for the amorphous (p_1), bcc (p_2), and fcc (p_3) phases, as well as composition (x_C) and temperature (T), ε and ε_c are constants relating to the structural and compositional interface thicknesses, respectively, and g_i^{GB} is an energy term, encapsulating the free energy of the crystallographic mismatch between phase i (2 for bcc and 3 for fcc) of a given control volume and that of its neighbors. This term may in fact be conceived to represent the local orientational order. The local energy density is given as follows:

$$f = \sum_{i=1}^{3} [h_i G_i(x_C, T) + p_i^2 (1-p_i)^2 W]$$

where G_i is the molar Gibbs free energy of phase i (1 for amorphous (liquid), 2 for bcc, and 3 for fcc) in the Fe–C system, W is the height of the energy barrier between two phases, and h_i is a function of the phase field variables given as follows:

$$h_i = \frac{p_i}{4} \left\{ 15(1-p_i)\left[1 + p_i - (p_k - p_j)^2\right] + p_i(9p_i^2 - 5) \right\}$$

in which i, j, and k designate different phases. The Gibbs free energies of the bcc, fcc, and amorphous phases as used in the present analysis are based on the thermodynamic model in the original work of Gustafson [44]. The mismatch energy, which represents the local orientational order, is defined in the discretized form as

$$g_i^{GB} = \tfrac{1}{2}(\varepsilon_i/\delta x)^2 \sum_n a^n p_i p_i^n E_i^n(\theta_i, \theta_i^n)$$

where ε_i is a constant, with the same dimension as ε and ε_c, relating to the energy of the boundary between two grains of phase i (bcc or fcc), δx is the grid spacing (cell size), θ_i is the orientation of phase i in the given cell, θ_i^n is that of its nth neighbor, E_i^n is a dimensionless function scaling with the value of the interfacial energy associated with the boundary between the grains of phase i and in the given cell and that in its nth neighbor, and a^n is a coefficient that equals unity and $\sqrt{2}/2$ for the nearest and the second nearest neighbors on a 2-D grid, respectively. The partial derivative of the discretized g_i^{GB} with respect to p_i is approximated by an expression similar to the right-hand side of the preceding equation where $p_i p_i^n$ is replaced by $(p_i + p_i^n)$, that is

$$g_i' \cong \tfrac{1}{2}(\varepsilon_i/\delta x)^2 \sum_n a^n (p_i + p_i^n) E_i^n(\theta_i, \theta_i^n)$$

In the present analysis, E_i^n is assumed to be independent of the location of the neighboring cell in a structured grid of square cells – which is always true for a 2-D system of fourfold symmetry – and calculated as follows: $E_i^n(\theta_i, \theta_i^n) = |\sin[2(\theta_i - \theta_i^n)]|$. A main characteristic of the discretized formulation of the grain boundary energy as used here is that the respective energy of crystallographic mismatch is attributed to the *boundaries* of the calculation cells. Thus, grain boundaries are manifested as sharp interfaces in the orientation field, though they can be diffuse in the phase or the concentration fields. Moreover, since each solid phase (bcc and fcc) is designated by an independent orientation variable, the grain boundary energy contribution can be present between grains of the same phase, and not necessarily between those of different phases. In this way, the grain boundary energy is decoupled from the interphase boundary energy.

Modeling the temporal evolution of the phase-field parameters follows the Allen–Cahn kinetic formulation [45–47]

$$\frac{\partial p_i}{\partial t} = -M_i \left(\frac{\partial f}{\partial p_i} - \varepsilon^2 \nabla^2 p_i + g_i' \right)$$

where M_i is the interface mobility, and g_i' is the derivative of the grain boundary energy contribution. Note that the discretized g_i' is nonzero only at the grain boundary, that is, for the cells adjacent to a discontinuity in the orientation field. Considering the constraint: $p_1 + p_2 + p_3 = 0$, the derivative of the local free energy with respect to the corresponding phase-field variable is obtained from the following relation:

$$\frac{\partial f}{\partial p_i} = 2W\left[p_i(1-p_i)(1-2p_i) - \tfrac{1}{3}\sum_{j=1}^{3}p_j(1-p_j)(1-2p_j)\right]$$
$$+ \sum_{j=1}^{3} h'_{ij} G_i(x_C, T)$$

where $h'_{ij} = -\tfrac{1}{2}h'_{ii} + \tfrac{15}{2}p_i^2(1-p_i)(p_k - p_j)$

and $h'_{ii} = \tfrac{5}{2}p_i^2\left[(p_j - p_k)^2(3p_i - 2) + (1-p_i)^2(3p_i + 2)\right]$

The temporal evolution of the concentration field is derived using the following equation:

$$\frac{\partial x_C}{\partial t} = \nabla \cdot M_c \nabla \left(\frac{\delta F}{\delta x_C}\right) = \nabla \cdot \frac{D}{\partial^2 f/\partial x_C^2} \nabla \left(\frac{\partial f}{\partial x_C} - \varepsilon_c^2 \nabla^2 x_C\right)$$

where M_c and D are the mobility and diffusivity of carbon, respectively. The first and second derivatives of the local free energy with respect to composition are obtained as follows:

$$\frac{\partial f}{\partial x_C} = \sum_{i=1}^{3} h_i \frac{\partial G_i(x_C, T)}{\partial x_C}; \quad \frac{\partial^2 f}{\partial x_C^2} = \sum_{i=1}^{3} h_i \frac{\partial^2 G_i(x_C, T)}{\partial x_C^2}$$

Combination of the discrete form of grain-boundary energy and the probabilistic algorithm as described earlier allows for the implementation of grain-boundary energy anisotropy, if required. It also allows for the formation of grain boundaries of finite thickness in phase field and zero thickness in the orientation field. Formation of orientationally sharp boundaries circumvents physically unjustifiable states of energy associated with diffuse interfaces, in addition to being consistent with the perception on the nature of grain boundaries in metals.

Another essential aspect of the approach explained earlier is the fact that the phase field model also predicts that the solute segregation on a bcc–bcc grain boundary can be so pronounced that this region can undergo a phase transformation from the initial solute-decorated bcc state into the fcc (austenite) phase. This means that according to this model, thin solid austenite films can in principle form at bcc steel grain boundaries provided that a sufficiently high equilibrium segregation of atoms promoting the fcc phase occurs and that reduced nucleation barriers are overcome by the disordered (and elastically stressed) state that characterizes martensitic (bcc, bct) grain boundaries (bct: body centered tetragonal). The elastic contribution to this problem is not included yet in the formulations outlined previously. However, elastic stresses should play an important role in this context specifically in the case when martensite-to-martensite grain boundaries undergo such local phase transformations, Figure 11.4.

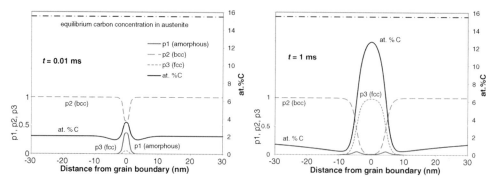

Figure 11.4 Modeling grain boundary segregation and austenite reversion in carbon steels by using a modified phase field approach. The figure shows two stages during the temporal evolution of the carbon concentration and the phase-field variables across the boundary of two bcc grains in Fe-2. at% C, held at 673 K, with a misorientation of 30°. The austenite layer grows to a maximum thickness of about 10 nm within the first millisecond. The variables indicate the disordered (amorphous) phase-field variable, p_1, which represents partial structural disorder at the interface, the fcc phase (p_3), the bcc phase (p_2), and the carbon content.

11.2.4
Interface Complexions at Grain Boundaries

Grain boundaries can assume a variety of structural, topological, chemical, and kinetic states, similar to bulk materials. These properties are each associated with a distinct thermodynamic state of that defect, hence, with its stability. Therefore, the state of an interface is specified by particular sets of intensive thermodynamic variables, interface inclinations, and crystallographic disorientation parameters. These states are similar to those of thermodynamic bulk phases and have associated interface equilibrium quantities (i.e., excess volume, entropy, and chemical adsorption). In the current context, where we aim at treating grain boundary segregation in steels, specifically, the chemical and phase states of the grain boundaries are of interest. However, it must be emphasized that grain boundaries are not conventional phases in the normal sense because a treatment of their thermodynamic state can obviously not be separated from the presence of their abutting bulk phases. In other words, a grain boundary does not exist without its neighboring grains [48–54].

The properties of grain boundaries not only change continuously with temperature and segregation state, but they also show discontinuities. Such transitions are likely to have an effect on corresponding transitions in the grain boundary energy and mobility. This behavior led to the notion of grain boundary phase transformations. This concept assumes that interfaces can be understood as two-dimensional phases, which can assume several distinct phase states each characterized by an individual free energy. This concept indicates that there will be a grain boundary phase transformation occurring whenever these free energy curves cross. At such interface phase transitions, properties can change discontinuously. Of course, the structural phase space for grain boundaries alone is huge, so that the ground-state structure and entropy can supposedly vary over a wide range of values and states. Complexity is increased when considering additionally the chemical state of the grain boundaries, such as for instance determined by grain boundary segregation.

To render these early concepts consistent with recent high-resolution transmission electron microscopy (HRTEM) observations, Harmer *et al.* [49–52] formulated an extension of the Gibbs definition of a bulk phase (which applies to identifiable volumes that differ in their structural state, composition, or symmetry, but not their geometrical features) to interfacial features.

They suggest designating these equilibrium features as interface complexions. More specifically, the concept assumes that grain boundaries are interface-stabilized phases (also referred to as interphases or complexions) that are chemically and structurally distinct from any of the abutting bulk phases. It should be noted that two different boundaries having the same complexion will not necessarily have the same exact atomistic structure, but rather will have similar characteristic equilibrium thermodynamic quantities.

Consistent with this, Baram *et al.* [48] extended the concepts of grain boundary structure and interface thermodynamics after examining interface structures that form between a gold particle in contact with a metal oxide surface. They observed that these nanoscale interface structures formed equilibrium phases that obey thermodynamic rules analogous to those for bulk phases. According to this work, an interface complexion can be considered as a separate interface phase, which can transform into different phases (complexions) with different properties by changing chemistry and heat treatment.

In classical interface theories, only the existence of three distinct interfacial phases has been discussed to occur at grain boundaries: These are intrinsic (undecorated) grain boundaries; Gibbs- or, respectively, Langmuir–McLean-type monolayer adsorption films [14–19] (or fractional monolayer occupation layers), and complete grain boundaries wetting films. Going beyond this classical structure and state concept, recent HRTEM work revealed the existence of additional types of impurity-containing intergranular films of equilibrium thickness in various ceramics and metals, as well as at metal–ceramic hetero-interfaces. More specifically, in a set of experiments, Dillon *et al.* [50] studied this phenomenon on alumina with controlled doping of well-adjusted impurity concentrations.

High-resolution microscopy revealed that grain boundaries could consist of one chemical species, or could have a single adsorbed (segregated) monolayer of a solute, or consist of an adsorbed bilayer of a solute, or multiple adsorbed layers, or thin intergranular wetting films of constant thickness. It was also observed that transitions from one complexion to another can be accompanied by a reduction in energy. From these observations the authors suggested that bulk phase diagrams can be modified to include equilibrium interface phases, or complexions [48].

The thermodynamic stability of these nanoscale impurity-containing intergranular films can be explained from a balance of various interfacial forces and by making analogies to the well-established surface premelting and prewetting theories. If grain boundaries and surface adsorptions are analogous, one may also expect that, in certain systems, interfacial phases can take on a discrete thickness, leading to the formation of distinct bilayer and trilayer interfacial phases [48–54].

These recent observations on the probably more complex nature of these multiple interface phase states (here referred to as complexions) are conceived to be highly relevant to the problem of segregation and, hence, included in this compilation. It should be noted that the level of solute segregation is – like temperature – a very effective and relatively easy-to-manipulate variable that determines the state and energy of a grain boundary (or likewise of a hetero-interface). It can be assumed that segregation and subsequent possible phase formation at a decorated grain boundary, as discussed earlier in the context of the phase-field model of grain boundaries, might proceed via a sequence of grain-boundary phase states before finally a new bulk phase is formed at the decorated grain boundary. However, the assumption of such a sequence of transition states with individual structural and energetic characteristics is speculative at this stage.

11.3
Atom Probe Tomography and Correlated Electron Microscopy on Interfaces in Steels

Quantitative experimental analysis of the thermodynamics of solute segregation at grain boundaries is very challenging owing to the high number of state variables involved. The crystallographic orientation change across a grain boundary can be quantified in terms of a rotation matrix (three independent orientation variables); two angles that characterize the grain boundary plane inclination for a given misorientation (two independent vector components); and one further variable for each chemical component. Local atomic in-plane relaxations are neglected here.

While the former five kinematical parameters (misorientation and interface plane inclination) are structural parameters that can be acquired by direct lattice imaging or the use of scanning electron microscopy (SEM) or transmission electron microscopy (TEM) diffraction tools for characterizing the abutting grains on either side of the grain boundary, the chemical characterization of the grain boundary segregation requires using energy dispersive spectroscopy (EDS), secondary ions mass spectrometry (SIMS), or atom probe tomography (APT), here listed in sequence of increasing spatial and chemical resolution [55–68].

In some cases, the spatial resolution of APT can even be high enough for crystallographic indexing, which gives access to the five kinematical grain boundary parameters (misorientation and interface plane inclination) as well as to all chemical variables [69–71]. However, this is confined to certain materials and has never been shown possible for highly alloyed steels. This means that the full characterization of all relevant grain boundary parameters (structural and chemical) normally requires the joint use of electron microscopy (SEM or TEM) and atom probe tomography in case that the best respective resolutions are desired (structural, chemical) [64].

Experiments setting the state of the art in this field currently pursue various types of approaches to identify both the structure and the chemical composition of grain boundaries in steels: In all cases, electron orientation microscopy has to be conducted prior to the atom probe experiment, as the latter one is a destructive technique. The samples are usually prepared electrochemically from rod-shaped geometries or via lift-out from the surface of a bulk material in a dual-beam focused ion beam (FIB) microscope (dual beam referring to an electron beam and an ion beam). The first approach goes along with less experimental effort but is nonsite-specific whereas the lift-out technique enables for sample preparation from specific regions containing grain boundaries as identified by SEM observations or electron backscatter diffraction (EBSD). The grain boundary type can be measured directly on the needle-shaped atom probe tips in the SEM/FIB by either EBSD or using diffraction techniques in the TEM.

An alternative approach lies in preparing atom probe tips directly from thin foils that have previously been analyzed in TEM. After conducting at first crystallographic or defect characterization in the TEM, a FIB lift-out can be done on TEM foils down to 100 nm thickness. But besides the challenging lift-out procedure, the disadvantage of preparing APT tips directly from thin TEM foils lies in residual stresses that can be released during preparation. This can lead to heavily curved and bent APT tips.

Figure 11.5 shows an example where TEM electron diffraction in a Philips CM 20 transmission electron microscope at an acceleration voltage of 200 kV is employed

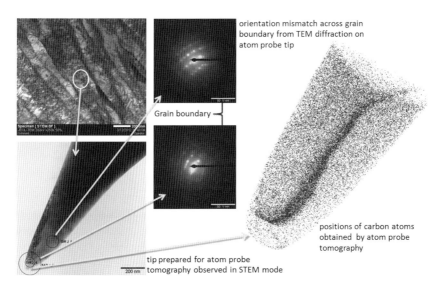

Figure 11.5 Example of a site-specific joint scanning transmission electron microscopy (STEM)–APT analysis of a grain boundary in a Fe–C–Mn martensite alloy. The dots indicate the positions of Mn atoms as measured by atom probe tomography. The joint analysis reveals strong segregation of carbon at the lath martensite grain boundary. The misorientation of this lath martensite interface amounts to about 7°. The shown depth of the evaporated volume was approximately 180 nm.

directly on an electropolished APT tip to characterize a lath martensite grain boundary before the field evaporation.

This sample and most of the other examples of atom probe tips for atomic-scale chemical analysis of steel discussed here are typically prepared in a standard two-step electrochemical polishing sequence. For this purpose, square rods with 0.3×0.3 mm^2 cross section and 20 mm length are cut from the bulk steel samples. In the first electrochemical polishing step, rods of square cross sections are then roughly sharpened using, for Fe–C or Fe–Mn steels, a solution of 9 vol% perchloric acid in glacial acetic acid. The second step is then conducted using a solution of 2 vol% perchloric acid in butoxyethanol.

Final sharpening for an optimal APT tip shape is conducted via FIB milling (in the cases discussed here using a FEI Helios Nanolab 600 dual-beam FIB). This step is carried out in order to obtain an initial tip radius of the APT samples below 50 nm. Finally, voltage- or laser-pulsed APT is performed for the examples presented here using a local electrode atom probe (LEAP 3000X HR^{TM}, Imago Scientific Instruments) at a specimen temperature of about 60–90 K. In voltage mode, the pulse fraction is usually set up at 0.2 with a constant pulse frequency of 200 kHz. In laser mode, beam energy of 0.2–0.4 nJ and 250 kHz laser frequency are commonly used. The detection rate amounts for many steels about 5 atoms per 1000 pulses. Data analysis can then be performed using the IVAS$^{®}$ software by Imago Scientific Instruments. The reconstruction of 3D atom maps is for the Fe–C and Fe–Mn steels presented here mostly done using an evaporation field constant of 33 V/nm, which corresponds to that of pure Fe at 77 K [1–5,68].

In either way, atom probe tomography is, in conjunction with electron microscopy, a powerful tool that can significantly contribute to our understanding and development of complex structural materials through the detailed analysis of interface structure and chemistry. The ability to probe the atomic-scale composition and structure of grain boundaries of complex structural alloys is instrumental in providing key insights into the atomistic mechanisms controlling microstructure, in providing routes for the development of better alloys, and in suggesting solutions to remediate problematic degradation and damage phenomena.

11.4
Atomic-Scale Experimental Observation of Grain Boundary Segregation in the Ferrite Phase of Pearlitic Steel

Cold-drawn near-pearlitic steel wires have a maximal tensile strength above 6 GPa, thus making them the strongest nanostructured bulk materials available [3,4,68]. They serve as wires in bridges, car tires, cables, and as strings in musical instruments. Also, pearlite, being a lamellar aggregate of ferrite and cementite, is an essential microstructure component in many complex steels.

The microstructural origin of the extreme strength of this material is not well understood. We observed that cold wire drawing not only strengthens pearlite via a Hall–Petch effect by gradually refining the lamellae structure, but also causes partial chemical decomposition of the constituent cementite and even a transition from crystalline to amorphous cementite [3,72].

The deformation-driven decomposition and associated microstructural changes of cementite are closely related to several other phenomena such as a strong redistribution of carbon and other alloy elements such as Si and Mn in both cementite and ferrite; a variation of the deformation accommodation at the phase interfaces due to a change in the carbon concentration gradient at the interfaces, mechanical alloying (i.e., shear-induced chemical mixing), and a further reduction of the deformability of cementite when rendered amorphous. Since these phenomena occur at the atomic scale, the understanding of the strengthening mechanisms of cold-drawn pearlitic wires and the specific role of the associated grain boundary segregation effects can only be improved on the basis of atomic-scale investigations as was outlined in the preceding section.

The carbon concentration in the ferrite increases with the wire drawing strain up to 3.47 and then saturates with further deformation. This high carbon excess content in the ferrite stems from dissolving cementite [3,4,68]. We observed the formation of dislocation subgrain boundaries in the ferrite lamellae and observed the segregation of carbon to the ferrite dislocations and to these subgrain boundaries already during wire deformation. In cementite, we found that the carbon concentration decreases with the thickness of the cementite lamellae. This finding allows one to establish a quantitative correlation between plastic deformation and cementite decomposition in cold-drawn pearlitic steel wires [3,4,68].

Other nanostructured metallic alloys that are produced by severe plastic deformation are highly susceptible to grain coarsening on heating due to large densities of dislocations and interfaces providing a high driving force for substructure coarsening. However, so far, only little attention has been paid to the thermal stability of the nanostructure of heavily cold-deformed pearlitic wires and its effect on the mechanical properties.

In this section, we therefore also discuss the changes in the nanostructure of a heavily cold-drawn hypereutectoid pearlitic steel wire on annealing in order to elucidate the mechanism leading to the ultrahigh strength of cold-drawn pearlitic steel wires. The wires studied here were subjected to an extreme deformation to a true strain of 6.02 and exhibited the highest tensile strength achieved to date (6.35 GPa) [4]. The tensile strength was then also measured as a function of the annealing temperature (varied between 423 and 723 K) and of the annealing time. The samples were characterized by means of APT in conjunction with TEM as outlined earlier.

APT results on the as-deformed, 473 K, and 673 K heat-treated samples are displayed in Figure 11.6. Two mutually rotated views are provided for each specimen. The atom maps of the as-deformed material (Figure 11.6a) reveal that a lamellar structure consisting of carbon-depleted (ferrite) and -enriched (cementite) regions still prevails after severe cold-drawing up to a strain of 6.02, although fragmentation of cementite lamellae is evident. The lamellar structure remains stable after annealing at 473 K for 30 min (Figure 11.6b and 11.7). No cementite spheroidization is observed yet. After heat treatment at 673 K for 30 min, strong

11 Interface Segregation in Advanced Steels Studied at the Atomic Scale

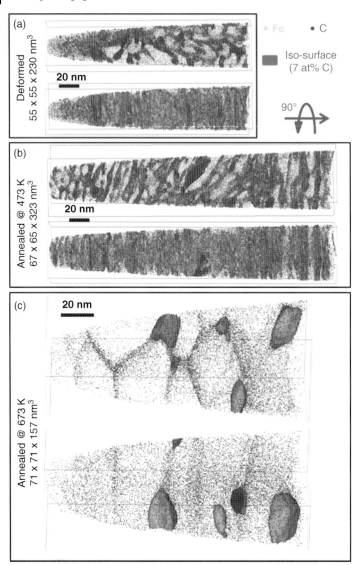

Figure 11.6 3D carbon atom maps of hypereutectoid pearlitic steel wires with a cold drawing strain of 6.02: (a) as-deformed state; (b) annealed at 473 K for 30 min; and (c) annealed at 673 K for 30 min. The isoconcentration surfaces for 7 at% carbon are shown in gray. Only 30% of all carbon and 0.5% of all iron atoms are displayed. Both the cross-sectional (top) and longitudinal views (bottom) of the wires are shown [3,4].

microstructural changes are detected: The lamellar pearlitic structure is no longer visible and instead a nearly equiaxed hexagonal subgrain structure has been formed inside the ferrite (Figure 11.6c). It can be observed that the subgrain boundaries in the ferrite are decorated with carbon atoms. The subgrains exhibit an average grain

Figure 11.7 Dislocation substructure in the carbon-enriched ferrite of a heavily deformed and annealed pearlite [4].

size of 30–40 nm, which is about 2 to 3 times higher than the interlamellar spacing in the as-deformed wire.

The formerly chemically and structurally decomposed cementite lamellae have during the 673 K annealing undergone spheroidization and are mainly located at the triple junctions of the ferrite subgrain structure. Closer observation of Figure 11.8 shows that the subgrains are not equiaxed, but rather elongated along the drawing direction with longitudinal sizes above 70 nm. It is noted that the reconstruction of the APT data shown in Figure 11.6c was done based on the final tip radius after APT measurement, and is hence reliable. The observation in Figure 11.8c and d suggests that the subgrain structure observed in the ferrite (Figure 11.6) is not due to recrystallization, as this mechanism would lead to equiaxed grains. The presence of dislocation arrays at low-angle grain boundaries of a heavily deformed and annealed (350 °C for 30 min) wire observed by TEM (Figures 11.7 and 11.8) gives further confirmation that recovery rather than recrystallization takes place during annealing.

Figure 11.8 shows TEM micrographs of pearlitic wires. Strong strain contrast due to heavy drawing can be observed in Figure 11.8a. The phase boundaries between ferrite and cementite are blurred and hardly visible. The observation in the cross section of the wire (Figure 11.8b) shows the typical ribbon-like lamellar structure around the wire axis. This so-called curling phenomenon is because the $\langle 110 \rangle$ fiber

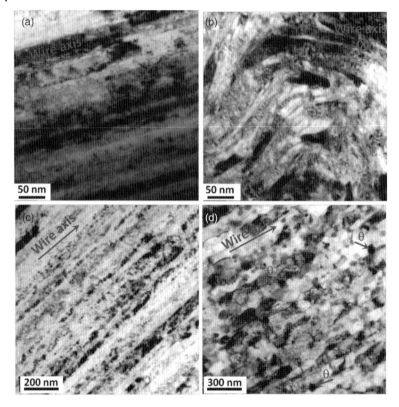

Figure 11.8 TEM images of hypereutectoid pearlitic steel wires. (a and b) As-deformed (true strain 6.02) states in longitudinal and transverse cross sections of wires, respectively. (c and d) As-annealed states at 523 and 723 K for 30 min, respectively. The arrows mark the wire axis, which is also parallel to the ferrite/cementite interfaces. Some globular cementite particles labeled as θ are marked [4].

texture developed during wire drawing confines further slip in each lamella to a plain strain state, and thus the compatibility of the neighboring grains can be maintained only by bending of the grains around one another. The observation of fragmentation of cementite in Figure 11.8b is consistent with the APT result shown at the bottom of Figure 11.6a. On annealing at 523 K for 30 min, the lamellar structure is still preserved and no recrystallization occurs (Figure 11.8c) [4]. After annealing at 723 K (see Figure 11.8d), the lamellar structure along the drawing direction is still visible, whereas cell/subgrain boundaries have formed in the ferrite and the cementite lamellae have undergone spheroidization. The ferrite cell/subgrains are elongated along the wire axis in agreement with the APT observation at 673 K (Figure 11.6c). In the direction perpendicular to the wire axis, the average subgrain size is below 80 nm. Figure 11.9 shows that carbon atoms strongly segregate to the ferrite subgrain boundaries when the sample is annealed at 673 K. The average carbon concentration inside the ferritic zones was measured to

11.4 Atomic-Scale Experimental Observation of Grain Boundary Segregation | 287

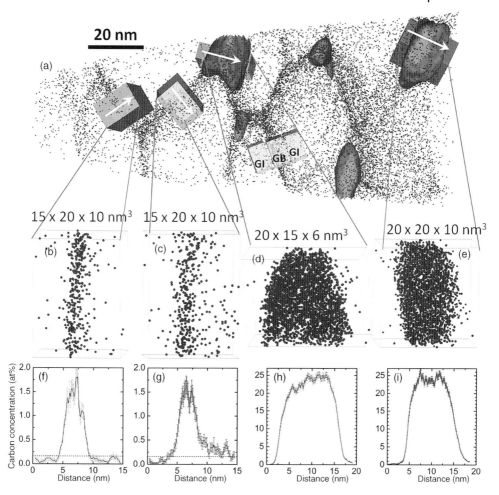

Figure 11.9 (a) Carbon atom map of the annealed (673 K for 30 min) hypereutectoid pearlitic steel wire with selected regions of interests (ROIs). The arrows mark the directions along which the concentration profiles are drawn. The gray-colored ROIs containing grain boundary (GB) and grain interior (GI) are shown as an example for determining the grain boundary excess of carbon (see text). (b–e) Carbon atom maps for the ROIs. The corresponding 1D carbon concentration profiles are displayed below each atom map (f–i). The average carbon concentration inside the ferrite grains is displayed as short-dashed lines in (f) and (g) [4].

be 0.163 +/− 0.057 at%, whereas it exceeds 1.5 at% at the grain boundaries, Figures 11.9f and g. Two regions of interest, cutting through the spheriodized cementite, are selected for a more detailed analysis of the carbon concentration. The measured values of about 25 at% C (Figure 11.9h and i) show that the carbon concentration in cementite has reached the stoichiometric value of 25 at% again after it was decomposed during the preceding wire drawing operation.

As was outlined in the theory section earlier, it is known that segregation of solute atoms at grain boundaries decreases the grain boundary energy, according to the Gibbs adsorption isotherm. From this analysis we conclude that the higher is the solute excess at the ferrite grain boundaries, the lower is the grain boundary energy.

The grain boundary excess content of carbon can be determined either from 1D concentration profiles (e.g., Figure 11.9f and g) or by direct counting of the number of carbon atoms per grain boundary area. With the former method, the carbon concentration in the ferrite subgrain interiors (as shown by the dotted lines in Figure 11.9f and g) is subtracted from the measured concentrations. Using the latter method, the number of carbon atoms within the selected region, for example, "GI" in Figure 11.9a is subtracted from the number of carbon atoms in region "GB," which has the same size as the domain "GI." Using the two methods, the grain boundary excess of carbon is determined to be $(5.30 +/- 0.73)$ atoms nm^{-2} $(8.80 \times 10^{-6}$ mol m$^{-2})$ and $(5.14 +/- 0.72)$ atoms nm^{-2} $(8.54 \times 10^{-6}$ mol m$^{-2})$, respectively.

According to the Gibbs formulation, a carbon excess of about 9×10^{-6} mol m^{-2} reduces the grain boundary energy and thus reduces the driving force for grain coarsening, even when annealing the heavily cold-drawn pearlitic steel sample at 723 K.

11.5
Phase Transformation and Nucleation on Chemically Decorated Grain Boundaries

11.5.1
Introduction to Phase Transformation at Grain Boundaries

As mentioned before, the observation of discontinuities in grain boundary properties and structures with changing composition or temperature has led to the concept of grain-boundary phase transformations. This perception assumes that grain boundaries are two-dimensional structural objects with a large possible variety in their respective atomistic structure, that there can be several such phases each of them having a specific free energy value, and that there will be a phase change whenever these free energy curves cross. At such phase transitions, properties can change discontinuously.

Transitions in the grain boundary state, structure, or phase state should not only occur when changing the temperature but they should be also possible when changing pressure, elastic load, or the chemical state. In this section we address the latter parameter, namely the change in the solute content that is segregated to a grain boundary. Here we shall address mainly martensite grain boundaries.

For this approach to work, the solutes used for segregation should fulfill three criteria. First, they should reveal a sufficiently high tendency to segregate at grain boundaries in the chosen matrix material in order to provide large concentration differences between matrix and interface after tempering, Figure 11.3. Second, the solute enrichment should affect the transformation temperature, that is, in the martensitic steels presented here it should, for instance, increase the transformation trend, that is, the chemical driving force, from martensite back to austenite (such as Ni, Mn, C, or N do). This would promote the start of martensite-to-austenite reversion particularly at such decorated martensite grain boundaries. Third, the solute element type should prefer segregation to the grain boundary over precipitation of a third phase. For instance, in the case of carbon, a competition typically exists among the segregation of carbon-to-martensite grain boundaries (Figure 11.3), the formation of carbides, and the trapping of carbon in the distortion fields surrounding dislocations [66]. For an advanced alloy design it is hence pertinent in such cases to add elements that prevent carbide formation, for example, Si. For Fe–Mn maraging steels a similar competition exists, since Mn has a strong tendency to segregate to grain boundaries and it can also form various intermetallic particles and decorate dislocations.

When these criteria are fulfilled and the difference in solute concentration is high enough between the matrix and the decorated grain boundary, local phase transformation at such an interface can be stimulated via changes in the thermodynamic state variables (temperature change, pressure change, or mechanical loading) [1,66]. For an estimate of the nucleation energy under such boundary conditions the interface energies involved must also be considered: These are the martensite-to-martensite grain boundary energy before the local phase transformation and the martensite-to-austenite interface energy after phase transformation. Also, it is important to note that in the case of martensite-to-austenite reverse transformation, a release in elastic energy of the transformed zone should be taken into consideration.

In practice, we realize this concept of promoting the onset of martensite-to-austenite phase transformation at decorated martensite grain boundaries in four steps:

We selected two types of alloy systems where at least one of the solute elements segregates to grain boundaries, namely Fe–Mn systems (e.g., Fe with 9–12 Mn (wt%)) [1] and Fe–Cr–C (Fe-13.6Cr-0.44C (wt%)) [66]. These systems meet the constraints outlined previously, that is, both Mn and C strongly segregate to defects in Fe and both elements reduce the transformation temperature from martensite to austenite, Figure 11.3 [24–31]. This promotes martensite reversion at relatively low temperatures (e.g., between 400 °C and 500 °C) [1,5,73,74]. Second, both alloys can be synthesized by casting, solidification, and thermomechanical homogenization, followed by austenitization and water quenching to obtain a martensitic (or martensitic–austenitic) matrix.

Third, both types of quenched systems (martensitic state) can be tempered into the diffusion regime (450 °C for Fe–Mn and 400 °C for Fe–Cr–C) such that strong segregation of Mn and C to the lath martensite grain boundaries occurs. In either case, the heat treatment takes place 100–150 K below the bulk re-austenitization temperature expected for the nondecorated grain interiors. The kinetics for grain boundary segregation follows in both cases the driving force exerted by the adsorption isotherm and the mobilities of the segregating elements in the martensite phase. In the Fe–Mn systems, we observe that the Mn enrichment at martensite–martensite grain boundaries can, at 450 °C (48 h), be as high as 27 at% as opposed to a content of only 4 at% in the martensite grain interior, Figure 11.10.

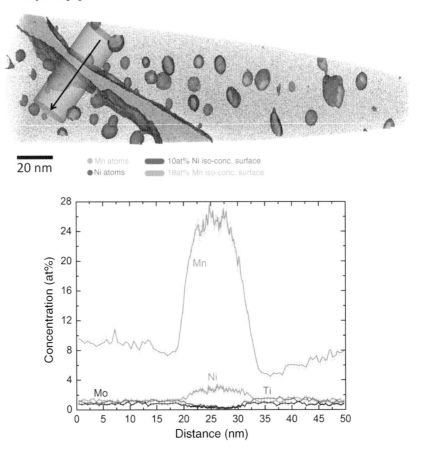

Figure 11.10 APT observation of the formation of Mn-rich austenite at a martensite interface in a Fe–Mn-based maraging steel (Fe-12Mn-2Ni-0.15Al-1Ti-1Mo-0.01C (wt%)) upon quenching and subsequent tempering at 450 °C for 48 h. The in-grain envelope zones in the APT map indicate intermetallic nanoparticles that form during tempering.

11.5.2
Grain Boundary Segregation and Associated Local Phase Transformation in Martensitic Fe-C Steels

Here we describe the segregation of carbon to martensite–martensite grain boundaries in a quenched and subsequently tempered Fe–Cr–C steel. On some of the decorated grain boundaries, the carbon segregation is so high that local phase transformation from martensite to austenite is promoted locally (martensite reversion), about 150 K below the bulk equilibrium re-austenitization temperature. The alloy is a martensitic stainless steel with chemical

11.5 Phase Transformation and Nucleation on Chemically Decorated Grain Boundaries

composition Fe-13.6Cr-0.44C (wt%; 1.4034, X44Cr13, AISI 420) [66]. The Ae3 temperature indicates that the incipient holding temperature for full austenitization should be above 800 °C. Full dissolution of chromium carbides in the austenite is achieved at about 1100 °C. Hence, the annealing conditions for homogenization before martensite formation via quenching were set to 1150 °C for 5 min. Dilatometry helped to identify the Ms temperature. Subsequently, a sequence of water quenching, tempering at 300 °C, 400 °C, and 500 °C, respectively, with different holding times was applied to study carbon redistribution, martensite-to-austenite reversion, and nanocarbide formation [66].

The volume fraction of austenite after the heat treatments (carbide dissolution annealing and tempering at 400 °C for 1, 2, 10 and 30 min) was measured by X-ray diffractometry (XRD), EBSD, and magnetic characterization (Feritscope MP30E-S). Thin foils were prepared using standard twin-jet electropolishing from the as-quenched material and the tempered samples. These samples were examined in a Philips CM 20 transmission electron microscope (TEM) to characterize the carbides and the formation of reverted austenite. Carbide characterization was also carried out using a carbon extraction replica technique and investigated by electron diffraction and energy dispersive spectroscopy (EDS) in the TEM. Needle-shaped APT samples were prepared by applying a combination of standard electropolishing and subsequent ion-milling with an FIB device. APT analyses were performed with a local electrode atom probe (LEAP^{TM} 3000X HR) in voltage mode at a specimen temperature of about 60 K. The pulse-to-base voltage ratio and the pulse rate were 15% and 200 kHz, respectively. Further experimental details can be found in [66].

After solid solution and subsequent water quenching, we found no retained austenite in the TEM foils, Figure 11.11a and b. This is in contrast to the results obtained from the EBSD maps, which showed retained austenite in the as-quenched state. We attribute this discrepancy between TEM and EBSD results to the fact that the as-quenched metastable retained austenite – when thinned for TEM analysis – is no longer constrained by the surrounding martensite and hence transforms into martensite.

Figures 11.11c and d give an overview of the nanoscaled elongated carbides formed during tempering: The carbides have an average length of 70 nm and an average width of 5 nm. After 30-min tempering, the average particle spacing is about 80 nm and the length 110 nm. The carbides after 1-min tempering at 400 °C were examined via carbon extraction replica. The diffraction patterns reveal that they have M_3C structure. This means that the formation of $M_{23}C_6$ carbides is suppressed at such a low tempering temperature. EDS analysis showed that the metal content in the carbide (M in M_3C) amounts to 74 at% Fe and 26 at% Cr, that is, the Cr/Fe atomic ratio is 0.35. Figure 11.11e shows the formation of a thin austenite layer that is located at a former martensite–martensite grain boundary. Figure 11.11f is a close-up view of a thin austenite zone that is surrounded by martensite. Electron diffraction analysis reveals that a Kurdjumov–Sachs

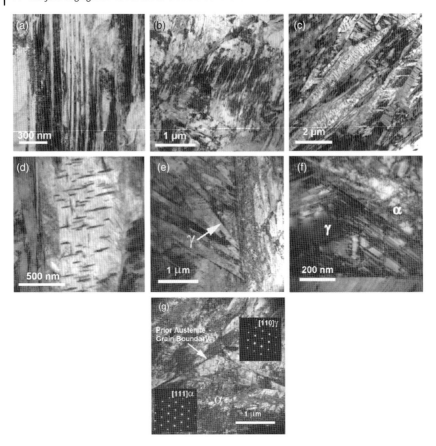

Figure 11.11 TEM images of as-quenched sample (only lath martensite was found: (a and b) TEM images of samples tempered at 400 °C for 1 min; (c) overview image of the very dense array of nanoscaled carbides that is formed during tempering; (d) in-grain view of the carbides; (e) overview image of the formation of a reverted austenite grain that is located at a former martensite–martensite grain boundary; (f) close-up view of reverted austenite that is surrounded by martensite; (g) electron diffraction analysis reveals that a Kurdjumov–Sachs growth orientation relationship exists between the martensite matrix and the reverted austenite [66].

orientation relationship exists between the martensite matrix and the thin austenite layer, Figure 11.11g.

The thin austenite film observed here in TEM might be either retained or reversed austenite. To determine more reliably which of the two kinds is observed, local atomic-scale chemical analysis was conducted via APT. The two types of austenite can then be distinguished in terms of their carbon content: Retained austenite has at first the nominal quenched-in C content (about 2 at% in the present case) of the alloy whereas reverted austenite has a higher C content (up to 9 at%) owing to local partitioning and kinetic freezing (drop in mobility once the C has partitioned from

11.5 Phase Transformation and Nucleation on Chemically Decorated Grain Boundaries

martensite into the austenite). However, we also have to account for the possibility that the retained austenite can have a higher C content as the lath martensite mechanism is slow enough to allow for some C diffusion out of the martensite into the retained austenite during quenching. This mechanism is referred to as self-tempering.

The local chemical compositions and their changes during 400 °C tempering of the martensite, austenite, carbides, and interface regions were also studied by atom probe tomography. Phase identification is in all cases achieved via the characteristic carbon contents of the present phases. Figure 11.12 shows the 3D atom maps after water quenching (Figure 11.12a), water quenching plus tempering at 400 °C for 1 min (Figure 11.12b), and water quenching plus tempering at 400 °C for 30 min (Figure 11.12c). Carbon atoms are visualized as gray dots and carbon iso-concentration surfaces in gray shade imaging for a value of 2 at%. This value corresponds to the nominal carbon concentration of the alloy of 0.44 wt%. The different phases (martensite, austenite, carbide) are marked. They were identified in terms of their characteristic carbon content and the TEM and EBSD data presented earlier. For more quantitative analyses, one-dimensional compositional profiles of carbon across the martensite–martensite and martensite–austenite interfaces were plotted (along cylinders marked in the 3D atom maps).

Figure 11.12a reveals that in the as-quenched condition the probed volume carbon is enriched along the martensite–austenite interface. The interface region, shown as composition profile in Figure 11.12a, reveals an average carbon concentration of about 1.90 at% in the austenite with strong local variations of about 0.98 at% in the abutting supersaturated martensite. The carbon concentration in the austenite nearly matches the nominal carbon concentration of the alloy. Some carbon clusters occur in both phases. The carbon concentration in these clusters is about 3 at%, that is, they are not carbides. In a thin interface layer of only about 5 nm, the carbon content is very high and reaches a level of 4–6 at%. In contrast to the variation in the carbon distribution, we found that the chromium content remained unchanged in the martensite, the interface, and the austenite.

After 1-min tempering at 400 °C after the quenching, a carbon-enriched austenite layer (15–20 nm width) was observed between two abutting martensite regions (Figure 11.12b). The thin austenite zone contains on average about 6.86 at% carbon whereas the martensite matrix contains only about 0.82 at% carbon. The identification of the phases in these diagrams follows their characteristic carbon content.

After 30-min tempering (Figure 11.12c), different carbon-enriched areas appear. They correspond to individual phases. The analyzed volume can be divided into two zones. The top region with low carbon content corresponds to martensite. The bottom zone with higher carbon content corresponds to austenite. Inside the martensitic region there are areas with very high carbon content (see arrow in Figure 11.12c). The carbon content is 25.1 at% in this particle indicating M_3C cementite stoichiometry. In the martensitic matrix surrounding the precipitate, the carbon content amounts to only 0.48 at%.

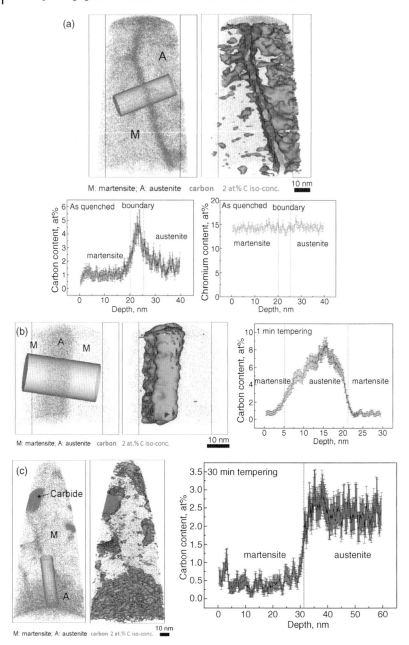

Figure 11.12 (a) 3-D reconstructions of the Fe–Cr–C sample after water quenching. The data clearly show that carbon redistribution already occurs during quenching. Cr redistribution does not occur. (b) Tempered at 400 °C for 1 min; (c) tempered at 400 °C for 30 min. The different phases are marked in the figure. Carbon iso-concentration surface (2 at%, corresponding to 0.44 wt%) and concentration profiles across the phase boundaries along the cylinder marked in the image) are also shown [66].

11.6
Conclusions and Outlook

We presented theoretical concepts and atomic-scale experimental analysis of grain boundary segregation in different types of steels. Detailed experimental analysis of grain boundary solutes requires the combined use of electron microscopy and atom probe tomography. Ideally, the electron microscopy characterization is conducted directly on the sample tips that were prepared for later tomographic atom probe analysis. This enables one to arrive at a segregation analysis that is specific for a certain type of grain boundary probed. Examples for the alloy systems Fe–C and Fe–Mn were given.

We also observed that for the specific case of high solute segregation concentrations on martensitic lath grain boundaries nanoscale martensite reversion into austenite at these grain boundaries is possible. This observation was made in both Fe–C and Fe–Mn systems. This is an important mechanism that allows one to decorate and transform lath martensite grain boundaries via simple quenching plus tempering. Mechanical characterization (not presented here but in [5] for Fe–Mn and in [66] for Fe–C) revealed that the so nanostructured martensite reveals pronounced increase in ductility. Another important consequence of high solute concentrations on grain boundaries particularly in nanostructured steels, such as pearlite, is the reduction in grain boundary energy and hence the decrease in the capillary driving forces that lead to competitive grain coarsening. This means that smart utilization of grain-boundary segregation can be of great benefit to stabilize nanostructured steels.

References

1 Dmitrieva, O., Ponge, D., Inden, G., Millán, J., Choi, P., Sietsma, J., and Raabe, D. (2011) Chemical gradients across phase boundaries between martensite and austenite in steel studied by atom probe tomography and simulation. *Acta Materialia*, **59**, 364–374.

2 Danoix, F., Bémont, E., Maugis, P., and Blavette, D. (2006) Atom Probe Tomography. I. Early stages of precipitation of NbC and NbN in ferritic steels. *Advanced Engineering Materials*, **8**, 1202–1205.

3 Li, Y.J., Choi, P.P., Borchers, C., Westerkamp, S., Goto, S., Raabe, D., and Kirchheim, R. (2011) Atomic-scale mechanisms of deformation-induced cementite decomposition in pearlite. *Acta Materialia*, **59**, 3965–3977.

4 Li, Y.J., Choi, P.P., Goto, S., Borchers, C., Raabe, D., and Kirchheim, R. (2012) Evolution of strength and microstructure during annealing of heavily cold-drawn 6.3 GPa hypereutectoid pearlitic steel wire. *Acta Materialia*, **60**, 4005–4016.

5 Raabe, D., Ponge, D., Dmitrieva, O., and Sander, B. (2009) Nano-precipitate hardened 1.5 GPa steels with unexpected high ductility. *Scripta Materialia*, **60**, 1141–1144.

6 Gottstein, G. and Shvindlerman, L.S. (1999) *Grain Boundary Migration in Metals – Thermodynamics, Kinetics, Applications*, CRC Press, Boca Raton, FL.

7 Molodov, D.A. (2001) Grain Boundary Boundary Character – A Key Factor for Grain Boundary Control. Recrystallization and Grain Growth, vol. **1** (eds G. Gottstein and D.A. Molodov), Springer, Aachen, pp. 21–38.

8 Rollet, A.D., Gottstein, G., Shvindlerman, L.S., and Molodov, D.A. (2004) Grain boundary mobility – a brief review. *Zeitschrift für Metallkunde*, **95**, 226–229.

9 Czubayko, U., Molodov, D.A., Petersen, B.C., Gottstein, G., and Shvindlerman, L.S.

(1995) An X-ray device for continuous tracking of moving interfaces in crystalline solids. *Measurements Science and Technology*, **6**, 947–952.

10 Upmanyu, M., Srolovitz, D.J., Shvindlerman, L.S., and Gottstein, G. (1999) Misorientation dependence of intrinsic grain boundary mobility: simulation and experiment. *Acta Materialia*, **47**, 3901–3914.

11 Furtkamp, M., Gottstein, G., Molodov, D.A., Semenov, V.N., and Shvindlerman, L.S. (1998) Grain boundary migration in Fe–3.5% Si bicrystals with [001] tilt boundaries. *Acta Materialia*, **46**, 4103–4110.

12 Li, C.H., Edwards, E.H., Washburn, J., and Parker, E.R. (1953) Stress-induced movement of crystal boundaries. *Acta Metallurgica*, **1**, 223–229.

13 Winning, M., Gottstein, G., and Shvindlerman, L.S. (2002) On the mechanisms of grain boundary migration. *Acta Materialia*, **50**, 353–363.

14 McLean, D. (1957) *Grain Boundaries in Metals*, Oxford Press, London.

15 Seah, M.P. and Hondros, E.D. (1973) Grain boundary segregation. *Proceedings of the Royal Society of London. Series A, Mathematical and Physical Sciences*, **335**, 191–212.

16 Hondros, E.D. (1965) The influence of phosphorus in dilute solid solution on the absolute surface and grain boundary energies of iron. *Proceedings of the Royal Society of London A*, **286**, 479–498.

17 Shea, M.P. (1975) Interface adsorption, embrittlement and fracture in metallurgy: A review. *Surface Science*, **53**, 168–212.

18 Hondros, E.D. and Seah, M.P. (1977) The theory of grain boundary segregation in terms of surface adsorption analogues. *Metallurgical Transactions*, **8**, 1363–1371.

19 Langmuir, I. (1918) he adsorption of gases on plane surfaces of glass, mica and platinum. *Journal of the American Chemical Society*, **40**, 1361–1368.

20 Gottstein, G., Shvindlerman, L.S., Molodov, D.A., and Czubayko, U. (1997) Grain boundary motion in aluminium bicrystals, in *Dynamics of Crystal Surfaces and Interfaces* (eds P.M. Duxbury and T.J. Pence), Plenum Press, New York, pp. 109–123.

21 Cahn, J.W. (1962) The impurity-drag effect in grain boundary motion. *Acta Metallurgica*, **10**, 789–798.

22 Lücke, K. and Detert, K. (1957) A quantitative theory of grain-boundary motion and recrystallization in metals in the presence of impurities. *Acta Metallurgica*, **5**, 628–637.

23 Sutton, A.P. and Ballufi, R.W. (1997) *Interfaces in Crystalline Materials*, Oxford University Press, USA.

24 Hoffman, S. (1990) Segregation of grain boundaries, in *Surface Segregation Phenomena* (eds P.A. Dowben and A. Miller), CRC Press, Boca Raton, FL, pp. 107–1134.

25 Hoffman, S. and Lejcek, P. (1996) Solute segregation at grain boundaries. *Interface Science*, **3**, 241–267.

26 Lejcek, P. and Hoffman, S. (2008) Thermodynamics of grain boundary segregation and applications to anisotropy, compensation effect and prediction. *Critical Reviews in Solid State and Materials Sciences*, **33**, 133–163.

27 Liu, F. and Kirchheim, R. (2004) Grain boundary saturation and grain growth. *Scripta Materialia*, **51**, 521–525.

28 Lejček, P. and Hofmann, S. (1995) Thermodynamics and structural aspects of grain boundary segregation. *Critical Reviews in Solid State and Materials Sciences*, **20**, 1–85.

29 Chapman, M.A.V. and Faulkner, R.G. (1983) Computer modelling of grain boundary segregation. *Acta Metallurgica*, **31**, 677–689.

30 Wynblatt, P.W. and Chatain, D. (2006) Anisotropy of segregation at grain boundaries and surfaces. *Metallurgical and Materials Transactions A – Physical Metallurgy and Materials Science*, **37**, 2595–2620.

31 Jiang, H. and Faulkner, R.G. (1996) Modelling of grain boundary segregation, precipitation and precipitate-free zones of high strength aluminium alloys–I. The model. *Acta Materialia*, **44**, 1857–1864.

32 Jiang, H. and Faulkner, R.G. (1996) Modelling of grain boundary segregation, precipitation and precipitate-free zones of high strength aluminium alloys–II. Application of the models. *Acta Materialia*, **44**, 1865–1871.

33 Lejček, P., Hofmann, S., and Paidar, V. (2003) Segregation based classification of [100] tilt grain boundaries in α-iron and its consequences for grain boundary engineering. *Acta Materialia*, **51**, 3951–3963.

34 Defay, R. and Prigogine, I. (1950) Surface tension of regular solutions. *Transactions of the Faraday Society*, **46**, 199–204.

35 Millett, P.C., Selvam, R.P., and Saxena, A. (2006) Molecular dynamics simulations of grain size stabilization in nanocrystalline materials by addition of dopants. *Acta Materialia*, **54**, 297–303.

36 Millett, P.C., Selvam, R.P., and Saxena, A. (2007) Stabilizing nanocrystalline materials with dopants. *Acta Materialia*, **55**, 2329–2336.

37 Kirchheim, R. (2002) Grain coarsening inhibited by solute segregation. *Acta Materialia*, **50**, 413–419.

38 Kirchheim, R. (2007) Reducing grain boundary, dislocation line and vacancy formation energies by solute segregation. I: Theoretical background. *Acta Materialia*, **55**, 5129–5138.

39 Raabe, D., Roters, F., Barlat, F., and Chen, L.-Q. (2004) *Continuum Scale Simulation of Engineering Materials – Fundamentals – Microstructures – Process Applications*, Wiley-VCH, Weinheim, ISBN: 3-527-30760-5.

40 Cahn, J.W. and Hilliard, J.E. (1958) Free energy of a nonuniform system. I. Interfacial free energy. *Journal of Chemical Physics*, **28**, 258–267.

41 Cahn, J.W. (1969) *The Mechanisms of Phase Transformations in Crystalline Solids* (eds P Wilkes and G Lorimer), The Institute of Metals, London.

42 Khachaturyan, A.G. (1983) *Theory of Structural Transformations in Solids*, John Wiley & Sons, Inc., New York.

43 Heo, T.W., Bhattacharyya, S., and Chen, L.-Q. (2011) A phase field study of strain energy effects on solute-grain boundary interactions. *Acta Materialia*, **59**, 7800–7815.

44 Gustafson, E. (1985) *International Journal of Thermophysics*, **6**, 395–409.

45 Allen, S.M. and Cahn, J.W. (1979) A microscopic theory for antiphase boundary motion and its application to antiphase domain coarsening. *Acta Metallurgica*, **27**, 1085–1095.

46 Zhu, J., Chen, L.-Q., Shen, J., and Tikare, V. (1999) Coarsening kinetics from a variable-mobility Cahn–Hilliard equation: Application of a semi-implicit Fourier spectral method. *Physical Review E*, **60**, 3564–3572.

47 Cahn, J.W. (1961) On spinodal decomposition. *Acta Metallurgica*, **9**, 795–801.

48 Baram, M., Chatain, D., and Kaplan, W.D. (2011) Nanometer-thick equilibrium films: The interface between thermodynamics and atomistics. *Science*, **332**, 206–209.

49 Dillon, S.J. and Harmer, M.P. (2007) Multiple grain boundary transitions in ceramics: A case study of alumina. *Acta Materialia*, **55**, 5247–5254.

50 Dillon, S.J., Harmer, M.P., and Rohrer, G.S. (2010) The relative energies of normally and abnormally growing grain boundaries in alumina displaying different complexions. *Journal of the American Ceramic Society*, **93**, 1796–1802.

51 Dillon, S.J. and Harmer, M.P. (2007) Direct observation of multilayer adsorption on alumina grain boundaries. *Journal of the American Ceramic Society*, **90**, 996–998.

52 Dillon, S.J., Tang, M., Carter, W.C., and Harmer, M.P. (2007) Complexion: A new concept for kinetic engineering in materials science. *Acta Materialia*, **55**, 6208–6218.

53 Harmer, M.P. (2011) The phase behavior of interfaces. *Science*, **332**, 182–183.

54 Ma, S., Meshinchi Asl, K., Tansarawiput, C., Cantwell, P.R., Qi, M., Harmer, M.P., and Luo, J. (2012) A grain boundary phase transition in Si–Au. *Scripta Materialia*, **66**, 203–206.

55 Lozano-Perez, S., Saxey, D.W., Yamada, T., and Terachi, T. (2010) Atom-probe tomography characterization of the oxidation of stainless steel. *Scripta Materialia*, **62**, 855–858.

56 Thuillier, O., Danoix, F., Gouné, M., and Blavette, D. (2006) Atom probe tomography of the austenite–ferrite interphase boundary composition in a model alloy Fe–C–Mn. *Scripta Materialia*, **55**, 1071–1074.

57 Danoix, F., Bémont, E., Maugis, P., and Blavette, D. (2006) Atom probe tomography I. early stages of precipitation of NbC and NbN in ferritic steels. *Advanced Engineering Materials*, **8**, 1202–1205.

58 Deschamps, F., Danoix, F., De Geuser, F., Epicier, T., Leitner, H., and Perez, M. (2011) Low temperature precipitation kinetics of niobium nitride platelets in Fe. *Materials Letters*, **65**, 2265–2268.

59 Danoix, F., Julien, D., Sauvage, X., and Copreaux, J. (1998) Direct evidence of cementite dissolution in drawn pearlitic steels observed by tomographic atom probe. *Journal of Materials Science and Engineering A – Structural Materials Properties Microstructure and Processing*, **250**, 8–13.

60 Sauvage, X., Copreaux, J., Danoix, F., and Blavette, D. (2000) Atomic-scale observation and modelling of cementite dissolution in heavily deformed pearlitic steels. *Philosophical Magazine A – Physics of Condensed Matter Structure Defects and Mechanical Properties*, **80**, 781–796.

61 Sauvage, X., Lefebvre, W., Genevois, C., Ohsaki, S., and Hono, K. (2009) Complementary use of transmission electron microscopy and atom probe tomography for the investigation of steels nanostructured by severe plastic deformation. *Scripta Materialia*, **60**, 1056–1061.

62 Hong, M.H., Reynolds, W.T.Jr., Tarui, T., and Hono, K. (1999) Atom probe and transmission electron microscopy investigations of heavily drawn pearlitic steel wire. *Metallurgical and Materials Transactions A – Physical Metallurgy and Materials Science*, **30**, 717–727.

63 Hono, K., Ohnuma, M., Murayama, M., Nishida, S., and Yoshie, A. (2001) Cementite decomposition in heavily drawn pearlite steel wire. *Scripta Materialia*, **44**, 977–983.

64 Miller, M.K., Cerezo, A., Hetherington, M.G., and Smith, G.D.W. (1996) *Atom Probe Field Ion Microscopy*, Oxford University Press, Oxford, UK.

65 Williams, C.A., Hyde, J., Smith, G.D.W.S., and Marquis, E.A. (2011) Effects of heavy-ion irradiation on solute segregation to dislocations in oxide-dispersion-strengthened Eurofer 97 steel. *Scripta Materialia*, **412**, 100–105.

66 Yuan, L., Ponge, D., Wittig, J., Choi, P., Jiminez, J.A., and Raabe, D. (2012) Nanoscale austenite reversion through partitioning, segregation and kinetic freezing: Example of a ductile 2GPa Fe–Cr–C steel. *Acta Materialia*, **60**, 2790–2804.

67 Ohsaki, S., Raabe, D., and Hono, K. (2009) Mechanical alloying and amorphization in Cu–Nb–Ag in situ composite wires studied by transmission electron microscopy and tomographic atom probe. *Acta Materialia*, **57**, 5254–5263.

68 Raabe, D., Choi, P.P., Li, Y.J., Kostka, A., Sauvage, X., Lecouturier, F., Hono, K., Kirchheim, R., Pippan, R., and Embury, D. (2010) Metallic composites processed via extreme deformation: Toward the limits of strength in bulk materials. *MRS Bulletin*, **35**, 982–991.

69 Moody, M.P., Gault, B., Stephenson, L.T., Haley, D., and Ringer, S.P. (2009) Qualification of the tomographic reconstruction in atom probe by advanced spatial distribution map techniques. *Ultramicroscopy*, **109**, 815–824.

70 Moody, M.P., Tang, F., Gault, B., Ringer, S.P., and Cairney, J.M. (2011) Atom probe crystallography: Characterization of grain boundary orientation relationships in nanocrystalline aluminium. *Ultramicroscopy*, **111**, 493–499.

71 Araullo-Peters, V.J., Gault, P., Shrestha, S.L., Yao, L., Moody, M.P., Ringer, S.P., and Cairney, J.M. (2012) Atom probe crystallography: Atomic-scale 3-D orientation mapping. *Scripta Materialia*, **66**, 907–910.

72 Borchers, C., Al-Kassab, T., Goto, S., and Kirchheim, R. (2009) Partially amorphous nanocomposite obtained from heavily deformed pearlitic steel. *Materials Science and Engineering A*, **502**, 131–142.

73 Nakada, N., Tsuchiyama, T., Takaki, S., and Hashizume, S. (2007) Variant selection of reversed austenite in lath martensite. *ISIJ International*, **47**, 1527–1532.

74 Nakada, N., Tsuchiyama, T., Takaki, S., and Miyano, N. (2011) Temperature dependence of austenite nucleation behavior from lath martensite. *ISIJ International*, **51**, 299–304.

12
Interface Structure-Dependent Grain Growth Behavior in Polycrystals

Suk-Joong L. Kang, Yang-Il Jung, Sang-Hyun Jung, and John G. Fisher

12.1
Introduction

Grain growth is commonly categorized into two types: normal grain growth (NGG) and abnormal grain growth (AGG, sometimes also called exaggerated grain growth). NGG exhibits a unimodal and an invariant relative grain size distribution (the relative grain size is the grain size divided by the mean grain size) with time [1–4]. On the other hand, AGG is characterized by a bimodal grain size distribution as a result of the rapid growth of a small number of large grains at the expense of the small matrix grains. Upon further annealing, however, the large abnormal grains impinge upon each other and the relative grain size distribution reverts to a narrow and unimodal one. The grain growth behavior can then become practically stagnant. Hence, the relative grain size distribution varies with annealing time for a system where AGG occurs [5–7].

Many investigations in solid–liquid two-phase systems [8–27] and also in single-phase systems [18,28–41] have shown that grain growth behavior is closely related to the interface morphology: NGG for rough (atomically disordered) interfaces and AGG or growth inhibition (stagnant grain growth, SGG) for faceted (atomically ordered) interfaces. The NGG behavior for rough interfaces has been explained in terms of diffusion-controlled growth since rough interfaces provide an unlimited number of sites for the easy attachment of atoms. The relative size distribution of grains is invariant with annealing time. For faceted interfaces, on the other hand, surface defects need to be present for growth of the grain [42–47]. Surface defects can form via two-dimensional atomic nucleation (2DN) or be present inherently (e.g, atomic steps formed by the presence of screw dislocations at the surface). The growth of faceted grains can then proceed with the attachment of atoms at the surface defects. Both 2DN mechanism and dislocation-assisted (DA) mechanism have a characteristic nonlinear relationship between the migration velocity of the interface and the driving force for migration [42,43,46,47], which results in a deviation from the normal behavior of grain growth.

The present chapter will discuss growth kinetics and behaviors with respect to the interface structures, that is, faceted and rough, for solid-state single-phase systems as well as solid–liquid two-phase systems. The equilibrium shape of a crystal (grain) will first be described, and the mechanisms and kinetics of interface migration (single crystal growth) will then be explained. Based on the interface migration mechanisms and kinetics, a grain growth model (mixed control model) will be presented and grain growth behavior will be discussed together with experimental results. The principle of microstructural evolution, which has been deduced from theoretical as well as experimental results, will be described.

12.2
Fundamentals: Equilibrium Shape of the Interface

12.2.1
Equilibrium Crystal Shape

The surface free energy of a crystal is an integral of the form

$$\int \gamma(\hat{n}) \, dS \tag{12.1}$$

extended over the surface S of the body, where the specific surface energy γ is a function of the orientation of the unit outward normal $\hat{n}$ at each surface point. The equilibrium shape of a crystal (ECS) can be determined by minimizing the surface free energy for a fixed volume, following the procedure called the Wulff construction [48].

If the surface energy is isotropic with respect to the orientation, the equilibrium shape is spherical; otherwise, it is partially or fully faceted. The equilibrium shape of a crystal is also related to the step free energy of the crystal surface [49], which is the excess energy of a step of a two-dimensional nucleus formed on the corresponding face of a crystal. When the step free energy is high ($\sigma \geq 1.0 \, h \, \gamma$, where σ is the step free energy and h is the step height), the equilibrium shape of the crystal is an angular polyhedron. As the step free energy decreases, the equilibrium crystal shape changes to a round-edged polyhedron with flat planes and rounded corners and edges. If the step free energy decreases to zero for all the surface of a crystal, the equilibrium shape is a sphere [49].

It is well documented that the step free energy decreases as temperature increases [50–53]. A common expression for the variation of step free energy with temperature takes the form

$$\sigma \approx \sigma_0 \exp\left(-\frac{M}{\sqrt{T_R - T}}\right) \tag{12.2}$$

where M is a nonuniversal constant and T_R is the roughening transition temperature. At a sufficiently low temperature, the γ-plot (the polar plot of the surface energy) is cusped in certain symmetry directions, leading to the appearance of facets in the equilibrium crystal shape. As the temperature is raised, step energies decrease, causing blunting of cusps and making the corresponding facets shrink.

The roughening transition at which each facet finally disappears takes place at the roughening temperature T_R of the corresponding planar interface. Different symmetry directions have different roughening temperatures. Above each T_R, the corresponding region of the equilibrium crystal shape becomes smoothly rounded.

12.2.2
Equilibrium Boundary Shape

Cahn and Hoffman [54,55] proposed a vector function $\xi(\hat{n})$, called the Cahn–Hoffman vector, as an alternative to the scalar function $\gamma(\hat{n})$ with which the free energy of anisotropic surfaces is commonly described. The Cahn–Hoffman formalism provides a direct way to elucidate the role of anisotropy on capillary pressure and its role on particle shape determination. Cahn and Hoffman [54,55] showed that the interface satisfies the condition

$$\nabla_s \cdot \xi = -\Delta F \tag{12.3}$$

where ∇_s is the surface divergence on the interface and ΔF is the bulk free energy difference across the interface. They also showed that the equilibrium shape is given by $r = -2\xi/\Delta F$, that is, the envelope of the ξ-vectors, which is equivalent to the Wulff construction.

The condition for local equilibrium at a triple junction with anisotropic grain boundary energies can be obtained using ξ-vectors. The capillary vector for a boundary is formed by combining the two scalar quantities $\gamma(\hat{n})$ and $(d\gamma/d\theta)$ with unit vectors normal and parallel to the plane of the boundary, respectively. Note that $\xi_\perp \times \hat{l}$ is a tension in the plane of the boundary and is a vector perpendicular to the line intersection of the three boundaries $\hat{l}$. Similarly, the vector perpendicular to the boundary has the form of $\xi_\parallel \times \hat{l}$, which is a torque force. For a triple junction, the equilibrium condition is expressed as

$$\left(\xi^i + \xi^j + \xi^k\right) \times \hat{l} = 0 \tag{12.4}$$

with the three boundaries denoted as i, j, and k, respectively

The grain boundary lying between two triple junctions also can be constructed using the same equilibrium conditions. If the plot of boundary energy (referred to as an ξ'-plot in this text) is angular with round edges, the boundary consists of rough segments connected to flat segments. If the ξ'-plot has a discontinuity in slope when the flat plane meets the rough plane, there are forbidden orientations of grain boundary planes. Cho [56] proposed the concept of gross equilibrium shape referred to as the ξ'' plot, which is a polar plot of the averaged energy in the directions from the reference coordinates for the whole misorientation, to describe various types of grain boundaries. Each plot of grain boundary energy was obtained by overlapping two gross ξ-plots of two grains with respect to their misorientation. Once the ξ'-plot or ξ''-plot is obtained, both the plots can define the shape of the grain boundaries.

12.3
Grain Growth in Solid–Liquid Two-Phase Systems

12.3.1
Growth Mechanisms and Kinetics of a Single Crystal in a Liquid

The growth mechanism and kinetics of a crystal in a liquid matrix is highly dependent on the surface structure of the crystal. Figure 12.1 shows the two extreme types of equilibrium shape of a crystal in a liquid: a sphere and a polyhedron with sharp edges and corners. The spherical grain has a rough (atomically disordered) surface whereas the polyhedral grain has a flat (atomically ordered) surface. For an atomically rough surface, there is no energy barrier for atom attachment. On the contrary, atom attachment on a flat surface results in an increase in the surface energy because more broken bonds with high bond energy are created on the surface. Therefore, an atom that attaches to a flat surface can soon detach unless it can move to a low energy site.

This structure-dependent atom attachment behavior results in a difference in growth kinetics with respect to the driving force for crystal growth [7,42–47,57]. For a spherical crystal, the growth rate is determined by the solute flux at the surface because the interface reaction of atom attachment does not require any energy change and occurs easily. However, a crystal with a flat surface grows by dissimilar and complex mechanisms.

12.3.1.1 Diffusion-Controlled Crystal Growth
When a crystal is growing (or dissolving) in a liquid, the diffusion current of solute across an imaginary spherical surface S around the crystal must be equal to the rate

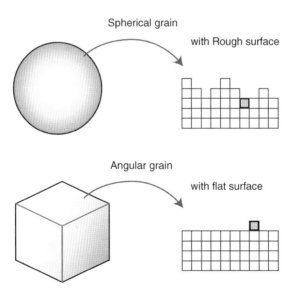

Figure 12.1 Two types of equilibrium shape and surface structure of a crystal.

of mass change of the crystal. Therefore,

$$\oint_S J \cdot dA = 4\pi l^2 D \nabla C = 4\pi r^2 \frac{dr}{dt} \tag{12.5}$$

where $J\,(=-D\nabla C)$ is the diffusion current per unit area, D is the diffusion coefficient, ∇C is the solute concentration gradient in the liquid, l is the mean distance of an imaginary surface from the grain center, r is the radius of the crystal, and dr/dt is the rate of crystal growth (dissolution). Integration by parts of Eq. (12.5) leads to

$$\frac{dr}{dt} = D \frac{C_a - C_r}{r} \tag{12.6}$$

where C_a is the solute concentration at a long distance away from the crystal.

12.3.1.2 Interface Reaction-Controlled Crystal Growth

For a polyhedral grain with flat surfaces, growth proceeds by lateral growth from atomic steps present at the surface. The possible step sources are two-dimensional nuclei, screw dislocations and twin re-entrant edges. If a crystal is cut not exactly parallel to a low-index zone axis, a vicinal surface with a set of equally spaced ledges and kinks results. It was suggested by Burton et al. [42] that the growth of such a crystal involves the following steps: (a) adsorption or desorption of atoms at the surface, (b) surface diffusion, and (c) transfer to the kink positions in monoatomic ledges on the surface. The theory proposed by Burton et al. [42] accounted for crystal growth from a vapor. For the growth of a faceted crystal from a liquid, it is assumed that growth of the step occurs via the bulk diffusion of solute atoms through the liquid matrix only [44]. The atoms that absorb at a kink or ledge site are assumed to be stable, but those that absorb on a flat surface are too unstable to stay there and soon desorb.

Growth from a Flat Surface Atom attachment onto a flat surface is unstable because of an increase in surface energy; the attached atoms tend to dissolve again into the liquid matrix. However, once a sufficiently large nucleation island forms, it generates an energetically stable ledge and kink sites along the circumference of the nucleus. Therefore, the growth of a faceted grain can proceed with 2DN [7,42–47,57], if there are no intrinsic atomic steps at the surface.

The change in Gibbs free energy for the formation of a two-dimensional nucleus with radius a and step height h is expressed as

$$\Delta g = 2\pi a \sigma + \pi a^2 h \Delta g_v \tag{12.7}$$

where $\Delta g_v\,(= g_s - g_l)$ is the difference in volumetric free energy between the solid (g_s) and the liquid (g_l), and σ the free energy for step generation (the step free energy), which is assumed to be isotropic. The activation energy for the formation of a critical two-dimensional nucleus, which can continuously grow, is expressed as

$$\Delta g^* = \frac{-\pi \sigma^2}{h \Delta g_v}. \tag{12.8}$$

The rate of 2DN is expressed as [43–45]

$$\dot{N} = n^* n_0 v \exp\left(-\frac{\Delta g_m}{kT}\right) \exp\left(-\frac{\Delta g^*}{kT}\right). \qquad (12.9)$$

Here, n^* is the number of atoms in a position near to a critical nucleus, n_0 is the number density of atoms in the liquid, v is the vibration frequency of the atoms in the liquid, and Δg_m is the activation energy for jumping across the interface.

Note that the value of $n_0 \exp\left(-\frac{\Delta g^*}{kT}\right)$ is the number density of the nuclei with radius a^*, the critical size of a stable two-dimensional nucleus, from the van't Hoff isotherm. The variation of $\dot{N}$ with Δg_v is drastic. For a sufficiently low Δg_v (as an absolute value), $\dot{N}$ is negligible. With increased Δg_v, however, $\dot{N}$ increases abruptly, as can be seen in Eq. (12.9).

The 2DN mechanism exhibits complex growth situations that themselves depend on the kinematics of ledge motion and nucleation. For an angular grain with a high step free energy, the nucleation rate is much slower than the rate of ledge propagation v_e. In this case, that is, $\dot{N} \ll v_e$, a single critical nucleus forms and spreads over the entire surface before the next nucleation event takes place. However, if the nucleation rate is relatively fast, for example, as in high supercooling in crystal growth, a large number of two-dimensional nuclei form at random positions at the surface before a layer is completed or on top of already growing two-dimensional islands. In both cases, every nucleation event results in growth normal to the surface by a distance equal to the step height. The growth kinetics by the 2DN mechanism is directly proportional to the nucleation rate $\dot{N}$, the grain facet size A, and the step height h, for the mono-nucleation of a two-dimensional island, that is, the growth rate $dr/dt = hA\dot{N}$. For the model of polynucleation of a monolayer, the growth rate is expressed as $dr/dt \approx h(\pi \dot{N} v_e^2)^{1/3}$ [57].

Growth from a Screw Dislocation If the polyhedral grain contains dislocations that intersect the solid–liquid interface, the problem of creating new interfacial steps can be circumvented. When a grain grows from a screw dislocation on its surface, the ledges associated with the dislocation wrap themselves into a spiral. The limit of curvature of the spiral can be determined by the Gibbs–Thomson relation, that is, the curvature of a ledge in equilibrium with the supersaturated liquid matrix [43]. Thus at the dislocation origin, the radius of curvature a_c is derived as

$$a_c = \frac{-\gamma}{\Delta g_v} = \frac{-\sigma}{h \Delta g_v} \qquad (12.10)$$

from Eq. (12.7). In Eq. (12.10), the step free energy is assumed to be isotropic. However, it is more typical to consider the anisotropic step free energy for an angular grain and then a_c becomes the smallest straight length of the low-index ledge adjacent to the dislocation [43].

Most theoretical aspects of the spiral growth mechanism were first reported in the classic paper of Burton et al. [42]. Assuming steady-state spiral rotation, the spacing λ

between spiral ledges emanating from a dislocation is approximately

$$\lambda = 4\pi a_c. \tag{12.11}$$

For polygonized spirals, on the other hand, it is estimated to be in the range of $5a_c$ to $9a_c$ [58].

For a crystal surface with equally spaced ledges due to the presence of screw dislocations, the growth rate v_e of the ledges on a terrace is expressed as [42,43,57]

$$v_e = \frac{dx}{dt} = 2\sqrt{2}\frac{D}{r}(C_a - C_e)\tanh\frac{\lambda}{\sqrt{2}r}. \tag{12.12}$$

As the spiral mechanism leads to the growth of a grain in the direction of its surface normal, the growth rate of a crystal is expressed as [42,43,57]

$$\frac{dr}{dt} = \frac{hv_e}{\lambda} = 2\sqrt{2}\frac{D}{r}(C_a - C_e)\frac{h^2\Delta g_v}{4\pi\sigma}\tanh\frac{4\pi\sigma}{\sqrt{2}rh\Delta g_v} \tag{12.13}$$

with the substitutions of $\lambda = 4\pi a_c$ and $a_c = -\sigma/hg_v$. In Eq. (12.13), the term $D(C_a - C_e)/r$ is the growth rate governed by diffusion (Eq. (12.6)).

Equation (12.13) predicts the same physical behavior as the previously developed equations do both for a solid–vapor system [42] and for a solid–liquid system [57]. Similar to the 2DN mechanism, it is assumed that the step free energy is high and the supersaturation is low in order for a grain to grow by the screw dislocation-assisted mechanism only. Therefore, in the case of $\sigma/h\Delta g_v \gg 1$, that is, for a small driving force, the growth rate is, in the first approximation, proportional to the square of Δg_v as

$$\frac{dr}{dt} \propto \Delta g_v^2. \tag{12.14}$$

Growth from a Re-Entrant Edge Two flat surfaces meeting at an angle smaller than 180°, for example, twin planes, form a re-entrant edge. The kinetics of atom attachment at a ledge with many kinks is much easier than that on a flat surface and is expressed as Eq. (12.12). As the nucleus formed at a re-entrant edge is a portion of a two-dimensional nucleus, the free energy change upon the formation of a critical nucleus at a re-entrant edge can be expressed as [59]

$$\Delta g = 2\pi a\sigma \langle L \rangle - \pi a^2 h \Delta g_v \langle S \rangle + \gamma(A_1 - A_2) \tag{12.15}$$

where $\langle L \rangle$ and $\langle S \rangle$ are structure factors, $(A_1 - A_2)$ is the change in surface area due to the formation of the nucleus, and γ is the surface energy equivalent to σ/h for angular grains. Eq. (12.15) is then expressed as

$$\Delta g = 2\pi a\sigma \langle L \rangle - \pi a^2 h \Delta g_v \langle S \rangle + 2\pi a\sigma \wp \langle A \rangle \tag{12.16}$$

where $\wp$ is a constant depending on the angle of the re-entrant edge. Equation (12.16) leads to the activation energy for nucleation to be

$$\Delta g^* = \frac{\pi\sigma^2}{h\Delta g_v} \cdot \frac{(\langle L \rangle - \wp\langle A \rangle)^2}{\langle S \rangle} \equiv \varsigma \cdot \Delta g_{2D}^*. \tag{12.17}$$

The activation energy for the nucleation at a re-entrant edge differs from that for 2DN by the structure factor ς [60]. A calculation for different geometries shows that the structure factor varies from ∼0.3 to ∼0.1 for a variation in angle of the re-entrant edge from 144° to 108° [60]. It is clear that nucleation on a ledge site is much easier because the value of the structure factor ς is much smaller than 1 in most cases.

12.3.1.3 Mixed Controlled Growth of a Faceted Crystal

Figure 12.2 plots schematically the growth kinetics of a crystal for the various growth modes described earlier. The growth kinetics of Eq. (12.6) (diffusion mechanism), Eq. (12.9) (2DN mechanism), and Eq. (12.13) (screw dislocation-assisted mechanism) were used for the plot in Figure 12.2. As the growth of a crystal in a liquid matrix is a result of serial processes of the diffusion of atoms through the matrix and the attachment of atoms at the crystal surface (interface reaction), the time consumed in the growth of a crystal, $1/v$, is the sum of the times consumed for diffusion, $1/v_D$, and for interface reaction, $1/v_R$. Therefore, the overall growth rate v takes the form [5,7]

$$v = \frac{v_R v_D}{v_R + v_D} \tag{12.18}$$

and is governed by the slower process.

For a spherical crystal, theoretically, as there are numerous sites for easy attachment of atoms, the diffusion process governs the overall kinetics, as schematically shown by a dashed line in Figure 12.2, irrespective of the magnitude of the driving force. In contrast, for the growth of a faceted crystal, while the diffusion kinetics in a liquid matrix is similar to that for a spherical grain (Eq. (12.6)), the interface reaction kinetics is a nonlinear function of the driving force, as schematically shown by dotted lines following Eqs. (12.9) and (12.13). The growth of a faceted crystal will be governed by the interface reaction for low driving forces and by diffusion for high driving forces, as schematically shown by solid lines in Figure 12.2. In the case of

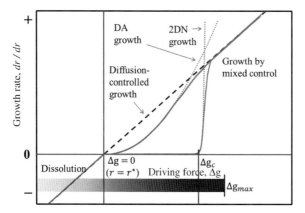

Figure 12.2 Schematic plot of the growth rate of a crystal in a liquid as a function of the driving force for different growth modes, showing the mixed control model of grain growth.

dissolution, however, the rate is always governed by diffusion, irrespective of the crystal shape, because there is essentially no energy barrier for the detachment of atoms from the corners and multilayer dissolution can occur [7,43,45]. This kinetic model of mixed control for growth and diffusion control for dissolution [5,7], the mixed control model of grain growth, can explain various grain growth behaviors and the resultant microstructural evolutions in two-phase systems, as described in the following section.

The growth kinetics of a round-edged polyhedral grain with flat surfaces and rounded edges and corners may be affected not only by the driving force but also by the continuity of the surface normal [45]. The growth of the flat surfaces, however, is thought to be governed by interface reaction via the 2DN mechanism for the case of a discontinuous surface normal [45] or by the surface-defect-assisted mechanism for the case of a continuous surface normal. Experimental observations have shown that the growth of round-edged grains is governed by the interface reaction kinetics, although the continuity of the surface normal was not identified [19,23,61].

As shown in Figure 12.2, the growth rates of the 2DN mechanism and spiral growth mechanisms cannot exceed the rate of the diffusion mechanism because of the supply limit of atoms by diffusion. Under a high driving force and with a high growth rate, kinetic roughening of facets, that is, random attachment of atoms at any point on the facet, can occur [62]. The estimated width of a diffuse step is approximately equal to the radius of a critical nucleus. Peteves and Abbaschian [63] showed experimentally the diffusion-limited growth of an initially faceted Ga single crystal by measuring its growth rate under various supercoolings.

12.3.2
Grain Growth Behavior

The grain growth behavior in two-phase systems can be predicted and explained, utilizing the mixed control model of grain growth in Figure 12.2. In a polycrystal, as each grain has its own driving force for growth (or dissolution), there is a range of values of the driving force for growth in the system, as schematically shown in Figure 12.2. The largest grain has the maximum driving force for growth (Δg_{max}) and the smallest grain has the maximum driving force for dissolution. The driving force for grain growth arises from the difference in the capillary pressure among grains, irrespective of their interface structure. The capillary driving force for a grain of size $2r$ is expressed as

$$\Delta g(r, r^*; t) = 2\gamma V_m \left(\frac{1}{r^*} - \frac{1}{r} \right) \qquad (12.19)$$

where γ is the interfacial energy, V_m is the molar volume, and r^* is the critical radius of grains neither growing nor shrinking, which is time variant. (In the case of a faceted crystal, r in Eq. (12.19) represents the distance from the center of the crystal to a flat surface and γ the specific energy of the surface.)

For a system where the growth and dissolution of grains are governed by the diffusion of atoms through the matrix, and hence their rates are linearly proportional to the driving force, normal grain growth (stationary in terms of the variation of the relative grain size distribution with annealing time) occurs, as Lifshitz and Slyozov [1], and Wagner [2] analyzed. Any deviation from the kinetics of diffusional growth, that is, any deviation from the linearity between the growth rate and the driving force, must result in nonnormal (nonstationary) growth behavior.

12.3.2.1 Stationary Grain Growth in Systems with Spherical Grains

In a polycrystal with numerous grains, each grain has its own value of solute solubility in the liquid. According to the Gibbs–Thomson equation, the equilibrium concentration C_r of solute in front of the surface of a grain with radius r is expressed as

$$C_r = C_\infty \left(1 + \frac{2\gamma V_m}{RTr}\right) \quad (12.20)$$

where C_∞ is the concentration of the solute in the liquid in equilibrium with a grain of an infinite size. In such a system, grains having solubilities lower than the average solubility in the system will grow and those having higher solubilities will shrink (dissolve). Lifshitz and Slyozov [1] and Wagner [2] considered diffusion-controlled grain growth for an infinitely dispersed system, that is, a system with a solid volume fraction of zero (LSW theory). From Eq. (12.6)

$$\frac{dr}{dt} = D\frac{C_{r^*} - C_r}{r} = \frac{2D\gamma V_m C_\infty}{RTr}\left(\frac{1}{r^*} - \frac{1}{r}\right) \quad (12.21)$$

where C_{r^*} is the solute concentration in the liquid (the solute concentration of the critical sized grain of radius r^*, which is neither growing nor shrinking). The grains larger than a critical size r^* grow at the expense of smaller ones. The critical size varies with time as grain growth takes place.

As the driving force for growth of a grain of radius r is expressed as Eq. (12.19), Eq. (12.21) is transformed into

$$\frac{dr}{dt} = \frac{2D\gamma V_m C_\infty}{RTr} \cdot \frac{\Delta g_v}{2\gamma} = \frac{2DV_m C_\infty}{RTr} \cdot \Delta g_v. \quad (12.22)$$

Therefore, the growth (or shrinkage) kinetics is directly proportional to the driving force, if the effect of the gradient of the driving force (which is marginal) is ignored. Here Δg_v is the volumetric driving force.

The solution of Eq. (12.22) for a spherical system gives a kinetic equation of

$$\bar{r}_t^3 - \bar{r}_0^3 = \frac{8}{9}\frac{D\gamma V_m C_\infty}{RT}t \equiv K_{LSW}t. \quad (12.23)$$

This equation is the well-known cubic law of the LSW theory. Another important result of this solution is that a steady state, where the relative grain size distribution is invariant, is reached from any initial grain size distribution as the annealing time increases.

Later, LSW theory was modified by many researchers to consider a finite volume fraction of solid grains [3,64–68]. For a steady state, all of these works report simple kinetic equations of grain growth and also unimodal and time-invariant (stationary) relative grain size distributions with respect to the annealing time. The main difference among the different works was the absolute rate of growth with different solid volume fractions [69]. The proportionality constant K in Eq. (12.23) increases with increasing solid volume fraction (ϕ), as given by

$$K(\phi) = K_{LSW} f(\phi) \tag{12.24}$$

with $f(\phi)$ being a system-independent function of the volume fraction, while the basic $t^{1/3}$ kinetics of the LSW theory are maintained. This consequence is natural because the concentration gradient at the particle–matrix interface becomes steeper as ϕ increases. According to the modified LSW theories, the theoretical distribution of particle sizes broadens rapidly at small values of ϕ [3,69] fitting the experimental data reasonably well [70].

While most theories and model calculations make predictions for the steady-state regime, a few attempts [71–73] have been made to describe the transient behavior of coarsening, that is, the evolution of the system prior to reaching the steady-state regime. Deviations from the steady state were well described, in particular for coarsening at the beginning; however, the predicted growth behavior was not very different from normal and the formation of abnormal grains was impossible.

12.3.2.2 Nonstationary Grain Growth in Systems with Faceted Grains

As explained in Section 3.1.2, the growth rate of a (partially) faceted crystal is not linearly proportional to the driving force, showing a nonlinear region in a plot of the growth rate versus the driving force. As a result, the growth behavior must deviate from normal behavior. The detailed behavior (specific type) of nonstationary grain growth is expected to be governed by the effect of the nonlinear region on the overall growth behavior at the time of observation. The effect of the nonlinear region can be predicted by considering the critical driving force for appreciable growth (Δg_c) relative to the maximum driving force for growth (Δg_{max}) in the system.

The maximum driving force for growth depends on the average grain size and the size distribution of grains. As the average grain size decreases for the same relative grain size distribution, the maximum driving force increases. For the same average size, broadening of the size distribution results in an increase in Δg_{max}. The critical driving force for appreciable growth Δg_c can be expressed as [42,43,45]

$$\Delta g_c = \frac{\pi \sigma^2}{kTh} (\ln \psi n_0)^{-1} \tag{12.25}$$

for two-dimensional nucleation and growth (2DNG). Here $\psi = n^* v \exp\left(\frac{\Delta g_m}{kT}\right)$ and n_0 is the number density of atoms in the liquid. When the growth of faceted grains occurs with the assistance of intrinsic surface defects, in particular screw dislocations, the definition of Δg_c is ambiguous. Nevertheless, the growth rate is still expressed as a nonlinear function of the driving force. The concept of the contribution of the nonlinear region to overall growth behavior can be applied to predict and explain the growth behavior in faceted systems [7,47]. Note that even in the presence of screw dislocations at the surface, the growth kinetics can still be governed by 2DN

[74,75]. For the 2DN mechanism, as the growth kinetics is directly proportional to the nucleation rate $\dot{N}$ (Eq. (12.9)), the equation of the nonlinear curve in Figure 12.2 can be written as [7,21]

$$\frac{dr}{dt} = A \exp\left(-\frac{B}{1/r^* - 1/r}\right) \quad \text{for} \quad \Delta g < \Delta g_c \tag{12.26}$$

and

$$\frac{dr}{dt} = \frac{C}{r}\left(\frac{1}{r^*} - \frac{1}{r}\right)\left(1 + \beta(\phi)\frac{r}{r^*}\right) \quad \text{for} \quad \Delta g > \Delta g_c. \tag{12.27}$$

Here, $A (= hA\psi n_0)$ and $B (= \pi \sigma^2 / 6kTh\gamma)$ are constants, if polynucleation occurs [44,57]. $\beta(\phi)$ is a function of the liquid volume fraction (ϕ) and C $(= 2\gamma V_m D_f C_\infty / RT)$ is a constant depending on the material [1–3].

There have been a few attempts to predict the microstructural evolution in systems where the grain growth is governed by mixed control [7,76]. Kang et al. [76] calculated numerically the abnormal growth of faceted grains for a system with infinitely dispersed two-dimensional grains, along the same framework as Wynblatt and Gjostein [5]. Jung et al. [7] predicted various growth behaviors using Eqs. (12.26) and (12.27) as functions of various physical and processing parameters. They demonstrated nonstationary grain growth for systems with faceted three-dimensional grains and developed the principle of microstructural evolution [7,47,77].

The principle is, in fact, a reflection of the coupling effect of the maximum driving force for grain growth Δg_{max} and the critical driving force for appreciable growth Δg_c as schematically shown in Figure 12.3. Figure 12.3 depicts several conditions for different types of grain growth behavior at the time of observation. Although the conditions were chosen for a constant Δg_{max} (constant grain size and distribution) and variable Δg_c (variable σ and thus variable equilibrium shapes of the grain), similar situations can be obtained for a constant Δg_c and variable Δg_{max}.

a) $\Delta g_c = 0$: Normal grain growth (NGG)
   The modified LSW theories as well as the LSW theory predict normal grain growth.
b) $0 \ll \Delta g_c \ll \Delta g_{max}$: Pseudonormal grain growth (PNGG)
   The contribution of the nonlinear region to the overall behavior is not significant compared with that of the linear region. As a result, the growth behavior is not very different from normal behavior (pseudonormal).
c) $0 \ll \Delta g_c \leq \Delta g_{max}$: Abnormal grain growth (AGG)
   As only a small number of grains have driving forces larger than the critical driving force for appreciable growth, only those grains grow rapidly, resulting in the formation of abnormal grains. The matrix grains, on the other hand, hardly grow.
d) $\Delta g_c \gg \Delta g_{max}$: Stagnant grain growth (SGG)
   In this case, none of the grains grow appreciably, showing a stagnation of grain growth. Although the apparent growth behavior is stagnant, inappreciable growth occurs during annealing because the growth rate is not absolutely zero.

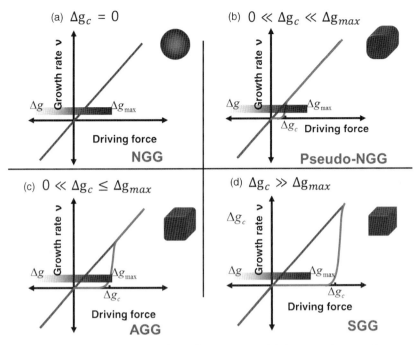

Figure 12.3 Schematic showing conditions for various growth behaviors at the time of observation of a system with the same Δg_{max} for different values of Δg_c (different equilibrium shapes of grains).

The different types of grain growth behavior, except for normal, are time dependent, that is, nonstationary, because Δg_{max} decreases, in general, as the annealing time increases. For a system where pseudonormal grain growth occurs at the beginning, AGG followed by SGG can appear with an increased annealing time due to the reduction of Δg_{max} via overall grain growth. In a system that exhibits SGG, AGG can occur during extended annealing (incubated AGG) as the relative growth rate of the largest grain, although inconsiderable, is the highest and, later, Δg_{max} can be larger than Δg_c [7,21].

The processing parameters that affect Δg_c include temperature, dopant, and oxygen partial pressure. As temperature increases, the step free energy decreases and hence Δg_c decreases, with the grains showing a tendency to rounding (roughening). Grain shape also changes with addition of dopants and with changing oxygen partial pressure, indicating a change in Δg_c. For some recently studied systems, the effects of dopant and oxygen partial pressure on grain shape and grain boundary structure were observed to be related to the variation of the total vacancy concentration [13,41]. As the total vacancy concentration was increased by any means, the grain shape tended to become rounded. This result is similar to surface roughening with the formation of thermal vacancies with increasing temperature.

The developed principle of microstructural evolution described earlier has been supported and confirmed by many experimental studies for various systems

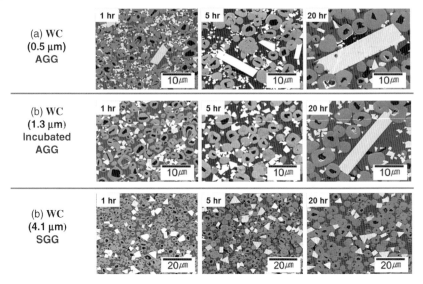

Figure 12.4 Microstructural evolution in 70(25TiC-75WC)-30Co (wt%) samples sintered at 1450 °C for various time periods [21]. Rounded gray grains: (Ti,W)C grains with a TiC core (black spot); faceted white grains: WC grains; dark region: liquid matrix.

[8–27,78,79] with changing initial particle size [9,18,21], changing temperature [19], dopant addition [10–15,17,20,24,26,79], and changing oxygen partial pressure [13,16,18,25,27]. All the observed, grain growth behaviors could be well explained using the concept of the coupling effect of Δg_c and Δg_{max}. The generality of the proposed principle of microstructural evolution has thus been experimentally supported.

Probably, the most compelling support for the applicability of this principle can be found in a previous study for the TiC–WC–Co system [21]. This system is an ideal model system that demonstrates the dependency of grain growth behavior on the shape of grains in a liquid matrix. In the same liquid matrix, two types of grain shape coexist: rounded (Ti, W)C grains and faceted WC grains, as shown in Figure 12.4. The growth of (Ti, W)C grains follows the cubic law, showing normal grain growth behavior. In contrast, the growth of WC grains exhibits nonstationary behavior. As Δg_{max} decreases with an increase of the initial WC grain size, the growth behavior changes from abnormal and incubated abnormal to stagnant, in agreement with the predictions from the principle of microstructural evolution [7,47,77].

12.4
Grain Growth in Solid-State Single-Phase Systems

12.4.1
Migration Mechanisms and Kinetics of the Grain Boundary

Grain growth in a single-phase polycrystal is a collective result of the migration of grain boundaries. Since the movements of grain boundaries are correlated with

12.4 Grain Growth in Solid-State Single-Phase Systems

other grain boundary networks, it is difficult to clarify the mechanism of boundary migration. An individual grain does not grow in an average or constant environment. Adjacent grains share common boundaries resulting in an ensemble that is topologically connected. In addition, the characteristics of atomic jumps between grains will result in different types of grain growth behavior, that is, stationary and nonstationary grain growth.

Equilibrium at the grain boundary junctions requires certain conditions to be satisfied for the angles at which the three boundaries intersect [54,55]. For simplicity, if all grain boundaries in a polycrystal are assumed to have the same grain boundary energy independent of grain boundary orientation, each of the grain boundaries makes an angle of 120° at the triple junction. Similarly, the angle at the four-grain corner should be 109.47°, even though the faces and edges can be curved in various ways [80]. This was conjectured by Plateau (called Plateau's rule), and had been proven to be true [81]. The atoms on both sides of a curved grain boundary are under different pressures across the boundary. If the atoms are in local equilibrium, the pressure difference is

$$\Delta P = \gamma_b \left(\frac{1}{R_1} + \frac{1}{R_2} \right) \tag{12.28}$$

where γ_b is the grain boundary energy and R_1 and R_2 are the two principal radii of curvature of the grain boundary. The effect of the pressure difference caused by a curved boundary is to create a difference in free energy Δg_v that drives the atoms across the boundary. The atoms in the shrinking grain detach themselves from the lattice on the high-pressure side of the boundary and relocate themselves on a lattice site on the low-pressure side of the boundary (the growing grain).

In order for an atom to be able to break away from a grain, it must acquire an activation energy by thermal activation. The effective flux of atoms J between grains is defined as in the form of

$$J = n^* n_0 \nu \exp\left(-\frac{\Delta g_m}{kT}\right) \frac{\Delta g_v}{kT} \tag{12.29}$$

where n^* is the number of atoms at the grain boundary per unit area, n_0 is the number density of atoms in the grain, ν is the vibration frequency, and Δg_m is the activation energy for jumping across the grain boundary [82,83].

Therefore, the boundary velocity $v = J \cdot (V_m/N_A)$ is expressed as

$$v = \frac{n^* n_0 \omega}{kT} \nu \exp\left(-\frac{\Delta g_m}{kT}\right) \cdot \Delta g_v = M \cdot \Delta g_v = M \cdot \gamma_b \left(\frac{1}{R_1} + \frac{1}{R_2} \right) \tag{12.30}$$

where ω is atomic volume and M the mobility of the boundary. If the boundary velocity is expressed in terms of the absolute mobility $B(=D/kT)$, and the chemical potential gradient across the boundary (driving force), Eq. (12.30) is rewritten as [6,84]

$$v = B \cdot F = B \cdot \gamma_b \left(\frac{1}{R_1} + \frac{1}{R_2} \right) \frac{\omega}{\delta} \tag{12.31}$$

where δ is the boundary thickness. This equation is the commonly used migration rate of the boundary for continuous growth kinetics. Both Eqs. (12.30) and (12.31) indicate that v is proportional to the capillary driving force. If the effect of the triple and quadruple junction energies is ignored – this condition is satisfied unless the grain size is less than a few tens of nanometers [85,86] – the well-known parabolic (or square) law is deduced from Eq. (12.30) or (12.31).

$$\bar{r}_t^2 - \bar{r}_0^2 = KT \tag{12.32}$$

where K is the proportionality constant.

For an impure as well as a pure system, it has been assumed that the intrinsic boundary mobility is constant irrespective of the driving force. This means that diffusion of atoms across the boundary governs the migration kinetics. The only difference between the cases of pure and impure systems is the reduction of the driving force by impurity drag at the boundary [6,87]. The apparent mobility in Eq. (12.31) is reduced, but the mechanism of the boundary migration was assumed to be the same. The effect of second-phase particles also appears as a reduction of the driving force for boundary migration due to the presence of the particles at the boundary [6,83,88].

Considering the migration mechanism of a faceted solid–liquid interface, the migration mechanism of a faceted (atomically ordered) boundary is expected to be different from that of a curved (atomically disordered) boundary. Based on the TEM observation that the grain boundaries of an Al–Cu alloy with {111} plane steps underwent spiral growth, Gleiter [89,90] suggested the step growth mechanism as the migration mechanism of grain boundaries. He also derived a nonlinear kinetic equation, adopting a similar procedure to that of Burton et al. for the growth of a single crystal in a matrix [42]. The equation predicts that under a low driving force, the migration rate is proportional to $(\Delta g_v)^2$ for a boundary with a low density of steps and to Δg_v for a boundary with a high density of steps.

The step growth process of grain boundaries has been observed in several high resolution TEM studies [91–94]. Kizuka [91] observed that the migration of a [210] tilt grain boundary in ZnO and the wurzite–zincblende phase boundary proceeded by the shifting of ledges. Song [92] observed periodic boundary structures in alternating symmetric–asymmetric segments along with atomic level facets in a high angle (Σ51b) CSL grain boundary. Merkle et al. [93,94] observed atomic shuffling (boundary movement in a jerky fashion) at the boundary formed between (001) and (011) Au bi-crystals. These observations are strong experimental evidence for the step growth mechanism (or atom shuffle mechanism) of a faceted boundary. The migration of a faceted boundary is expected to be nonlinear with respect to the driving force as the step growth mechanism is phenomenologically similar to the interface reaction mechanisms, in particular the 2DN mechanism, of a faceted solid–liquid interface.

Recently, the nonlinear migration of a faceted boundary has been observed in a model experiment using bi-layer samples of single crystals and polycrystals with different grain sizes [95]. Figure 12.5 plots the measured growth distances of {100}

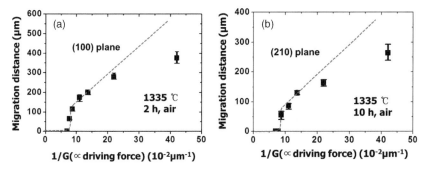

Figure 12.5 Plots of the average migration distances of BaTiO$_3$ single crystals into BaTiO$_3$ polycrystals of different average grain sizes as a function of the inverse of the average grain size, which is proportional to the driving force [95]. (a) {100} crystals after 2 h of air annealing and (b) {210} crystals after 10 h of air annealing at 1335 °C.

and {210} BaTiO$_3$ single crystals into BaTiO$_3$ polycrystals as a function of the inverse of the average grain size, which is proportional to the driving force, after annealing the bi-layer samples at 1335 °C for 2 and 10 h, respectively. Note that the average grain size of the polycrystals was fixed by a prior heat treatment and did essentially not change during single crystal growth. The measurement demonstrates the presence of a critical driving force for appreciable migration of the boundary: essentially no migration and appreciable migration below and above the critical driving force, respectively. Above the critical driving force, the migration distance deviates from linearity (the dotted line) most probably due to the accumulation of solutes at the migrating boundary. These experimental results demonstrate the nonlinear migration behavior of a faceted boundary, i.e. the mixed control of the boundary migration, similar to that of a faceted solid–liquid interface. It is, therefore, concluded that the principle of microstructural evolution for two-phase systems is applicable to understanding and predicting the microstructural evolution in single-phase systems as well.

12.4.2
Grain Growth Behavior

Many experimental studies on grain growth in metallic as well as ceramic single-phase systems show a correlation between grain growth behavior and grain boundary structure: NGG for rough boundaries and AGG or SGG for faceted boundaries [18,28–38,40,41]. The AGG or SGG in systems with faceted boundaries has been well explained by the coupling effect between the critical driving force for appreciable growth (Δg_c) and the maximum driving force for grain growth (Δg_{max}) (the principle of microstructural evolution), similar to that in a system with solid–liquid interfaces, AGG occurs for $\Delta g_{max} \geq \Delta g_c$, while SGG occurs for $\Delta g_{max} \ll \Delta g_c$.

The critical driving force Δg_c can vary with changing temperature, dopant addition, and changing atmosphere, in particular oxygen partial pressure [28,37,38,41], as in the case of a solid–liquid interface. With the reduction of Δg_c the boundary tends to become rough. With roughening of the boundary, that is, with the reduction of Δg_c, the grain growth behavior was observed to change from AGG to NGG (PNGG) in many systems, such as Al_2O_3 with MgO addition [31], $BaTiO_3$ with oxygen partial pressure change and donor doping [18,28,37,38,41], Ni-base superalloy with temperature increase [29], and Ag with temperature increase [33]. The roughening transition of faceted boundaries with temperature increase is easily understandable as a result of the increase in thermal vacancy concentration, and hence an increased entropy contribution. The roughening transition with doping and oxygen partial pressure change has also been observed to be related to an increase in total vacancy concentration [41]. Theoretical calculations also predict the roughening transition of the boundary with an increased total vacancy concentration in the bulk [96,97].

Various types of microstructural evolution can appear depending on the variation of Δg_c with respect to Δg_{max}, as in the solid–liquid two-phase systems. An example can be found in a study on the microstructural evolution in $BaTiO_3$ at 1250 °C under different oxygen partial pressures [18]. As the oxygen partial pressure (P_{O_2}) decreased from 0.2 to 10^{-19} atm, the fraction of faceted boundaries decreased from 100% to 10%, showing a reduction in Δg_c. Various types of microstructures were obtained after sintering for 50 h, as shown in Figure 12.6. At $P_{O_2} = 0.2$ atm, {111} twins formed [34] and {111} twin-assisted abnormal grain growth took place. The shape of the abnormal grains is elongated along {111} planes. The abnormal grains contained {111} twins while {111} twins were absent in the fine matrix

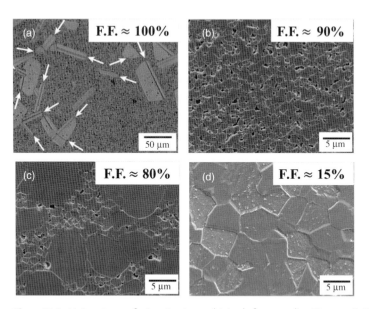

Figure 12.6 Various types of microstructures obtained after annealing Ti-excess $BaTiO_3$ powder compacts at 1250 °C for 10 h (a) and for 50 h ((b)–(d)) under different oxygen partial pressures: (a) 0.2 atm, (b) 2×10^{-17} atm, (c) 8×10^{-18} atm, and (d) 4×10^{-19} atm [38]. The percentage of faceted boundaries (FF) is also noted in each micrograph.

grains. This means that Δg_c decreased to below Δg_{max} when {111} twins were present. At $P_{O_2} = 2 \times 10^{-17}$ atm {111} twins did not form and grain growth was suppressed (SGG), as shown in Figure 12.6b. This result implies that the reduction of Δg_c with the P_{O_2} reduction is not as large as that caused by the presence of {111} twins. Further reduction of P_{O_2} to 8×10^{-18} atm, however, induced the formation of abnormal grains without {111} twins, as shown in Figure 12.6c. The shape of abnormal grains in this sample, however, is equiaxed because of the absence of {111} twins, in contrast to that in the sample at $P_{O_2} = 0.2$ atm. As most of the boundaries were roughened at $P_{O_2} = 4 \times 10^{-19}$ atm, the growth behavior was quite normal (Figure 12.6d). These results demonstrate that with the reduction of Δg_c for a given system, the growth behavior can change from SGG and AGG to PNGG and NGG, following the prediction of the principle of microstructural evolution.

12.5 Concluding Remarks

The kinetics of grain growth in both single-phase and two-phase (solid–liquid) polycrystals has been described in terms of the structure of the interfaces. Both solid–solid interfaces (grain boundaries) and solid–liquid interfaces can be rough or faceted on an atomic scale. For rough interfaces, grain growth is limited by diffusion across the interface from the shrinking to the growing grain. The grain growth rate increases linearly with the driving force for grain growth and normal grain growth (NGG) occurs. For faceted interfaces, grain growth is limited by interface reactions taking place on the surface of the growing grain, showing a nonlinear dependence of growth rate on the driving force, if the driving force is smaller than a critical value. These interface reactions include 2D nucleation and growth and spiral growth around screw dislocations for solid–liquid interfaces, and step growth for grain boundaries, which might occur via 2D nucleation and growth or defect-assisted growth. The mixed control model of grain growth (interface reaction and diffusion) has been introduced and the general principle of microstructural evolution has been developed. The principle states that the growth behavior in a polycrystal is a result of the coupling effect between the critical driving force for appreciable boundary migration Δg_c and the maximum driving force for growth Δg_{max}. Different types of grain growth, normal grain growth (NGG), pseudonormal grain growth (PNGG), abnormal grain growth (AGG), and stagnant grain growth (SGG) can take place depending on the relative values of Δg_c and Δg_{max}. A variety of grain growth behaviors observed in single- and two-phase polycrystals in both ceramic and metallic systems supports the generality of the developed model and the principle.

Acknowledgment

This work was supported by a National Research Foundation of Korea (NRF) grant funded by the Korean government (MEST) (No. 2012-0005707) and Samsung

Electro-Mechanics Co. Ltd. through the Center for Advanced MLCC Manufacturing Processes.

References

1 Lifshitz, I.M. and Slyozov, V.V. (1961) The kinetics of precipitation from supersaturated solid solutions. *Journal of Physics and Chemistry of Solids*, **19**, 35–50.
2 Wagner, C. and Elektrochem, Z. (1961) Theorie der alterung von Niederschlägen durch Umlösen (ostwald-reifung). *Zeitschrift für Elektrochemie, Berichte der Bunsengesellschaft für physikalische Chemie*, **65**, 581–591.
3 Ardell, A.J. (1972) The effect of volume fraction on particle coarsening: theoretical considerations. *Acta Metallurgica*, **20**, 61–71.
4 Atkinson, H.V. (1988) Theories of normal grain growth in pure single phase systems. *Acta Metallurgica*, **36**, 469–491.
5 Wynblatt, P. and Gjostein, N.A. (1976) Particle growth in model supported metal catalysts: I. Theory. *Acta Metallurgica*, **24**, 1165–1174.
6 Kang, S.-J.L. (2005) *Sintering: Densification, Grain Growth, and Microstructure*, Elsevier Butterworth-Heinemann, Oxford.
7 Jung, Y.I., Yoon, D.Y., and Kang, S.-J.L. (2009) Coarsening of polyhedral grains in a liquid matrix. *Journal of Materials Research*, **24**, 2949–2959.
8 Kang, S.-J.L. and Han, S.M. (1995) Grain growth in Si_3N_4-based materials. *MRS Bulletin*, **20**, 33–37.
9 Park, Y.J., Hwang, N.M., and Yoon, D.Y. (1996) Abnormal growth of faceted (WC) grains in a (Co) liquid matrix. *Metallurgical and Materials Transactions A*, **27**, 2809–2819.
10 Yoo, Y.S., Kim, H., and Kim, D.Y. (1997) Effect of SiO_2 and TiO_2 addition on the exaggerated grain growth of $BaTiO_3$. *Journal of the European Ceramic Society*, **17**, 805–811.
11 Oh, K.S., Jun, J.Y., Kim, D.Y., and Hwang, N.M. (2000) Shape dependence of the coarsening behavior of niobium carbide grains dispersed in a liquid iron matrix. *Journal of the American Ceramic Society*, **83**, 3117–3120.
12 Choi, K., Hwang, N.M., and Kim, D.Y. (2002) Effect of grain shape on abnormal grain growth in liquid-phase-sintered $Nb_{1-x}Ti_xC$–Co alloys. *Journal of the American Ceramic Society*, **85**, 2313–2318.
13 Chung, S.Y., Yoon, D.Y., and Kang, S.-J.L. (2002) Effects of donor concentration and oxygen partial pressure on interface morphology and grain growth behavior in $SrTiO_3$. *Acta Materialia*, **50**, 3361–3371.
14 Park, C.W. and Yoon, D.Y. (2002) Abnormal grain growth in alumina with anorthite liquid and the effect of MgO addition. *Journal of the American Ceramic Society*, **85**, 1585–1593.
15 Wallace, J.S., Huh, J.M., Blendell, J.E., and Handwerker, C.A. (2002) Grain growth and twin formation in 0.74PMN-0.26PT. *Journal of the American Ceramic Society*, **85**, 1581–1584.
16 Jang, C.W., Kim, J., and Kang, S.-J.L. (2002) Effect of sintering atmosphere on grain shape and grain growth in liquid-phase-sintered silicon carbide. *Journal of the American Ceramic Society*, **85**, 1281–1284.
17 Lee, H.R., Kim, D.J., Hwang, N.M., and Kim, D.Y. (2003) Role of vanadium carbide additive during sintering of WC-CO: Mechanism of grain growth inhibition. *Journal of the American Ceramic Society*, **86**, 152–154.
18 Jung, Y.I., Choi, S.Y., and Kang, S.-J.L. (2003) Grain-growth behavior during stepwise sintering of barium titanate in hydrogen gas and air. *Journal of the American Ceramic Society*, **86**, 2228–2230.
19 Cho, Y.K., Yoon, D.Y., and Kim, B.K. (2004) Surface roughening transition and coarsening of NbC grains in liquid cobalt-rich matrix. *Journal of the American Ceramic Society*, **87**, 443–448.
20 Fisher, J.G., Kim, M.S., Lee, H.Y., and Kang, S.-J.L. (2004) Effect of Li_2O and PbO additions on abnormal grain growth in the $Pb(Mg_{1/3}Nb_{2/3})O_3$-35 mol% $PbTiO_3$

system. *Journal of the American Ceramic Society*, **87**, 937–942.

21 Yoon, B.K., Lee, B.A., and Kang, S.-J.L. (2005) Growth behavior of rounded (Ti,W)C and faceted WC grains in a Co matrix during liquid-phase sintering. *Acta Materialia*, **53**, 4677–4685.

22 Sano, T. and Rohrer, G.S. (2007) Experimental evidence for the development of bimodal grain size distributions by the nucleation-limited coarsening mechanism. *Journal of the American Ceramic Society*, **90**, 211–216.

23 Moon, K.S. and Kang, S.-J.L. (2008) Coarsening behavior of round-edged cubic grains in the $Na_{1/2}Bi_{1/2}TiO_3$–$BaTiO_3$ system. *Journal of the American Ceramic Society*, **91**, 3191–3196.

24 Motohashi, T. and Kimura, T. (2008) Formation of homo-template grains in $Bi_{0.5}Na_{0.5}TiO_3$ prepared by the reactive-templated grain growth process. *Journal of the American Ceramic Society*, **91**, 3889–3895.

25 Heo, Y.H., Jeon, S.C., Fisher, J.G., Choi, S.Y., Hur, K.H., and Kang, S.-J.L. (2011) Effect of step free energy on delayed abnormal grain growth in a liquid phase-sintered $BaTiO_3$ model system. *Journal of the European Ceramic Society*, **31**, 755–762.

26 Moon, K.S., Rout, D., Lee, H.Y., and Kang, S.-J.L. (2011) Effect of TiO_2 addition on grain shape and grain coarsening behavior in $95Na_{1/2}Bi_{1/2}TiO_3$–$5BaTiO_3$. *Journal of the European Ceramic Society*, **31**, 1915–1920.

27 Fisher, J.G., Choi, S.Y., and Kang, S.-J.L. (2011) Influence of sintering atmosphere on abnormal grain growth behaviour in potassium sodium niobate ceramics sintered at low temperature. *Journal of the Korean Ceramic Society*, **48**, 641–647.

28 Lee, B.K., Chung, S.Y., and Kang, S.-J.L. (2000) Grain boundary faceting and abnormal grain growth in $BaTiO_3$. *Acta Materialia*, **48**, 1575–1580.

29 Lee, S.B., Yoon, D.Y., and Henry, M.F. (2000) Abnormal grain growth and grain boundary faceting in a model Ni-base superalloy. *Acta Materialia*, **48**, 3071–3081.

30 Lee, S.B., Hwang, N.M., Yoon, D.Y., and Henry, M.F. (2000) Grain boundary faceting and abnormal grain growth in nickel. *Metallurgical and Materials Transactions A: Physical Metallurgy and Materials Science*, **31**, 985–994.

31 Park, C.W. and Yoon, D.Y. (2000) Effects of SiO_2, CaO_2, and MgO additions on the grain growth of alumina. *Journal of the American Ceramic Society*, **83**, 2605–2609.

32 Koo, J.B. and Yoon, D.Y. (2001) Abnormal grain growth in bulk Cu - The dependence on initial grain size and annealing temperature. *Metallurgical and Materials Transactions A: Physical Metallurgy and Materials Science*, **32**, 1911–1926.

33 Koo, J.B. and Yoon, D.Y. (2001) The dependence of normal and abnormal grain growth in silver on annealing temperature and atmosphere. *Metallurgical and Materials Transactions A: Physical Metallurgy and Materials Science*, **32**, 469–475.

34 Lee, B.K. and Kang, S.-J.L. (2001) Second-phase assisted formation of {111} twins in barium titanate. *Acta Materialia*, **49**, 1373–1381.

35 Park, C.W., Yoon, D.Y., Blendell, J.E., and Handwerker, C.A. (2003) Singular grain boundaries in alumina and their roughening transition. *Journal of the American Ceramic Society*, **86**, 603–611.

36 Cho, Y.K., Kang, S.-J.L., and Yoon, D.Y. (2004) Dependence of grain growth and grain-boundary structure on the Ba/Ti ratio in $BaTiO_3$. *Journal of the American Ceramic Society*, **87**, 119–124.

37 Choi, S.Y. and Kang, S.-J.L. (2004) Sintering kinetics by structural transition at grain boundaries in barium titanate. *Acta Materialia*, **52**, 2937–2943.

38 Jung, Y.I., Choi, S.Y., and Kang, S.-J.L. (2006) Effect of oxygen partial pressure on grain boundary structure and grain growth behavior in $BaTiO_3$. *Acta Materialia*, **54**, 2849–2855.

39 Holm, E.A. and Foiles, S.M. (2010) How grain growth stops: A mechanism for grain-growth stagnation in pure materials. *Science*, **328**, 1138–1141.

40 Bäurer, M., Shih, S.J., Bishop, C., Harmer, M.P., Cockayne, D., and Hoffmann, M.J. (2010) Abnormal grain growth in undoped strontium and barium titanate. *Acta Materialia*, **58**, 290–300.

41 An, S.M. and Kang, S.-J.L. (2011) Boundary structural transition and grain growth

behavior in BaTiO$_3$ with Nd$_2$O$_3$ doping and oxygen partial pressure change. *Acta Materialia*, **59**, 1964–1973.

42 Burton, W.K., Cabrera, N., and Frank, F.C. (1951) The growth of crystals and the equilibrium structure of their surfaces. *Philosophical Transactions of the Royal Society of London Series A, Mathematical and Physical Sciences*, **243**, 299–358.

43 Hirth, J.P. and Pound, G.M. (1963) *Condensation and Evaporation, Nucleation and Growth Kinetics*, Pergamon Press, Oxford, pp. 77–148.

44 Howe, J.M. (1997) *Interfaces in Materials, Atomic Structure, Thermodynamics and Kinetics of Solid-Vapor, Solid-Liquid and Solid-Solid Interfaces*, John Wiley & Sons, Inc., New York, pp. 256–268.

45 Yoon, D.Y., Park, C.W., and Koo, J.B. (2001) The step growth hypothesis for abnormal grain growth, in *Ceramic Interfaces* (eds H. Yoo and S.-J.L. Kang), Institute of Materials, London, **2**, pp. 3–21.

46 Jo, W., Kim, D.Y., and Hwang, N.M. (2006) Effect of interface structure on the microstructural evolution of ceramics. *Journal of the American Ceramic Society*, **89**, 2369–2380.

47 Kang, S.-J.L., Lee, M.G., and An, S.M. (2009) Microstructural evolution during sintering with control of the interface structure. *Journal of the American Ceramic Society*, **92**, 1464–1471.

48 Wulff, G. (1901) On the question of speed of growth and dissolution of crystal surfaces. *Zeitschrift Fur Krystallographie Und Mineralogie*, **34**, 449–530.

49 Rottman, C. and Wortis, M. (1984) Equilibrium crystal shapes for lattice models with nearest-and next-nearest-neighbor interactions. *Physical Review B*, **29**, 328–339.

50 Kosterlitz, J.M. and Thouless, D.J. (1973) Ordering, metastability and phase transitions in two-dimensional systems. *Journal of Physics C: Solid State Physics*, **6**, 1181–1203.

51 Kosterlitz, J.M. (1974) The critical properties of the two-dimensional xy model. *Journal of Physics C: Solid State Physics*, **7**, 1046–1060.

52 Beijeren, H. (1977) Exactly solvable model for the roughening transition of a crystal surface. *Physical Review Letters*, **38**, 993–996.

53 Beijeren, H. and Nolden, I. (1987) The roughening transition. Structure and dynamics of surfaces II, in *Topics in Current Physics* (eds W. Schommers and P. Blanckenhagen), Springer, Berlin, **43**, p. 259.

54 Hoffman F D.W. and Cahn, J.W. (1972) A vector thermodynamics for anisotropic surfaces: I. Fundamentals and application to plane surface junctions. *Surface Science*, **31**, 368–388.

55 Cahn, J.W. and Hoffman, D.W. (1974) A vector thermodynamics for anisotropic surfaces: II. Curved and faceted surfaces. *Acta Metallurgica*, **22**, 1205–1214.

56 Cho, Y.K. (2003) Interface Roughening Transition and Grain Growth in BaTiO3 and NbC-Co, Ph. D. thesis, KAIST, Daejeon, Republic of Korea, 149–200.

57 Peteves, S.D. and Abbaschian, R. (1991) Growth kinetics of solid–liquid Ga interfaces: Part II. Theoretical. *Metallurgical Transactions A*, **22**, 1271–1286.

58 Bennema, P. (1984) Spiral growth and surface roughening: developments since Burton, Cabrera and Frank. *Journal of Crystal Growth*, **69**, 182–197.

59 Kang, M.K., Yoo, Y.S., Kim, D.Y., and Hwang, N.M. (2000) Growth of BaTiO3 seed grains by the twin-plane reentrant edge mechanism. *Journal of the American Ceramic Society*, **83**, 385–390.

60 Jung, Y.I. (2006) Effect of Grain Boundary Structure on Grain Growth in BaTiO3 below the Eutectic Temperature, Ph.D. thesis, KAIST, Deajeon, Republic of Korea, 48–56.

61 Sheldon, B.W. and Rankin, J. (2002) Step-energy barriers and particle shape changes during coarsening. *Journal of the American Ceramic Society*, **85**, 683–690.

62 Cahn, J.W., Hillig, W.B., and Sears, G.W. (1964) The molecular mechanism of solidification. *Acta Metallurgica*, **12**, 1421–1439.

63 Peteves, S.D. and Abbaschian, R. (1991) Growth kinetics of solid–liquid Ga interfaces: Part I. Experimental. *Metallurgical Transactions A*, **22**, 1259–1270.

64 Brailsford, A.D. and Wynblatt, P. (1979) The dependence of ostwald ripening

kinetics on particle volume fraction. *Acta Metallurgica*, **27**, 489–497.

65 Davies, C.K.L., Nash, P., and Stevens, R.N. (1980) The effect of volume fraction of precipitate on Ostwald ripening. *Acta Metallurgica*, **28**, 179–189.

66 Tsumuraya, K. and Miyata, Y. (1983) Coarsening models incorporating both diffusion geometry and volume fraction of particles. *Acta Metallurgica*, **31**, 437–452.

67 Voorhees, P.W. and Glicksman, M.E. (1984) Ostwald ripening during liquid-phase sintering effect of volume fraction on coarsening kinetics. *Metallurgical Transactions A, Physical Metallurgy and Materials Science*, **15**, 1081–1088.

68 Marsh, S.P. and Glicksman, M.E. (1996) Kinetics of phase coarsening in dense systems. *Acta Materialia*, **44**, 3761–3771.

69 German, R.M. and Olevsky, E.A. (1998) Modeling grain growth dependence on the liquid content in liquid-phase-sintered materials. *Metallurgical and Materials Transactions A: Physical Metallurgy and Materials Science*, **29**, 3057–3067.

70 Kang, C.H. and Yoon, D.Y. (1981) Coarsening of cobalt grains dispersed in liquid copper matrix. *Metallurgical Transactions A, Physical Metallurgy and Materials Science*, **12**, 65–69.

71 Chen, M.K. and Voorhees, P.W. (1993) The dynamics of transient Ostwald ripening. *Modelling and Simulation in Materials Science and Engineering*, **1**, 591–612.

72 Snyder, V.A., Alkemper, J., and Voorhees, P.W. (2000) The development of spatial correlations during Ostwald ripening: A test of theory. *Acta Materialia*, **48**, 2689–2701.

73 Snyder, V.A., Alkemper, J., and Voorhees, P.W. (2001) Transient Ostwald ripening and the disagreement between steady-state coarsening theory and experiment. *Acta Materialia*, **49**, 699–709.

74 Bennema, P. and Van Der Eerden, J.P. (1987) *Morphology of Crystals, Part A* (ed. I. Sunagawa), Terra Scientific Publishing Company, Tokyo, pp. 1–75.

75 Kang, M.K., Yoo, Y.S., Kim, D.Y., and Hwang, N.M. (2000) Growth of $BaTiO_3$ seed grains by the twin-plane reentrant edge mechanism. *Journal of the American Ceramic Society*, **83**, 385–390.

76 Kang, M.K., Kim, D.Y., and Hwang, N.M. (2002) Ostwald ripening kinetics of angular grains dispersed in a liquid phase by two-dimensional nucleation and abnormal grain growth. *Journal of the European Ceramic Society*, **22**, 603–612.

77 Kang, S.-J.L., Jung, Y.I., and Moon, K.S. (2007) Principles of microstructural design in two-phase systems. *Materials Science Forum*, **558–559**, 827–834.

78 Schreiner, M., Schmitt, T., Lassner, E., and Lux, B. (1984) On the origins of discontinuous grain growth during liquid phase sintering of WC-Co cemented carbides. *Powder Metallurgy International*, **16**, 180–183.

79 Bateman, C.A., Bennison, S.J., and Harmer, M.P. (1989) Mechanism for the role of magnesia in the sintering of alumina containing small amounts of a liquid phase. *Journal of the American Ceramic Society*, **72**, 1241–1244.

80 Smith, C.S. (1964) Some elementary principles of polycrystalline microstructure. *Metallurgical Reviews*, **9**, 1–48.

81 Almgren, F.J. and Taylor, J.E. (1976) The geometry of soap films and soap bubbles. *Scientific American*, **235**, 82–93.

82 Humphreys, F.J. and Hatherly, M. (1996) *Recrystallization and Related Annealing Phenomena*, Elsevier, Oxford, pp. 114–120.

83 Gottstein, G. and Shvindlerman, L.S. (1999) *Grain Boundary Migration in Metals: Thermodynamics, Kinetics, Applications*, Taylor & Francis, Florida, pp. 125–133.

84 Rahaman, M.N. (2008) *Sintering of Ceramics*, CRC Press, Boca Raton, FL, pp. 119–121.

85 Gottstein, G. and Shvindlerman, L.S. (2006) Grain boundary junction engineering. *Scripta Materialia*, **54**, 1065–1070.

86 Streitenberger, P. and Zöllner, D. (2011) Evolution equations and size distributions in nanocrystalline grain growth. *Acta Materialia*, **59**, 4235–4243.

87 Cahn, J.W. (1962) The impurity-drag effect in grain boundary motion. *Acta Metallurgica*, **10**, 789–798.

88 Manohar, P.A., Ferry, M., and Chandra, T. (1998) Five decades of the Zener equation. *ISIJ International*, **38**, 913–924.

89 Gleiter, H. (1969) The mechanism of grain boundary migration. *Acta Metallurgica*, **17**, 565–573.

90 Gleiter, H. (1969) Theory of grain boundary migration rate. *Acta Metallurgica*, **17**, 853–862.

91 Kizuka, T. (1999) Atomic processes of grain-boundary migration and phase transformation in zinc oxide nanocrystallites. *Philosophical Magazine Letters*, **79**, 417–422.

92 Song, S.G. (1999) In situ high-resolution electron microscopy observation of grain-boundary migration through ledge motion in an Al–Mg alloy. *Philosophical Magazine Letters*, **79**, 511–517.

93 Merkle, K.L. and Thompson, L.J. (2001) Atomic-scale observation of grain boundary motion. *Materials Letters*, **48**, 188–193.

94 Merkle, K.L., Thompson, L.J., and Phillipp, F. (2002) Collective effects in grain boundary migration. *Physical Review Letters*, **88**, 225501.

95 An, S.M., Yoon, B.K., Chung, S.Y., and Kang, S.-J.L. (2012) Nonlinear driving force-velocity relationship for the migration of faceted boundaries. *Acta Materialia*, **60**, 4531–4539.

96 Imaeda, M., Mizoguchi, T., Sato, Y., Lee, H.S., Findlay, S.D., Shibata, N., Yamamoto, T., and Ikuhara, Y. (2008) Atomic structure, electronic structure, and defect energetics in [0 0 1](310)Σ5 grain boundaries of $SrTiO_3$ and $BaTiO_3$. *Physical Review B*, **78**, 245320.

97 Lee, H.S., Mizoguchi, T., Mistui, J., Yamamoto, T., Kang, S.-J.L., and Ikuhara, Y. (2011) Defect energetics in $SrTiO_3$ symmetric tilt grain boundaries. *Physical Review B*, **83**, 104110.

13
Capillary-Mediated Interface Energy Fields: Deterministic Dendritic Branching
Martin E. Glicksman

13.1
Introduction

Dendritic crystal growth is the ubiquitous mode of solidification in casting and fusion welding. The dense branching patterns of alloy dendrites establish important as-cast length scales governing microsegregation and microstructure evolution. Establishing a basic understanding of dendritic growth requires, per force, analysis of the branching mechanism and verification of the dendrite pattern evolution details.

Contemporary ideas explaining dendritic morphogenesis and growth kinetics are currently based on stochastic physics, including selective noise amplification and perturbation-induced morphological instability [1]. Two such theoretical approaches describing dendritic growth remain in general acceptance: (1) *marginal stability*, developed in the mid-1970s [2,3], which contends that dendrite tips grow at their limit of stability, perhaps controlled by a spectrum of noisy fluctuations that continually perturb the dendrite's tip and selectively amplify into a train of branches [4]; and (2) *microscopic solvability*, a theory developed in the 1980s that seeks discrete steady–state solutions for smooth (branchless) tips with anisotropic interfacial energy [5,6]. The stable dendrite solution in this theory is believed to be the discrete state with the maximum velocity [7].

One major study purporting to support noise-induced branching measured dendrites of NH_4Br growing in a thin cell containing a supersaturated aqueous solution [8]. Branch coarsening, which is induced by polydispersity in the dendritic length scales, was observed to influence *all* wave numbers in the recorded branching spectrum, thus making it difficult to ascertain whether the peak frequencies, centered near 0.1 Hz, and deemed "responsible for branching", were also simply corrupted by the subsequent coarsening.

Other studies show that noise-based theories fail at quantitative descriptions of dendritic growth kinetics in well-characterized nonfaceting materials, such as high-purity (7–9s) succinonitrile, a low-anisotropy BCC crystal, and pure (5–9s) pivalic acid anhydride, an anisotropic FCC crystal. For example, predictions based on solvability theory were checked in experiments conducted in the early 1990s by

Muschol et al., who found solvability incapable of predicting the correct dendritic scaling laws, growth kinetics, or branching patterns [9].

Additional quantitative checks were finally performed in the late 1990s. These efforts were specifically designed to test dendritic growth theories that were based on stochastic physics, namely marginal stability and solvability. Those experiments were conducted in microgravity over a period of several years and produced a large data archive based on results obtained from the first and second space flights of NASA's *Isothermal Dendritic Growth Experiment* (IDGE). A thorough review of these microgravity data was published recently by LaCombe et al., who concluded that some of the key theoretical predictions, which were based on either of these conventional noise-based theories of dendritic growth, could not be confirmed on the basis of the IDGE data [10].[1]

The third IDGE space flight, USMP-4, fortuitously also provided videos of the melting kinetics on which the initial ideas for this paper are based [11,12]. These videos led to later studies of conduction-limited melting in microgravity and stimulated a re-evaluation of dendritic growth based on deterministic (noise-free) processes [13,14]. The present chapter provides both a brief outline of these experimentally based observations of melting, from which the current ideas on noise and interfacial capillarity evolved, along with a concise, explicative summary of theoretical progress on capillary-mediated dendritic branching.

13.2
Capillary Energy Fields

13.2.1
Background

Capillarity ranks as a weak, second-order energy source in crystal–melt transformations, and is usually ignored when formulating the local Stefan energy balance for a moving interface. To help quantify this statement, consider the ratio of the interfacial energy density, $\gamma_{s\ell}$, to the volumetric latent heat, $\Delta H_f / \Omega_s$, where ΔH_f and Ω_s are the molar latent heat and solid volume, respectively. Their ratio defines an interface "capillary length",

$$\lambda_{\text{cap}} \equiv \frac{\Omega_s \gamma_{s\ell}}{\Delta H_f} \tag{13.1}$$

Given that volumetric latent heats for many materials are relatively large, so $\Delta H_f / \Omega_s \approx 10^9 \, \text{J/m}^3$, and that corresponding crystal–melt interfacial energy densities

---

[1] The IDGE comprised a series of microgravity solidification experiments flown by NASA during the period 1994–1997 aboard the space shuttle *Columbia*. Microgravity experiments known as the IDGE series were designed to observe and measure the kinetics and morphology of dendritic growth under pure-conduction heat transfer [15–20].

are small, so $\gamma_{s\ell} \approx 0.1 \, \text{J/m}^2$, one finds that the characteristic "capillary" length of ordinary materials to be circa $\lambda_{cap} \approx 0.1$ nm, which is a length scale far smaller than those in dendritic structures, which exhibit tip radii in the range $0.1 \leq R_{tip} \leq 100 \, \mu\text{m}$. Thus, the importance of considering capillary energies along with the large volume-based phase transformation energy released near a dendrite tip may be gauged by the dimensionless product $\lambda_{cap} \mathcal{H}_{tip}$. Here, $\mathcal{H}_{tip}$ is defined as the mean curvature of the dendrite tip, namely,

$$\mathcal{H}_{tip} = \frac{1}{R_{tip}} = \frac{1}{2}\left(\frac{\partial A}{\partial V}\right) \qquad (13.2)$$

which provides a measure of the change of the local $\ell \rightarrow s$ (liquid-to-solid) volume transformation, V, to the corresponding change in interface area change, A. Based on the scale of dendritic tip radii, R_{tip}, the energetic contribution from capillarity would not be expected to exceed about one part per thousand of the total. Not surprisingly, therefore, capillary energies are ignored in conventional solidification theories as far as their sensibly contributing to the total transformation energy. An excellent review of the fundamental conservation laws, including capillary effects considered here, for solidification interfaces evolving under nonequilibrium conditions may be found in [21].

13.2.2
Melting Experiments

The first instance, to the author's knowledge, when a solid–liquid phase transformation proceeding under rigorous thermal control was observed to be influenced quantitatively, by what eventually proved to be the addition of energy derived from capillarity, occurred during the third IDGE space flight in late 1997. Video data show pivalic anhydride (PVA) crystallites *melting* under microgravity conditions and reveal what proved to be an extremely interesting and important phenomenon [12].[2)]

Figure 13.1 shows several video frames (captured at 30 fps) of PVA dendrites melting progressively. The distant melt remains heated about 1.8 K above PVA's melting point (35.9 °C). When a properly oriented crystallite (long axis in the plane-of-focus) became sufficiently isolated to avoid thermal interactions with neighbors, its axial (C/A) ratio was measured, frame-by-frame, as melting progressed toward total extinction (see Figure 13.1a).

---

2) At the termination of each of the more than 100 cycles of growing PVA dendritic crystals during the third IDGE, the dendrites produced during the prior growth cycle had to be destroyed by melting. The melting that completed each experimental cycle was required to allow initializing the subsequent cycle with a melt totally devoid of any remaining crystals. Video recordings of both freezing and melting in microgravity were taken and transmitted back to Earth for analysis via the Space Shuttle's K-band antenna.

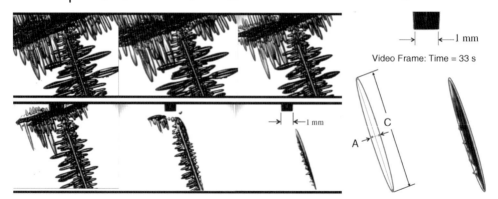

Figure 13.1 (a) Progressive melting of PVA dendrites in microgravity. Video frames from Cycle 04, IDGE, and USMP-4, *Columbia*. Individual crystallites remain motionless in the melt, despite the density difference between the solid and liquid phases. (b) Measurement of the C/A ratio of an isolated ellipsoidal crystallite of PVA captured in the lower-right frame of the adjacent video frame. Elliptical boundary and the crystallite's axial ratio are drawn and measured automatically using image analysis software. Crystallites selected for shape analysis remain motionless in the melt and must be oriented within the plane of focus, allowing true length measurements.

13.2.3
Self-Similar Melting

The potential theory [23] describing quasistatic melting of ellipsoidal needle crystals shows that with constant far-field superheating of the melt, melting proceeds self-similarly [11]. Self-similar melting occurs in this instance because the heat flux entering the poles, Φ_{pole}, of an ellipsoidal crystal with an axial ratio of C/A is itself exactly C/A times the heat flux entering the equator, Φ_{eq}. This result – that is, concentrating the heat flux from the melt at the crystal's poles – is called the "lightning rod" effect of electrostatics, or, in this instance, heat conduction, which may be expressed as

$$\frac{\Phi_{pole}}{\Phi_{eq}} = C/A \tag{13.3}$$

Self-similar melting, described by Eq. (13.3), is a phase transformation process that remains independent of the crystal's size, or volume, and should continue over time until extinction occurs, and the crystal completely disappears.[3] Experimental proof of observing self-similar melting therefore consists of showing that the axial ratio, C/A, remains constant during melting.

Axial ratio measurements versus melting time data for PVA crystallites were obtained from the IDGE videos. Figure 13.2 shows what one observes. These data

---

3) Self-similarity during melting to full extinction also assumes that the ellipsoidal shape remains compact over time, and that the remaining melting time is too short to allow development of a Rayleigh-type unduloidal instability that leads to pinch-off.

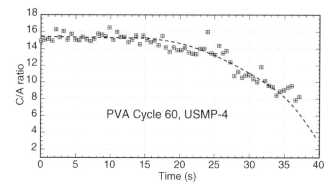

Figure 13.2 Quasistatic melting of an ellipsoidal PVA crystallite observed in microgravity. Axial ratio data extracted from video frames captured in Cycle 60, IDGE, and USMP-4. The first 15 s of melting show a constant $C/A = 15.5 \pm 1$, confirming initial self-similar melting behavior of a relatively long ($C \geq 1$ cm) needle-shaped crystal. The axial ratio then decreases to $C/A \approx 9$ over the next 20 s as melting continues, then rapidly plummets toward unity as the shrinking crystallite approaches extinction in the form of a sphere.

indeed show self-similar melting occurring as the needle crystal melts initially and slowly decreases in size and volume from $C \approx 10$ mm, to about $C \approx 5$ mm. One would posit that as a crystal shrinks to smaller and smaller sizes, self-similarity should become even easier, at least in principle, as the shrinking crystallite becomes increasingly "isolated" from its surroundings. But, contrary to this reasonable hypothesis, one sees instead that the polar regions begin to melt faster than the equatorial regions, violating self-similarity, and causing the axial C/A ratio to fall at an increasing rate toward unity.

13.2.4
Influence of Capillarity on Melting

A computational model was used to check whether or not the accelerated melting observed at the poles of ellipsoidal needle crystals was caused by capillarity (S. Salon, 2005, Department of Electrical Engineering, Rensselaer Polytechnic Institute, Troy, NY, USA, private communication). The Laplace heat field surrounding, and within, an isolated, slowly melting ellipsoid was computed using an accurate finite difference code. Capillarity was introduced by imposing the (isotropic) Gibbs–Thomson temperature distribution over the ellipsoid's curved solid–melt interface. The thermal conductivities for the solid and melt phases were assumed to be equal in the computations, which is a good approximation for PVA. The isotropic form of the Gibbs–Thomson equilibrium temperature distribution at any location along the ellipsoidal interface, $\vec{r}$, is

$$T_e(\vec{r}) = T_m\left(1 - 2\lambda_{cap}\mathcal{H}_{int}(\vec{r})\right) \qquad (13.4)$$

where the mean curvature of the interface is $\mathcal{H}_{int} = \frac{1}{2}(1/R_{11} + 1/R_{22})$, and R_{11} and R_{22} are the local *principal* radii of curvature that diagonalize the local curvature matrix of the ellipsoidal interface. T_m is the thermodynamic melting point at a flat solid–liquid interface, where $\mathcal{H}_{int} = 0$; and the material's characteristic length, λ_{cap}, appearing in Eq. (13.4), is defined in Eq. (13.1).

If $\lambda_{cap} = 0$, that is, capillarity is absent in the computations, the calculated polar flux, Φ_{pole}, was found to be *precisely* C/A times larger than the equatorial flux, Φ_{eq}, in agreement with quasistatic potential theory. However, if $\lambda_{cap} \neq 0$, the ratio of the polar-to-equatorial fluxes rose above unity, corresponding to our experimental observation that C/A decreased toward unity as the volume diminished. See again Figure 13.2. Heat flux data derived from the finite difference melting code with capillarity imposed on ellipsoids with various C/A ratios are displayed in Figure 13.3. Although a relatively large (but observed) $C/A \approx 13$ would be required for the excess melting flux at the poles to reach 10% – that is, capillarity adding 10% more energy flux to the poles than does the surrounding hot melt alone – but, as explained in the next section, even much smaller capillary-mediated energies prove important in the case of freezing crystallites.

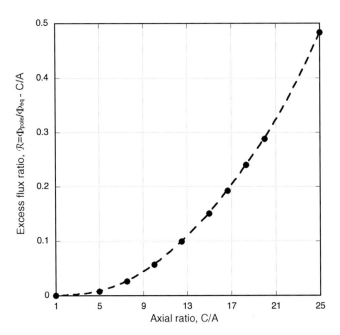

Figure 13.3 Excess polar-to-equatorial flux ratio, R, plotted as a function of the axial ratio, C/A. As a crystal's shape becomes increasingly anisotropic, or needle-like, the excess flux ratio contributed by capillarity increases rapidly. When, however, a melting crystal eventually becomes a sphere ($C/A = 1$), the excess polar flux is contributed by isotropic capillarity vanishes.

13.3
Capillarity-Mediated Branching

Observations of late-stage crystallite melting in microgravity support the idea of accelerated melting of ellipsoids, which is caused by additional capillary-mediated energies directed from the flatter (warmer) equatorial areas to the curved (cooler) polar areas. Initially stable, self-similar melting kinetics, by external heat conduction from the surrounding melt to larger ellipsoidal crystals (where interfacial capillarity remains sufficiently weak) eventually gives way to accelerated polar melting and falling C/A ratios, as crystallites shrink toward extinction as spheres.

Melting experiments such as described here suggest an interesting companion question: How might capillarity-mediated energy affects *freezing* interfaces such as occur in dendritic crystals?

13.3.1
Local Equilibrium

For both brevity and simplicity, we analyze here the equilibrium temperature distribution along an interface in two dimensions, with its momentary shape chosen as the semi-infinitely long parabola $y/a = \sqrt{1/2 - x/a}$, scaled in the Cartesian coordinates x and y by the length a. The parabola extends from $-\infty \leq x/a \leq 1/2$. Here the scale factor is chosen as $a = 2R_{\text{tip}}$, where R_{tip} is the in-plane radius of curvature of the parabola's tip. The (dimensionless) curvature of this parabolic interface, κ, at any arbitrary point on the interface, $x/a = \eta$, is given by

$$\kappa = \frac{2}{[2(1 - 2\eta) + 1]^{3/2}} \tag{13.5}$$

13.3.2
Gibbs–Thomson–Herring Interface Potential

In order to analyze the branching process in dendrites solidifying in a supercooled melt, the crystal–melt surface energy density, $\gamma_{s\ell}$, and the strength of its fourfold in-plane anisotropy, will be specified. Specifically,

$$\gamma_{s\ell} = \gamma_0 (1 + \varepsilon \cos 4\varphi) \tag{13.6}$$

where γ_0 is the average modulus of the energy density over all interface orientations, which are specified by the local normal angle, φ, along the parabolic interface. The dendrite tip normal orientation, $\varphi = 0$, is aligned with the $\eta = (x/a)$-axis. The amplitude, or strength, of the fourfold anisotropy is specified by $0 \leq \varepsilon \leq 1/15$.[4]

---

[4] Beyond an amplitude of $\varepsilon = 1/15$, the interfacial energy density becomes nonconvex, leading to facet formation, that is, missing orientations. This significant complication in the mechanism of dendritic branching will not be dealt with in this paper.

The normal angle along the parabola is the following function of the scaled interface coordinate η

$$\varphi = \pm \arctan\left[\left(-2\sqrt{\frac{1}{2}-\eta}\right)^{-1}\right] + \frac{\pi}{2} \tag{13.7}$$

with the plus sign for the interface in the upper half-plane, and the minus sign for the lower half-plane.

The local equilibrium temperature along an interface subject to anisotropic interfacial energy is given by the Gibbs–Thomson–Herring (GTH) equation [24,25]

$$T_{int} = T_m\left[1 - (\gamma_{s\ell} + \gamma_{\varphi,\varphi})\left(\frac{\Omega}{a\,\Delta H_f}\right)\kappa(\varphi)\right] \tag{13.8}$$

with $\gamma_{\varphi,\varphi}$ denoting the second angular derivative of $\gamma_{s\ell}$ with respect to the normal angle, φ, which provides the anisotropic "torque term." Inserting Eq. (13.6) into Eq. (13.8) and carrying out the indicated differentiations yields the local interfacial equilibrium temperature as

$$T_{int}(\varphi;\varepsilon) = T_m\left[1 - (1 - 15\varepsilon\cos 4\varphi)\left(\frac{\gamma_0\,\Omega}{a\,\Delta H_f}\right)\kappa(\varphi)\right] \tag{13.9}$$

Subtracting T_m from both sides of Eq. (13.9) and then dividing through by T_m yields the dimensionless GTH capillary potential, ϑ_{int}, defined here as the function

$$\vartheta_{int} \equiv \frac{T_{int}(\varphi;\varepsilon) - T_m}{T_m} = \left[-(1 - 15\varepsilon\cos 4\varphi)\left(\frac{\gamma_0\,\Omega}{a\,\Delta H_f}\right)\kappa(\varphi)\right] = \Gamma(\varphi)\kappa(\varphi). \tag{13.10}$$

Equation (13.10) is plotted in Figure 13.4 against the scaled running coordinate for several values of the anisotropy strength, ε. The capillary potential, which is everywhere negative, approaches zero as the interface flattens, in accord with $T_{eq} \to T_m$ as $\eta \to -\infty$, and the curvature $\kappa \to 0$. For isotropy ($\varepsilon = 0$), the potential falls steadily toward the tip of the parabola, where it achieves its extreme value $\vartheta_{int} = -2$. Anisotropy, however, causes the interface potential to become nonmonotone, so the tip is no longer at the minimum potential, but actually *rises* in its potential. When the allowed anisotropy achieves its maximum value, $\varepsilon = 1/15$, the tip potential rises to zero, which represents the bulk melting point, T_m. These behaviors demonstrate the relatively complicated behavior of the GTH potential, even for as simple a shape as a parabolic interface.

13.3.3
Tangential Gradients and Fluxes

The fact that the interface potential varies along a nonequilibrium interface shape – chosen in this instance as a parabola – suggests that local tangential gradients exist directed along the solid–liquid interface [21]. If the interface conductivity, k_{int}, is

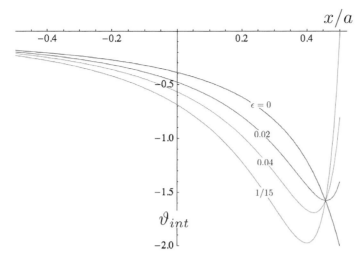

Figure 13.4 Gibbs–Thomson–Herring potentials, ϑ_{int}, along a parabolic interface, plotted against the dimensionless running variable $\eta = x/a$. The parameter ε denotes the strength of the interfacial energy anisotropy, aligned with the growth axis as one of the four maxima in $\gamma_{s\ell}$. The parabola's tip, located at $x/a = 1/2$, is at the lowest potential for $\varepsilon = 0$. Nonzero anisotropies can produce nonmonotone behavior. At the limit of fourfold energy convexity, where faceting occurs, $\varepsilon = 1/15$, and the potential at the parabolic tip rises to $\vartheta_{int} = 0$, where correspondingly, $T_{eq} \to T_m$.

nonzero,[5] tangential energy fluxes, $\vec{f}_\sigma$, associated with the capillary field, will be present in accord with Fourier's law for energy conduction along the interface,

$$\vec{f}_\sigma = -\left(\frac{\gamma_0 k_{int} \Omega T_m}{a^2 \Delta H_f}\right) \vec{\nabla}_\sigma [\vartheta_{int}] \tag{13.11}$$

where the tangential gradient operator, $\vec{\nabla}_\sigma [\vartheta_{int}]$, is the dimensionless (arc length) derivative of the interface potential. The *dimensionless* tangential energy flux, $\vec{\Phi}_\sigma = -\vec{\nabla}_\sigma [\vartheta_{int}]$ may now be found with the aid of Eq. (13.11). One sees that $\vec{\Phi}_\sigma = [(a^2 \Delta H_f)/(\gamma_0 k_{int} \Omega T_m)] \vec{f}_\sigma$, which for a parabolic interface yields the expression,

$$\vec{\Phi}_\sigma = \frac{2\left((3-4\eta)^2(-5+12\eta) - 15\varepsilon(67+4\eta(-79+68\eta+48\eta^2))\right)}{(-3+4\eta)^{9/2}\sqrt{-\frac{3}{2}+(5-4\eta)\eta}}. \tag{13.12}$$

13.3.4
Capillary-Mediated Energy

If the capillary-mediated energy fluxes along the parabolic interface are either convergent or divergent, they accumulate, or remove, respectively, net energy.

---

5) The interface conductivity, k_{int} (watts/K), is a little known transport property, akin to the interface diffusivity and interface viscosity. As real crystal-melt interfaces are known to have finite thicknesses, it is a reasonable assumption therefore that $k_{int} \neq 0$ [21,22].

The details of the energy released from point-to-point, acting either as a "source", or as a "sink", depend on the divergence of the vector fluxes given in Eq. (13.11). The interfacial energy source/sink density $(J/m^2/s)$ is easily determined from Eq. (13.11) using the (negative) tangential divergence operator, namely,

$$-\frac{1}{a}\vec{\nabla}_\sigma \cdot \vec{f}_\sigma = \left(\frac{\gamma_0 k_{int}\Omega T_m}{a^3 \Delta H_f}\right)\vec{\nabla}_\sigma 2[\vartheta_{int}] \quad (13.13)$$

Equation (13.13) may be nondimensionalized as the net capillary-mediated energy density, $E(\eta)$, deposited, or removed, from the interface.

$$E(\eta) = -\vec{\nabla}_\sigma \cdot \vec{\Phi}_\sigma = \vec{\nabla}_\sigma 2[\vartheta_{int}] \quad (13.14)$$

where one notes in Eq. (13.14) that the negative divergence (convergence) of the dimensionless interfacial flux, $\vec{\Phi}_\sigma$, equals the surface Laplacian of the interface potential. This relationship is of major importance in interpreting the predicted physical results.

Carrying out the indicated operation in Eq. (13.14), using Eq. (13.12), yields, after some algebraic manipulation, the local net energy addition or subtraction along the parabolic interface,

$$E(\eta) = \frac{8(3-4\eta)^2(-17+60\eta) - 120(263+4\eta(-507+4\eta(173+60\eta)))\varepsilon}{(3-4\eta)^{13/2}} \quad (13.15)$$

The capillary energy density, $E(\eta)$, is plotted in Figure 13.5 versus the dimensionless coordinate, $\eta = x/a$. The energy initially is negative – indicating a capillary-mediated *sink* – from $x/a \geq -\infty$ until $x/a \geq 0.2$, beyond which the sign changes and it becomes a *source*. Energy sources climb in magnitude rapidly to their

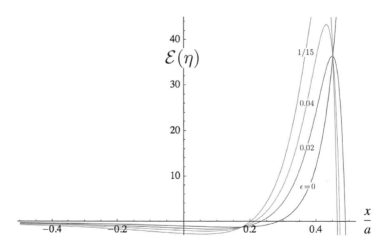

Figure 13.5 Local interfacial energy source/sink, $E(\eta)$, from the negative divergence of the tangential capillary-mediated fluxes. Curves plotted for several strengths, ε, of the fourfold anisotropy. The ascending and descending roots of these energy curves, and the sign of their slopes, provide *necessary* and *sufficient* conditions for branching.

peak-positive values near $x/a \approx 0.4$, and then fall to zero again closer to the tip, again changing sign, except for the case of isotropy, $\varepsilon = 0$, for which $E(\eta)$ continuously rises to the tip location at $x/a = 0.5$.

13.4 Branching

13.4.1 Stefan Balance

We have shown to this point that small amounts of energy are deposited or removed from a parabolic interface because of capillary-mediated fluxes, reflecting *rapid* atomic-scale local equilibrium. Notwithstanding their small values, these usually neglected energies can be accounted in the local macroscopic interface Stefan balance. That is, capillary-mediated energy source terms, $E(\eta)$ either add to, or subtract from, the usual latent heat source term, $v_n \Delta H_f / \Omega$, at a moving solid–liquid interface. Here v_n is the normal freezing velocity of the interface. The Stefan balance, also includes the conduction of heat, normal to the interface, via the slowly varying Ivantsov thermal field in the melt plus any macroscopic gradients encountered in the solid, specifically,

$$v_n \left(\frac{\Delta H_f}{\Omega}\right) + E(\eta) \left(\frac{k_{\text{int}} T_m}{a^2}\right) = k_\ell \nabla T_\ell + k_s \nabla T_s \quad (13.16)$$

where k_ℓ and k_s are the thermal conductivities of the bulk liquid and solid phases.

Equation (13.16) makes clear that capillary-mediated *sources*, $E(\eta) \geq 0$, would reduce the interface velocity, v_n, slightly, whereas capillary-mediated *sinks*, $E(\eta) \leq 0$, would increase v_n slightly. This remains true provided that the macroscopic gradient fields on the right-hand side of Eq. (13.16) change slowly relative to the rapidly responding local equilibrium potential, ϑ_{int}, the Laplacian of which is $E(\eta)$.

13.4.2 Zeros of the Surface Laplacian

The modulation over time of the normal interface velocity by capillary-mediated energy has a local characteristic where the surface Laplacian of the interface potential vanishes along the interface. This corresponds to the point where the vector divergence of the capillary-mediated heat flux also vanishes. The interface becomes retarded or accelerated on either side of such a zero, as suggested schematically in Figure 13.6.

The Laplacian of the GTH potential involves the fourth derivative of the tip shape. Such a high-order derivative might possibly explain why numerical models have not connected branching to capillarity fields – it is too subtle an effect to be casually observed. Moreover, the local net energy induced by capillarity changes sign where this Laplacian vanishes, corresponding to $E(\eta) = 0$. Interface rotation ("tilting") and

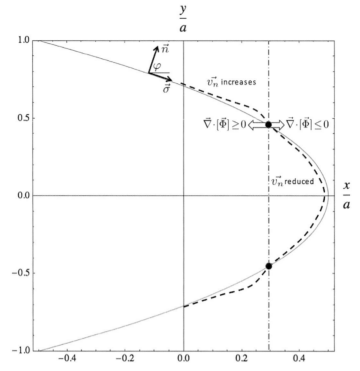

Figure 13.6 Dynamic influence of a zero in the surface Laplacian of the interface potential. The interface rotates (tilts clockwise in the upper half-plane, and tilts anticlockwise in the lower half-plane) about the vanishing point of the potential's surface Laplacian, causing the local curvatures to sharpen behind, and flatten nearer to the tip. These curvature changes then couple with the external (Ivantsov) field and induce branch formation. Moreover, rotation occurs *only* if the slope of the $E(\eta)$ curve is positive. Other zeros closer to the tip that cross the axis with negative slopes would not be effective in producing a branch, because their chirality is unsuitable for coupling with the exterior field in the melt. Compare the location and slopes of the zeros on the $E(\eta)$ curves shown in Figure 13.5.

branch formation occur, however, only if the slope $dE(\eta)/d\eta \geq 0$, which provides the required chirality for the tilting motion to evolve into a side branch by coupling to the exterior Ivantsov field. Thus, one finds that the *necessary* condition for dendritic branch formation is a zero in the fourth derivative of the dendrite tip shape, whereas the *sufficient* condition for subsequent side-branch growth depends on the sign of the fifth-order derivative of the tip shape.

13.5
Dynamic Solver Results

The dynamic evolution of several closed starting dendritic shapes were studied by J. Lowengrub and Shuwang Li, who used a low-noise, high-precision, Greens function solver [26,27]. Their dynamic solver allows direct checks on the correspondence of pattern evolution with the rotation points predicted from the capillary-mediated theory described in this chapter (S. Li, 2011, Department of Applied Mathematics, Illinois Institute of Technology, Chicago, IL, USA, private communication).

Initial efforts to detect capillary-mediated interface rotation and accurately locate subsequent branching required the dynamic evolution of ellipses with different eccentricities and surface energy anisotropies [14]. These initial elliptical "crystals" grew into a surrounding "melt" with either a specified heat flux removal from the far field or a fixed supercooling imposed away from the crystal. The simulations showed the development of "noise-free" dendrites [13].

The dynamically observed first rotation points for a 2:1 ellipse, with fourfold anisotropy of strength $\varepsilon = 0.005$, are shown in Figure 13.7. Here the elliptical

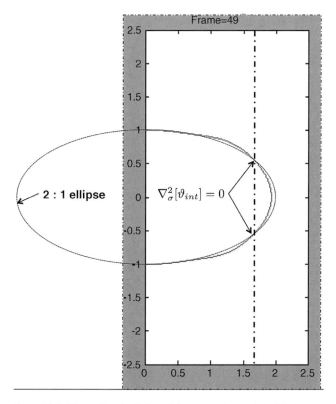

Figure 13.7 The surface Laplacian of the interface potential along a 2:1 ellipse is predicted to vanish at $x/a \approx 1.69$, as indicated by the vertical dash-dot line, for $\varepsilon = 0.005$. Here a is the semi-major axis of the starting ellipse. The elliptical starting shape appears here as the broken curve. An early video frame (segmented curve) of the dynamic evolution of this shape shows that the normal velocity of the tip region is retarded slightly over time, causing the local tip curvature to flatten, whereas for $x/a < 1.69$ the adjacent interface accelerates and the local curvatures sharpen. The countervailing interface acceleration and retardation cause the interface to tilt clockwise about the upper rotation point, and anticlockwise about the lower rotation point, stimulating side-branch formation.

starting shape is superimposed with an early frame taken from the dynamic solver. At the scaled position $x/a \approx 1.69$, which is determined analytically from Eq. (13.14), the superimposed starting shape and the dynamically evolved dendritic interface show the expected interface rotation, with the interface slightly retarded (flattened) ahead of the rotation point, and slightly sharpened just aft of it, in close agreement with Eq. (13.16).

13.6
Conclusions

1) Capillarity at a crystal–melt interface provides a weak energy field, the energy from which is ignored in the usual Stefan interface balance. Capillarity, as traditionally used in crystal growth theory, provides only a boundary condition on the robust transport fields controlling the $\ell \rightarrow s$ phase transformation.
2) We treat the GTH temperature distribution as a physical energy field capable of exerting gradients, fluxes, and a divergent energy conduction field along the interface. This capillary field reacts rapidly to local changes in the interface curvature as the dendrite grows subject to the macroscopic (Ivantsov) conduction field that is responsible for the transport of latent heat through the melt.
3) Analysis shows that small amounts of capillary-mediated energy are added close to the tip and removed aft of the tip. The unique point separating regions of energy addition from regions of energy loss is found to correspond with the vanishing of the surface Laplacian of the GTH potential.
4) The influence of capillarity on dendritic branching is deterministic, with the *necessary* condition for branching being a zero in the surface Laplacian of the GTH potential, which is equivalent to the vanishing of the fourth derivative of the tip shape. The associated *sufficient* condition for branching is a positive fifth derivative of the shape, producing the correct rotation chirality. These high-order, rather subtle criteria, explain why this intrinsic mechanism for branching has remained obscure.
5) Capillary-mediated energy influences the interface Stefan balance and causes the local normal velocity to decrease where energy is added, and increase where energy is removed. The slight bias in retarding or accelerating the normal velocity, respectively, flattens or sharpens the local curvatures surrounding the vanishing point of the surface Laplacian of the potential where interface rotation or tilting occurs. The changes in interface curvature lead to side-branch formation as local gradients in the melt become either locally enhanced or reduced.
6) A low-noise dynamic solver confirms the location for the initial rotation points predicted from the capillary field analysis for various closed elliptical shapes subject to varying anisotropy. A more complete discussion of the dynamic simulation data gathered to date will be published elsewhere.

Acknowledgments

The author thanks both the Alexander von Humboldt Foundation, Bonn, Germany, and the Allen S. Henry Chair, Florida Institute of Technology, Melbourne, Florida, USA, for providing financial support to participate in the Günter Gottstein Honorary Symposium, prompting preparation of this chapter. Thanks are also due to Professor John Lowengrub, Mathematics Department, University of California, Irvine, CA, and Professor Shuwang Li, Department of Mathematics, Illinois Institute of Technology, Chicago, IL, for applying their integral equation solver that provides independent dynamical checks on the capillary-mediated theory presented here.

References

1 Glicksman, M.E. (2011) Chapter 13, Dendritic Growth, in *Principles of Solidification*, Springer USA, New York, pp. 305–343.

2 Langer, J.S. and Müller-Krumbhaar, H. (1977) Stability effects in dendritic crystal growth. *Journal of Crystal Growth*, **42**, 11–14.

3 Langer, J.S. and Müller-Krumbhaar, H. (1978) Theory of dendritic growth: 1. Elements of a stability analysis. *Acta Metallurgica*, **26**, 1681–1687.

4 Pieters, R. and Langer, J.S. (1986) Noise-driven sidebranching in the boundary layer model of dendritic solidification. *Physical Review Letters*, **56**, 1948–1951.

5 Kessler, D., Koplik, J., and Levine, H. (1986) Steady-state dendritic growth. *Physical Review A*, **33**, 3352–3357.

6 Kessler, D. and Levine, H. (1986) Stability of dendritic crystals. *Physical Review Letters*, **57**, 3069–3072.

7 Kessler, D. and Levine, H. (1988) Pattern selection in 3-dimensional dendritic growth. *Acta Metallurgica*, **36**, 2693–2706.

8 Dougherty, A., Kaplan, P.D., and Gollub, J.P. (1987) Development of side branching in dendritic crystal growth. *Physical Review Letters*, **58**, 1652–1655.

9 Muschol, M., Liu, D., and Cummins, H.Z. (1992) Surface tension anisotropy measurements of succinonitrile and pivalic acid – comparison with microscopic solvability theory. *Physical Review A*, **46**, 1038–1050.

10 LaCombe, J.C., Koss, M.B., and Glicksman, M.E. (2007) Tip velocities and radii of curvature of pivalic acid dendrites under convection-free conditions. *Metallurgical and Materials Transactions A – Physical Metallurgy and Materials Science*, **38**, 116–126.

11 Glicksman, M.E., Lupulescu, A., and Koss, M.B. (2003) Melting in microgravity. *Journal of Thermophysics and Heat Transfer*, **17**, 69–76.

12 Lupulescu, A., Glicksman, M.E., and Koss, M.B. (2005) Conduction-limited crystallite melting. *Journal of Crystal Growth*, **276**, 549–565.

13 Glicksman, M.E., Lowengrub, J.S., Li, S., and Li, X. (2007) A deterministic mechanism for dendritic solidification kinetics. *JOM*, **59**, 27–34.

14 Glicksman, M.E. (2012) Mechanism of dendritic branching. *Metallurgical and Materials Transactions A – Physical Metallurgy and Materials Science*, **43**, 391–404.

15 Glicksman, M.E., Koss, M.B., and Winsa, E.A. (1994) Dendritic growth velocities in microgravity. *Physical Review Letters*, **73**, 573–576.

16 Glicksman, M.E., Koss, M.B., Bushnell, L.T., LaCombe, J.C., and Winsa, E.A. (1995) Space flight data from the isothermal dendritic growth experiment. *Advances in Space Research*, **16**, 181–184.

17 Koss, M.B., LaCombe, J.C., Tennenhouse, L.A., Glicksman, M.E., and Winsa, E.A. (1999) Dendritic tip velocities and radii of

curvature in microgravity. *Metallurgical and Materials Transactions A – Physical Metallurgy and Materials Science*, **30**, 3177–3190.

18 LaCombe, J.C., Koss, M.B., and Glicksman, M.E. (1999) Nonconstant tip velocity in microgravity dendritic gravity. *Physical Review Letters*, **83**, 2997–3000.

19 Corrigan, D.P., Koss, M.B., LaCombe, J.C., de Jager, K.D., Tennenhouse, L.A., and Glicksman, M.E. (1999) Experimental measurements of sidebranching in thermal dendrites under terrestrial gravity and microgravity conditions. *Physical Review E*, **60**, 7217–7223.

20 Glicksman, M.E. (2008) Solidification research in microgravity. *ASM Handbook*, **15**, 398–401.

21 Caroli, B., Caroli, C., and Roulet, B. (1984) Nonequilibrium thermodynamics of the solidification problem. *Journal of Crystal Growth*, **66**, 575–585.

22 Guggenheim, E.A. (1940) Thermodynamics of interfaces in systems of several components. *Transactions of the Faraday Society*, **36**, 397–412.

23 Morse, P.M. and Feshbach, H. (1953) *Methods of Theoretical Physics*, **2**, McGraw-Hill, New York, p. 1284.

24 Herring, C. (1951) *Physics of Powder Metallurgy* (ed. W.E. Kingston), McGraw-Hill, New York, p. 143.

25 Glicksman, M.E. (2011) Chapter 13, in *Principles of Solidification*, Springer USA, New York, p. 333.

26 Li, S., Lowengrub, J., Leo, P., and Cristini, V. (2005) Nonlinear morphological control of growing crystals. *Journal of Crystal Growth*, **277**, 578–592.

27 Li, S., Lowengrub, J., and Leo, P. (2005) Nonlinear morphological control of growing crystals. *Physica D*, **208**, 209–219.

Part III
Advanced Experimental Approaches for Microstructure Characterization

Microstructural Design of Advanced Engineering Materials, First Edition. Edited by Dmitri A. Molodov.
© 2013 Wiley-VCH Verlag GmbH & Co. KGaA. Published 2013 by Wiley-VCH Verlag GmbH & Co. KGaA.

14
High Angular Resolution EBSD and Its Materials Applications
Claire Maurice, Romain Quey, Roland Fortunier, and Julian H. Driver

14.1
Introduction: Some History of HR-EBSD

Standard electron backscatter diffraction (EBSD) has been used for about 20 years to measure the local orientations of crystalline materials and has now become a common feature of many, if not most, modern materials laboratories. The principle consists of applying a "stationary" electron beam to a tilted sample in a scanning electron microscope to generate a backscattered Kikuchi pattern that is captured by a phosphor screen and then transformed, via a lens assembly and charge-coupled device CCD camera, to a digitized image that can be computer analyzed to obtain the orientation of the diffracting volume. There have been many reviews of the technique, for example, in Refs. [1–3]. Using standard equipment, the accuracy of orientation measurement is often considered to be of the order of $0.5°$.

High angular resolution EBSD (commonly abbreviated to HR-EBDSD) basically employs the same type of equipment and methods but aims to measure lattice orientations to accuracies of about $10^{-2°}$. There are two major advantages for materials science (and geological) applications: (i) the evaluation of the misorientations associated with crystal defects, phase boundaries and so on to much greater accuracy than hitherto, and thereby the development of better models for their formation and evolution; (ii) the measurement of the distortion of the crystal lattice associated with elastic strains (either residual or generated *in situ*). The latter problem turns out to be a much more challenging task and therefore requires very careful analysis of the potential errors and suitable correction procedures. This chapter will consider HR-EBSD techniques for both applications with a particular emphasis on the problem of strain measurement.

The first measurement of elastic strain from EBSD diagrams was performed in the early 1990s by Troost *et al.* [4] from the tetragonal distortion of epitaxial $Si_{1-x}Ge_x$ films grown on a silicon (1 0 0) substrate. One component of strain was calculated from the shift of a zone axis measured by cross-correlation between two patterns (one is considered as reference, i.e., with no strain). The authors announced a strain sensitivity of 1×10^{-3}. Wilkinson [5] adopted a similar approach and claimed a strain sensitivity of 2×10^{-4}. Over the last 10 years, this cross-correlation method

Microstructural Design of Advanced Engineering Materials, First Edition. Edited by Dmitri A. Molodov.
© 2013 Wiley-VCH Verlag GmbH & Co. KGaA. Published 2013 by Wiley-VCH Verlag GmbH & Co. KGaA.

has been continuously improved. By considering the measured shifts of two zone axes, four components of the displacement gradient tensor could be determined (Wilkinson [6]), then extended to eight components from the shifts of four zone axes (Wilkinson *et al.* [7]). The remaining ninth component corresponds to hydrostatic dilatation to which zone axis position is insensitive since it only changes the widths of Kikuchi bands.

Currently, this method is not only applied to zone axes but more generally to regions of interest (ROI). Any direction captured on the phosphor screen can be used to determine strains, and the cross-correlation windows are not necessarily centered on zone axes. As a consequence of digital image processing improvements (cross-correlation and filtering), this *pattern shift method* enables a strain sensitivity of 1.3×10^{-4} according to Wilkinson *et al.* [7].

As discussed in more detail below, numerical simulations of EBSP can also provide much valuable information, and the first true multibeam dynamical simulation of an EBSP was published in 2007 by Winkelmann *et al.* [8]. The use of simulated patterns as a benchmark for HR-EBSD was then proposed by Maurice and Fortunier [9], initially with a simpler geometrical diffraction model (but improved to allow for intensity variations across the K bands). Multibeam dynamical simulations of elastically deformed crystals are now becoming more widely available.

Of course, strain sensitivity is important but the real test concerns the true accuracy of the strain measurements as obtained by these techniques. This aspect was first addressed by Villert *et al.* [10] who applied the pattern shift method to simulated patterns of numerically strained crystals to evaluate the strain components, as a function of both their imposed values and the microscope parameters including sample geometry. An accuracy of 10^{-4} on the strain tensor components could be obtained but the microscope projection parameters (pattern center and detector distance) were shown to have a very strong influence; in some cases quite small errors in the relative PC position can induce large errors in the strain measurement – afterward known as the pattern center problem. This has been the subject of much recent research.

An alternative to the relative pattern shift method was also proposed in 2008 by Maurice and Fortunier [9]: the three-dimensional (3D) Hough transformation. This absolute method is based on the fact that Kikuchi band edges are (slightly curved) hyperbolae and not straight lines. A 3D Hough transform uses two parameters to describe a reflector plane normal and the third parameter is the associated Bragg angle. In principle, treating the bands via the mathematical formulation of conics allows one to use more of the information that is in fact present in the EBSP.

14.2
HR-EBSD Methods

14.2.1
Basic Geometry of HR-EBSD

To obtain high angular accuracy requires either a very precise analysis of the Kikuchi diffraction bands (absolute method) or cross-correlation of the current pattern with that of a reference pattern. The cross correlation method has been the most widely employed so far and basically consists of applying digital image correlation methods to two Kikuchi patterns to obtain their relative shifts in selected ROIs. From the shifts, one obtains the material deformation gradient tensor by standard optimization techniques (such as least-square fitting), based on the following fundamental relations for EBSD geometry.

An EBSD pattern is basically the gnomonic projection of the crystal lattice cell geometry (recalling that it projects a sphere onto a tangent plane, here the phosphor screen). As such, any elastic distortion of the probed material (point S in Figure 14.1) will be mapped onto the EBSP. It should be recalled that an elastic deformation transforms, in general, an initially strain-free cubic cell into a triclinic cell as illustrated in Figure 14.2. The angular deviations of the lattice planes give rise to angular variations of the K bands and zone axes in the EBSP. These are the basic shifts that are measured by HR-EBSD. The quantities (b/a, c/a, alpha, beta, gamma) are then directly related to the deviatoric strain tensor.

The aim of HR-EBSD is then to quantify the material distortion between a reference configuration and a "deformed" configuration, by comparing the EBSPs collected from these two probed volumes. Consider, in Figure 14.1, an infinitesimal element $\vec{dX}$ in the reference crystal cell; this particular direction is mapped to point P on the phosphor screen (i.e., $\vec{SP} = \vec{r} = \alpha \cdot \vec{dX}$): the gnomonic projection is

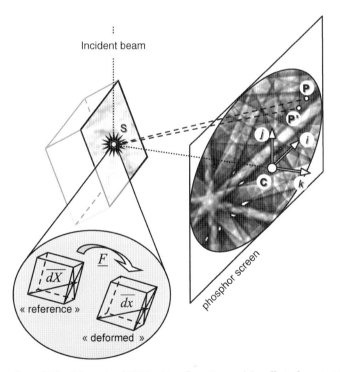

Figure 14.1 Schematic of EBSD pattern formation and the effect of a material deformation gradient tensor $\underline{F}$ displacing point P to P' on the phosphor screen.

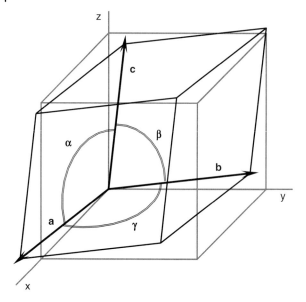

Figure 14.2 Schematic of the elastic deformation of an initial cubic cell to triclinic.

expressed in the screen reference frame as $\vec{SP} = (x_p - X^*, y_p - Y^*, Z^*)^T$ where (x_p, y_p) are the coordinates of pixel P in the EBSP image (from the lower left corner) and (X^*, Y^*, Z^*) are the pattern center coordinates (denoted C).

From basic continuum mechanics, the reference element $\vec{dX}$ is transported to $\vec{dx} = \underline{F} \cdot \vec{dX}$ in the deformed configuration, where $\underline{F}$ is the material deformation gradient tensor. The direction collinear to $\vec{dx}$ is now projected to point P' such that $\vec{SP'} = \vec{r} + \vec{q} = a' \cdot \vec{dx}$ ($\vec{q}$ denotes the translation or shift vector $\vec{PP'}$). Using the gnomonic projection, the scalar factor a' can be determined from the fact that the third coordinate of $\vec{SP'}$ is equal to Z^* (point P' is on the screen), hence $a' = Z^* / (\vec{dx} \cdot \vec{k})$ where $\vec{k}$ is the unit vector normal to the screen (a similar relation holds for the scalar factor a). Therefore, the mapping transformation from the reference EBSP to the current EBSP is simply written as:

$$\vec{r}' = \vec{r} + \vec{q} = \frac{Z^*}{(\underline{F} \cdot \vec{dX}) \cdot \vec{k}} \underline{F} \cdot \vec{dX}$$

Multiplying the numerator and the denominator of the right-hand term by the scalar factor a, this is equivalent to

$$\vec{r}' = \vec{r} + \vec{q} = \frac{Z^*}{(\underline{F} \cdot \vec{r}) \cdot \vec{k}} \underline{F} \cdot \vec{r} \tag{14.1}$$

Equation 14.1 is the fundamental transformation formula of HR-EBSD that relates the shift $\vec{q}$ of a given direction $\vec{r}$ produced on the screen to the material deformation gradient tensor $\underline{F}$ (Villert et al. [10], Maurice et al. [11]). Clearly, this is not a linear relation between the shifts and the strains, although it is often treated as such in elementary analyses. Equation (14.1) is used to identify the tensor F by standard numerical optimization schemes from a number (typically 20 or 30) of shifts determined by cross-correlation of the corresponding ROIs. The deformation gradient tensor is then decomposed into an elastic strain tensor and a rotation tensor. Modern cross-correlation techniques enable shifts to be measured to 1/20 of a pixel. Details on subpixel registration techniques can be found in [10].

14.2.2
EBSP Numerical Simulations

EBSPs can be simulated for a wide variety of crystal structures and microscope parameters and thereby benchmark the experimental patterns. In particular, the patterns can be simulated for elastically strained lattices under different stress states. Although Bragg's law simulations have been used in the past [9,10], the best simulations are obtained using the many-beam dynamical theory [12] of EBSP formation, as explained by Winkelman et al. [8]. Over the last 2 years, an in-house parallelized implementation of many-beam dynamical EBSD has been extensively used in our group. This particular software allows simulating EBSPs of elastically distorted lattices at typical no-binning resolutions (e.g., 1344×1024 pixels for NordlysII cameras) in about 2 to 5 min when run on a 64 processor PC cluster [11]. The computation time depends on the number of atoms of the lattice cell, the number of chosen reflectors and convergence criteria. In the current version, a 2-min calculation is obtained for an FCC material (copper), that is, 4 atoms/cell, 537 reflectors, an omega perturbation factor of 40, a termination criterion of 95% (see [8,12] for the definition of these factors). The projection parameters are typical of an experimental setup: pattern center located at ($X^* = 0.5$, $Y^* = 0.7$, and $Z^* = 0.5$) expressed in fractions of the pattern dimensions (width for X^* and Z^*, height for Y^*) from the lower left corner of the pattern. Figure 14.3 compares an experimental pattern obtained on a Ge single crystal with two numerically simulated patterns, the first based on a simple geometrical Bragg law model and the second based on many-beam dynamical theory. It is clear that the latter more closely reproduces the experimental intensity distribution. See Britton et al. [13] for a critical review of a cross-correlating experimental with different flavors of simulated patterns.

14.2.3
Practical Problems

Figure 14.4 illustrates schematically the sample-phosphor screen-camera configuration for which small variations of the backscattered Kikuchi bands from the sample are reproduced on the CCD in order to be analyzed. HR-EBSD requires

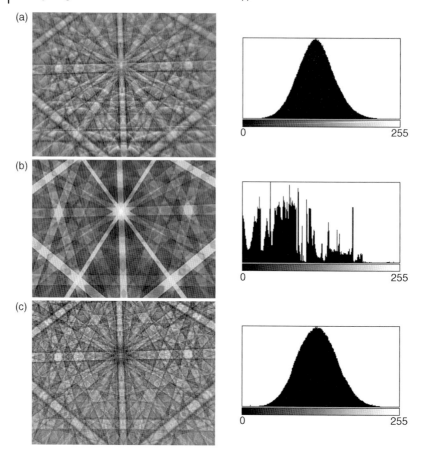

Figure 14.3 Example of numerically simulated EBSPs of germanium in (b) and (c) compared with the experimental pattern (a). (b) Bragg's law simulation using weighted Kikuchi band intensities [10] and (c) is a multibeam dynamical simulation. On the right are the corresponding intensity distributions.

extremely accurate geometrical calibrations of the electron diffraction equipment, particularly the pattern center position and the camera lens distortions.

14.2.3.1 Lens Distortion

All CCD cameras have some lens distortion and optics misalignment that can have a significant influence on HR-EBSD analyses (although usually negligible for standard EBSD). The problem has been treated in some detail by Mingard *et al.* [14]. A standard method for quantifying lens distortion is to image a planar calibration artifact, often a chessboard of alternating black and white squares. By replacing the phosphor with a printed chessboard pattern and shining a light through the paper, the grid can be directly observed by the CCD and distortions measured from a comparison of real and captured images. Figure 14.5 shows this method schematically.

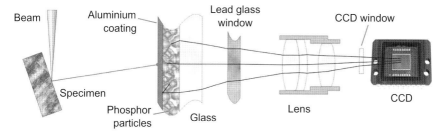

Figure 14.4 Schematic of diffracted beam to CCD chip via the camera assembly. adapted from Austin Day, private communication

The major distortion components are the center offset, rotation, and radial distortion that are conveniently characterized by Brown's model parameters [15]. They should remain constant for any given camera and lens set-up. Once these parameters are measured, the acquired EBSPs can then be "undistorted" by numerically re-mapping the EBSPs using the inverse model and the particular lens distortion parameters; this is done before any other processing such as ROI cross-correlation. Conversely, lens distortions can be artificially introduced in the EBSP simulations to quantify their influence on strain measurements [13].

14.2.3.2 Orientation Gradients

Orientation gradients are present in most crystals and their amplitude depends strongly on the material history; near-perfect crystals possess orientation variations of less than 1° over a millimeter but the grains of plastically deformed metals may vary in orientation by several degrees at the micron scale. Until recently, and for the cross-correlation method, it was considered that orientation gradients between the current and reference patterns would have little or no influence on the HR-EBSD results. However, it has now been shown by Maurice *et al.* [11], and subsequently confirmed by Britton and Wilkinson [16], that orientation differences as small as 1° can lead to strain measurement errors of several hundred microstrains. A typical

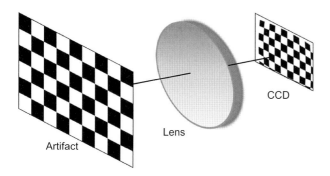

Figure 14.5 Schematic of lens distortion calibration.

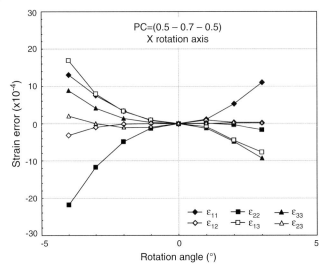

Figure 14.6 Measured strain errors for individual strain component errors $\left(\varepsilon_{ij}^{measured} - \varepsilon_{ij}^{imposed}\right)$ as a function of the applied finite rotation angle around the x sample axis. Data obtained with the standard infinitesimal strain model on many-beam numerically simulated crystals.

plot of strain error as a function of rotation angle between current and reference EBSP is given in Figure 14.6. It can be seen that for a rotation of 3°, the strain errors can be of order 10^{-3} that implies that some previous published results on plastically deformed crystals are simply incorrect. These results, as typified by Figure 14.6, were obtained by applying the cross-correlation method to many-beam dynamical EBSP simulations of rotated and elastically deformed crystals as described above in Section 14.2.2.

Maurice et al. [11] demonstrated that these errors are due to the large rotations distorting the shifts leading to poor quality cross-correlation peaks; the errors in the measured shifts $\vec{q}$ induce errors in the subsequent analysis of the strains. A relatively simple method of correcting for these orientation gradients has been proposed and consists of rotating the reference pattern toward that of the current pattern by the known difference on orientation (or equivalently rotating the current pattern the other way [16]). Figure 14.7 illustrates the principles of the method: the reference EBSP on the left is numerically remapped by a finite rotation R corresponding to the (approximate) rotation measured by standard EBSD between reference and current patterns. The infinitesimal elastic distortion is then identified from the "residual shifts" between the "rotated reference" and the "current" pattern by the standard cross-correlation procedure. A word of caution is required concerning the remapping procedure as it involves the pattern center and lens distortion correction.

Since all the parameters are known for the simulated patterns, then the errors of the shifts can be immediately determined for both the standard (un-rotated reference) and rotated reference methods. Figure 14.8 shows the major reduction, by two

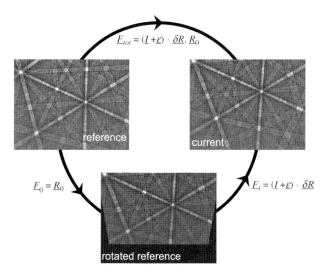

Figure 14.7 Sketch showing how the total deformation tensor can be decomposed into an infinitesimal elastic distortion and a finite rotation (e.g., with a 15° rotation around A, with PC located at (0.5–0.7–0.5)). The finite rotation is used to compute the "rotated reference" pattern for cross-correlation with the current pattern [11].

orders of magnitude, of the shift errors that can be achieved by this rotated reference technique (RRT). Clearly, this new method opens up the possibility of accurate elastic strain analysis in dislocated crystals characterized by orientation gradients.

14.2.3.3 Pattern Center

The fundamental equation of HR-EBSD, Eq. (14.1) above, explicitly involves the source point position, or pattern center (PC). The gnomonic projection being

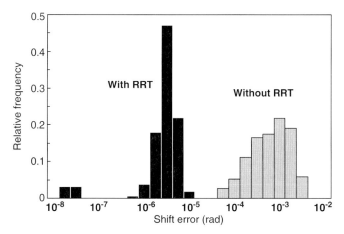

Figure 14.8 Shift error distributions (on log scale) for a case study of 4° rotation about the x-axis showing the large improvement of shift measurements using the RRT method [11].

nonconformal, accurate angular measurements can only be performed if the projection center is known. The sensitivity of HR-EBSD to the PC position has been studied by Villert et al. [10], by means of simulated patterns. It has been demonstrated that the "absolute" PC position has little influence on the results (e.g., an error of 1% in either component of the PC yields an error of $\sim 2.5 \times 10^{-5}$ on the strain components). This conclusion should be taken with caution when the remapping procedure, based on Eq. (14.1), is applied. If the source point used for remapping is not accurate, geometrical distortions are artificially introduced in the remapped pattern, subsequently interpreted as phantom strain and rotation components (see Britton and Wilkinson [16]). We recall that the PC position in standard EBSD is determined at each point by an algorithm of Krieger Lasson [17] and an accuracy of about ± 5 pixels that is inadequate for HR-EBSD.

It has also been demonstrated that the "relative" PC position (i.e., the PC change between the reference and current pattern, due, e.g., to beam shift) is a very critical point.

Figure 14.9, taken from Villert et al. [10], gives an example of the error on 2 displacement gradient components induced by relative PC variations along the x-axis (PC_x). Using a screen of 1000 pixels, it follows that one pixel error in PC_x gives a d_{yz} error of 2×10^{-3}; this is obviously prohibitive for elastic strain measurements. Such relative PC displacements during beam scanning can in principle be, and effectively are, compensated for by simple geometrical considerations. However, the relative PC position becomes a real problem if one wants to use a simulated pattern as reference, see [13].

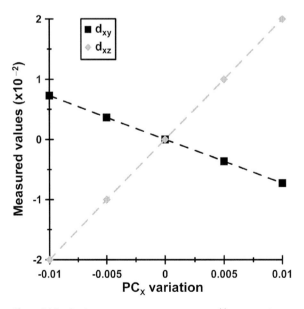

Figure 14.9 Strain measurement errors assessed by comparing measured and applied components of d induced by PC_x variations (in units of screen fractions), from [10].

Methods to determine the PC to the required accuracy have been proposed: an improved moving screen method by Maurice et al. [18], an improved shadow casting technique by Mingard et al. [14], and the K band edge parallelism [19]. However, for the moment these techniques have not yet completely solved the problem. The moving screen method consists of acquiring 2 EBSPs at different zooms to find their convergence point, and it has been shown by numerical EBSP simulations for different screen distances that the method should be sufficiently accurate, if, that is, the zoom direction could be perfectly controlled. In fact current camera systems do not have unique zoom directions because of mechanical imperfections. The improved shadow casting technique [14] has been demonstrated to give a PC accuracy of about 1/3 pixel that is a significant improvement but still on the limit for accurate HR-EBSD.

14.3
Applications

14.3.1
Low-Angle Grain Boundaries

Since the angular resolution can be of order $1/100°$, the method lends itself immediately to the accurate analysis of low misorientations in crystals. Figure 14.10 illustrates an example taken from a Ni-based superalloy single crystal that contains a substructure of low-angle boundaries developed during directional solidification; the misorientations are much less than $1°$, typically about $0.1°$. The left orientation map is given by standard (FEGSEM: field emission gun scanning electron microscope) EBSD and the right image is provided by HR-EBSD using the same data set; the improvement is obvious showing that substructures can now be analyzed with the same, or better angular, resolution as TEM.

14.3.2
Pure Elastic Strains

As mentioned above, the first application of HR-EBSD by Troost et al. [4] concerned elastic strains in Si–Ge thin films and this field is still of major importance [18]. A validation test of the method as used for measuring elastic strains was carried out by Villert et al. [10] on an elastically bent Si crystal. The crystal was elastically strained by four-point bending in the SEM (JEOL 6500 FEGSEM equipped with a HKL Technology EBSD device using the Nordlys II CCD camera and Channel 5 acquisition software). To reduce PC errors, a stage scanning map was done for 100 point with a grid step distance of $7\,\mu m$. At each point, a high-definition EBSD pattern was recorded (8-bit gray level bitmap image with 1344×1024 pixels, i.e., no-binning). The center point of the map is considered as an unstrained reference. Every other EBSD pattern is compared to this reference.

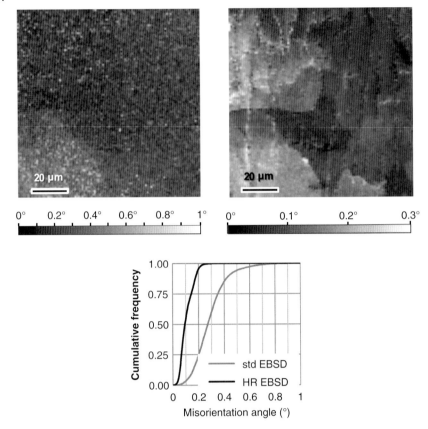

Figure 14.10 EBSD misorientation maps of a Ni-based superalloy single crystal "AM1." The maps give the misorientation with respect to the average orientation, (although with different misorientation angle scales), (a) standard EBSD map and (b) HR-EBSD map. Figure 14.5c plots out the cumulative misorientation angle frequencies for these two misorientation measurements.

Figure 14.11 shows the measured elastic strains through the section of the crystal which, from simple beam theory, should be tensile along the outer surface (along the length direction y) and compressive on the inner surface. The HR-EBSD elastic strain data are compared to a finite element (FE) simulation of the beam distortion using anisotropic elastic theory. It can be seen that the principal strains are in good agreement with the values expected from standard elasticity. Furthermore, the shear strains should be zero and in fact the majority of the measured shear strains are less than 10^{-4}. The rigid body rotation should also be zero and this is confirmed to within 10^{-4} for ω_{xy} and ω_{xz}, although ω_{yz} is slightly larger.

An application to elastic strains in thin Si–Ge plots (or mesas) deposited at high temperature on a (1 1 0) silicon wafer is shown in Figure 14.12. In this case, the elastic strain varies over short distances within each 2 μm wide plot. A beam scan of 100×100 points with a 100 nm step was performed on the area outlined in the

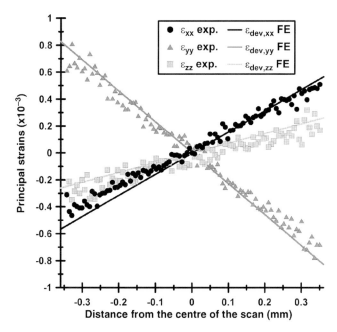

Figure 14.11 Test of elastic strain measurement on a bent Si crystal. The sample directions are x through the thickness, y along the length, and z along the short transverse direction (the incident electron beam is on the xy plane), from [10].

figure and the strains evaluated as before using the adjacent un-deformed Si as a reference. The measured normal and shear strain components were found to be in good agreement with FE simulations assuming that the elastic strains are due to thermal distortion, function of the local (Ge) composition [10]. Figure 14.12 gives an example of the ε_{zz} (normal) strain distribution.

The method has recently been applied to the "elastic" strains within grains of an *in situ* tensile-deformed austenitic stainless steel by Ojima *et al.* [20]. An example of the measured tensile strain maps over several grains at increasing (macroscopic) loads is shown in Figure 14.13. There are clear differences in strains within grains but care is required in interpreting the results since the highest stress (Figure 14.13d) corresponds to a macroscopic strain of about 8%, which is well into the plastic regime.

The problems created by dislocation substructures are obviously not so critical in ceramic materials so that local stresses can be measured with more confidence. Villanova *et al.* [21] have recently measured the residual (cooling) stresses in a yttria-stabilized zirconia layer of a solid oxide fuel cell) (SOFC), using the X-ray and HR-EBSD techniques. The macroscopic biaxial compressive stresses of the X-ray measurements were recovered by HR-EBSD in the 10 grains examined and their significant local variations due to elastic anisotropy could then be determined.

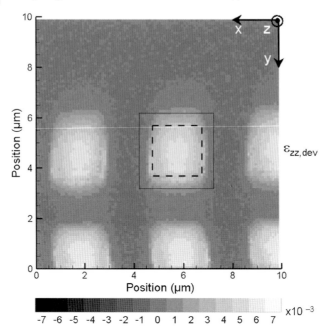

Figure 14.12 ε_{zz} (normal strain) map from HR-EBSD analysis of a SiGe–Si sample.

14.3.3
Crystal Defects and Lattice Misorientations: GND Analysis

The lattice curvatures that develop in (elasto) plastically deformed crystalline materials are usually accommodated by geometrically necessary dislocations (GNDs). The first theoretical analysis of the relations between lattice curvature and GND density was formulated by Nye [22] and complemented further by Ashby [23] and Kröner [24]. HR-EBSD has been used by several groups for GND analyses, see for example, [25–27].

The relation between lattice curvature and GND density is given in terms of Nye's dislocation density tensor α by

$$\alpha = \mathrm{curl}(\beta^e) \tag{14.2}$$

where β^e is the elastic part of the displacement gradient and can be expressed as the sum of the lattice elastic strain ε and the lattice rotation ω.

In principle, HR-EBSD gives access to both ε and ω. However, the elastic strains are usually neglected compared to the lattice rotations so that α is taken as curl (ω) with

$$\mathrm{curl}(\omega) = \begin{pmatrix} \dfrac{\partial \omega_{13}}{\partial x_2} - \dfrac{\partial \omega_{12}}{\partial x_3} & \dfrac{\partial \omega_{23}}{\partial x_2} & -\dfrac{\partial \omega_{32}}{\partial x_3} \\ -\dfrac{\partial \omega_{13}}{\partial x_1} & \dfrac{\partial \omega_{12}}{\partial x_3} - \dfrac{\partial \omega_{23}}{\partial x_1} & \dfrac{\partial \omega_{31}}{\partial x_3} \\ \dfrac{\partial \omega_{12}}{\partial x_1} & -\dfrac{\partial \omega_{21}}{\partial x_2} & \dfrac{\partial \omega_{32}}{\partial x_1} - \dfrac{\partial \omega_{31}}{\partial x_2} \end{pmatrix} \tag{14.3}$$

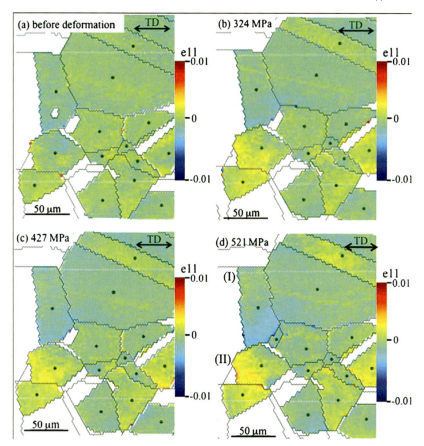

Figure 14.13 Elastic strain maps via the Wilkinson (small strain) method of an austenitic stainless steel deformed *in situ*, from [20]. ε_{11} is along the tensile direction. (a) before deformation, (b) at 324 MPa, (c) at 427 MPa, and (d) at 521 MPa.

It should be noticed that the expression involves derivatives in the three directions of space, which cannot all be evaluated by conventional 2-D electron microscopy—this has been referred to as the "opacity problem" [28]. In most studies, these derivatives are taken as zero as does in fact occur in specific configurations such as uniaxial compression or wedge indentation of symmetrically-oriented single crystals [28,29]. In other cases, the result is taken as a lower bound of dislocation density.

Man *et al.* [30] describe in detail how the derivatives in curl (ω) can be estimated from EBSD orientation maps. With the data aligned on a 2-D rectangular grid, the derivatives along the two directions of the grid are evaluated from the orientations of neighboring pixels along these directions. Given two neighboring pixels along the direction x, of orientations g_1 and g_2, and separated by a distance Δl, the expression of the derivative along x is,

$$\frac{\partial \omega}{\partial x} = \frac{1}{\Delta l} \ln(g_2^{-1} g_1) \text{ where } \ln(a) = \frac{\theta}{2\sin\theta}(a - a^{-1}) \quad (14.4)$$

g_1 and g_2 must be close to each other in orientation space (i.e., crystal symmetry must be taken into account) and θ is the angular part of rotation a. As described by Man et al. [30], this expression has the advantage that the result lies in the tangent space of orientation space (the space of skew-symmetric tensors).

Nye's tensor $\boldsymbol{\alpha}$ can also be related to the population of geometrically necessary dislocations, herein reduced to a finite number N of dislocation families,

$$\alpha = \sum_{s=1}^{N} \rho^s (b^s \otimes t^s) \quad (14.5)$$

where b^s is the Burgers vector, t^s is the tangent line direction and ρ^s is the density (in m^{-2}). As described by Arsenlis and Parks [31] for a face-centered cubic lattice with $\{111\}\langle 110 \rangle$ slip systems, and dislocations considered to be either of pure edge or pure screw types, 18 different dislocation families should be considered.

To determine the dislocation densities from lattice curvature, Eq. (14.5) can be rewritten as a system of linear equations: $A[\rho] = \Lambda$ where Λ contains the components of Nye's tensor $\boldsymbol{\alpha}$, $[\rho]$ is the dislocation density vector, and A is populated by the geometric parameters of the dislocation families. For an FCC lattice, as the problem is underdetermined, an additional criterion must be used to determine a unique solution for $[\rho]$. The latter can be obtained by looking for the minimum of a given function. Using the L^2 norm of $[\rho]$ has the advantage of providing an analytic solution, and has been used, for example, by Field et al. [32]. The physical argument for using such a norm is not obvious. On the other hand, the L^1 norm

$$\rho = \sum_{s=1}^{N} \rho^s \quad (14.6)$$

represents a quantity of interest in materials applications: the total GND density. A variant consists of minimizing the total dislocation energy, which requires weighting the densities of the different families by the line energy [26]. In both cases, the problem can be solved using the standard linear optimization techniques, for example, the simplex algorithm.

In the following, we apply the above method (with the L^1 norm), to the analysis of the GND density distribution in a Ni single crystal of cube orientation deformed in plane strain compression. This configuration has been the subject of previous studies in our group, on the deformation mechanisms of cube-oriented crystals [33]. Two Ni single crystals were deformed to true strains of 0.4 and 0.85, respectively. Typical areas of the microstructure were mapped by high-angular EBSD. Figure 14.14 provides the orientation distributions at the two strains and the (macroscopic) stress–strain curve. The increase of orientation spread and the strain hardening are well-known, characteristic, features of plastic deformation and GND storage.

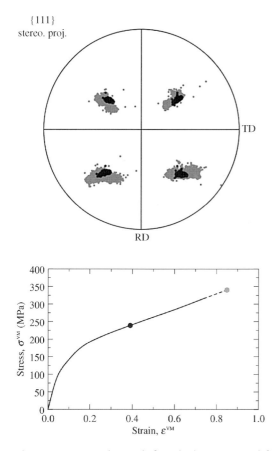

Figure 14.14 Ni single crystal of initial cube orientation deformed in plane strain compression at strains $\varepsilon = 0.4$ (black) and $\varepsilon = 0.85$ (gray). (a) Orientations and (b) stress–strain curve.

Orientation maps and GND density maps are provided in Figures 14.15 and 14.16, respectively. The banded microstructure obtained at a strain of 0.85 is the characteristic of the deformation of unstable cube crystals [32].

The average GND densities obtained from the maps have values of 0.8×10^{14} m^{-2} and 2.3×10^{14} m^{-2} at epsilon = 0.4 and 0.85, respectively.

The GND density values calculated from the EBSD measurements can be compared to the values of the (total) dislocation density expected from the strain hardening as shown by the stress–strain curve (Figure 14.14b). From dislocation theory, and to first order, the flow stress is related to the total dislocation density, ρ^{tot}, by

$$\sigma(\varepsilon) = \sigma^0 + \alpha.M.G.b\sqrt{\rho^{tot}}(\varepsilon) \tag{14.7}$$

where the hardening coefficient α is usually between 0.3 and 0.4, M is the Taylor factor $= \sqrt{6}$ for the cube orientation, G is the shear modulus $= 67$ GPa for Ni, and $b = 0.305$ nm.

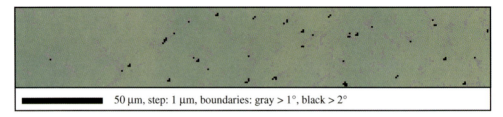
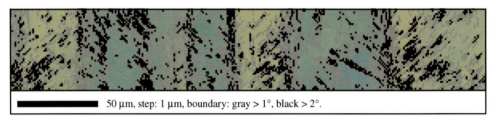

Figure 14.15 Orientation maps of a Ni single crystal of cube orientation deformed in plane strain compression. (a) $\varepsilon = 0.4$, (b) $\varepsilon = 0.85$. Rodrigues vector coloring. TD horizontal, RD vertical.

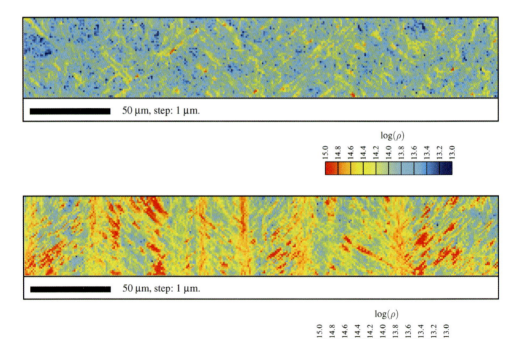

Figure 14.16 GND density maps of a Ni single crystal of cube orientation deformed in plane strain compression. (a) $\varepsilon = 0.4$ and (b) $\varepsilon = 0.85$.

Taking the crystal yield stress as about 30 MPa gives limiting values of the total dislocation density of:

Strain 0.4: ρ^{tot} between 1.1 ($\alpha = 0.4$) and 2.0 ($\alpha = 0.3$) $\times 10^{14}$ m^{-2} and
Strain 0.85: ρ^{tot} between 2.6 ($\alpha = 0.4$) and 4.7 ($\alpha = 0.3$) $\times 10^{14}$ m^{-2}.

The GND dislocation densities obtained by EBSD are of the same order of magnitude as the total dislocation densities expected from the stress–strain curve and tend to be closer to the lower bounds of the total dislocation densities. If we assume $\alpha = 0.3$ then the GND densities are about half the total dislocation densities in these two samples.

14.3.4
Crystal Defects and Elastic Strains

In the general case dislocations are, of course, associated with both lattice misorientations and localized elastic strains. The HR-EBSD method has not yet been extended to generalized, defect induced, rotations and elastic strains but we should like to give an example of elastic strains around a slip event in a lightly deformed 316 Fe–Ni–Cr fcc single crystal in Figure 14.17.

Figure 14.17 shows a linear defect (top left to bottom right) whose strain and rotation components are given in the standard sample reference frame in (a). However, slip is expected on $\{111\}\langle110\rangle$ and the line is indeed parallel to the slip plane trace of $(-1 -1\ 1)$ so the rotation components are then plotted out in the three possible slip coordinates with the three potential Burgers vectors (labeled D1–D3). The third column of Figure 14.17c shows rotations around the plane normal that are basically extinct. The first column shows rotations around the slip direction (// dislocation Burgers vector) and this component vanishes for the slip system D2. The second column shows rotations around the dislocation line: this component has maximum contrast in the slip system D2.

These observations are consistent with the rotation field expected around threading edge dislocations on the D2 system. This is a low stacking fault energy (SFE) metal so that edge dislocation pile-ups can be expected on the $\{111\}$ planes.

Dislocations have been identified in the SEM by channeling contrast in nitride thin films [34] but the above analysis appears to be a first example of dislocation type identification in a SEM directly from the strain fields using HR-EBSD.

14.4
Discussion

HR-EBSD is currently in a state of rapid evolution since new problems with the relative cross-correlation method have emerged recently and general solutions to these issues have not yet been proposed, although some ideas are being developed. In many cases, the problems (e.g., those related to the pattern center or orientation gradients) have been revealed by careful numerical validation tests using simulated patterns for which all the sample and projection parameters are known. It appears that, prior to these numerical validation tests, some "elastic" strain maps obtained by cross-correlation HR-EBSD have been published without adequate verification. In more detail, we can summarize the situation as follows:

First, it is clear that the cross-correlation method is, in general, valid when the material is only deformed elastically (although care should be exercised to correct for

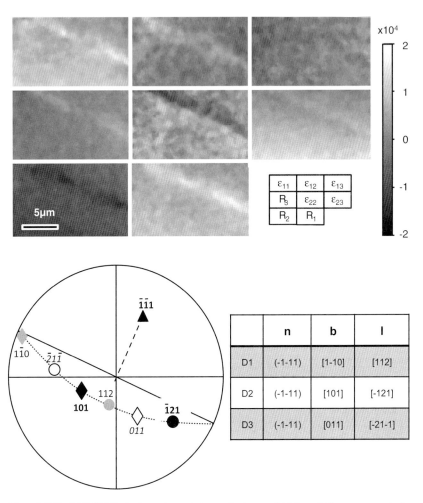

Figure 14.17 A HR-EBSD elastic strain analysis of a slip line in lightly deformed 316 austenitic stainless steel crystal; (a) map of the 8 strain and rotation components in the (standard) sample reference frame, (b) the principal crystallographic directions on a stereographic projection, and (c) the rotation components for the three possible Burger's vectors: column 1 rotations about ⟨110⟩, column 2 rotations about the dislocation line vectors, and column 3 rotations about the slip plane normal.

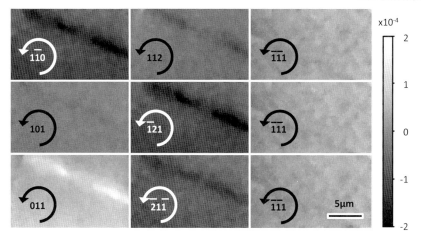

Figure 14.17 (Continued)

lens distortion and avoid major PC errors, that is, compensate for beam shift between the two patterns to be cross-correlated). Pure elastic strain measurements by the pattern shift HR-EBSD method have been validated by independent techniques for thin films, bent crystals, and oxide samples and shown to give satisfactory results—with the advantage of excellent spatial resolution.

Second, the application of HR-EBSD to more accurate misorientation measurements is another area where the technique can be employed directly – see, for example, the substructures developed in as-solidified superalloy crystals. Deformation substructures in particular are often characterized by low-angle dislocation cells whose misorientation distributions are important quantities but are difficult to determine by standard EBSD. Current methods involve much postprocessing of EBSD maps (Kuwahara filtering etc.) that significantly influences the misorientation distributions [35,36]. HR-EBSD could be a useful solution to this problem. It also provides better input misorientation data for GND analyses based on the hypothesis of a direct relation between misorientation and dislocation density, that is, by neglecting elastic strains. Note that misorientations are local, relative, measurements for which absolute values of the camera projection parameters are unnecessary.

The major problems concern the frequent situations where lattice rotations are combined with elastic strains as occurs in most dislocated crystals. In these cases, the standard cross-correlation method is severely limited both by the poor cross-correlation quality and the pattern center inaccuracy. A first step toward improving HR-EBSD measurements in these situations is the very recent introduction of the Rotated Reference technique that compensates for the poor cross-correlation quality due to lattice rotations between the reference and current patterns. Nevertheless, the pattern center problem is still not adequately solved so that artifacts might be introduced by the remapping step [13].

Table 14.1 Summary of diffraction techniques for elastic strain measurements.

Technique	Spatial resolution (μm)	Strain sensitivity	Comments
Neutron diffraction	500	5×10^{-5}	
Microbeam X-ray diffraction	100	10^{-4}	
SEM Kossel	3	3×10^{-4}	
Laue synchrotron radiation	1	10^{-4}	Synchrotron
SEM HR-EBSD	0.02	10^{-4}	
TEM CBED	0.02	10^{-4}	Thin foil

Another potential problem is of course the fact that a known reference point state might not be readily available for the analysis, in which case only spatial variations of stress and strain can be characterized.

The question now is whether more original EBSD analyses can provide the missing information. In this context, there is some on-going work on the development of an absolute HR-EBSD method using the 3D Hough transform, that is, a detailed analysis of the K band curvatures that depend, in part, on sample geometry. It is quite plausible that in the near future, the exact shape of the Kikuchi hyperbolae can be used to provide the missing PC information. This would be a major step forward since there is a critical requirement for resolving elasto-plastic problems at the local scale; clearly there is a very high potential for HR-EBSD in this respect. One should recall the general possibilities of strain measurements by diffraction at the local scale as shown in Table 14.1:

From Table 14.1 it is clear that only HR-EBSD and TEM convergent beam electron diffraction (CBED) possess a spatial resolution substantially better than the micron (both are about 20–30 nm). However, TEM suffers from the handicap of thin foil preparation that alters the strain distribution. HR-EBSD also requires good surface preparation but can be applied to selected zones of relatively large areas.

14.5
Conclusions

The current state of research on HR-EBSD is reviewed with the aim of developing valid applications for problems in the modern materials science.

It is shown that HR-EBSD is capable of measuring misorientations to about $1/100°$ and elastic strains to 10^{-4} but this requires very careful calibration of the EBSP acquisition system (i.e., the sample-phosphor screen-camera configurations, particularly, the pattern center position and the camera distortions).

Many-beam dynamical simulations of EBSPs of deformed crystals provide a very useful guide to both the possible errors that can occur and appropriate methods for reducing them.

The relative (pattern shift) method can be used for pure elastic strains, exemplified by the cases of bent Si single crystals and distortions in epitaxial thin films. It can also give high-quality data for dislocation boundary analyses (GNDs and LAGBs).

The problems of dislocated crystals characterized by orientation and strain gradients are discussed and some possible solutions are outlined. An original application to the strain fields of dislocation arrays in a low SFE stainless steel crystal is described.

It is concluded that HR-EBSD has a major potential to resolve many micromechanics characterization problems at the scale of tens of nanometers but that this entails very careful analysis of local crystal distortion, its influence on Kikuchi band diffraction and the acquired EBSP.

Acknowledgments

This work has been partially funded by the French National Agency (ANR) as part of the SAKE and AMOS projects (ANR-07- NANO-029 & ANR-10-NANO-015). The authors wish to thank Austin Day and Ken Mingard for many fruitful discussions on HR-EBSD. We should also like to dedicate this paper to the memory of Professor Claude Goux who passed away in 2012 and who, many years ago, was a strong proponent of developing high angular resolution techniques.

References

1 Humphreys, F.J. (2001) Review – Grain and subgrain characterisation by electron backscatter diffraction. *Journal of Materials Science*, **36** (16), 3833–3854.

2 Dingley, D. (2004) Progressive steps in the development of electron backscatter diffraction and orientation imaging microscopy. *Journal of Microscopy-Oxford*, **213**, 214–224.

3 Wright, S.I., Nowell, M., and Field, D.P. (2011) A review of strain analysis using electron backscatter diffraction. *Microscopy and Microanalysis*, **17**, 1–14.

4 Troost, K.Z., Vandersluis, P., and Gravesteijn, D.J. (1993) Microscale elastic-strain determination by backscatter kikuchi diffraction in the scanning electron-microscope. *Applied Physics Letters*, **62** (10), 1110–1112.

5 Wilkinson, A. (1996) Measurement of elastic strains and small lattice rotations using electron back scatter diffraction. *Ultramicroscopy*, **62**, 237–247.

6 Wilkinson, A.J. (2000) Advances in SEM-based diffraction studies of defects and strains in semiconductors. *Journal of Electron Microscopy*, **49** (2), 299–310.

7 Wilkinson, A.J., Meaden, G., and Dingley, D.J. (2006) High-resolution elastic strain measurement from electron backscatter diffraction patterns: New levels of sensitivity. *Ultramicroscopy*, **106** (4–5), 307–313.

8 Winkelmann, A., Trager-Cowan, C., Sweeney, F., Day, A.P., and Parbrook, P. (2007) Many-beam dynamical simulation of electron backscatter diffraction patterns. *Ultramicroscopy*, **107** (4–5), 414–421.

9 Maurice, C. and Fortunier, R. (2008) A 3D Hough transform for indexing EBSD and Kossel patterns. *Journal of Microscopy-Oxford*, **230** (3), 520–529.

10 Villert, S., Maurice, C., Wyon, C., and Fortunier, R. (2009) Accuracy assessment of elastic strain measurement by EBSD. *Journal of Microscopy-Oxford*, **233** (2), 290–301.

11 Maurice, C., Driver, J.H., and Fortunier, R. (2012) On solving the orientation gradient dependency of high angular resolution EBSD. *Ultramicroscopy*, **113**, 171–181.

12 Spence, J.C.H. and Zuo, J.M. (1992) *Electron Microdiffraction*, Plenum Press, New York, London.

13 Britton, T.B., Maurice, C., Fortunier, R., Driver, J.H., Day, A.P., Meaden, G., Dingley, D.J., Mingard, K., and Wilkinson, A.J. (2010) Factors affecting the accuracy of high resolution electron backscatter diffraction when using simulations. *Ultramicroscopy*, **110**, 1443–1453.

14 Mingard, K., Day, A.P., Maurice, C., and Quested, P. (2011) Towards high accuracy calibration of EBSD systems. *Ultramicroscopy*, **111**, 320–329.

15 Brown, D.C. (1966) Decentering distortion of lenses. *Photogrammetric Engineering*, **32** (3), 444–462.

16 Britton, T.B. and Wilkinson, A.J. (2012) High resolution electron backscatter diffraction measurements of elastic strain variations in the presence of larger lattice rotations. *Ultramicroscopy*, **114**, 82–95.

17 Krieger Lassen, N.C. (1999) Source point calibration from an arbitrary electron backscattering pattern. *Journal of Microscopy-Oxford*, **195**, 204–211.

18 Maurice, C., Dzieciol, K., and Fortunier, R. (2011) A method for accurate localisation of EBSD pattern centres. *Ultramicroscopy*, **111**, 140–148.

19 Basinger, J., Fullwood, D., Kacher, J., and Adams, B. (2011) Pattern centre determination in electron backscatter diffraction microscopy. *Microscopy and Microanalysis*, **17** (3), 330–340.

20 Ojima, M., Adachi, Y., Suzuki, S., and Tomota, Y. (2011) Stress partitioning behavior in an fcc alloy evaluated by the in situ/ex situ EBSD-Wilkinson method. *Acta Materialia*, **59**, 4177–4185.

21 Villanova, J., Maurice, C., Micha, J.-S., Bleuet, P., Sicardy, O., and Fortunier, R. (2012) Multi-scale measurements of residual strains in stabilized zirconia layer. *J. App. Cryst*, **45**(5), 926–935.

22 Nye, J. (1953) Some geometrical relations in dislocated crystals. *Acta Metallurgica*, **1**, 153–162.

23 Ashby, M.F. (1970) The deformation of plastically non-homogeneous materials. *Philosophical Magazine*, **21**, 399–424.

24 Kröner, E. (1962) Dislocations and continuum mechanics. *Appl. Mech. Rev.*, **15**, 599.

25 Gardner, C., Adams, B., Basinger, J., and Fullwood, D. (2010) EBSD-based continuum dislocation microscopy. *International Journal of Plasticity*, **26**, 1234–1247.

26 Wilkinson, A. and Randman, D. (2010) Determination of elastic strain fields and geometrically necessary dislocation distributions near nanoindents using electron back scatter diffraction. *Philosophical Magazine*, **90**, 1159–1177.

27 Karamched, P.S. and Wilkinson, A.J. (2011) High resolution electron backscatter diffraction analysis of thermally and mechanically induced strains near carbide inclusions in a superalloy. *Acta Materialia*, **59** (1), 263–272.

28 El Dasher, B., Adams, B., and Rollett, A. (2003) Viewpoint: experimental recovery of geometrically necessary dislocation density in polycrystals. *Scripta Materialia*, **48**, 141–145.

29 Kysar, J., Gan, Y., Morse, T., Chen, X., and Jones, M. (2010) High strain gradient plasticity associated with indentation into face-centered cubic single crystals: Geometrically necessary dislocation densities. *J. Mech. Phys. Solids*, **55**, 1554–1573.

30 Man, C.-S., Gao, X., Godefroy, S., and Kenik, E. (2010) Estimating geometric dislocation densities in polycrystalline materials from orientation imaging microscopy. *International Journal of Plasticity*, **26**, 423–440.

31 Arsenlis, A. and Parks, D. (1999) Crystallographic aspects of geometrically-necessary and statistically stored dislocation density. *Acta Materialia*, **47**, 1597–1611.

32 Field, D., Trivedi, P., Wright, S., and Kumar, M. (2005) Analysis of local orientation gradients in deformed single crystals. *Ultramicroscopy*, **103** (1), 33–39.

33 Basson, F. and Driver, J.H. (2000) Deformation banding mechanisms during plane strain compression of cube-oriented f.c.c. crystals. *Acta Mater.*, **48** (9), 2101–2115.

34 Trager-Cowan, C., Sweeney, F., Trimby, P.W., Day, A.P., Gholinia, A., Schmidt, N.-H., Parbrook, P.J., Wilkinson, A.J., and Watson, I.M. (2007) Electron backscatter diffraction and electron channeling contrast imaging of tilt and dislocations in nitride thin films. *Physical Review B – Condensed Matter and Materials Physics*, **75** (8), 085301.

35 Humphreys, F.J., Bate, P.S., and Hurley, P.J. (2001) Orientation averaging of electron backscattered diffraction data. *Journal of Microscopy*, **201** (1), 50–58.

36 Godfrey, A., Mishin, O.V., and Liu, Q. (2006) Processing and interpretation of EBSD data gathered from plastically deformed metals. *Materials Science and Technology*, **22** (11), 1263–1270.

15
4D Characterization of Metal Microstructures
Dorte Juul Jensen

15.1
Introduction

The Materials Science theme generally aims at linking materials processing with materials properties and performances. The key toward understanding this link is typically in the materials microstructure, and microscopy, recently in particular electron microscopy, is by far the most valuable experimental tool for microstructure investigations. All the microscope techniques are by nature 2D (x, y) techniques that limit the microstructure information and have led to misinterpretations and mistakes (e.g., [1]). Transmission electron microscopy (TEM) allows $2^1/_2$ D (x, y, thin z section) characterizations both by tomography and diffraction techniques (e.g., [2–4]). A third dimension, namely time, t, has been studied by electron microscope techniques, and the evolution of a specific microstructure during for example deformation or annealing has been studied by implementing stress rigs and furnaces in the microscope (e.g., [5–7]). Such measurements clearly reveal the microstructural evolution in the sample surface, but uncertainties of course always exist if the observations are affected by the free surface, and bulk phenomena occurring below the inspected surface may suddenly appear – for example, a nucleus may form below the surface and grow to be seen in the inspected surface [8]. So clearly 3D (x, y, z) and 4D (x, y, z, t) techniques are the way forward for many types of experiments.

Three-dimensional techniques are not new. Serial sectioning has been used for decades (e.g., [9,10]), x-ray methods have long been used for local bulk characterization (e.g., [11,12]), and neutron diffraction has been used to determine, for example, average textures and residual stresses in large bulk sample volumes (e.g., [13,14]).

What is new is that within the last 10–20 years; the scientific community has really realized the need for 3D and 4D techniques and has devoted significant efforts to advance the existing 3D methods and to develop new unique techniques. The goals in these efforts include: much easier, less manpower-requiring operations; better spatial resolution; nondestructive methods; fast measurements. This has led to a whole suite of modern 3D techniques, and 3D results are reported in

Microstructural Design of Advanced Engineering Materials, First Edition. Edited by Dmitri A. Molodov.
© 2013 Wiley-VCH Verlag GmbH & Co. KGaA. Published 2013 by Wiley-VCH Verlag GmbH & Co. KGaA.

more and more publications. The techniques behind the most frequently published 3D results are in two groups: advanced serial sectioning and x-ray methods. The serial sectioning methods encompass semiautomatic and fully automatic mechanical sectioning with microstructural inspection in optical as well as scanning electron microscopes (SEMs) (e.g., [15–18]), high resolution focused ion beam (FIB) sectioning in the scanning microscope (e.g., [19,20]) and sectioning by lasers [21]. The new x-ray methods include three-dimensional x-ray diffraction (3DXRD) for fast nondestructive orientation measurements (e.g., [22–24]) and 3D polychromatic x-ray microdiffraction for high spatial resolution orientation measurements (e.g., [25–27]) as well as a whole range of x-ray tomography techniques (e.g., [28–31]) and techniques where slits [32] and collimators [33] are used to get the third dimension.

15.2
4D Characterizations by 3DXRD – From Idea to Implementation

The vision was to develop a technique allowing nondestructive measurements in selected local micrometer-sized volumes within the bulk of cubic millimeter to cubic centimeter samples. It was motivated by the need to study local deformation and recrystallization phenomena occurring in the bulk *in situ*, while samples were deformed or annealed. For example, for deformation it is not known how the grains inside a polycrystalline sample break up and rotate in crystallographic orientation. For recrystallization, it is remarkable that real key phenomena are not known, for example, where do the nuclei form and with which crystallographic orientations, how do the annealing-induced boundaries surrounding the nuclei and recrystallizing grains migrate and interact with all the deformation-induced dislocation boundaries and what parameters are of importance here.

A basic requirement for developing a technique to fulfill the described vision is a beam that can penetrate deeply (>1 mm) in materials and which is so intense that even diffraction signals from micrometer-sized volume elements can be distinguished from the background. High energy x-rays from synchrotrons provide such a beam. When the energy of the x-rays is about 80 keV, the penetration depth (defined as the depth at which the transmitted intensity is 10%) is 5 mm in steel and 4 cm in aluminum. A very intense beam on the sample is obtained by focusing.

The first paper describing the idea behind the 3DXRD microscope and the first set of preliminary data measured at HASYLAB in Hamburg, Germany, was published already in 1995 [22]. A permanent installation of a 3DXRD microscope at the European Synchrotron Radiation Facility (ESRF) in Grenoble, France was decided in 1998 and the first real 3DXRD microscope was commissioned during the summer 1999 and has been in regular operation ever since. During the recent years, it has been upgraded to obtain finer spatial resolution including moving it to a new experimental hutch.

15.2.1
The 3DXRD Microscope

The principal components of the 3DXRD microscope are sketched in Figure 15.1. The incoming beam is a monochromatic focused x-ray beam tunable to energies in the range 45–100 keV. The focusing is obtained by means of a bent Laue crystal and/or multilayer mirrors [34,35].

The sample tower (Figures 15.1 and 15.2) allows translations (x, y, z), rotation (ω), and tilting of the sample with an absolute accuracy of 1 μm. Detectors (Figure 15.1) can be translated in all directions (x, y, z). Several charge coupled device (CCD) detectors are available, covering a range of pixel sizes from 1.2 to 225 μm. Detectors with a small pixel size are typically placed close to the sample for optimal spatial resolution, while those with a large pixel size are placed far from the sample for strain characterization. Tools to control the sample environment include a cryostat, a tensile machine, and several furnaces.

The incoming monochromatic beam penetrates the sample, and all microstructural volume elements within the illuminated gauge volume that fulfill the Bragg condition will generate diffracted beams. The incoming beam is confined by focusing, possibly in combination with slits. It determines the gauge volume, which is typically a plane or a volume within the sample. To probe the complete crystal structure within the illuminated layer or volume, the sample is rotated (ω) around the axis perpendicular to the incoming beam (Figure 15.2).

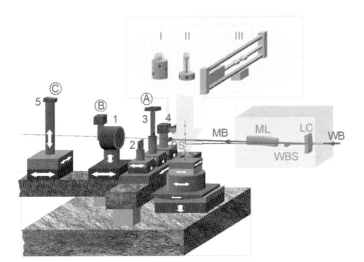

Figure 15.1 Sketch of the principal components of the 3DXRD, where WB is white beam; LC, bent Laue crystal; ML, curved multilayer; WBS, white beam stop; MB, two-dimensional microfocused monochromatic beam; and BS, monochromatic beam stop. Sample environment: (I) cryostat; (II) furnace; (III) 25 kN stress rig. Detector bank: (1) large-area detector; (2) conical slit system; (4) high-resolution area detector; and (5) optional detector system [37].

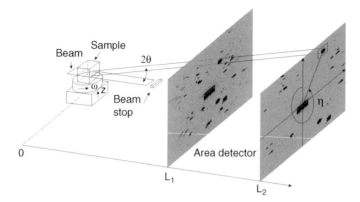

Figure 15.2 Sketch of the tracking procedure. Diffraction patterns (examples shown in gray) are acquired on a high-resolution two-dimensional detector positioned at several distances to the sample. A so-called far-field detector is placed further away from the sample (see detector No. 1 in Figure 15.1) [37].

A critical issue is how the position of the diffracting volume within the gauge volume is determined. The conventional approach for this is to limit the gauge volume to a point by slit(s) and scan the sample with respect to this point. However, in general, this approach is too slow for dynamic studies. Hence, it is replaced in the 3DXRD concept by a tracking method that allows diffraction from large gauge volumes [36,37]. By recording the diffraction pattern at two or more distances between the sample and the detector(s) (L_1, L_2, ..., and so on in Figure 15.2), it is possible to extrapolate a straight line through the corresponding diffraction spots at different distances all the way to the sample and determine the position of the diffracting volume element (Figure 15.2).

The crystallographic orientations of individual grains are determined directly from the positions of the spots recorded on the detector, using the information from all the ω settings. This determination is not possible if there is too much overlap of the spots in the diffraction patterns. Fewer diffracting grains is the solution, which may be achieved by inserting a conical slit [36] between the sample and the detector, by making the beam size smaller, or by reducing the sample size.

At present, the mapping precision is $\sim 1\,\mu m \times 1\,\mu m \times 0.5\,\mu m$, while microstructural elements down to 70 nm can be detected provided they have a sufficient crystallographic orientation difference to avoid overlap of diffraction spots on the detector. However, the x, y, z positions of these small elements cannot be determined at present to better than the mapping precision. For further details see [37].

The aforementioned describes the classical 3DXRD set-up, which is now implemented not only at beamline ID11 at ESRF but also at sector 1-ID at the advanced photon source (APS) in Chicago, USA and at High Energy Materials Science beamline (HEMS) at the radiation source PETRA-III in Hamburg, Germany.

X-ray tomography combined with 3DXRD is another version of the classical 3DXRD, which is referred to as diffraction contrast tomography (DCT) [29,30].

15.2 4D Characterizations by 3DXRD – From Idea to Implementation

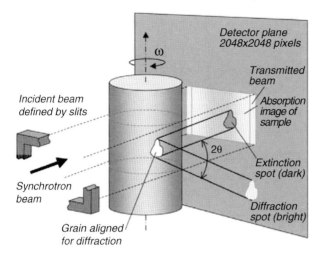

Figure 15.3 Sketch of the DCT set-up. The middle part of the detector (shown in light gray) records usual tomography data, whereas the outer parts record diffraction data similar to 3DXRD [29,30].

The DCT set-up is sketched in Figure 15.3. It only uses the near field detector. The incident beam (and thus the directly transmitted beam) will illuminate only the central part of the detector (shown in light gray in Figure 15.3). Thereby standard absorption contrast tomography is observed in this small central part of the detector, whereas the outer regions are used to record diffraction spots (see Figure 15.3). For further details, see [29]. Compared to the classical 3DXRD instruments, the DCT set-up is more robust and currently 3D maps of grain structures with the highest spatial resolution are made in this way. On the other hand, the DCT set up is less versatile and in particular there are fewer options for crystallography [38].

15.2.2
Another Approach for 3D Mapping of Crystallographic Contrast Microstructure by Synchrotron x-rays

Another instrument called the polychromatic x-ray microdiffraction technique is in operation at APS sector 34-ID [25–27]. This instrument is, however, significantly different from the 3DXRD microscope; it operates at lower x-ray energies, typically applies a polychromatic microbeam, and performs wire scanning to get the three-dimensional information. This means that the APS polychromatic x-ray microdiffraction technique has much better spatial resolution but shorter penetration depths and much worse time resolution than the 3DXRD. Typical deformation microstructures can be beautifully resolved in 3D by the polychromatic x-ray microdiffraction technique, but it is typically better to perform *ex-situ* (rather than *in situ*) experiments on this instrument.

15.3
Examples of Applications

The 3DXRD microscope has been used for a range of different applications. Figure 15.4 shows key examples. As will be described, various experimental 3DXRD methodologies can be chosen. The simplest is to measure the diffracted intensity from selected structural elements (such as grains or subgrains) as a function of time while exposing the sample to stimuli such as stress and temperature. This provides data on volume kinetics of individual structural elements with the highest possible time resolution of seconds or subseconds. The most comprehensive 3DXRD method allows *in-situ* measurements of complete three-dimensional microstructural maps with typical time resolutions of tens of minutes or even hours. These modes of operation will

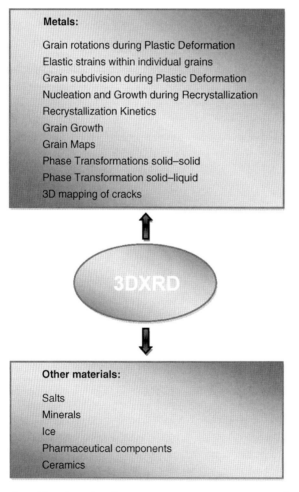

Figure 15.4 Examples of 3DXRD experiments covering a range of materials. In the figure the experiments for metallic materials are highlighted at the top.

be illustrated in the following where examples of applications will be given. These examples will focus mostly on recrystallization but also 3DXRD results for plastic deformation, grain growth, and phase transformation shall be briefly reviewed.

15.3.1
Recrystallization Studies

15.3.1.1 Nucleation

Nucleation of recrystallization is very difficult to study experimentally; the nuclei are small and few, so large data sets are typically needed to get statistically sound data. Furthermore, if the issue is to study nucleation sites or orientation relationships between nuclei and parent site in the deformed sample, static 2D measurements suffer from the so-called lost-evidence problem [39], namely that when a nucleus is observed with a 2D (x, y) technique, it is impossible to see what was there before the nucleus formed so the exact nucleation site is unknown. Also results obtained by *in-situ* annealing measurements in, for example, a SEM (x, y, t) may be problematic to interpret due to surface effects.

In order to predict the recrystallization texture and understand the various nucleation mechanisms, it is important also to know which crystallographic orientations the nuclei have at the various nucleation sites. It is generally assumed that nuclei form with orientations present at the nucleation sites in the deformed state. However, other publications report on formation of nuclei with new orientations that are not seen in the deformed microstructure (e.g., [8,40–42]). With the classical 2D or the serial sectioning 3D method one can only guess which orientations were present at the exact nucleation sites before the annealing.

Only 4D investigations can quantify if nuclei may form with orientations different from those present at the nucleation sites prior to annealing. 3DXRD has been used for such investigations [43,44] and it was shown that nuclei with new orientations do form and their orientations cannot be explained by twinning up to second order (see Figure 15.5). The data were analyzed for relationships between the new nuclei

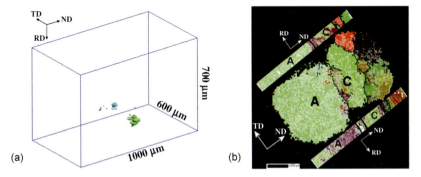

Figure 15.5 (a) Position of six nuclei in Al cold rolled 30% and annealed to the beginning of recrystallization. (b) Plane section and side sections of the microstructure as characterized by 3DXRD and EBSP, respectively, with nuclei positions marked by numbers [44].

orientations and the rotation axis across deformation induced-dislocation boundaries in the deformed parent grains. The results suggest that the origin for the formation of nuclei with new orientations relates to the misorientation distribution in the deformed parent grain and that the misorientation axes between a nucleus and the parent grain may be identical to a misorientation axes across dislocation boundaries within that grain [44].

15.3.1.2 Grain Boundary Migration

During recrystallization the grains grow by boundary migration through the deformed matrix. Whenever two grains impinge, the boundary motion will stop. Further migration is defined as grain growth. Generally, it is supposed that the velocity of boundary motion, v, is proportional to the driving force F as expressed by the following relationship:

$$v = MF \tag{15.1}$$

where M is the mobility [1]. For recrystallization, the driving force F is assumed to be given by the stored energy in the deformed microstructure [1] that typically is orders of magnitude larger than the driving force related to boundary curvature during grain growth. While several grain growth experiments have confirmed the validity of Eq. (15.1) (e.g., [45,46]) much less is done for recrystallization. The few recrystallization experiments focusing on this issue have all been done by static stereological 2D characterizations that either confirm Eq. (15.1) (e.g., [47,48]) or found a power–law relationship between v and F [49], that is, another simple relationship.

In-situ 2D investigations during recrystallization as well as *ex-situ* 2D characterizations of the same surface area after a series of consecutive annealing treatments, however, reveal very complex boundary migration and velocities that locally deviate significantly from that predicted using average F and M values in Eq. (15.1) [7,50]. Whether this is a real effect or a consequence of the 2D character of the experiment cannot be proven. Four-dimensional (x, y, z, t) experiments are required.

The migration of a boundary surrounding a single recrystallizing grain through a weakly deformed single crystal matrix has been studied *in situ* by 3DXRD [51]. These measurements also reveal a very complex boundary migration process. A number of snapshots from the *in-situ* film are shown in Figure 15.6. It is found that

1) Flat facets may form, which migrate forward at a constant rate for extended periods of time.
2) The nonfaceted segments of the boundary do not migrate at a constant rate through the relatively homogeneous deformed matrix of a single crystal. On the contrary, segments of the boundary move forward for a while then stop, move again and so on (stop–go motion).
3) Local protrusions and retrusions typically form locally on the migrating boundary. After these measurements were published [51], attention has also been given to optical and electron micrographs of partly recrystallized structures, and many

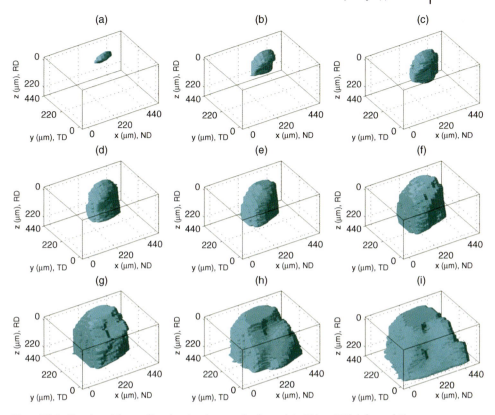

Figure 15.6 Storyboard from a film showing the growth of a nuclei within a 42% deformed Al single crystal of {1 1 0} ⟨0 0 1⟩ orientation. Nucleation was stimulated by hardness indents. The annealing temperature was varied in the range 270–300 °C [51].

of these do show clear protrusions and retrusions along the recrystallization boundaries (see e.g., Figure 15.7).

As these results are observed directly in 4D, they cannot be due to experimental artifacts. Standard recrystallization theories and models (e.g., Eq. (15.1)) do not describe the 4D results well and have to be revised. For this purpose, the 3DXRD measurements have to be supplemented by detailed microscopical investigations of the deformed microstructure in front of the migrating boundary in order to understand what is going on [6]. Also, an experiment has been performed using the 3D polychromatic x-ray microdiffraction method at APS. As mentioned earlier, this technique has a much better spatial resolution than the 3DXRD and thus allows direct observations of effects of the local deformed microstructure on migration of recrystallization boundaries, which is considered instrumental for establishment of improved recrystallization theories and models.

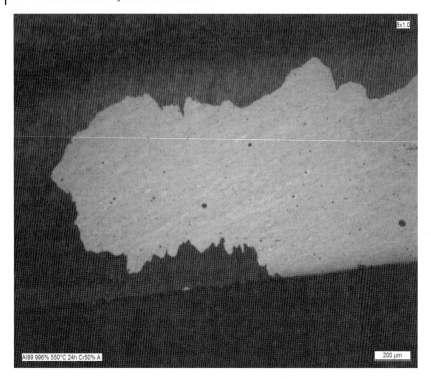

Figure 15.7 Optical micrograph of partly recrystallized aluminum (RD ND plane). Protrusions and retrusions are seen on the boundaries surrounding the recrystallizing grain.

15.3.1.3 Recrystallization Kinetics

The aforementioned examples have used the full grain mapping mode of 3DXRD. The fast mode, where only diffracted intensities (not peak shapes) are recorded, has been used to follow recrystallization kinetics.

Traditionally, recrystallization kinetics is characterized by measuring the volume fraction of recrystallized material, V_1, in a series of partly recrystallized samples annealed for different times at a given temperature by optical or electron microscopy. 3DXRD kinetic experiments for 90% cold-rolled (cr) Al and Cu cr 92% have shown that growth rates of newly nucleated grains vary significantly from grain to grain as well as change with time [52]. Recent measurements for Al cr 50% and Al cr 98% have also shown big variations in growth rate from grain to grain [53,54].

In the Al cr 50% experiment, the annealing temperature was changed rapidly during a kinetic measurement, whereby determination of apparent activation energy Q for the individual grains is possible [53]. As shown in Figure 15.8, it was found that Q varies significantly and a wide distribution of activation energies was observed for the ensemble of grains. This result supports detailed calorimetric measurements during isothermal annealing of Cu cr 92%, where it was shown that the experimental data were much better fitted if a spectrum of activation energies (enthalpies) was assumed instead of a single value [55].

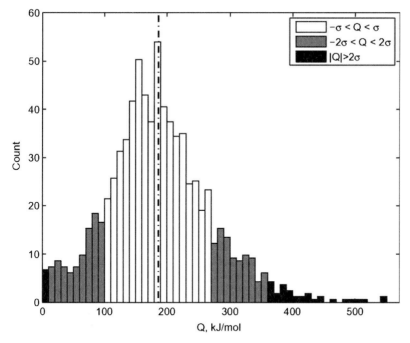

Figure 15.8 Distribution of grain-averaged activation energies. The mean of the distribution, $\langle Q \rangle = 187$ kJ mol^{-1} is indicated by the vertical line. The standard deviation is $\sigma = 82.9$ kJ mol^{-1}, and the 1 and 2σ confidence intervals are indicated by the color of the bars [53].

Possible effects of the fact that each grain has its own kinetics have been simulated for a series of size/growth rate distributions [56]. An example of the simulated kinetic results is given in Figure 15.9 showing that such distributions can significantly affect the overall (sample averaged) recrystallization kinetics. As seen, significant deviations from the straight black line representing standard assumptions are observed. Also the recrystallized grain size distribution is very strongly affected by the grain-to-grain variations in growth rates [56].

15.3.2
Other 4D Studies

15.3.2.1 Plastic Deformation

During plastic deformation, the individual grains in a polycrystalline sample break up and the crystal lattice rotate (e.g., [57]). X-ray line broadening experiments have been extensively used to estimate the size of coherently scattering domains, which may be related to the cell size and thus give information about the grain break up. In these types of measurements, ensembles of grains are measured simultaneously, which may lead to complications when analyzing the data. These complications are avoided by 3DXRD measurements focusing on individual (bulk) grains during *in-situ* tensile or compressive deformation [58,59].

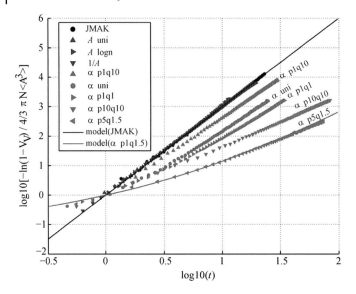

Figure 15.9 Normalized recrystallization kinetic curves averaged over grains with time dependent sizes $r = At^{1-\alpha}$, where A and α are distributions covering the range from almost delta functions to uniform distributions. The black full line is for the analytical solution according to the JMAK (Johnson, Mehl, Avrami, and Kolmogorov) model [56].

By 3DXRD, it is shown that the dislocation structures typically seen in TEM are not "graveyards," where the dislocations rest after the deformation is terminated, but they do develop during straining. Furthermore, the measurements document development of largely strain-free cells surrounded by cell walls with much higher dislocation density [59,60]. The data are used as basis for evaluation of existing deformation structure models and for development of improved models [59].

The crystal lattice rotation can also be followed *in situ* by 3DXRD during tensile deformation (e.g., [61]). By this type of measurements, it is shown for many metals that the rotations are determined by the individual grain orientation and to a far less extent by its neighbors. These data provide a unique database for critical evaluation of texture models for deformation of polycrystals.

15.3.2.2 Grain Growth

Grain growth subsequent to recrystallization should be an obvious and not too complicated experiment to perform using 3DXRD. However, so far not many papers with such data have been published. Challenges have shown to relate to getting the annealing times and temperatures right to allow mapping of many grains over enough annealing steps. An example, however, is shown in Figure 15.10. Here grain growth was followed during five annealing steps, whereby the initially 491 grains were reduced to 49 within the selected gauge volume [62]. Data of this type may be used for evaluation of models for grain growth, for identification of important microstructural parameters determining which grains grow and which shrink and for analysis of topological changes.

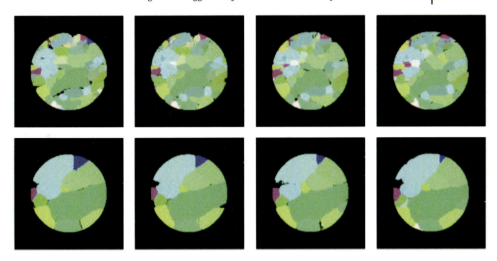

Figure 15.10 Grain growth in Al-0.1% Mn. A 3D sample volume of diameter 700 μm and height 350 μm was characterized after five annealing steps. The initial structure and that after the final annealing is shown for four layers within the sample volume [62].

15.3.2.3 Phase Transformation

Measurements of solid-state nucleation and grain growth during phase transformations are both extremely challenging, and 3DXRD does not currently fulfill all the requirements needed to fully quantify the processes [63]. However, the ferrite nucleation rate from austenite in medium carbon steel has been followed *in situ* by 3DXRD during slow continuous cooling [64,65]. When the measured rates are compared to classic nucleation model [66] predictions, it is found that the measured activation energies are orders of magnitude smaller than the predicted ones, and revisions to the model are currently underway.

In the papers from the Delft group (e.g., [64,65]) also the growth of the individual grains have been followed. The data were compared to the well-established theory of Zener for diffusion-controlled growth. For some grains good agreement was found, whereas others deviate from the theory. For a further discussion see [65].

Whereas the aforementioned results were obtained just watching the evolution in diffraction peaks (i.e., fast measurements resembling those for recrystallization kinetics mentioned earlier), full 4D mapping of cementite particle has also recently been achieved by phase contrast tomography [67] and the structural changes during solid–liquid phase transformation in a Cu Al alloy have been followed by diffraction contrast tomography (DCT) and successfully compared with phase field modeling [68].

15.4
Challenges and Suggestions for the Future Success of 3D Materials Science

In this section, challenges and suggestions for future successes, not only of 3DXRD, but of 3D and 4D experimental methods in general shall be discussed.

Indication of success for 3D/4D characterizations already exist: (i) results have been published, which prove that 2D is not enough and some new physical mechanisms have been suggested based on 3D/4D data, (ii) several of the experimental techniques have been further advanced and can now be considered as relatively mature, and (iii) the 3D/4D community is growing from the initial stage when it was mostly the group(s) behind a given technique, who was using that technique.

There is still, however, a long way to go before 3D/4D will be widely used. This is not unusual, and history has shown that some degree of commercialization is often needed. The electron back scattering pattern (EBSP) techniques for example was developed over many years in individual universities and national laboratories but not until two companies were created focusing on the EBSP method and in particular the data analysis; EBSP became a "need to have" technique in most materials science laboratories in academia and industry. Clearly of course the two companies realized correctly that the EBSP technique fulfills a general scientific need of sufficient importance to base a business model on. And the companies contributed significantly to the field by developing easy, user friendly, and fast experimental measurements and even more important, they developed systems for easy data analysis and beautiful data presentation.

Little commercialization has yet been realized for the 3D/4D methods, except from FIB in the SEM and the Robo-Met.3D technique that is developed by an American company and based on mechanical serial sectioning combined with optical microscopy. This is not because of lack of scientific need for the 3D/4D methods – some users may have been scared off because of the present complexities with the methods, but the need is there.

Major obstacles and suggestions for the future encompass:

There are many principally different 3D/4D methods; hence, one data analysis system cannot be developed that serves them all. The raw data analysis and the alignment procedure needed for example to get serial sections in register are different. Hence, this part of the data analysis clearly has to be developed by the instrument builders or in very close collaboration with these. From thereon, however, the 3D/4D data analysis and representation could (should) be united. The data sets are large, which may be a challenge in its own right, but other communities – in particular the game industries do handle far larger sets, and it might be an idea to learn from them. The whole 3D/4D materials society would benefit, if a company would take up this challenge (or the society could unite to work on just one representation system). Another challenge here is that in present days' publication channels, it is hard to visualize 3D/4D data. This would mean that in the papers we cannot focus on the pretty 3D pictures but on specific scientific points; which may actually be an advantage for this field.

The 3D/4D community is scattered, and the scientific focus is broad. It might be an advantage to keep the community close together via establishments of networks. This should be networks focusing on technique development, on science-technique questions and possibilities and as discussed above on data representation and visualization. Web-based sharing of information and tools may be a way forward.

Some of the 3D/4D methods (e.g., 3DXRD) only operate at large international facilities. Here it is either very costly to get beamtime or if the usual peer-reviewed beamtime application procedure is used, it is hard to get beamtime – success rates below 10% are not unusual and in all cases the time from idea to experiment is very long. This clearly limits the widespread use of these 3D/4D methods, and it is expected that this will always be the case in spite of the fact that more large scale facilities are being built. One might think of making some of the measurements faster and more standard, so some kind of block allocation system might be put into place as parts of the biology community has done. But generally, an advice is only to use the 3D/4D techniques at the large-scale facilities when it is really needed, that is, the results cannot be obtained in any other way and always be very well prepared with optimal and well-characterized samples for the 3D/4D experiment at the large scale facility.

Data analysis takes months, if not years. This is in particular true for the 3D/4D technique at the large scale facilities. Some *on-line* data visualization is extremely important, which can guide the next measurements and the user can avoid losing some of the precious beamtime on useless data. On-line visualization has been successfully implemented on some of the beamlines, for example, on APS 34-ID for the 3D polychromatic x-ray microdiffraction technique. Easier data analysis may come as a compromise between constantly improving the experimental techniques to reach new possibilities and keeping to some fixed standards, which can be used for series of measurements. Also, here establishment of communities between technique developers and scientific users could help getting this balance right.

15.5
Concluding Remarks

All experimental techniques have potentials and limitations and so does the 3DXRD technique. The main potentials are that the measurements are nondestructive and fast, that is, they allow full 4D characterizations. Limitations are as follows.

1) The technique is unique; it is constantly under development and is being used for many different types of measurements – (which is actually also a plus). A "newer ending" and substantial effort is thus required to develop and update the necessary data analysis software. In the future, a good balance between easier, more standard, yet scientifically important measurements (which benefit from faster and easier data analysis with already developed software) and forefront new "never done before" type of measurements have to be established.
2) The spatial resolution is limited and spot overlap may represent a problem, in particular in strongly textured or deformed metals. Work is underway to develop a new 3DXRD mode somewhat similar to dark field imaging in the electron microscope, whereby even small cells/subgrains may be differentiated from the "background" of the rest of the sample and mapped with very high precision.

In the field of recrystallization, 3D and 4D methods have given important new information that is neither consistent nor included in present day models. To advance modeling, experimental characterization is essential. Of course the best suite of experimental techniques has to be chosen for the actual problem. A combination of many 2D measurements with a few selected 3D/4D characterizations may often be the right choice to ensure statistical solid representative data and at the same time to avoid that serious mistakes and neglects are being made because of lack of information in the third and fourth dimension [69].

Acknowledgments

The Danish National Research Foundation is acknowledged for funding the Danish-Chinese Center for Nanometals, within which this review has been written.

References

1 Spanos, G., Rowenhorst, D.J., Chang, S., and Olson, G.B. (2010) 3D Characterization of Microstructures in Metallic Systems by Serial Sectioning. Proceedings of the 31st Risø International Symposium on Materials Science. Eds N. Hansen *et al.* 159–170.

2 Barnard, J.S., Sharp, J., Tong, J.R., and Midgley, P.A. (2006) High-resolution three-dimensional imaging of dislocations. *Science*, 313, 319–1319.

3 Barnard, J.S., Sharp, J., Tong, J.R., and Midgley, P.A. (2006) Three-dimensional analysis of dislocation networks in GaN using weak-beam dark-field electron tomography. *Philosophical Magazine*, 86, 4901–4922.

4 Liu, H.H., Schmidt, S., Poulsen, H.F., Godfrey, A., Liu, Z.Q., Sharon, J., and Huang, X. (2011) Three-dimensional orientation mapping in the transmission electron microscope. *Science*, 332, 833–834.

5 Zhang, X.D., Godfrey, A., Winther, G., Hansen, N., and Huang, X. (2010) In-Situ TEM Compression of Submicron-Sized Single Crystal Copper Pillars. Proceedings of the 31st Risø International Symposium on Materials Science. Eds. N. Hansen *et al.* 489–496.

6 Zhang, Y., Godfrey, A., and Juul Jensen, D. (2011) Local boundary migration during recrystallization in pure aluminium. *Scripta Materialia*, 64, 331–334.

7 Anselmino, E. (2007) Microstructural effects on grain boundary motion in AlMn alloys. PhD Thesis, Delft University of Technology

8 Sabin, T.J., Winther, G. and Juul Jensen, D. (2003) Orientation Relationships Between Recrystallization Nuclei at Triple Junctions and Deformed Structures. *Acta Materialia*, 51, 3999–4011.

9 Mangan, M.A., Lauren, P.D., and Shiflet, G.J. (1997) Three-dimensional reconstruction of Widmanstatten plates in Fe – 12.3 Mn – 0.8 C. *Journal of Microscopy*, 188, 36–41.

10 Vandermeer, R.A. and Gordon, P. (1959) Edge-nucleated, growth controlled recrystallization in aluminium. *Transactions of the Metallurgical Society AIME*, 215, 577–588.

11 Bowen, D.K. (1984) Application of Synchrotron X-ray Methods for the Characterization of Deformed and Recrystallized Microstructures. Proceedings of the 5th Risø International Symposium on Materials Science. Eds. N. Hessel Andersen *et al.* 1–17.

12 Dadson, A.B.C. and Doherty, R.D. (1992) Transmission pseudo-kossel (TK) studies of the structure of hot-deformed (dynamically recovered) polycrystalline

aluminium. *Acta Metallurgica et Materialia*, **40**, 345–352.

13 Juul Jensen, D. and Kjems, J.K. (1983) Apparatus for dynamical texture measurements by neutron diffraction using a position sensitive detector. *Textures and Microstructures*, **5**, 239–251.

14 Lorentzen, T. and Ibsø, J.B. (1995) Neutron-diffraction measurements of residual strains in offshore welds. *Materials Science and Engineering A*, **197**, 209–214.

15 Spowart, J.E. (2006) Automatic serial sectioning for 3D analysis of microstructures. *Scripta Materialia*, **55**, 5–10.

16 Rowenhorst, D.J., Lewis, A.C., and Spanos, G. (2010) Three-dimensional analysis of grain topology and interface curvature in a beta–titanium alloy. *Acta Materialia*, **58**, 5511–5519.

17 Sükösd, Z., Hanneson, K., Wu, G.L., and Juul Jensen, D. (2007) 3D spatial distribution of nuclei in 90% cold rolled aluminium. *Materials Science Forum*, **558–559**, 345–350.

18 Zhang, Y.H., Juul Jensen, D., Zhang, Y.B., Lin, F.X., Zhang, Z.Q., and Liu, Q. (2012) Three-dimensional investigation of recrystallization nucleation in a particle-containing Al alloy. *Scripta Materialia*, **67**, 320–323.

19 Uchic, M.D., Groeber, M.A., Dimiduh, D.M., and Simmonds, J.P. (2006) 3D microstructioral characterization of nickel superalloys via serial-sectioning using dual beam fib-sem. *Scripta Materialia*, **55**, 23–28.

20 Zaefferer, S., Wright, S.I., and Raabe, D. (2008) Three-dimensional orientation microscopy of a focused ion beam-scanning electron microscope: A new dimension of microstructure characterization. *Metallurgical and Materials Transactions A – Physical Metallurgy and Materials Science*, **39**, 374–389.

21 Echlin, M.L.P., Mottura, A., Torbet, C., and Pollock, T.M. (2012) A new tribeam system for three-dimensional multimodal materials analysis. *The Review of Scientific Instruments*, **83**, 023701.

22 Poulsen, H.F. and Juul Jensen, D. (1995) Synchrotron Radiation Diffraction: A Novel Tool for Recrystallization Studies in Bulk μm Sized Local areas. Proceedings of the 16th Risø International Symposium on Materials Science: Microstructural and Crystallographic Aspects of Recrystallization. Edited by N. Hansen *et al*. 503–508.

23 Poulsen, H.F., Juul Jensen, D., and Vaughan, G.B.M. (2004) Three-dimensional x-ray diffraction microscopy using high energy x-rays. *MRS Bulletin*, **29**, 166–169.

24 Juul Jensen, D., Kvick, A., Lauridsen, E.M., Lienert, U., Margulies, L., Nielsen, S.F., and Poulsen, H.F. (2000) *Proceedings of the Materials Research Society Symposium*, **590**, 227–240.

25 Ice, G.E., Larson, B.C., Tischler, J.Z. Liu, W. and Yang, W. (2005) X-ray microbeam measurements of subgrain stress distributions in polycrystalline materials. *Materials Science and Engineering: A*, **399**, 43–48.

26 Larson, B.C., Yang, W., Ice, G.E., Budai, J.D., and Tischler, T.Z. (2002) Three dimensional X-ray structural microscopy of submicrometre resolution. *Nature*, **415**, 887–890.

27 Ice, G.E., Larson, B.C., Yang, W., Budai, J.D., Tischler, J.Z., Pang, J.W.L., Barabash, R.I., and Liu, W. (2005) Polychromatic X-ray microdiffraction studies of mesoscale structure and dynamics. *Journal of Synchrotron Radiation*, **12**, 155–162.

28 Baruchel, J., Maire, E., and Buffiere, J.Y. (2000) *X-ray Tomography in Materials Science*, Hermes Science, London.

29 Ludwig, W., Reischig, P., King, A., Herbig, M., Lauridsen, E.M., Johnson, G., Marrow, T.J., and Buffiere, J.Y. (2009) Three-dimensional grain mapping by X-ray diffraction contrast tomography and the use of Friedel pairs in diffraction data analysis. *Review of Scientific Instruments*, **80**, 33905–33909.

30 Ludwig, W., Cloetens, P., Härtwig, J., Baruchel, J., Hamelin, B., and Bastie, P. (2001) Three-dimensional imaging of crystal defects by "topo-tomography". *Journal of Applied Crystallography*, **34**, 602–607.

31 Baruchel, J., Cloetens, P., Hartwig, J., Ludwig, W., Mancini, L., Pernot, P., and Schlenker, M. (2000) Phase imaging using

highly coherent X-rays: Radiography, tomography, diffraction topography. *Journal of Synchrotron Radiation*, **7**, 196–201.

32 Bunge, H.J., Wcislak, L., Klein, H., Garbe, U., and Schneider, J.R. (2003) Texture and microstructure Imaging in six-dimensions with high-energy synchrotron radiation. *Journal of Applied Crystallography*, **36**, 1240–1255.

33 Wroblewski, T., Clauss, O., Crostack, H.-A., Ertel, A., Fandrich, F., Genzel, Ch., Hradil, K., Ternes, W., and Woldt, E. (1999) A new diffractometer for materials science and imaging at HASYLAB beamline G3. *Nuclear Instruments and Methods in Physics Research Section A*, **428**, 570–582.

34 Lienert, U., Schulze, C., Honkimaki, V., Tschentscher, T., Garbe, S., Hignette, O., Horsewell, A., Lingham, M., Poulsen, H.F., Thomsen, N.B., and Ziegler, E. (1998) Focusing optics for high-energy X-ray diffraction. *Journal of Synchrotron Radiation*, **5**, 226–231.

35 Lienert, U., Poulsen, H.F., Honkimaki, V., Schulze, C., and Hignette, O. (1999) A focusing multilayer analyser for local diffraction studies. *Journal of Synchrotron Radiation*, **6**, 979–984.

36 Nielsen, S.F., Wolf, A., Poulsen, H.F., Ohler, M., Lienert, U., and Owen, R.A. (2000) A conical slit for three-dimensional XRD mapping. *Journal of Synchrotron Radiation*, **7**, 103.

37 Poulsen, H.F. (2004) *Three-Dimensional X-ray Diffraction Microscopy*, Springer, Berlin.

38 Ludwig, W., King, A., Reischig, P., Herbig, M., Lauridsen, E.M., Schmidt, S., Proudhon, H., Forest, S., Cloetens, P.D., Rolland du Roscoat, S., Buffière, J.Y., and Marrow, T.J., and Poulsen, H.F. (2009) New opportunities for 3D materials science of polycrystalline materials at the micrometer length scale by combined use of X-ray diffraction and X-ray imaging. *Materials Science and Engineering; A*, **524**, 69–76.

39 Duggan, B. (1996) The Problem of Lost Evidence. Term Discussed at the ICOTOM 11 Conference, Xian, China.

40 Paul, H., Driver, J.H., and Jasienski, Z. (2002) Shear banding and recrystallization nucleation in a Cu-2% Al alloy single crystal. *Acta Materialia*, **50**, 815–830.

41 Skjervold, S.R. and Ryum, N. (1996) Orientation relationships in a partially recrystallized polycrystalline AlSi alloy. *Acta Materialia*, **44**, 3407–3419.

42 Wu, G.L. and Juul Jensen, D. (2007) Orientations of recrystallization nuclei developed in columnar-grained Ni at triple junctions and a high-angle grain boundary. *Acta Materialia*, **55**, 4955–4964.

43 Larsen, A.W., Poulsen, H.F., Margulies, L., Gundlach, C., Xing, Q.F., Huang, X., and Juul Jensen, D. (2005) Nucleation of recrystallization observed *in situ* in the bulk of deformed metal. *Scripta Materialia*, **53**, 553–557.

44 West, S.S., Schmidt, S., Sorensen, H.O., Winther, G., Poulsen, H.F., Margulies, L., Gundlach, C., and Juul Jensen, D. (2009) Direct non-destructive observation of bulk nucleation in 30% deformed aluminum. *Scripta Materialia*, **61**, 875–878.

45 Gottstein, G. and Shvindlerman, L.S. (1992) On the true dependence of grain boundary migration rate on drive force. *Scripta Metallurgica*, **27**, 1521–1526.

46 Viswanathan, R. and Bauer, G.L. (1973) Kinetics of grain boundary migration in copper bicrystals with (0 0 1) rotation axes. *Acta Metallurgica*, **21**, 1099–1109.

47 Vandermeer, R.A. (1965) Dependence of grain boundary migration rates on driving force. *Transaction of AIME*, **233**, 265–267.

48 Vandermeer, R.A., Juul Jensen, D., and Woldt, E. (1997) Grain boundary mobility during recrystallization of copper. *Metallurgical and Materials Transactions*, **28**, 749–754.

49 Rath, B.B. and Hu, H. (1969) Effect of driving force on the migration of high-angle tilt grain boundaries in aluminium bicrystals. *Transactions of the Metallurgical Society of AIME*, **245**, 1577–1585.

50 Zhang, Y.B., Godfrey, A., Liu, Q., Liu, W., and Juul Jensen, D. (2009) Analysis of the growth of individual grains during recrystallization in pure nickel. *Acta Materialia*, **57**, 2631–2639.

51 Schmidt, S., Nielsen, S.F., Gundlach, C., Margulies, L., Huang, X., and Juul Jensen, D. (2004) Watching the growth of bulk grains during recrystallization of deformed metals. *Science*, **305**, 229–232.

52 Lauridsen, E.M., Poulsen, H.F., Nielsen, S.F., and Juul Jensen, D. (2003) Recrystallization kinetics of individual bulk grains in 90% cold-rolled aluminium. *Acta Materialia*, **51**, 4423–4435.

53 Poulsen, S.O., Lauridsen, E.M., Lyckegaard, A., Oddershede, J., Gundlach, C., Curfs, C., and Juul Jensen, D. (2011) In-situ measurements of growth rates and grain averaged activation energies of individual grains during recrystallization of 50% cold-rolled aluminium. *Scripta Materialia*, **64**, 1003–1006.

54 Wu, G.L. and Juul Jensen, D. (2012) In-situ measurements of annealing kinetics of individual bulk grains in nanostructured aluminium. *Philosophical Magazine*, **92**, 3381–3391.

55 Krüger, P. and Woldt, E. (1992) The use of an activation energy distribution for the analysis of the recrystallization kinetics of copper. *Acta Metallurgica et Materialia*, **40**, 2933–2942.

56 Godiksen, R.B., Schmidt, S., and Juul Jensen, D. (2007) Effects of distribution of growth rates on recrystallization kinetics and microstructure. *Scripta Materialia*, **57**, 345–348.

57 Hansen, N. and Juul Jensen, D. (2011) Deformed metals – structure, recrystallization and strength. *Materials Science and Technology*, **8**, 1229–1240.

58 Ungar, T., Ribarik, G., Balogh, L., Salem, A.A., Semiatin, S.L., and Vaughan, G.B.M. (2010) Burgers vector population, dislocation types and dislocation densities in single grains extracted from a polycrystalline commercial-purity Ti specimen by X-ray line-profile analysis. *Scripta Materialia*, **63**, 67–72.

59 Pantleon, W., Wejdemann, C., Jakobsen, B., Lienert, U., and Poulsen, H.F. (2010) Advances in characterization of deformation structures by high resolution reciprocal space mapping. Proceedings of the 31st Risø Int. Symp. on Materials Science (eds. N. Hansen et al.) 79–100.

60 Jakobsen, B., Poulsen, H.F., Lienert, U., Almer, J., Shastri, S.D., Sørensen, H.O., Gundlach, C., and Pantleon, W. (2006) Formation and subdivision of deformation structures during plastic deformation. *Science*, **312**, 889–892.

61 Winther, G., Margulies, L., Schmidt, S., and Poulsen, H.F. (2004) Lattice rotations of individual bulk grains. *Acta Materialia*, **52**, 2863–2872.

62 Schmidt, S., Olsen, U.L., Poulsen, H.F., Sørensen, H.O., Lauridsen, E.M., Margulies, L., Maurice, C., and Juul Jensen, D. (2008) Direct observation of 3D grain growth in Al-0.1% Mn. *Scripta Materialia*, **59**, 491–494.

63 Juul Jensen, D., Offerman, S.E., and Sietsma, J. (2008) 3DXRD characterization and modelling of solid-state transformation processes. *MRS Bulletin*, **33**, 621–629.

64 Offerman, S.E., van Dijk, N.H., Sietsma, J., Grigull, S., Lauridsen, E.M., Margulies, L., Poulsen, H.F., Rekveldt, M.T., and van der Zwaag, S. (2002) Grain nucleation and growth during phase transformations. *Science*, **298**, 1003.

65 Offerman, S.E., van Dijk, N.H., Sietsma, J., Lauridsen, E.M., Margulies, L., Grigull, S., Poulsen, H.F., and van der Zwaag, S. (2004) Solid-state phase transformation involving solute partitioning: Modeling and measuring on the level of individual grains. *Acta Materialia*, **52**, 4757–4766.

66 Lange, W.F.III, Enomoto, M., and Aaronson, H.I. (1988) The kinetics of ferrite nucleation at austenite grain-boundaries in Fe–C alloys. *Metallurgical Transactions A.*, **19**, 427–440.

67 Kostenko, A., Sharma, H., Gözde Dere, E., King, A., Ludwig, W., van Oel, W., Stallinga, S., van Vliet, L.J., and Offerman, S.E. (2012) Three-dimensional morphology of cementite in steel studied by X-ray phase-contrast tomography. *Scripta Materialia*, **67**, 261–264.

68 Aagesen, L.K., Johnson, A.E., Fife, J.L., Voorhees, P.W., Miksis, M.J., Poulsen, S.O., Lauridsen, E.M., Marone, F., and Stampanoni, M. (2010) Universality and self-similarity in pinch-off of rods by bulk diffusion. *Nature Physics*, **6**, 796–800.

69 Robertson, I.M. et al. (2011) Towards an integrated materials characterization toolbox. *Journal of Materials Research*, **26** (11), 1341–1383.

16
Crystallographic Textures and a Magnifying Glass to Investigate Materials
Jürgen Hirsch

16.1
Introduction

Crystallographic texture is an essential feature of the microstructure, besides more obvious and easier perceptible grain and precipitation structures. Orientations of crystallites are formed in a very specific way by almost any microstructure evolution process involved in material processing, that is, casting, forming, and annealing. Directional effects during solidification, plastic deformation, or recrystallization significantly affect the textures whose sound analysis can statistically give specific information on the processes, allowing deeper insight into the local and global microstructure and variables of their formation [1,2].

The texture describes the whole set of grain orientations of crystalline materials, which is an essential feature of the microstructure, however, is not easy to visualize [3,4]. Based on the origin of characteristic textures and the knowledge of basic principles of their formation and correlation to microstructure, texture data can be used to draw conclusions on these processes. Furthermore, certain properties of texture data are of special advantage for this purpose, such as the statistical nature of X-ray texture data or the local information of EBSD data [5].

In metals, any process forms the texture in a very specific way, either during casting, forming, or annealing. The texture depends on the corresponding (orientation dependent) formation mechanisms, which are oriented glide on selected slip planes during plastic deformation or oriented nucleation and/or growth selection effects of the newly formed grains, either during solidification in casting or during recrystallization in hot deformation or annealing. So besides alloy composition and constitution, the microstructure evolution is determined during processing (e.g., casting, extrusion, hot and cold rolling, and annealing) by the specific parameters applied. This can be used to investigate such mechanisms, based on underlying physical principles. The advanced simulation tools available can be applied in both directions. In many industrially processed materials, strong crystallographic effects occur, sometimes unintended, but sometimes introduced intentionally and often strictly controlled to achieve specific (orientation dependent) advantageous material properties such as strength, formability, permeability, etching characteristics, and so on [6,7]. In these cases the demand for best material quality and optimum performance the control of texture is of key importance. It is analyzed by established measuring techniques (e.g., automatic X-ray diffraction goniometers or REM with EBSD equipment) and evaluation methods available and applied for industrial R&D [8,9]. Hence, the systematic knowledge on textures is growing and their advanced control during industrial processing, such as rolling, annealing, and extrusion for many processes and products like aluminum [10].

16.2
Texture Evolution and Exploitation of Related Information in Metal Processing

In principle, only those mechanisms can be observed and analyzed by textures that lead to notable and characteristic orientation changes. Here the size and reproducibility of texture effects need to be considered carefully, which for industrially fabricated material is usually quite good. Slight variations in industrial production parameters can cause notable and clearly distinguishable texture changes. Modern measuring techniques and evolution methods allow an accurate and detailed analysis of even minor effects, but for laboratory experiments the conditions must always be fixed accurately to ensure their exploitation by textures.

The main processes of metal production and related orientation effects are as follows.

- *Casting:* directional solidification by growth selection.
- *Forming:* deformation on the selected slip system with specific local geometry effects.
- *Annealing:* recrystallization by oriented nucleation and/or growth.

In the latter, oriented nucleation plays a role besides oriented growth. The interpretation of both mechanisms and their contribution caused a long dispute among scientists in texture and recrystallization science [11] in early days of texture research due to limited accuracy and missing local resolution. However, the dispute has been resolved by the modern equipment and advanced simulation methods, revealing evidence for both and attributing their exclusive role only in very special cases.

Since many years now textures have been analyzed in great detail for all kind of materials and processes. For aluminium the main texture effects are listed in Table 16.1. They are described in [6,7] together with their underlying physical mechanisms and affected properties and their importance for certain applications, such as the anisotropy of strength and formability, surface formation, and chemical (e.g., etching) behaviour.

16.2 Texture Evolution and Exploitation of Related Information in Metal Processing | 389

Table 16.1 Main texture components in aluminum processing.

1 Casting:	(1a) random textures (DC-cast with grain refinement)
	(1b) ⟨1 0 0⟩ fiber texture (directionally solidified pure Al)
2 Deformation:	(2a) ⟨1 1 1⟩/⟨1 0 0⟩ fiber in uniaxial tension
	(2b) ⟨1 1 0⟩ fiber in uniaxial compression
	(2c) β-fiber in plane strain compression (e.g., rolling)
3 Recrystallization:	(3a,b) for uniaxial deformed aluminum
	(cases 2a and 2b: fiber type textures = transformed deformation fibres)
	(3c) for cold rolled and annealed aluminum sheet (case 2c): Cube, R (in three variants), PSN: P, ND rotated Cube, Q orientations depending on alloy composition and processing conditions (ingot casting/homogenization, hot–cold rolling, annealing)
	(PSN: particle stimulated nucleation, ND: normal direction)

16.2.1
Casting

During casting, the solidification process of crystalline material generates characteristic texture effects caused by epitaxial growth of crystalline grains on certain surfaces. The effect is further enhanced by heterogeneous nucleation, whereas homogeneous nucleation generates randomly oriented grains. For the latter, growth selection prefers the fastest growing crystallites during subsequent "oriented" solidification. In pure aluminum a ⟨1 0 0⟩ direction is preferred that therefore generates a strong ⟨1 0 0⟩ fiber texture in directional solidification [12]. The direction of the ⟨1 0 0⟩ fiber axis in turn indicates the solidification direction in a unidirectional cast sample. The observed tilting angle α of the fiber axis taken from a pole figure (Figure 16.1) helps us to analyze the corresponding solidification geometry for melt spun metal cast on a fast rotating wheel [13] and is another method to determine the important solidification rate of the metal strip on the rotating wheel by geometrical considerations [9]. Furthermore, the texture reveals information about other effects of the solidification process, for example, nucleation on the wheel surface and microstructural changes in different layers of the strip [13].

16.2.2
Deformation

Plastic deformation of crystalline material is caused by shear on selected crystallographic planes in selected directions. Hence, quite strong orientation effects can develop, depending on the slip system geometry and imposed deformation. Making the use of resulting texture effects is an important method in geology research to derive the deformation history of geological formations from the geometrical conditions in rocks [14]. In crystalline metallic materials, plastic deformation is caused by oriented dislocation glide on few crystallographic slip systems, demonstrated easily in the deformation geometry observed in single crystals.

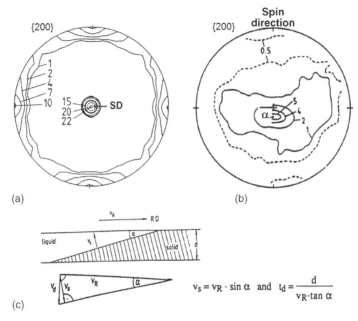

Figure 16.1 Textures of directionally solidified face centered cubic (FCC) metals showing a ⟨1 0 0⟩ fiber texture. (a) Unidirectionally solidified pure aluminum ({2 0 0} pole figure, solidification direction (SD) – center of pole figure); (b) melt spun copper alloy ({2 0 0} pole figures, angle α indicates orientation shift from the sheet normal); (c) effective solidification rate and time estimated from the tilting angle α (Figure 16.1b).

In polycrystalline face-centered cubic (FCC) metals like aluminum sufficient (>5) slip systems are available to accommodate any deformation imposed. The specific selection of active slip systems out of the 12 available ones are predicted [15] which then allow us to predict the resulting texture evolution [16].

For a uniaxial deformation (such as tensile testing or wire drawing) a double fiber texture (with major ⟨1 1 1⟩ and minor ⟨1 0 0⟩ orientations in the elongation direction) is formed (Figures 16.2 and 16.3 a). This is also predicted by the classical theory [15]. The correlation of this texture type to any specific deformation effects, however, is limited due to only one crystallographic direction indicated. For plane strain deformation (like sheet rolling) a typical FCC rolling texture with well-defined orientations along the β-fiber occurs (Table 16.2, Figure 16.3b) and can be analyzed with more background information in more depth when predicted quantitatively [16,17].

Figure 16.3 shows the pole figure of the cross section of an extruded, that is, highly deformed (but unrecrystallised) Al–Li alloy. The extruded profile was an L-shaped bar [18] showing a characteristic double fiber texture on the left side of the cross section with Figure 16.3a showing a uniaxially deformed quadratic area and Figure 16.3b the typical plane-strain- (rolling-) type texture on the right side the bar, deformed into a flat geometry. Here the texture effects are drastically

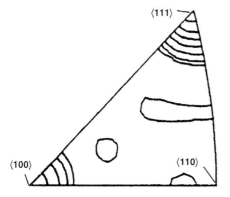

Figure 16.2 Texture of unidirectionally deformed (drawn) aluminum wire (inverse pole figure of the drawing direction).

different, revealing the (local) different geometry of deformation and related materials flow. The corresponding textures also generate drastic differences in final properties. Besides anisotropy significant strength variations (>20%) occurred in the underaged material depending on the differences in texture and the inhomogeneity of dislocation glide on the selected slip systems due to their specific geometry.

16.2.2.1 Textures Revealing Details on the Mechanisms of Deformation

In hexagonal closed packed (HCP) magnesium, glide on the only "basal" slip plane creates a strong texture that drastically affects properties and forming behavior [19].

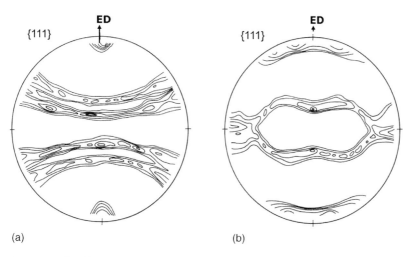

Figure 16.3 {1 1 1} pole figures obtained on different sections of an extruded Al–Li profile (a) ⟨1 1 1⟩ and ⟨1 0 0⟩ fiber texture; (b) plane strain texture. (ED: extrusion direction).

Table 16.2 Main orientations in rolled and annealed FCC sheet (Miller indices $\{hkl\}\langle uvw\rangle$ and Euler angles $\varphi_1, \Phi, \varphi_2$).

Name	Miller indices $\{hkl\}\langle uvw\rangle$	Euler angles $\varphi_1, \Phi, \varphi_2$
(a) plane strain (rolling) deformation texture components – as listed in Table 16.1, 2c		
C	$\{112\}\langle 111\rangle$	90°, 35°, 45°
S	$\{124\}\langle 211\rangle$	63°, 31°, 60°
B	$\{011\}\langle 211\rangle$	35°, 45°, 0°
(b) corresponding recrystallization texture components – as listed in Table 16.1, 3c		
Cube	$\{001\}\langle 100\rangle$	0°, 0°, 0°
R	$\{124\}\langle 211\rangle$	63°, 31°, 60°
P	$\{011\}\langle 122\rangle$	70°, 45°, 0°
Q	$\{013\}\langle 231\rangle$	45°, 15°, 10°
Goss	$\{011\}\langle 100\rangle$	0°, 45°, 0°

In FCC metals like copper or aluminum, the selection of slip systems for plastic deformation is usually more homogeneous and in plane strain (i.e., homogeneous rolling) deformation the typical "copper-type" rolling texture is formed (Figure 16.3b and Figure 16.4). It consists of a characteristic orientation tube (called "β-fiber") composed of three texture components – C ($\{112\}\langle 111\rangle$ (Euler angles ($\varphi_1, \Phi, \varphi_2 = 90°, 35°, 45°$)), S ($\{123\}\langle 634\rangle$ (59°, 37°, 63°)) and B ($\{011\}\langle 211\rangle$ (35°, 45°, 90°)) [16]. The texture intensity in general increases with the degree of deformation, so comparative texture studies can be used to estimate this parameter.

The selected slip system and the plane strain or rolling texture evolution can be described by Taylor-type theories [15]. However, always more than the sufficient (five) slip systems are present that creates some ambiguity of glide effects. But the

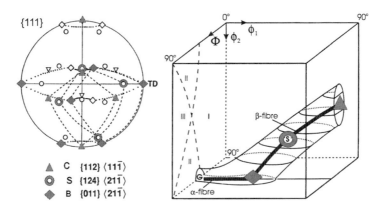

Figure 16.4 Rolling texture in FCC metals – α and β-fibers. (a) Ideal orientations in $\{111\}$ pole figure and (b) α- and β-fibers in the three-dimensional orientation space (Euler space in Bunge's notation [3]).

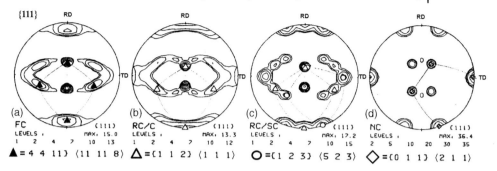

Figure 16.5 Simulated FCC rolling textures under various boundary conditions for plane strain deformation: full constraints (FC), relaxed constraints (RC), and no constraints (NC). (a) FC "Taylor" model = ideal plane strain; (b) RC "lath" model = relaxed ε^{NR} "C"-shear; (c) RC "pancake" model = relaxed ε^{NR+NT} "SC"-shear; (d) NC "SACHS" model = all shears relaxed.

textures vary only slightly due to the ambiguity in slip system selection [16]. Deformation texture variations much more depend on local boundary conditions of the deforming grain. So in highly rolled (i.e., textured) sheet finally less than five slip systems are sufficient to achieve the imposed strain. The "relaxed constraints" models predict specific variations in stable rolling texture peaks (Figure 16.5) along the β-fiber and some characteristic variation in their exact orientation in the Euler angle space: the shift of the "FC Taylor" orientation {4 4 11}⟨11 11 8⟩ (90°, 27°, 45°) (Figure 16.5a) to the C orientation {1 1 2}⟨1 1 1⟩ (90°, 35°, 45°), that is, by a 8° rotation around the transverse direction (TD) (Figure 16.5b), formation of a second rolling texture component "S" {1 2 3}⟨5 2 3⟩ (55°, 37°, 63°), together with the C peak (Figure 16.5c) and the formation of the pure B orientation {0 1 1}⟨2 1 1⟩ (35°, 45°, 0°), (with a minor Goss {0 1 1}⟨1 0 0⟩ (0°, 45°, 0°), when no constraints are applied (Figure 16.5d).

Advanced simulation models like the "GIA model" [20] include certain grain interactions effects in their slip system selection and so are able to predict the corresponding textures to a high accuracy. So based on the model predictions experimental texture data can now be used to compare different materials and analyze their individual deformation mechanisms. In Figure 16.6 a the β-fiber orientation density maxima $f(g)$ versus φ_2 and in Figure 16.6b their exact orientation (i.e., $f(g)$ orientation as φ_1 and Φ versus φ_2) are plotted for 95% cold rolled FCC metals, that is, aluminum, copper, and Cu alloys. They show the characteristic texture transition from the copper-type to the brass-type rolling texture with decreasing stacking fault energy (SFE). The comparison of the corresponding exact β-fiber orientations with theoretical predictions, like the FC Taylor model (included in Figure 16.6b) reveals information about the local strain geometry [20]. Also the activated slip systems can quantitatively be analyzed making use of the detailed model predictions of their exact orientations [16]. The active slip systems in FCC sheet rolling are plotted in Figure 16.7 for the main orientations along the β-fiber in a stereographic projection of their {1 1 1} slip

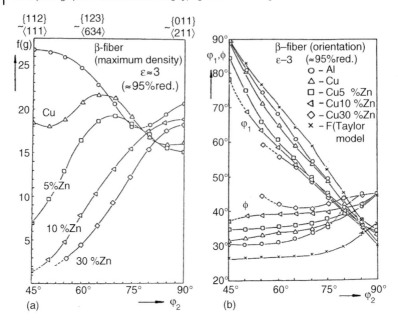

Figure 16.6 (a) Measured orientation density maxima $f(g)$ along the β-fiber and (b) their exact orientation in the Euler space (φ_1 and Φ) as a function of φ_2 for different FCC metals and alloys after 95% reduction by cold rolling (including the fiber position "F" theoretically predicted by the "full constraints" (FC) Taylor model).

planes (open symbols) and ⟨1 1 0⟩ slip directions (filled symbols), connected by full (or broken) lines:

A) Two coplanar slip systems with the common slip plane (1 11) : (1 1 1) [−1 0 1] (A1) and (1 1 1) [0–11] (A2) and
B) Two co-directional (cross) slip systems with the common slip direction [1 1 0] : (2 1 1)[1 1 0] (B1) and (1 1 1) [1 1 0] (B2)

For the B orientation only two slip systems A1 and B1 are active, while all four are active for the C orientation (triangles in Figure 16.7), symmetrically arranged in each pair, A1, 2 and B1, 2. Slip on both system pairs causes a rotation around the transverse direction TD (⟨1 1 0⟩, called τ-fiber), each pair with an opposite sign. The difference between models FC (Figure 16.5a) and RC/C (Figure 16.5b) is the different activation of the two opposing systems. For the experimentally observed average orientation at $\Phi = 30°$ an equal amount of slip on the Systems A1, 2 and B1, 2 (a slip ratio = 0.5) is required, resulting in an ε^{NR} "C" shear of 25% of the unit thickness strain ε^{T}. No shear allowed in the FC model ($\Phi = 27.3°$) requires 50% more slip on the co-directional Systems B1, 2. The S orientation deforms in the relaxed constraints "SC" mode, also with four slip systems but in a slightly different combination. For the S orientation positioned at (63°, 30′, 65°) the ratio of slip

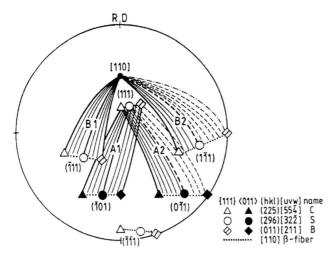

Figure 16.7 The slip systems active during plane strain deformation (rolling) in FCC metals with an established stable rolling texture (β-fiber). The slip planes {1 1 1} are plotted as open symbols and the slip directions [1 1 0] as full symbols for the main rolling texture orientations C, S, and B (copper, S, and brass components, respectively).

system activations is B1: (A1, B2): A2 ≈ 10 : 7 : 3. So the S orientation deforms on the same two slip systems A1 and B1 (active solely in the B orientation), with additional two slip systems (A2 and B2, additionally active in the C orientation) being able to balance the ε^B shear. (Figure 16.7) but allowing some S shear ε^S.

16.2.2.2 Stacking Fault Energy Analysis in Rolling Texture Formation

For decreasing SFE in FCC metals characteristic texture changes occur due to a drastic change in deformation mechanism. For highly rolled α-brass or silver the characteristic brass-type rolling texture is dominant that only shows the B orientation of the β-fiber (Figure 16.4) and some minor Goss (or G) orientation {0 1 1}⟨1 0 0⟩ (0°, 45°, 90°; predicted in Figure 16.5d). The transition from the copper to the brass-type rolling texture depends on the SFE, strain, and temperature in a rather complex manner [22]. Based on texture results the transition can be plotted as variation of volume fractions for the main texture components for various homogeneous copper alloys for a constant strain of 95% rolling reduction (Figure 16.8). With the scale of alloy content chosen in a way that the texture components are closest, this evaluation provides a measure for the SFE that very well agrees with other measuring techniques like line broadening observed in TEM (Figure 16.8b).

16.2.2.3 Surface Shearing and Lubrication Effects

As already shown in Figure 16.3, the extrusion process texture formation during deformation strongly depends on the (local) strain geometry imposed. In sheet rolling an important effect is surface lubrication and friction that can be analyzed by the characteristic texture effects involved. Aluminum is very reactive with the roll

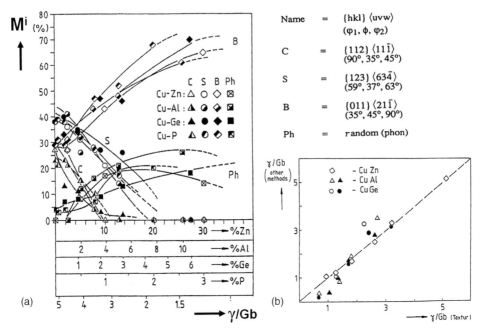

Figure 16.8 Evaluation of stacking fault energy (SFE) by texture effects [22]. (a) Volume fractions of rolling texture components in various copper alloys cold rolled to 95% reduction, plotted as a function of the alloy content and the normalized SFE (b) comparison with conventional methods of SFE determination.

surface and requires special lubrication and under specific friction conditions the rolling texture changes at the surface into a predominant ND-rotated cube orientation $\{0\,0\,1\}\,\langle1\,1\,0\rangle$ (90°, 0°, 45°) "CubeND" and a $\{1\,1\,1\} = $ ND "γ-fiber" texture [22]. The depth of the influenced shear deformation zone can be visualized by the height and variation of the corresponding texture peaks. In Figure 16.9, the texture peaks of a rolled aluminum sheet (with low ratio of roll contact length t_c to sheet thickness t: $t_c/t < 1$) are plotted for different depths of the sheet layers (indicated by the index of S running from $S = 1$ (surface) and $S = 0$ (sheet mid-thickness) as ODF intensities $f(g)$ along the τ-fiber. In this plot, the CubeND (located at $\Phi = 0°$) and their intensity variation is plotted together with all orientations with a common $\langle 1\,1\,0 \rangle$ axis parallel to the transverse direction TD ($\varphi_1 = 90°$, $\varphi_2 = 45°$, plotted along Φ, indicating the rotation angle from CubeND). The experimentally (stable) rolling texture orientation $\{2\,2\,5\}\,\langle 5\,4\,4\rangle$ C* (at $\Phi = 30°$) is strongest at the sheet center where plane strain deformation prevails (here measured at $S = 0.19$, i.e., near mid-thickness), but its intensity $f(g)$ continuously decreases closer to the surface (i.e., increasing S) where the surface shear texture CubeND emerges. (The locations of C* and CubeND are indicated on top line of Figure 16.9). This evaluation helps us to analyze the friction conditions and the development of the shear zone and measures to affect it – like lubrication media and rolling conditions.

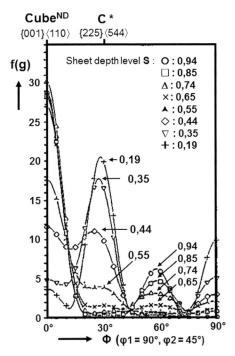

Figure 16.9 Texture ODF intensity $f(g)$ plotted for CubeND (0 0 1) ⟨1 1 0⟩ (90°, 0°, 45°) and along the τ-fiber (Φ at $\varphi_1 = 90°$ and $\varphi_2 = 45°$), that is, orientations with a common ⟨1 1 0⟩ axis parallel to the (long) transverse direction (TD) of the sheet.

16.2.3
Recrystallization

16.2.3.1 Oriented Nucleation, Oriented Growth and in situ Recrystallization in Aluminum Alloys

Recrystallization occurs by the mechanisms of local nucleation of new grains and the subsequent motion of high angle boundaries into the surrounding matrix. Both mechanisms imply strong changes in texture and therefore, in turn, texture can help us to analyze the corresponding underlying mechanisms [22]. Even small effects that might change any of these (local) conditions influence the resulting recrystallization texture and lead to texture modifications or in some cases even totally different texture components that can be statistically evaluated. Figure 16.10 shows typical examples of recrystallization textures (components are listed in Table 16.2) of highly rolled and annealed aluminum alloys.

Here the cube orientation {0 0 1} ⟨1 0 0⟩ is usually the prominent texture component. In some cases, a strong R orientation (near {1 2 3} ⟨6 3 4⟩) also appears [25] (Figure 16.10a), which is quite similar to the main rolling texture component S (see Figure 16.4). However, its intensity and exact position can vary significantly that has

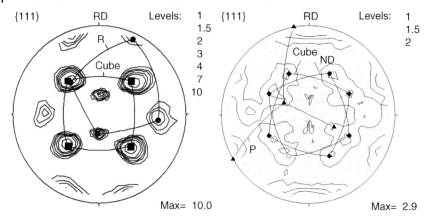

Figure 16.10 Recrystallization textures in cold rolled and annealed aluminum alloy sheet. (a) Cube and R texture in 95% rolled commercial purity aluminum (Al 99.5%) (b) ND rotated Cube and P texture in age hardening Al–Mg–Si–Cu alloy for automotive sheet [24].

been analyzed by detailed texture analysis (Figure 16.11) [26]. The typical shift in orientation is caused by a change in recrystallization mechanism from preferred nucleation (high φ_1 values, near the C rolling texture) to oriented growth processes (lower φ_1 values, near the 40° $\langle 1\,1\,1 \rangle$ rotated S* rolling texture) [23]. Even a third mechanism of extended recovery (called "in-situ recrystallization") has been identified this way when no new grains are formed (i.e., neither nucleation nor growth takes place) and the rolling texture is maintained in all its details (intensity and exact position). This is the case in Figure 16.11 at 360 °C, where the R orientation is near the S rolling texture. This is valid for the whole texture [26].

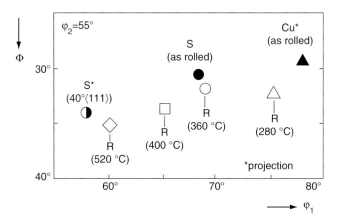

Figure 16.11 Exact central position of the R recrystallization texture components, with systematic peak shifts indicating different recrystallization mechanisms as a function of annealing temperature after high cold rolling reductions [23].

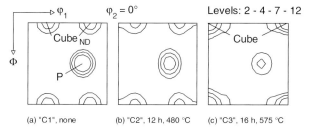

Figure 16.12 Recrystallization textures ($\varphi_2 = 0°$ section) in Al–Mg–Si sheet with systematic variations of Cube and PSN texture components due to different ingot preannealing.

16.2.3.2 Recrystallization in Particle Containing Al Alloys

In Al alloys, small amounts of alloy elements can cause drastic changes in the recrystallization behavior when particles are present strongly influencing recrystallization nucleation and so affecting the final recrystallization texture. The shift of the Cube component to its ND rotated variant $\{0\,0\,1\}\langle 3\,1\,0\rangle$ and the occurrence of a strong $\{0\,1\,1\}\langle 1\,1\,1\rangle$ (55°, 45°, 0°) P texture indicates this change in mechanism [27]. Figure 16.12 shows this texture variation caused by different preannealing treatments of the as-cast ingots prior to cold rolling and solution annealing [28]. For this texture transition observed in industrial Al–Mg–Si (6xxx) alloys particle-stimulated nucleation (PSN) [29] is the recrystallization mechanism generating a weak recrystallization texture with P and the "ND-rotated Cube" components. The two main peaks are very close to orientation accumulations predicted from the rolling texture by a complete $40°\langle 1\,1\,1\rangle$ transformation [27] so the texture is interpreted as a preferred growth selection process taking place out of the large spectrum of potential nuclei around particles, for example, in early stages of recrystallization [30]. However, it still competes with the dominant cube nuclei, so the final texture depends on the relative amounts of (successful) nucleation at either orientation site.

In Figure 16.12, the exact cube orientation together with its RD-scatter increases at the expense of the PSN (P) orientations with increasing ingot homogenization temperature and time. This affects the particles formed during early processing, including the supersaturation level that inhibits early growth of preferred (Cube) nuclei by simultaneous fine precipitation. The decrease of Cube^{ND} and P (the R orientation – not shown in Figure 16.12 – appeared to scale with the cube orientation) reveals a decrease in the efficiency of PSN with increasing precipitation of Mg_2Si-particles during annealing [31]. The dispersed particles impede PSN more strongly than nucleation at the cube bands, so nuclei emerging from the cube bands (i.e., Cube – and Cube^{RD} oriented grains) and the grain boundaries (i.e., R oriented grains) prevail in the recrystallization texture.

16.3 Summary

Crystallographic textures are an essential feature of the microstructure and can be very sensitive measures for their evolution being strongly influenced by many important parameters in metal processing. So in all the main steps of metal fabrication processes, that is, casting, plastic deformation, and annealing, the corresponding physical mechanisms, that is, solidification, dislocation glide, and recrystallization are controlled by the crystalline structure of metals. This correlation with the resulting preferences in crystallographic orientation leads to the formation of typical textures that in many cases reveal details of the underlying mechanisms.

Solidification, deformation, and annealing textures of typical FCC materials have been analyzed to investigate the corresponding main process parameters and provide information about the physical mechanisms involved. Systematic investigation of the complex dependences of textures on the different material and process parameters will further open this new field of material characterization. The established texture measurement facilities properly used in fundamental as well as in applied research laboratories will help us to establish textures as a suitable method to analyze physical mechanisms or as an investigative tool for process and quality control in industrial practice.

So besides their important role in affecting the anisotropy of the mechanical properties, texture data can also be applied to characterize different processes of material production and help us to better understand, control, and improve related fabricating processes.

Acknowledgment

Many fruitful discussions with texture experts in Hydro Aluminium R&D and my former colleagues at the Institut für Allgemeine Metallkunde and Metallphysik der RWTH Aachen are gratefully acknowledged.

References

1 Bunge, H.J. (1988) *Directional Properties of Materials*, DGM Informationsgesellschaft Verlag, Oberursel.
2 Tewari, A., Suwas, S., Srivastava, D., Samajdar, I., and Haldar, A. (eds) (2012) *Textures of Materials (ICOTOM 16)*, Trans Tech Publications, Dürnten-Zürich.
3 Bunge, H.J. (1981) *Mathematical Methods of Texture Analysis*, Butterworths, London.
4 Hirsch, J. and Lücke, K. (1988) Description and presentation methods for textures. *Textures and Microstructures*, **8–9**, 131–151.
5 Randle, V. and Engler, O. (2000) *Introduction to Texture Analysis: Macrotexture, Microtexture and Orientation Mapping*, Gordon and Breach, Amsterdam.
6 Hirsch, J. (2005) Texture and anisotropy in industrial applications of aluminium alloys. *Archives of Metallurgy and Materials Quart*, **50**, 21–34.
7 Hirsch, J. (2012) Textures in industrial processes and products. *Materials Science Forum*, **702–703**, 18–25.
8 Hirsch, J. (1988) New applications of textures analysis for industrial processing, in *Proceedings of the 8th International Conference on Textures of Materials, Santa Fe, New Mexico, September 20–25, 1987 (ICOTOM 8)* (eds

J.S. Kallend and G. Gottstein), Metallurgical Society Publication, Warrendale, PA, pp. 1011–1016.

9 Hirsch, J. (1990) Crystallographic texture analysis as a new method for material characterization, in *Nondestructive Characterization of Materials IV* (eds J.F. Bussière, R.E. Green, and C.O. Ruud), Plenum Publishing, New York, pp 429–438 (4th International Symposium on Non-Destructive Characterization of Materials, Annapolis, USA).

10 Hirsch, J. (ed.) (2006) *Virtual Fabrication of Aluminium Products*, Wiley-VCH, Weinheim.

11 Doherty, R.D., Gottstein, G., Hirsch, J., Hutchinson, W.B., Lücke, K., Nes, E., and Willbrandt, P.J. (1988) Report of panel on recrystallization textures: mechanism and experiments, in *Proceedings of the 8th International Conference on Textures of Materials, Santa Fe, New Mexico, September 20–25, 1987 (ICOTOM 8)* (eds J.S. Kallend and G. Gottstein), Metallurgical Society Publication, Warrendale, PA, pp. 563–572.

12 Hirsch, J., Nes, E., and Lücke, K. (1987) Rolling and recrystallization textures in directionally solidified aluminium. *Acta Metallurgica*, **35** (2), 427–438.

13 Eucken, S., Hirsch, J., and Hornbogen, E. (1988) Texture and microstructure of meltspun shape memory alloys. *Textures and Microstructures*, **8–9**, 415–426.

14 Wenk, R. and Kocks, F. (1985) *Introduction to Modern Texture Analysis*, Academic Press, Orlando.

15 Taylor, G.I. (1938) Plastic strain in metals. *Journal of the Institute of Metals*, **62**, 307–324.

16 Hirsch, J. and Lücke, K. (1988) Simulation and interpretation of experiments on the basis of Taylor-type theories. *Acta Metallurgica*, **36** (11), 2883–2904.

17 Van Houtte, P., Li, S., and Engler, O. (2006) Simulation of deformation textures: a review of latest grain-interaction models, in *Virtual Fabrication of Aluminium Products* (ed. J. Hirsch), Wiley-VCH, Weinheim, pp. 177–188.

18 Tempus, G., Scharf, G., and Calles, W. (1987) Influence of extrusion process parameters on the mechanical properties of Al–Li-extrusions. *Journal de Physique Colloques*, **48** (C3), 187–194.

19 Avedesian, M. and Baker, H. (eds) (1999) *ASM specialty handbook, Magnesium and Magnesium Alloys*, ASM International, Materials Park, OH.

20 Crumbach, M., Pomana, G., Wagner, P., and Gottstein, G. (2001) A taylor type deformation texture model considering grain interaction and material properties, in *Recrystallization and Grain Growth* (eds G. Gottstein and D.A. Molodov), Springer, Berlin, pp. 1053–1060.

21 Hirsch, J.R. (1990) On the correlation of deformation textures and microstructure. *Material Science and Technology*, **6** (11), 1048–1057.

22 Hirsch, J., Lücke, K., and Hatherly, M. (1988) The influence of slip inhomogeneities and twinning. *Acta Metallurgica*, **36** (11), 2905–2927.

23 Hirsch, J. (1988) *Untersuchung der primären Rekristallisation und ihrer Mechanismen mit Hilfe der Texturanalyse*, Habilitation-thesis, RWTH Aachen, Aachen.

24 Engler, O. and Hirsch, J. (1996) Recrystallization textures and plastic anisotropy in Al–Mg–Si sheet alloys. *Materials Science Forum*, **217–222**, 479–486.

25 Hirsch, J. (1986) Recrystallization of FCC metals as investigated by ODF analysis, in *Annealing Process: Recovery, Recrystallization and Grain Growth, Proceedings of the 7th Risø International Symposium on Metallurgy and Materials Science* (ed. N. Hansen), Risø National Laboratory, Roskilde, pp. 361–366.

26 Hirsch, J. and Lücke, K. (1985) The application of quantitative texture analysis for investigating continuous and discontinuous recrystallization processes of Al-0.01 Fe. *Acta Metallurgica*, **33** (19), 1927–1938.

27 Hirsch, J. (1990) Recrystallization and texture control during rolling and annealing in Al Alloys, in *Recrystallization '90* (ed. T. Chandra), TMS Publication, Warrendale, PA, pp. 759–768.

28 Hirsch, J. and Engler, O. (2001) Recrystallization texture in industrial aluminium sheet, in *Recrystallization and Grain Growth* (eds G. Gottstein and D.A. Molodov), Springer, Berlin, pp. 731–740.

29 Humphreys, F.J. (1977) The nucleation of recrystallisation at second phase particles in aluminum. *Acta Metallurgica*, **25** (11), 1323–1344.
30 Engler, O., Löchte, L., and Hirsch, J. (2007) Through-process simulation of texture and properties during the thermo-mechanical processing of aluminium sheets. *Acta Materialia*, **55**, 5449–5463.
31 Engler, O. and Hirsch, J. (2009) Control of recrystallisation texture and texture-related properties in industrial production of aluminium sheet. *International Journal of Materials Research*, **100** (4), 564–575.

Part IV
Applications: Grain Boundary Engineering and Microstructural Design for Advanced Properties

17
The Advent and Recent Progress of Grain Boundary Engineering (GBE): In Focus on GBE for Fracture Control through Texturing
Tadao Watanabe

17.1
Introduction

On a special occasion of the Professor Gottstein honorary symposium, the present author has a great honor and privilege to dedicate this chapter to him on our long challenging research project "grain boundary engineering" (GBE). This project has been performed since the 1980s to endow polycrystalline structural and functional materials with desirable bulk properties and high performance, by controlling detrimental effects or by enhancing beneficial effects of grain boundaries. Professor Gottstein and his co-workers have been deeply involved in fundamental studies of the evolution of microstructure, texture, grain boundary structure, and properties in metallic materials for many years. They have greatly contributed to the development and establishment of basic knowledge of "microstructural control" in connection with texture evolution and grain-boundary-related phenomena. The present author must confess that he has learned much and was stimulated by outstanding view and insight given by Professor Gottstein that experimental work on bicrystals with a single boundary is important to understanding of macro- and microscopic aspects of texture: that is, characteristic feature of texture and mechanism of texture evolution, in connection with microscopic observations on the generation and migration of grain boundaries during grain growth [1]. Professor Gottstein and co-workers have been deeply involved in important subjects along the direction indicated by his early statement and view, in the past four decades up to now [2]. The present author has much benefited from personal contact and collaboration with Professor Gottstein for many years, in proposing and developing the concept of "grain boundary design and control" or "grain boundary engineering" for structural and functional materials, through this longstanding project since the 1980s [3], until recently [4]. This chapter is written to demonstrate that the advent and recent progress of GBE was based on the fundamental knowledge of structure-dependent grain boundary properties obtained by many dedicated research studies through their bicrystal experiments and on quantitative and statistical analyses of grain boundary microstructures in polycrystalline materials which became possible by recent development of a new technique of electron backscatter diffraction (EBSD) in a scanning

Microstructural Design of Advanced Engineering Materials, First Edition. Edited by Dmitri A. Molodov.
© 2013 Wiley-VCH Verlag GmbH & Co. KGaA. Published 2013 by Wiley-VCH Verlag GmbH & Co. KGaA.

electron microscope (SEM)/orientation imaging microscopy (OIM) by Adams and co-workers [5].

17.2
Historical Background

17.2.1
Demand for Strong and Tough Structural Materials

In the history of human civilization and its development, materials have always been a key issue for sustaining and supporting our lives and activities in order to prosper human societies, as previously discussed by Smith [6]. From a historical perspective of engineering materials, there has been always a demand for strong and tough materials, as evidenced by a change of several periods named after widely used materials, from "the stone age" in the ancient time, to "the bronze age," further to "the iron age" which seems still continuing and dominantly supporting our current human society. Nevertheless, it is evident that we are now in the midst of "the semiconductor age" which is also supporting our human lives and activities through recent development of modern electronics and information technology (IT). Looking at historical changes of such practical materials, the present author as a materials scientist thinks: "What is a very key issue of materials development necessary for sustaining human societies and activities, among many demands made upon metallurgists and foundry men in old times, and nowadays materials scientists and engineers." In the author's personal opinion, the control of material fracture, particularly brittle fracture, is the most important and urgent issue, because material fracture often causes serious accident and enormous disaster, occurring suddenly and rapidly beyond human prediction, as seen from accidents of small-scale machines to large-scale aircraft, space shuttle, nuclear power station due to brittle fracture of engineering materials, and furthermore, earthquake caused by brittle fracture of rock plate occurring beneath ground in nature beyond human prediction.

According to the discipline of materials science and engineering (MSE) which has been almost established during the past half century since the 1960s, the subject of "microstructure control" is well recognized as a very centering issue of development of high performance structural or functional materials. Many researchers have been deeply involved in establishment of MSE, as discussed by Cohen [7], and more recently by Cahn [8]. Almost all engineering materials are polycrystalline aggregates consisting of a huge number of grains with their size ranging from mm to nm order, connecting to neighbors at grain boundaries and/or interphase boundaries so that "microstructural control" needs to be understood from the viewpoint of grain- and interphase-boundary-related phenomena. In recent years, the size and shape of individual grains, and also the character of individual grain boundaries (including interphase boundaries in multiphase material), grain boundary connectivity, have been precisely determined for polycrystalline samples of metallic, semiconductor, and ceramic materials by a fully computerized SEM-EBSD/OIM technique without much difficulty. Accordingly, it is not difficult to consider that conventionally called "microstructure" in polycrystalline materials [9–11] can be more precisely termed as "grain boundary microstructure" characterized by newly introduced several grain-boundary-related parameters such as the grain boundary character distribution (GBCD) and grain boundary connectivity, discussed later, in addition to the conventional microstructural parameter, that is, grain size. On the other hand, polycrystalline materials can be produced by using different processing methods developed so far: (i) solidification from the melt or casting and unidirectional solidification, (ii) thermomechanical processing based on recrystallization and grain growth of plastically deformed material, (iii) sintering of powder compacts, (iv) vapor and electrolytic deposition and so on, depending on the state, type, shape, and size of raw material and finishing material. We need to know the details of characteristic features of grain boundary microstructures for a large variety of polycrystalline materials produced by different processing methods under different conditions to endow them with desirable bulk properties and high performance.

17.2.2
Long Pending Materials Problem and Dilemma

In the history of materials design and development, the mankind has always made ever increasing effort to obtain stronger and more fracture-resistant/tough materials. This is evident from examples of most popular metallic materials like iron-based alloys and steels [12,13] and other engineering materials [14]. There has been enormous demand for engineering materials with higher strength and higher fracture resistance, that is, fracture toughness. However, with all enormous efforts, there still remains a long pending problem and dilemma that a stronger material tends to become more brittle and break more easily, resulting in inevitable loss of ductility and fracture toughness due to intergranular fracture. It is widely recognized that strengthening in polycrystalline metallic, intermetallic, and ceramic materials, tends to reduce ductility and toughness. As a materials scientist, the author must insist that any effort for strengthening a structural material with ductility loss has an inevitable danger and lack of reliability in engineering applications. This is truly against the engineering purpose for design and development of high strength structural materials with higher fracture toughness and reliability. How can we solve this long pending problem and dilemma?

Nowadays it is well known that the presence of grain boundaries (here after including interphase boundaries) dominantly controls deformation and fracture in single- and multiphase materials, as discussed by Hondros and McLean [15], so that we cannot understand the mechanical properties of polycrystalline materials without basic knowledge of roles of grain boundaries in yielding deformation and fracture [16]. According to the established discipline of MSE, the presence of grain boundaries strongly affects the motion of dislocations in plastic deformation and fracture of crystalline solids so that an increase of the density of grain boundaries, or grain refinement, is effective for strengthening of polycrystalline materials, as experimentally conformed very early by Hall and Petch, and the grain size effect on the flow stress and fracture stress is described by the well-known "Hall–Petch relation" [17,18]. Up to now, grain size refinement is a standard engineering technique, named "grain boundary strengthening" or "polycrystalline strengthening," which has attracted many researchers being involved in strengthening of conventional polycrystalline materials [19], more recently extending to the extreme case of polycrystalline materials, that is, nanocrystalline materials [20].

Since the year 1986 when a nanocrystalline material was first reported by Gleiter and co-workers [21], basic studies of nanocrystalline materials have been extensively performed, drawing an ever-increasing attention of materials researchers and engineers. A drastic decrease in grain size (d) has been achieved from mm to μm order for conventional polycrystalline materials up to 10–100 nm order bringing about a significant increase of deformation strength in nanocrystalline materials by development of new processing methods, as discussed in detail by Valiev and Langdon [22]. Indeed, grain boundaries are recognized as sources strength as well as other boundary-related bulk properties which conventional polycrystalline materials never possess, as comprehensively discussed by Gleiter [23,24] so far. From the viewpoint of the pending problem and dilemma mentioned before, unfortunately, nanocrystalline materials are still in the conflict of the dilemma, without getting rid of the problem of "loss of ductility." For the purpose of solving the pending problem of "loss of ductility and toughness" associated with grain refinement, the elimination of grain boundaries was once seriously discussed, for example, by Gilman for several types of engineering materials such as high-temperature materials [25] and semiconductor electronic materials. Indeed the elimination of grain boundaries can reduce detrimental effects associated with the presence of grain boundaries, by engineering effort of complete elimination through the production of single crystalline engineering material. Nevertheless, we should note one important thing that this way of the single crystal approach will naturally and inevitably lose beneficial effects produced by the presence of grain boundaries, for example, grain boundary strengthening. How can we compromise and overcome this dilemma?

Now we should seriously think and try to find a new possibility that enables us to control detrimental effect and also to enhance beneficial effects, in order to endow polycrystalline materials with desirable bulk properties over a wide range of grain size. What is the reason why grain size refinement causes a loss of ductility and fracture toughness? Can we now fully understand and explain the effect of grain boundary microstructure on fracture processes in polycrystalline and nanocrystalline materials? In the present author's opinion, until recently, grain size refinement for strengthening has been performed without sufficient recognition of possible changes of "grain boundary structure and character" or "grain boundary microstructure" with decreasing grain size. Probably "grain boundary microstructure" in nanocrystalline materials will more strongly depend on the processing method because of the presence of an extremely high density of grain boundaries. Here let me remind the readers of an interesting paper by A. W. Armstrong entitled "The strengthening or weakening of polycrystals due to the presence of grain boundaries" [26]. From the title of this chapter, we recognize that there are two possibilities of the grain size effect which can produce either strengthening or weakening by controlling grain size in polycrystalline materials. Here, let us just think about the reason why the plots of experimental data of the flow stress or fracture stress almost always tend to deviate downward, as seen from the Hall–Petch curves for metallic materials with decreasing grain size to submicron size. In fact, structure-dependent intergranular fracture is considered as a major origin of intrinsic brittleness in polycrystalline and nanocrystalline materials. Is it reasonable to consider that the effectiveness of grain boundaries as dislocation barrier or dislocation source remains unchanged over a wide range of grain sizes from mm to nm, irrespective of materials processing and condition?

17.2.3
Origin of Heterogeneity of Grain Boundary Phenomena

We often recognize in microstructural observations that grain boundary phenomena never take place simultaneously at every boundary in a polycrystal. If we carefully observe any case of grain boundary phenomena, we can easily realize that some boundaries are preferential sites, but some other boundaries are immune to the boundary phenomena. Figure 17.1 shows an example of heterogeneity of grain boundary phenomena, that is, high temperature deformation and fracture in a polycrystalline sample of iron–tin alloy during creep deformation at 973 K [27]. Some boundaries could extensively slide and migrate, but some others little or less, resulting in the heterogeneity of grain boundary sliding and intergranular fracture occurring in a polycrystal.

It was the end of the 1940s when the difference in the propensity to grain boundary phenomena began to be recognized and seriously discussed by Smith in the relation of microstructure to the grain boundary energy [28] and further to the character and structure of grain boundaries, in order to clarify the relation between the structure and properties of grain boundaries. Careful experimental studies with

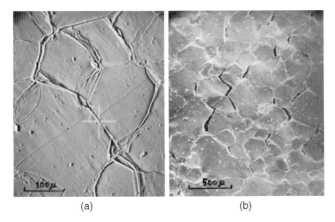

Figure 17.1 Heterogeneity of occurrence of grain boundary phenomena at individual grain boundaries during high temperature creep deformation in iron–tin alloy. (a) Grain boundary sliding and migration, (b) intergranular fracture due to sliding and migration.

metal bicrystals were performed later by pioneering researchers, Aust, Chalmers, and others, and extensive fundamental data on structure-dependent grain boundary properties were collected as seen from the literature [29–31], as examined later. In fact, the heterogeneity of grain boundary (GB) microstructure characterized by the occurrence and the distribution of different types of grain boundaries is a primary source of characteristic features of GB phenomena. They occur and appear in a more complicated manner controlled by responsible mechanisms which are normally based on the percolation-controlled process of single or coupled GB phenomena in two-dimensional (2D) or three-dimensional (3D) polycrystalline material systems.

17.3
Basic Concept of Grain Boundary Engineering

17.3.1
Beneficial and Detrimental Effects of Grain Boundaries

Grain boundaries are the most important "microstructural element" in polycrystalline materials of single phase and multiphase. They can exist generally irrespective of the kind of material whether being metallic, semiconductor, ceramic materials, and of crystal structure whether cubic or noncubic crystal structure. The atomic structure of grain boundaries is the source of a large variety of structure-dependent boundary properties due to the atomistic level of structural order, because GB atomic structures can be often more distorted and even disordered in comparison with the crystal lattice of the grain interior. Grain boundaries are very versatile from the viewpoints of their structure/character and geometrical configurations in polycrystalline materials. Moreover, grain boundaries can change their character and structures by interaction with other defects and by a structural transformation due to temperature, pressure, and other external fields. From the reasons mentioned above, grain boundaries have a higher intrinsic and extrinsic potential originating from their versatility and flexibility. Why should we not utilize such a higher potential, versatility, and flexibility of grain boundaries for the development of advanced materials through manipulating GB microstructures?

As discussed later in detail, it has been almost established that the activity of individual grain boundaries can be controlled by manipulating GB character and structure, in order to produce optimum GB microstructure and to result in desirable boundary-related bulk properties of polycrystalline materials. High-energy random boundaries with excess strain energy can play dominant roles as preferential sites for metallurgical phenomena in polycrystalline materials. It is reasonable to consider that the effectiveness and the magnitude of influence of individual grain boundaries may vary from boundary to boundary, depending on the type and the degree of disorder of the atomic structure of individual boundaries. We have already made enormous research efforts to establish the relationship between the structure and properties of grain boundaries, as the fundamental knowledge of structure-dependent boundary properties, obtainable from experimental and theoretical studies mostly on bicrystals. Now we should effectively utilize fundamental knowledge and accumulated data for design and development of different kinds of materials from metals and alloys, semiconductors to ceramics. The structure-dependent properties of grain boundaries have been studied over the past half century. We can access those useful data on the structure-dependent properties of grain boundaries in reference books and proceedings of conferences on related topics [2,30–33]. Effects of grain boundaries on the bulk properties of polycrystalline materials are simply classified into two categories. One is such that the presence of grain boundaries causes some degradation of the bulk properties and performance of polycrystalline materials. The other is such that the presence of boundaries brings about some improvement and enhancement of bulk properties, even generation of a new function and property. So, we may define the former as the "detrimental effect" and the latter as the "beneficial effect" of grain boundaries. The presence of grain boundaries (hereafter including interphase boundaries) can affect beneficial and detrimental effects on the bulk properties of polycrystalline materials. We can separate detrimental effects from beneficial effects by the precise control of GB microstructure. If it is possible to suppress the "detrimental effect" and conversely to increase the "beneficial effect" as much as we could in the course of their fabrication, we would be able to endow a polycrystalline material of ordinary composition or the same composition with much better material performance than the original material.

17.3.2
Basic Concept of Grain Boundary Engineering

Now, let us briefly discuss the basic concept of GBE, initially proposed by the present author as "grain boundary design and control" in the early 1980s [3,34,35] based on the fundamental knowledge of structure-dependent GB properties in metals and alloys, recently established through extensive bicrystal work performed since the early 1950s [36,37]. Fortunately we could utilize quantitative experimental data on structure-dependent GB properties in metal bicrystals, previously reported in the proceedings of symposia on the topics [29,30]. Looking at historical development of the approach of GBE, it is evident that theoretical and experimental studies of GB structure and properties were performed stimulating to each other, resulting in a rapid progress of our basic knowledge. In particular, the theories of atomic structure of grain boundaries based on the structure-unit model have provided clues to explain the experimental data on structure-dependent GB properties obtained from real materials. Development of experimental techniques for observations and characterization of grain boundaries such as field ion microscopy (FIM), transmission electron microscopy (TEM), high resolution transmission electron microscopy (HRTEM), and more recently SEM-EBSD/OIM, electron energy loss spectroscopy (EELS) greatly helped those researchers who actively worked in the GB research field and could meet a strong demand and desire to reveal the relation between the structure and properties of grain boundaries. The present author has also enjoyed his experience of timely advents of a new experimental technique or theoretical tool.

Indeed, "Necessity is the mother of invention," as confirmed in the course of the advent and development of GBE. "Mutual interaction and stimulation" between the theoretical and experimental approach were essential for researchers to obtain useful insight into understanding of GB structure-controlled bulk properties based on structure-dependent GB phenomena.

Table 17.1 presents a brief history of the research field of the structure and properties of grain boundaries, interfaces, and related fields during the past one

Table 17.1 A brief history of development of the research field of structure and properties of grain boundaries, interfaces and of related fields during the past one century.

1900–1940s

    Amorphous Cement Theory (Rosenhain-Ewen, 1912), Good Fit-Bad Fit Model (Mott, 1948)

    Coincidence-Site-Lattice (CSL) Model (G. Friedel 1920, Kronberg-Wilson 1949)

    Transition-Lattice Theory (Hargreaves-Hill, 1929), Disordered Group Model (Ke, 1949)

    Geometrical and Topological Approach to GB microstructure (C. S. Smith, 1948)

1950s–1960s

    Dislocation Theory of Low-angle GBs (Read-Shockley, 1952, Amelincks, 1957)

    Boundary Structure and Properties in Bicrystals (Chalmers-Aust, R. W. Cahn)

    Thermodynamics of GBs (J. W. Cahn, 1956), First Book on GBs (D. McLean, 1957)

    Geometrical and Mathematical Approach to CSL (Brandon, Ranganathan, 1966)

    FIM, TEM Observations (Brandon, Ryan-Suiter, Smith, Ralph-Jones, Gleitcr)

    O-Lattice Theory (Bollmann, 1968)

1970s–1980s

    HREM of GB Structure (Schober-Balluffi-Bristowc, Sass, Smith-Pond-King, Ishida-lchinose, Bourret-Bacman, Ruhle)

    Bicrystal Work in Metals (extensively in France, Russia, Japan)

    Extension of CSL model to HCP, Non-cubic crystals (Bruggemen-Bishop, Grimmer-Warrington)

    Computer Calculations (Biscondi, Vitek-Sutton, Wolf, Doyama-Kohyama)

    Nanocrystalline (Gleiter)

    Interface Approach to Phase Transformation (Hillert, Aaronson-Enomoto-Purdy, Maki-Furuhara)

1990s–2000

    Microscale Texture Analysis (Lücke-Gottstein, Bunge-Esling)

    SEM-EBSD/OIM (Dingley-Adams-Wright-Kunze, 1991–1993)

    GB Microstructure and Properties in Polycrystallinc Materials (Aust-Palumbo-Erb, Ralph-Howell-Jones-Randle, Grabski, Priesster, Watanabe-Kokawa-Tsurekawa)

    Bicrystal Berhaviour (Metals: Gottstein-Shvindllerman-Straumal-Molodov-Winning, Paidar-Lejcek, Mori- Monzen-Kato-Miura, Ceramics: Sakuma-lkuhara)

    Triple-Junction Behavior (Gottstein-Shvindlerman, King, Aust-Palumbo)

    Nanocrystalline Materials by ECAP Processing (Valiev-Langdon-Nemoto-Horita-Furukawa)

century. It is evident from this table that the objective of the basic study of microstructure changed from macroscopic to microscopic level, and after the 1950s to 1960s, experimental and theoretical approaches on atomic level were performed in accordance with the advent of electron microscopy. During the period 1970s to 1980s, a rapid progress in the science of GB and interface was made which enabled us to establish the fundamental knowledge of GB and interface science. Then since the 1980s, several attempts were made by which a transformation of "Science" to "Engineering" of GB and interfaces was intended to understand the effect of microstructure upon the bulk properties of polycrystalline materials. Theoretical and experimental tools for statistical analyses of "GB microstructure" were strongly required. To meet this requirement, SEM-EBSD/OIM was timely developed by Adams and co-workers in the early 1990s.

On the other hand, as basic studies for the control of intergranular fracture and brittleness, several systematic investigations of intergranular fracture were made by using orientation-controlled bicrystal samples of metals and alloys, particularly in order to reveal the effect of GB structure on fracture and to find some clue for the control of intergranular fracture and brittleness, as shown in the early review [27]. A systematic study of fracture processes performed in the light of structure-dependent GB phenomena provided us much useful information to solve the long pending problem and dilemma, as discussed in this chapter. We have already begun to collect some fruits, although it is still premature. The state-of-the-art of GBE for structural materials as well as functional materials seems very successful and promising, indicating a new direction toward further achievements and establishment of GBE, as seen from recent proceedings of international conferences and special issues of international journals on grain boundaries and interphase boundaries in crystalline materials [38–42].

17.3.3
Structure-Dependent Grain Boundary Properties in Bicrystals

In 1864 when Sorby first reported on the grain structure of a polycrystalline material, the heterogeneity of microstructure, particularly indicating the difference in degree of corrosion and fracture at individual grain boundaries, was demonstrated on early photomicrograph for oxygen-rich blister steel, by using an optical microscope with magnification power of only eight times [43]. Grain boundaries surrounding larger grains were more heavily corroded and more easily broken in the polycrystalline sample. In order to quantitatively and systematically study possible effects of GB structure and character on GB properties and phenomena (corrosion, migration, sliding, segregation, precipitation, fracture, etc.), orientation-controlled bicrystal samples were produced, sometimes with much difficulties, for quantitative studies of structure-dependent GB properties by experiments. The relative orientation of adjoining grains was systematically controlled by using specially designed graphite boats for bicrystal growth. A tilt- or twist-type boundary can be selectively introduced into a bicrystal sheet or rod of metals, depending on the application by the crystal growth method, or the diffusion-bonding method using two single crystals fixed in a specific relative orientation arrangement. Pure tilt-type, pure twist-type, and mixed-type boundaries can be artificially introduced in metal bicrystals with systematically controlled misorientation angles about a given rotation axis of crystal orientation, since the very early time around 1950 of the basic study of grain boundaries [29,30,44]. So far the following methods for bicrystal growth have been invented: (a) Chalmers' technique based on solidification from the melt under temperature gradient, (b) strain annealing, (c) diffusion bonding of two single crystals, (d) arc zone melting technique, (e) floating zone technique, and so on [45]. Practically, there are more or less some difficulties of bicrystal growth even for metals and alloys so that much care is necessary for successful bicrystal growth. For example, in the most popular case of Chalmers' technique based of crystal growth from the melt, it is necessary to avoid the oxidation during growth, by using the vacuum condition or inert gas atmosphere. In the case of metal with a high vapor pressure, bicrystal growth is necessary to be carried out in inert gas positive pressure to avoid oxidation and vaporization, as is the case of magnesium and zinc bicrystals. In general, bicrystal growth of metals of higher melting temperature is more difficult than those of low melting temperature. Probably this is the primary reason why experimental work on bicrystals of tin, lead, and aluminum has been extensively performed from the early stage of GB studies until recently. Designing a boat (graphite or alumina) for bicrystal growth is a key point to produce such a bicrystal sheet or rod whose boundary character is precisely controlled as exactly as expected with an accuracy of orientation control, within a few degrees. Moreover, it is often experienced that the GB plane is curved or facetted, not planar or straight in the bicrystal specimen. Some special effort is needed to obtain a planar GB in a grown bicrystal sheet or rod, for example, by putting a straight ridge on the inside surface of the graphite boat in order to lead a straight GB along the growth direction. Moreover, it is now possible to produce even ceramics bicrystals (such as alumina, zirconia) with well-controlled boundary character by the diffusion-bonding technique for the quantitative investigation of structure-dependent GB properties, as reported by Shibata et al. [46].

So far experimental work on metal bicrystals has been extensively performed to reveal structure-dependent GB properties involved in metallurgical phenomena such as GB migration, sliding, diffusion, segregation, corrosion, and fracture since the 1950s [30,36]. A brief introduction of an example taken from the literature on the GB phenomenon in systematically orientation-controlled bicrystals will help readers in understanding the characteristic features of structure-dependent GB properties. Here, to demonstrate an example of structure-dependent GB properties, we look at the result from our early work on GB sliding in $\langle 10\bar{1}0 \rangle$ symmetrical tilt zinc bicrystals [47]. It is well known that GB sliding plays key roles in intergranular fracture during creep deformation in metals and alloys at high temperatures, as previously discussed on GB sliding-induced intergranular creep fracture in detail in several review papers [27,48,49].

Figure 17.2 shows the misorientation dependence of GB sliding observed for the $\langle 10\bar{1}0 \rangle$ symmetrical tilt zinc bicrystals at different temperatures. As seen from Figure 17.2a, the amount of sliding (observed at 300 min after the onset of the sliding test) increases with increasing misorientation angle θ, more rapidly beyond

17.3 Basic Concept of Grain Boundary Engineering

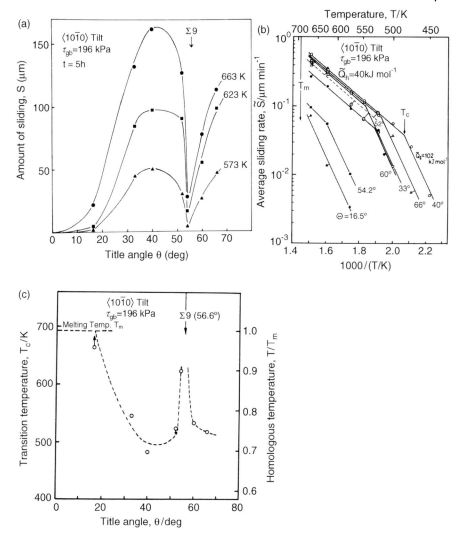

Figure 17.2 Structure-dependent grain boundary sliding in $\langle 10\bar{1}0 \rangle$ symmetrical tilt zinc bicrystals [47]: (a) misorientation dependence of the amount of sliding at different temperatures, (b) change in the temperature dependence of sliding due to a grain boundary structural transformation, and (c) misorientation dependence of the critical temperature for grain boundary phase transformation. Note that much higher critical temperatures are observed for the $\Sigma 9$ coincidence boundary and low-angle boundaries, suggesting a higher stability of grain boundary structure at low temperatures.

$\theta > 10°$, indicating a peak around $\theta = 40°$, then a deep cusp around 57° which well corresponds to the 56.6° $\langle 10\bar{1}0 \rangle$ orientation, theoretically predicted as $\Sigma 9$ near-coincidence boundary. It is evident that sliding in the $\langle 10\bar{1}0 \rangle$ symmetrical tilt zinc bicrystals is very difficult at low-Σ coincidence orientation. It is well known that GB sliding plays important roles in creep deformation, superplasticity, and fracture in polycrystalline materials at high temperatures. Figure 17.2b exhibits the temperature dependence of the sliding rate for the $\langle 10\bar{1}0 \rangle$ symmetrical tilt bicrystals with different misorientation angles. It should be noted that the temperature dependence of the average sliding rate changes at a certain critical temperature, depending on the misorientation angle. It was also found that the critical temperature is higher for the Σ9 near-coincidence boundary while high-angle random boundaries show much lower critical temperatures. Accordingly, as seen from Figure 17.2c, the critical temperature becomes higher for the Σ9 near-coincidence boundaries and also very likely for the low-angle boundary whose critical temperature must be above the melting temperature T_m. The observed change of sliding behavior due to temperature has been ascribed to a GB structural transformation. With increasing temperature, GB structure transforms from a low-temperature structure to a high-temperature structure due to an increase of entropy associated with increasing temperature. It may be said that low-angle boundary and low (Σ9) near-coincidence boundary possess higher thermal stability compared with high angle random boundaries. Similar misorientation dependence has been observed for other GB properties and boundary phenomena.

Figure 17.3 shows a schematic presentation of two types of structure-dependent GB properties: coincidence boundaries show whether they possess peaks or cusps around specific coincidence orientations of low Σ. This is because the type A structure-dependent boundary properties are due to the activity of the GB: high-angle/high-energy random boundaries are potential site for metallurgical phenomena, while low-energy boundaries such as low-Σ coincidence boundaries

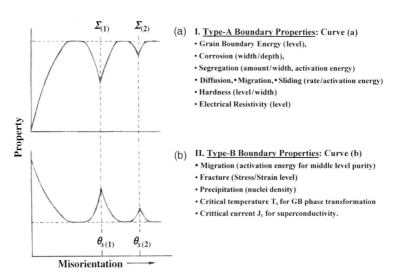

Figure 17.3 Two types of structure-dependent grain boundary properties, depending on grain boundary energy: (a) Type A: more significant at higher energy boundaries, (b) Type B: more significant at lower energy boundaries, particularly at low-Σ coincidence boundaries with a special misorientation angle, Θ_Σ [4].

and low-angle boundaries with smaller misorientation angle (can <7°) are not. On the other hand, the type B structure-dependent properties are generated by the inactivity or resistance of the GB to metallurgical phenomena. Accordingly, low-energy grain boundaries exhibit such properties. Simply speaking, the type A and type B structure-dependent boundary properties show completely opposite features of their misorientation dependence, as demonstrated in Figure 17.3a and b, respectively. It has been found that the misorientation dependence of the fracture stress and fracture strain in metal bicrystals generally shows type B feature [50]. This suggests the possibility that a polycrystalline material with high strength and high fracture toughness can be realized by controlling the GB microstructure, particularly by introducing a higher fraction of low-energy strong boundaries, or reversely by eliminating high-energy weak random boundaries, as shown soon later.

17.3.4
Structure-Dependent Fracture Processes in Polycrystals

In situ observations are very useful and powerful for the basic understanding of microstructure-controlled metallurgical processes occurring in polycrystals, particularly for fracture processes. Probably SEM *in-situ* observations of structure-dependent intergranular fracture processes in bismuth-doped copper and beta brass polycrystalline specimens performed by the present author and co-workers in the early 1980s were first in the literature [3,51]. All grain boundaries in coarse-grained tensile specimens were characterized by the SEM – electron channeling patterns (ECPs) technique prior to fracture testing with a SEM-tensile stage. The whole fracture path after final fracture is shown in Figure 17.4a. The initial crack formed at random boundary (R) on the bottom edge then propagated in the mixed fracture mode of intergranular and transgranular (cleavage) fracture up to the triple junction where fracture mode changed from transgranular to intergranular fracture along random boundary, further proceeded simultaneously partly in transgranular and intergranular fracture mode. It was found that crack formed and propagated preferentially at weak random boundaries, and when the propagating intergranular crack meets the low-energy fracture-resistant boundary such as low-Σ coincidence boundary, fracture mode changes to transgranular fracture, as shown in Figure 17.4c.

The change of fracture mode was found to depend on the direction of the propagating crack and the stress condition whether the GB in front of the crack is aligned perpendicular to the tensile stress direction. Characteristic features structure-dependent fracture processes in a polycrystal obtained from *in-situ* observations are schematically shown in Figure 17.5. It was revealed that fracture processes in polycrystals take place in different manners, depending on the type of grain boundaries and their connectivity. High-energy random boundaries are preferential sites for crack nucleation and break easily, while low-energy special boundaries such as low-Σ coincidence boundaries (including $\Sigma 1$ or low-angle boundary) will not break, indicating a higher resistance to fracture at low and high temperatures and in various environments, as revealed by early work by the

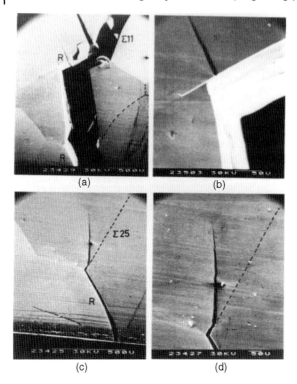

Figure 17.4 SEM observations on crack propagation for liquid-gallium-induced fracture of a beta brass polycrystal [51]. Note the change in fracture mode between intergranular and transgranular fracture, depending on the type of grain boundary in front of the propagating crack and the type of triple junction. (a), (c) Successive observations; (b), (d) higher magnifications of (a) and (c).

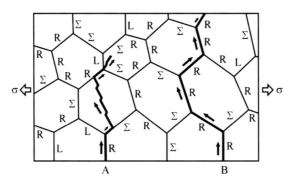

Figure 17.5 Schematic representation of grain boundary structure-dependent fracture processes in polycrystal [3]. Path A: combined process of intergranular and transgranular fracture, Path B: typical intergranular fracture. The overall fracture behavior is controlled by the percolation of crack propagation in the 2D or 3D grain boundary network of polycrytals with different grain boundary microstructures.

author's group [48], and further confirmed by more recent studies [49,50]. It is easily understood that when random boundaries are connected to each other at triple junctions throughout a specimen, the crack can propagate continuously along random boundaries and fracture occurs in a typical brittle fracture mode controlled by dominant intergranular fracture without much plastic deformation in the grain interior (Case B). On the other hand, when the random boundary and low-Σ coincidence boundary connect at triple junction, the crack which initially nucleated at the random boundary will not propagate along the low-energy boundary so that fracture mode has to change from intergranular to transgranular fracture mode. Nevertheless, if the crack meets other random boundary after penetration by transgranular fracture in the grain interior, again it propagates along the random boundary. Thus, the whole fracture process in a polycrystal specimen is due to the mixture of intergranular and transgranular fracture, spending much higher fracture energy necessary for plastic deformation in the grain interior, resulting in an increase of fracture resistance and fracture toughness (Case A). Thus, fracture processes and fracture mode in polycrystals strongly depend on the distribution of random boundaries and special boundaries and their connectivity, which are salient features of the "GB microstructure," quantitatively characterized by a modern SEM-EBSD/OIM technique developed by Adams et al. [5].

17.3.5
Parameters to Characterize Grain Boundary Microstructure

Until recently, the grain boundary character distribution (GBCD) and GB connectivity have been widely used to quantitatively characterize the GB microstructure in polycrystalline metallic materials since the early 1980s when the present author proposed these parameters and others [3]. These were determined by using the SEM-EBSD-ECP technique [52], and more recently by SEM-EBSD/OIM, for metallic, ceramic, and semiconductor materials that received different processing, see the reviews on GBE by the present author and co-workers [53,54]. In fact, it is very important for our understanding of the microstructure-dependent bulk properties of polycrystalline materials to clarify what type of GB and to what extent different types of boundaries can occur depending on metallurgical factors, such as the kind of material, crystal structure, composition, processing method and condition, and so on. Now we have reached the stage of understanding and prediction of some general features of the GBCD and its relationship with the bulk properties of polycrystalline materials, as discussed in the recent reviews on GBE [52–54] although much remains to be explored by more systematic experiments based on OIM analysis.

The grain size has been generally used as the primarily important parameter to quantitatively characterize the microstructures, irrespective of different geometrical arrangements and configurations of grain boundaries in 1D, 2D, and 3D polycrystals. However, at the beginning of the 1980s several new parameters were introduced which could quantitatively characterize the GB microstructure taking consideration into not only grain size but also the difference of character and structure of individual grain boundaries and their statistical distribution in a polycrystal. So far, the following parameters were proposed and repeatedly discussed to relate the GB microstructure to the bulk properties of polycrystalline materials: (i) GBCD, (ii) GB connectivity, (iii) GB inclination distribution, (iv) boundary triple junction character distribution, (v) GB chemistry distribution, in addition to the conventional parameter (vi) grain size, that is, GB density. It is interesting to clarify how these parameters can properly characterize and describe a large variety of GB microstructures in different kinds of polycrystalline materials produced by different processing methods.

17.4
Characteristic Features of Grain Boundary Microstructures

It may be useful to briefly examine important features of the GB microstructures observed and quantitatively described by the GBCD and GB connectivity in order to relate the bulk properties of polycrystalline materials. Firstly, let us examine how the GBCD depends on the grain size for polycrystalline metallic materials produced thermomechanically by annealing of their deformed state resulting in recrystallization and subsequent grain growth, and nanocrystalline materials produced by electro- or vapor-deposition methods.

17.4.1
Grain Size Dependence of Grain Boundary Microstructure

Until recently, the effect of grain boundaries on the bulk properties of polycrystalline materials has always been discussed in connection with the average grain size, as a standard approach to various kinds of GB effects. Here it should be pointed out that "the average grain size" physically means "the average spacing" between grain boundaries, simply assuming a random distribution of grain boundaries without any consideration into the special structural feature of individual grain boundaries. Accordingly, "the average grain size" is obtained by simply calculating under another tacit assumption that every GB plays a similar role without exhibiting any difference between them, completely neglecting the difference in the character and structure of individual grain boundaries. So far, almost all discussions on the bulk properties of polycrystalline materials from the viewpoint of just "grain size effect" ignore any intrinsic nature of individual grain boundaries and their structure-dependent boundary properties well documented by a bicrystal experiment. It is very suspicious whether the tacit assumption of the grain size-independent effectiveness of grain boundaries is reasonable, without any examination in the light of recent knowledge of structure-dependent GB properties.

17.4.2
Change in Grain Boundary Energy during Grain Growth

In fact, during the 1980s it was found that the type of grain boundaries and GBCD can change during annealing and grain growth, namely with grain size, in polycrystalline metallic materials produced by conventional thermomechanical processing [55–57] or rapid solidification and subsequent annealing [58]. The possibility of change in GB energy during grain growth was pointed out by Grabski and co-workers on the basis of their extensive TEM observations of grain boundaries [59,60] and predicted by Watanabe based on a simple regular polygon grain shape model of grain growth [61]. It is very likely that the type and character of grain boundaries can change with grain size in metallic polycrystalline materials, depending on material processing and condition, as experimentally confirmed later [62]. During the annealing of a deformed polycrystal sample, low-angle cell boundaries tend to transform to high-angle boundaries having a higher mobility for the later stage of grain growth. On the other hand, initially existing high-angle grain boundaries have to absorb lattice dislocations from the grain interior changing their boundary misorientation angle or character during the stage of recovery and primary recrystallization, as similarly taking place at sliding grain boundaries in creep deformation at high temperature, as observed by Kokawa *et al.* in aluminum [63]. Thus, the character and structure of grain boundaries can flexibly change due to their interaction with other boundary or lattice defects like dislocations and point defects, again depending on the type and structure of boundary. High-energy random boundaries can more easily absorb lattice dislocations and point defects than special low-energy boundaries. In the author's opinion, the significance of the flexibility of GB structure has not yet been fully recognized in previous discussions on metallurgical phenomena in the literature. Probably this is one of the reasons why the previous discussions on GB effects were mostly restricted to the grain size effect.

It is worth mentioning that a unique experimental technique for sample preparation based on deposition of smoke of MgO particles was developed by Chaudhari and Matthews in order to study the relation between GB structure and energy [64]. They produced a cluster of twist boundaries and determined the distribution of misorientation angles on the {001} surface, that is, for ⟨001⟩ twist boundaries. Preferential occurrence of the ⟨001⟩ twist-type $\Sigma 5$, $\Sigma 13$, $\Sigma 17$, $\Sigma 25$ coincidence boundaries was observed, probably because of their low energies, as predicted by possible coincidence orientations for the ⟨001⟩ rotation. This suggests that some specific low-Σ coincidence boundaries may occur preferentially due to their energy prevalence, in a sharply textured polycrystal of well-defined texture type. A similar approach was attempted to clarify the relation between GB structure and energy, and the effect of solute segregation on the relation by using a single crystal ball sintering method developed by Gleiter and co-workers [65,66]. The orientation distribution of single crystal balls of vapor-deposited samples of silver alloys or copper was analyzed by the X-ray diffraction method after annealing. The observation of high intensities at specific diffraction angles indicates the preferential occurrence of special low-energy coincidence boundaries due to the relative rotation of single crystal balls during annealing, as predicted by the rotation axis vs. angle relation for the ⟨111⟩ rotation in an fcc crystal.

17.4.3
Relation between Grain Boundary Character Distribution and Grain Size

Surprisingly, since the 1980s until recently, only a few systematic experimental investigations have been reported on the relationship between the GBCD and grain size for polycrystalline metallic materials produced by thermomechanical processing from the initial polycrystalline sample, not from the single crystal sample without grain boundaries, although the SEM-EBSD/OIM technique has been extensively used. The present author and co-workers have made a systematic investigation of the GBCD in polycrystals produced by thermomechanical processing of single crystals with different initial orientations, as first attempted for aluminum [67] and then recently for austenitic stainless steel [68] and molybdenum [69], as discussed later. Now we have simple questions to be answered: what kind of GB microstructure can occur in ordinary polycrystalline samples with ordinary grain size (cm–μm) and nanocrystalline samples with nm grain size, at least three orders of magnitude smaller than that of ordinary grain size, for the same material? Can we predict quantitatively the relationship between the GBCD and grain size from the currently available knowledge of Materials Science and Engineering?

Let us examine early experimental data on the relation between the GBCD and grain size in metallic materials. Figure 17.6 shows the fraction of special low-Σ coincidence boundaries ($\Sigma < 29$ including low-angle/$\Sigma 1$) as a function of the average grain size for bulky polycrystalline samples of Al, Fe, Fe-based bcc alloys, W, Mo, Ni$_3$Al produced by thermomechanical processing of deformed (rolled or extruded) polycrystalline samples and also for thin ribbon samples of bcc Fe–6.5 mass% Si alloy produced by rapid solidification from the melt and subsequent annealing [58]. Firstly, it is evident that the frequency of low-Σ coincidence boundaries tends to decrease with grain size; conversely the fraction of high-angle random boundaries increases during grain growth for most of bulky metallic materials produced by thermomechanical processing, although there is a large scatter among the data for

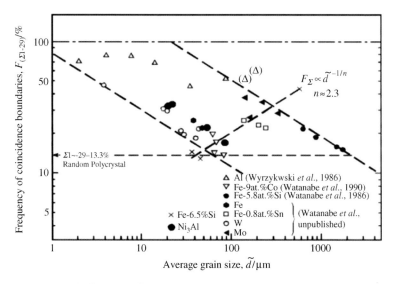

Figure 17.6 The frequency of low-Σ coincidence boundaries ($\Sigma < 29$) as a function of the grain size for bulky polycrystalline of metals and alloys produced thermomechanically and for thin ribbons of an Fe–6.5mass%Si alloy produced by rapid solidification and subsequent annealing [62].

different materials and test conditions. Grabski also reported that the frequency (fraction) of special boundaries decreases with increasing grain size or with annealing temperature in hydrostatically extruded aluminum [56]. The observations suggest that low-energy grain boundaries which initially formed during the primary recrystallization transform to high-energy random boundaries which possess a higher mobility for migration and can easily absorb lattice dislocations as a preferential site for atomic diffusion and also source and sink of vacancies and dislocations. On the other hand, in the case of Fe–6.5mass% Si ribbons produced by rapid solidification from the melt and subsequent annealing [58,70], the frequency of low-Σ coincidence boundaries increases with increasing grain size, quite oppositely to the case of those bulky samples produced by thermomechanical processing. This is due to the difference in the driving force of GB migration and the mechanism of grain growth between bulky samples produced by thermomechanical processing and thin ribbon samples produced by rapid solidification and subsequent annealing. The stored strain energy and GB energy are the origin of the driving force for GB migration and grain growth in the former, while the orientation-dependent surface energy is the driving force in the latter. The evolution of coarse-grained thin Fe–6.5mass %Si alloy ribbons with a very sharp texture of {100} or {110} was observed, depending on annealing temperature which affects the orientation-dependent surface energy in the Fe–Si alloy system [58,70].

In the case of the evolution of GB microstructure in polycrystalline metallic materials produced by thermomechanical processing, several systematic experimental studies have been performed with techniques for precise orientation determination such as the SEM-ECP or SEM-EBSD/OIM technique for bulky polycrystalline samples of aluminum [67], stainless steel [68], and molybdenum [69], by applying a more precise thermomechanical processing, that is, by annealing deformed single crystals with different initial orientations, in order to avoid some possible effect of the orientation distribution of grain structure existing in an initial polycrystal sample. It is reasonable to expect that the usage of single crystals with different initial orientations would more clearly reveal the key factors and crystal orientation affecting deformation and recrystallization, resulting in the evolution of simple GB microstructure.

Figure 17.7 shows the result of the first experimental investigation of GB microstructure for aluminum polycrystals produced thermomechanically from single crystals, conducted by the author and co-workers in the early 1980s [67]. The characterization of GB microstructure was carried out with the SEM selected area channeling pattern (SACP) technique, more than 10 years before the development of modern computer-assisted SEM-EBSD/OIM [51]. We notice several interesting features of GB microstructure, from the viewpoint of the fraction of different types of grain boundaries: The fraction of low-Σ coincidence boundaries, low-angle boundaries, and high-angle random boundaries depends on the initial orientation of the single crystal and the amount of compressive prestrain (60%, 70%, 80%) before annealing, particularly being significant for the single crystal with near the {111} orientation. It was found that the fraction of low-Σ coincidence boundaries (20%–31%) is much higher than that (13.6%) predicted for a random polycrystal, as well as for low-angle

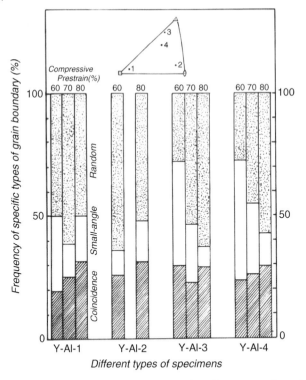

Figure 17.7 The frequency of occurrence for different types of grain boundaries in high purity (99.9998%) aluminum polycrystals produced by thermomechanical processing from single crystals with different initial orientations [67].

boundaries. As a result of this, the fraction of high-energy random boundaries ranged from 28% to 64%, being quite different from the theoretical value (86.4%) for a random polycrystal. It was revealed that the characteristic feature of GB microstructure in polycrystalline aluminum with fcc crystal structure and high stacking fault energy can be quite different from that predicted for a random polycrystal and can be likely controlled by the condition of thermomechanical processing, as one of polular pure metals. Thus, it may be reasonable to consider that the feature of GB microstructure can be more complicated by involvement of other factors controlling the evolution of GB microstructure in ordinary thermomechanically processed materials starting from initially polycrystalline solid. This was the reason why similar systematic experimental work was performed recently by the author's group using single crystals with different initial orientations in austenitic stainless steel (SUS304L) with fcc structure and low-stacking fault energy [68]. Several new findings have been obtained which are worth noting in relation to the present subject of GBE.

It should be mentioned that a significant difference in the GBCD has been observed between bcc metals/alloys with high-stacking fault energy such as iron, alpha iron-based alloys mentioned above, and molybdenum (including fcc metals like aluminum) and fcc metals/alloys with low-stacking fault energy. As seen from Figure 17.8, in the

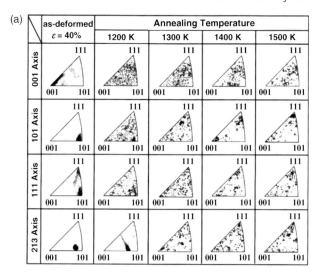

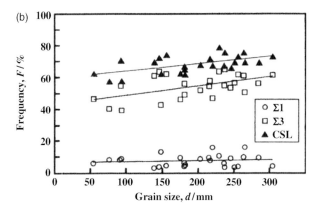

Figure 17.8 The frequency of occurrence for low-angle (Σ1), twin boundary (Σ3), and the overall frequency for low-Σ coincidence (CSL) boundaries (Σ < 29) in 304 L stainless steel polycrystals produced by thermomechanical processing from single crystals with different initial orientations [68]. (a) The types and the sharpness of texture, depending on the initial orientation of single crystal and annealing temperature; (b) relationship between the fraction of low-Σ coincidence boundaries and the mean grain size.

case of a 304L-austenitic stainless steel polycrystalline sample produced by themomechanical processing from single crystals with different initial orientations [68], the frequency of Σ3 and other low-Σ coincidence boundaries tends to increase with increasing grain size. The dominant occurrence of Σ3 coincidence boundaries (so-called twin boundary) and other Σ3-related coincidence boundaries due to multiple twinning during annealing should be noted, although the frequency of low-angle/Σ1 boundaries remains unchanged with increasing grain size. In fact, it has been experimentally evidenced by Lim and Raj that preferential occurrence of Σ3-related coincidence boundaries in descending order of Σ value is due to multiple twinning which can produce $\Sigma 3^n$ coincidence boundaries (n: integer as 2, 3, 4, for $\Sigma 9$, $\Sigma 27$, $\Sigma 81$) selectively and preferentially, during annealing in nickel [71]. As mentioned above, the relation between the GBCD and grain size strongly depends on fundamental mechanisms of GB migration and grain growth so that more systematic investigation into the relation between the GBCD and grain size is necessary for precise control of GB microstructure in various types of polycrystalline materials with a conventional range of grain size and more urgently for nanocrystalline materials, under deeper consideration into metallurgical factors, such as crystal structure (bcc, fcc, hcp, and others), stacking fault energy, and impurity effect. Especially, we need a new insight into structure-dependent GB energy, migration, structural transformation, and orientation/temperature-dependent surface energy, as pointed out by Aust [72]. Unfortunately there has been a huge lack of fundamental knowledge of the factors controlling the evolution of GB microstructure even in single-phase materials so that we need to urgently collect reliable experimental data for both single-phase and multiphase materials including interphase boundaries, for future application of GB/interfacial engineering to advanced materials technology.

17.5
Relation between Texture and Grain Boundary Microstructure

Here, let us discuss the relation between texture and GB microstructure in polycrystalline materials, particularly, some theoretical bases of the GBCD in textured polycrystals, and the recent knowledge of GB microstructures obtained from recent experimental studies using SEM-EBSD/OIM, in addition to some early work by SEM-SACP since the end of the 1980s up to now.

17.5.1
Theoretical Basis of GB Microstructure in Textured Polycrystals

From the basic knowledge of GB characterization, a localization of grain orientations on a specific orientation means that the connection of neighboring grains produces a higher fraction of pure tilt-type or twist-type boundaries than the theoretical prediction based on the statistical approach for a random polycrystal by Mackenzie [73] for crystals with cubic structure, and with other crystal structures [74,75]. From the viewpoint of structure-dependent GB properties observed by the bicrystal experiment, the results of the disorientation distribution from statistical analyses are not sufficient because any structural effect of GB is little taken into consideration in these crystallographic-geometrical analyses. So it should be noted that theoretical data on the disorientation (misorientation) distribution are normally likely different from the data obtained from experimental determination. How should we overcome this gap between the theoretical approach and the experimental one for textured polycrystalline materials? There is a simple guideline: Nature is very economic even in microstructural evolution in polycrystalline materials, without spending much energy in generation of grain boundaries in most metallurgical processing. Namely, grain boundaries of lower energy tend to be formed much easily. As a result of this fundamental nature, those grain boundaries that possess lower energy originating from their atomistic structures tend to occur with a higher frequency than that for a random polycrystal. It is not difficult to imagine that a higher fraction of special boundaries such as low-angle boundaries, high-angle special boundaries like coincidence, or plane-matching type boundaries if the grain orientation distribution sharply localizes around specific crystallographic orientation(s), that is, generation of a sharp texture, in a polycrystalline sample.

17.5.2
Characteristic Features of GB Microstructures in Textured Polycrystals

The development of a new technique of computer-assisted automatic SEM-EBSD/OIM was achieved by Adams and co-workers for microscale orientation analysis and GB characterization [5]. This has enabled us to perform a rapid and quantitative characterization of microscale texture and GB microstructure in polycrystalline samples with a wide range of grain size from mm to nm. So far, it has been clarified by experiments that, strictly speaking, real polycrystalline materials never possess ideally a random grain orientation distribution. In reality, they inevitably possess some localization of the grain orientation distribution originating from the initial state of crystalline solid and applied processing method and condition. Accordingly it is necessary to precisely characterize the GB microstructure together with texture analysis based on the grain orientation distribution. Up to the end of the 1980s, it was quite difficult and time consuming to characterize individual grain boundaries by determining the orientation relationship between neighboring grains by the X-ray method and even by TEM for bulky polycrystalline samples, in connection with the local grain orientation distribution, that is, microscale texture, except for very coarse grained one. However, the advent of the SEM-EBSD/OIM technique in the early 1990s drastically changed the situation. It enabled us to obtain quantitative data on the distributions of GB character, GB connectivity, grain size, and shape of individual grains, as well as statistical data on the grain orientation distribution (texture) in a polycrystal without much difficulty. Nevertheless, in my opinion, the importance of GB microstructure has not been fully recognized yet even though the useful data can be obtained from SEM-DBSD/OIM analysis, but unfortunately not effectively used for understanding grain-boundary-related bulk properties. Accordingly the present author must emphasize that the recognition of the importance of GB microstructure is the first requirement for understanding the physical meaning of microscale texture associated with grain boundaries. This is of particular importance for nanocrystalline materials with grain size down to a few tens of nm, in order to fully understand their unique boundary-related bulk properties, as demonstrated by our recent studies of GBE for several nanocrystalline structural and functional materials for controlling high-cycle fatigue in electrodeposited Ni–P alloy [76], segregation-induced embrittlement in Ni–S alloy [77] and for the high performance electronic connect property in sputtered and annealed gold thin film [78].

Since the early 1990s, the importance of microscale texture likely began to be recognized for understanding of the bulk properties of polycrystalline materials [79]. Until recently, it was not so clear how the evolution and characteristic features of texture are related to GB microstructure so that much has not been discussed about the origin of newly observed characteristic features of texture and GB microstructures, that is, between overall and local feature of microstructure, which is our primary concern here. However, if we carefully check the literature, we can find a certain number of interesting theoretical and experimental studies previously performed by Howell and co-workers in the early 1970s by using TEM to reveal the relation between texture and grain boundaries in a tungsten swaged rod [80]. A high frequency of grain boundaries with ordered structure was found in swaged tungsten rods and then a systematic study of the relation between texture and GB structure was further performed by TEM, focusing on the relationship between texture type and the misorientation function in tungsten [81]. It was also found that the occurrence of a high fraction of the {110} plane-matching-type special boundaries plays an important role in the evolution of texture in tungsten with bcc crystal structure. The idea of a plane-matching-type boundary was originally proposed by Pumphrey in the 1970s [82], but unfortunately the important role of this type of GB has not been fully recognized. In fact, the plane-matching-type GB is considered as a one-dimensional coincidence boundary defined by a coincidence axial direction (CAD), which is more loosely coincidence-sited than the ordinary two-dimensional coincidence site lattice (CSL) condition, as discussed by Warrington and Boon [74].

More recently, a role of the migration of the $\{110\}_\alpha/\{111\}_\gamma$ plane-matching-type interphase boundary has been discussed by the present author and co-workers for α/γ phase transformation in iron-based alloy [84,85]. It is interesting to investigate the evolution of texture and GB microstructure during phase transformation in connection with the migration of such a special boundary as the plane-matching-type interphase boundary, together with the plane-matching grain boundaries, from the viewpoint of GB and interface engineering.

Quite recently, Kobayashi et al. [69] have performed a basic study of the evolution of texture and GB microstructure in molybdenum polycrystals produced thermomechanically by using single crystals with different initial surface orientations of ⟨100⟩, ⟨110⟩, ⟨111⟩, and ⟨112⟩, with the objective of controlling GB microstructures and intergranular fracture. The characterization of GB microstructure was carried out with the FE-SEM-EBSD/OIM technique. Here, we just briefly introduce some interesting findings, so for the detail, please refer to the original paper [69]. The single crystals were uniaxially compressed at room temperature up to 80%, and then annealed in vacuum at 1873 K ($0.65 T_m$, where T_m is the melting point) for 7.2 ks. OIM observations revealed that the microstructures produced by recrystallization from deformed molybdenum single crystals of different initial orientations are significantly different. In the ⟨100⟩, ⟨110⟩, and ⟨112⟩ specimens, a nonrecrystallized area in a partially recovered state was observed, but recrystallization had begun at the peripheral area of the specimens. The nonrecrystallized areas always had the ⟨100⟩ surface orientation. We look at some interesting findings from this work.

Figure 17.9a gives a summarized data on the inverse pole figures, the sharpness of grain orientation distributions, average grain sizes, and GBCDs, for grain clusters of

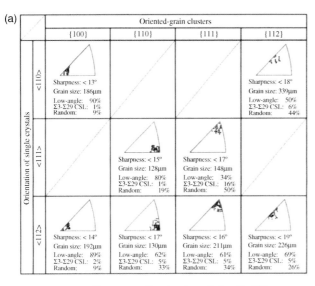

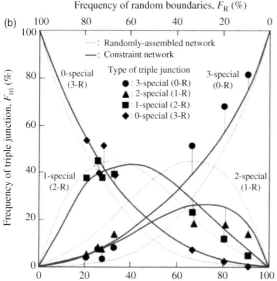

Figure 17.9 (a) Inverse pole figure, the average grain size and GBCD for distinctly oriented grain clusters in molybdenum polycrystals produced by thermomechanical processing from single crystals with different initial orientations [69]. In the pore figures, one data point indicates the surface orientation of a single grain. (b) Relationship between the frequency of different types of triple junctions and the frequency of low-Σ coincidence (CSL) boundaries ($1<\Sigma<29$) for molybdenum polycrystals produced by thermomechanical processing from single crystals with different initial orientations [69]. The curves drawn by thick or dotted lines indicate the prediction for the constraint grain boundary network or randomly assembled network, respectively.

different orientations which quantitatively characterize the GB microstructures in the recrystallized molybdenum samples. Generally speaking, the frequency of low-energy boundaries, such as low-angle and low-Σ coincidence boundaries, is closely related to the grain orientation distribution. The sharper the texture, the higher the frequency of low-energy boundaries. It is evident that the occurrence of different types of oriented-grain clusters strongly depends on the initial orientation of single crystals among $\langle 110 \rangle$, $\langle 111 \rangle$, and $\langle 112 \rangle$. The $\langle 112 \rangle$ specimen included {100}, {110}, {111}, and {112} grain clusters. Surface orientation in the {100} grain cluster was most sharply concentrated and the frequency of low-angle boundaries had the highest value of 89%. There was little difference between the frequencies of special (low-angle, low-Σ boundaries) for the {110}, {111}, and {112} oriented-grain clusters.

Next, it is interesting to examine the relationship between the GBCD and the triple junction distribution which plays a key role in percolation-controlled phenomena in the GB network, such as intergranular crack propagation in polycrystalline materials, in order to control intergranular fracture and to enhance the fracture toughness [86,87]. Figure 17.9b shows the relationship between GBCD and triple junction distribution plotted the data obtained for recrystallized molybdenum polycrystalline samples produced by thermomechanical processing from single crystals [69]. The thin and thick solid lines are the curves for a randomly assembled grain boundary network calculated by Fortier et al. [88,89] and for a crystallographically constraint network calculated by Fray and Schuh [90], respectively. The frequency of 3-special type triple junctions increased monotonically with increasing frequency of special (low-angle and CSL) boundaries, while the frequency of 0-special (3-R) type triple junctions that consisted of only random (R) boundaries decreased. It is evident that the experimental results are in much better agreement with the prediction for a crystallographically constraint (namely textured) network than that for a randomly assembled network.

Now let us examine the relationship between texture and GB microstructure, or the GBCD more precisely and quantitatively on the basis of experimental data obtained from early studies of GB microstructures in Fe–6.5mass%Si ribbons [58,70]. The present author would like to emphasize that there exists a close relationship between the type and the sharpness of texture and GB microstructure in textured polycrystalline materials, even if produced by different types of material processing methods. Although the studied Fe–6.5mass%Si ribbon samples were produced by such a special processing, as rapid solidification from the melt and subsequent annealing, not by popular thermomechanical processing, the evolution of a very sharp texture of {100} or {110} within about 5° deviation has provided us some important knowledge of GB microstructure in a typical case of a sharply textured polycrystal with a specific type of texture. The details should refer to the original papers [58,70].

Figure 17.10 shows the frequency of occurrence of coincidence boundaries as a function of the Σ value for sharply {100} and {110} textured samples, indicated by hatched bars. The theoretical prediction for a random polycrystal is given by open bars.

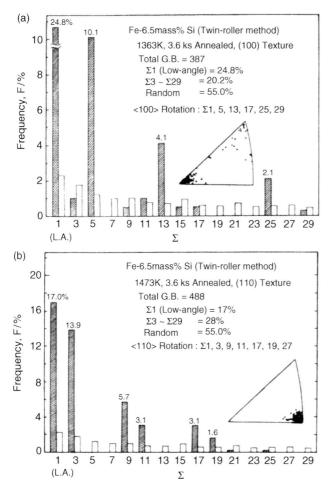

Figure 17.10 (a), (b) The frequency of coincidence boundaries as a function of Σ for sharply textured Fe–6.5mass%Si alloy ribbons produced by rapid solidification and subsequent annealing [58,70]. (a) {100} texture, (2) {110} texture, within 5° deviation of surface orientation of individual grains from $\langle 100 \rangle$ and $\langle 110 \rangle$, respectively. (c) The frequency of coincidence boundaries as a function of Σ for different polycrystalline materials produced differently [58]: the data were taken from the literature for nickel [71] and 304 steel [91] with fcc structure produced thermomechanically.

In the case of the {100} textured sample (Figure 17.10a) it is evident that those coincidence boundaries of low-Σ values such as 1, 5, 13, 25 occurred in much higher frequency in descending order of Σ value (almost 10 times higher than the theoretical prediction for a random polycrystal). On the other hand, in the case of the {110} textured sample, low-Σ coincidence boundaries occurred in higher frequency, in descending order of 1, 3, 9, 11,17, 19, as shown in Figure 17.10b. First, it is surprising to see that low-Σ coincidence boundaries occur very systematically as a function of the

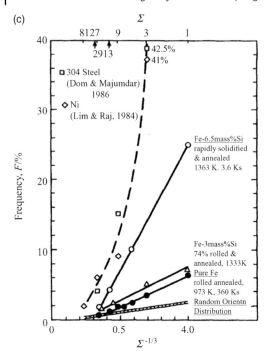

Figure 17.10 (Continued)

Σ value in descending order. Those Σ values which actually occurred in the studied two types of samples were found to be exactly possible coincidence orientations predicted for {100} and {110} rotation. Second, it was found that the observed frequencies of low-Σ boundaries obey the inverse cubic square root dependence on Σ, as indicated in Figure 17.10c for the {100} textured Fe–6.5mass%Si ribbon, together with the data from bulky samples of pure Fe and Fe–3%Si produced by ordinary thermomechanical processing based on rolling and annealing. The level of the frequency of coincidence boundaries for a random polycrystal is indicated by the curve at the bottom. We see that the level of the frequency of occurrence of low-Σ coincidence boundaries strongly depends on the processing method and condition used, as specified by the slope of the curve which crosses the abscissa around Σ29. The presence of the pivot point around Σ29 for the curves for Fe and Fe-based Si alloys with bcc structure may suggest that those coincidence boundaries which have Σ values larger than 29 are not different from high-angle random boundaries from the viewpoint of structure-dependent boundary properties controlled by the GBE. Moreover, as seen from the top curve in Figure 17.10c, it is worth noting that low-Σ coincidence boundaries occur with a much higher frequency for fcc metal and alloy such as nickel [71] and 304 stainless steel [91]. In fact, multiple-tinning plays an important role in selective formation of Σ^{3n} twin-related coincidence boundaries during annealing in fcc metals and alloys with low-stacking fault energy.

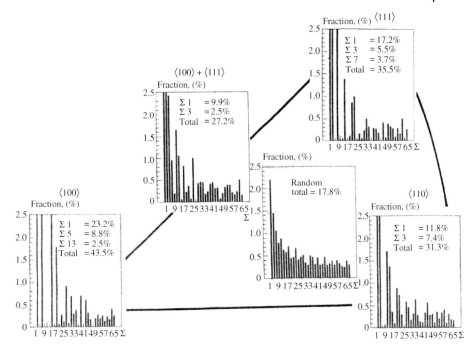

Figure 17.11 The frequency of low-Σ coincidence (CSL) boundaries (Σ < 65) obtained for different ⟨hkl⟩ textures for the sharpness within 5° deviation [92]. It should be remembered that not much difference will actually occur between the cases where the maximum value of Σ is taken either 29 or 65, for a random polycrystal: 13.6% for Σ < 29 [94], 17.8% for Σ < 65 [92].

Garbacz and Grabski [92] have theoretically predicted the frequency of low-Σ coincidence boundaries for differently textured polycrystalline samples with different types of texture and different degrees of grain orientation distributions and compared with experimental observations in Fe–6.5%Si alloy [58]. Their prediction for the {100} textured sample with 5° deviation is in a good agreement with experimental data on the Fe–6.5 mass%Si ribbons already shown in Figure 17.10a. A brief summary of their simulations is shown in Figure 17.11, which gives the prediction of the frequency of occurrence of low-Σ coincidence boundaries with different types of textures within 5° deviation angle, not only for the single type of texture such as {100}, {110}, {111}, but also for the mixed type composed of ⟨100⟩ and ⟨111⟩ texture. Unfortunately, until recently, there has not been much attempt to verify the close relation between characteristics of texture and GB microstructure which must dominantly control bulk properties (mechanical, physicochemical, electrical, and electromagnetic properties) in polycrystalline engineering materials, as discussed by the author and co-workers in their recent reviews on GBE [4,12,54]. In the author's opinion, the finding of the quantitative relation between texture and GB microstructure is the most important clue to understanding the origin of the unique bulk properties of sharply textured polycrystalline materials. Nevertheless, the current understanding of the relation between texture and GB microstructures is still premature and remains much to be studied, although several groups have made comparative studies of the GBCD between observed data and theoretical prediction for textured materials until recently [88–90,93–96]. We need to clarify by experiment how GB microstructure (at least defined by the GBCD and GB connectivity) affects the bulk properties of real polycrystalline materials. In recent years, we have just began to control possible effects of GB microstructure in the light of the GBCD and boundary connectivity on fracture, corrosion, and magnetization in differently textured structural materials, as well as functional materials like magnetic materials in which the interaction between different types of grain boundaries and domain wall motion dominates the magnetization and magnetic property, as reported by Yamaura *et al.* [97]. The author would like to emphasize that we need to study more systematically the relation between texture and GB microstructure, in order to endow polycrystalline materials' high performance and desirable bulk properties by GBE through texturing.

17.6
Grain Boundary Engineering for Fracture Control through Texturing

17.6.1
The Control of Microscale Texture and GB Microstructure

The computer-assisted SEM-EBSD/OIM technique has been widely used for precise and rapid orientation determination and microtexture analyses. This technique is also considered as the most powerful tool for analysis of GB microstructure for polycrystalline metallic, semiconductor, and ceramic materials with a wide range of grain size order from a few mm to 10 nm. Generally speaking, grain sizes are not so homogeneous and locally change in a polycrystalline material so that the length of individual grain boundaries surrounding neighboring grains also locally changes. The OIM technique installed well-programmed software can quantitatively determine the grain orientation distribution, the character of individual grain boundaries, as well as the length of individual grain boundaries observed on the scanned specimen surface, when taking reasonable step size for scanning. We can obtain quantitative data on microscale texture and characteristic features of GB microstructure, such as the GBCD and GB connectivity at triple junction, for bulky or thin polycrystalline samples without much difficulty. For polycrystalline samples with well-defined GB microstructure, bulk properties such as mechanical properties based on measurements of the flow and fracture stresses can be related to the GB microstructure. Then the design and control of GB microstructure is attempted by an appropriate processing method in order to achieve the GBE for desirable bulk properties and higher performance. Here let us examine an example of our recent attempt at the GBE for controlling intergranular brittleness in

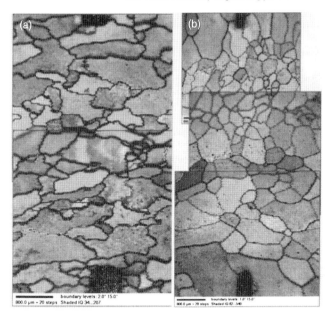

Figure 17.12 Grain boundary microstructures characterized by OIM for molybdenum polycrystals produced through different thermomechanical processing routes from sintered polycrystalline samples [53]. (a) Type I, swaged and annealed at 1873 K for 14.4 ks, (b) Type II, floating zone melted, forged and annealed at 1773 K for 14.4 ks.

polycrystalline molybdenum which was generally considered as being intrinsically brittle due to a high propensity to intergranular fracture [53,86]. The material used was sintered 99.999% molybdenum. Two types of specimens were prepared by thermomechanical processing: Type I was swaged into a rod of 10 mm diameter at 1773 K or 1873 K, and then thin sheets of about 500 μm thickness were cut from the swaged rod and annealed at 1773 and 1873 K for 14.4 and 28.8 ks. Type II was that sintered Mo was floating zone melted and solidified into a single crystal, and then forged to 5 mm thickness at 1593 K. Finally thin sheets of 500 μm thickness were cut out and then annealed under the same condition for Type I. The details of experimental procedures and results should be referred to the original papers [53].

Figure 17.12a and b shows OIM micrographs of GB microstructures observed for two types of sintered molybdenum specimens (Type I and Type II) that had received different routes of thermomechanical processing (swaged and annealed, or forged and annealed), in order to introduce different GB microstructures. The results of characterization of GB microstructure are shown in Figure 17.13 for both types of specimens, indicating the frequency of coincidence boundaries as a function of Σ, together with the results of texture analysis indicted by inserted pole figures. The OIM micrographs clearly demonstrate great differences in grain structure (rather uniform coarse-grained structure for Type I, but coarse-fine mixed heterogeneous

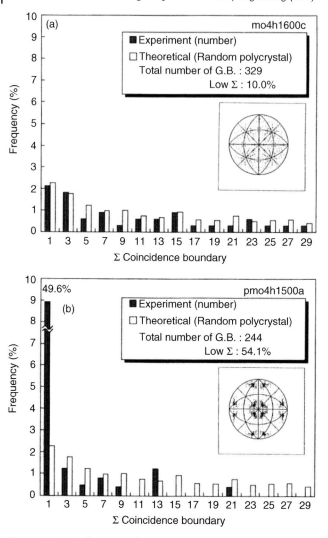

Figure 17.13 The frequency of coincidence boundaries as a function of Σ for sintered molybdenum polycrystals with grain boundary microstructures indicated in Figure 17.12a and b. Experimental data are compared with theoretical values for a random polycrystal [94].

grain structure for Type II) and in GB microstructure, between Type I and Type II specimens. The Type I specimen possesses almost random oriented grains and therefore the frequency of low-Σ coincidence boundaries (∼2%) is similar to that theoretically predicted for a random polycrystal indicated by white bars in Figure 17.13a, while the Type II specimen has sharply textured grains around {114}, resulting in the occurrence of a higher fraction (∼50%) of low-angle (Σ1) boundaries, as indicated in Figure 17.13b.

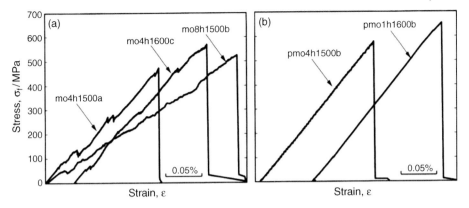

Figure 17.14 Stress–strain curves obtained by the four-point bending test at 77 K and strain rate 0.05 mm/min, for the sintered molybdenum polycrystals with well-defined grain boundary microstructures [53]: (a) Type I specimen with a random texture and (b) Type II specimen with a sharp texture. Note that jerky-shaped stress–strain curves were observed for Type I, while smooth stress–strain curves were observed for Type II.

17.6.2
GB Microstructure-Dependent Brittle Fracture in Molybdenum

For those two types of molybdenum specimens mentioned in the previous section, four-point bending tests were carried out at 77 K. Figure 17.14a and b depict stress–strain curves for the Type I specimen with a random texture and for the Type II specimen with a sharp texture, respectively. There was no definite sign of plastic deformation and brittle fracture for the two types of specimens as normally expected. Moreover, it should be noted that stress irregularities in the stress–strain curves are found for Type I, but such irregularities are scarcely recognized for Type II. The fracture stress for Type I is lower than that for Type II. The morphology of the fracture path observed by SEM clearly showed that fracture occurred in the mixed facture mode of transgranular and intergranular fracture in Type I, while transgranular (cleavage) fracture occurred dominantly exhibiting a rather straight crack path over many neighboring grains in the Type II specimen with a sharp texture. One of important findings from the previous experimental studies for polycrystalline molybdenum with well-defined GB microstructures is that the fracture stress is plotted following the Hall–Petch relation [98]. The slope of the Hall–Petch plot curve was found to increase with decreasing fraction of fracture-resistant low-Σ coincidence boundaries, as seen from Figure 17.15. A change of the Hall–Petch slope with the GBCD was reported by Wyrzykowski and Grabski for the flow stress in aluminum polycrystals which showed a rapid increase of the slope with increasing fraction of special boundaries beyond 50% [57]. Since the effect of an increase of the fraction of special boundaries appears opposite between the fracture stress and the flow stress, it is interesting to reveal the origin of such changes of the Hall–Petch slope observed differently for fracture stress in molybdenum and flow stress in

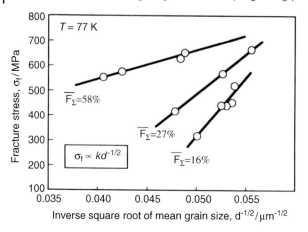

Figure 17.15 The Hall–Petch type plot of the fracture stress vs. the mean grain size for three different groups of polycrystalline molybdenum specimens with different levels of the fraction of low-Σ coincidence boundaries [98]. Note that the slope increases with decreasing fraction of low-Σ coincidence boundaries.

aluminum polycrystals. We need to explore structure-dependent GB deformation and fracture in the light of the structure-dependent interaction of grain boundaries with cracks and dislocations, as potential sites for lattice defects.

17.6.3
Percolation-Controlled and GB Microstructure-Dependent Fracture

A modeling of GB microstructure-dependent fracture processes in polycrystals has been previously performed by Lim and Watanabe [99] in order to predict the possibility of controlling intergranular brittle fracture and of improvement in the fracture toughness by GBE. The modeling was made for different cases of 3D equiaxed or elongated polycrystals made up of tetrakaidecahedron-shaped grains (for the detail, refer to the original paper). Here, we look at the result for the case of a 3D-random polycrystal composed of equiaxed grains. It is well known that the fracture paths in polycrystals are different depending on fracture mode, that is, the zig-zag path for intergranular fracture and the planar path for transgranular fracture mode. In the modeling, the toughness of a random GBCD was calculated as a function of the overall fraction of special low-energy fracture-resistant boundaries. One of the results obtained is shown in Figure 17.16, which indicates the toughness (G_R) of a random polycrystals with respect to the toughness of transgranular fracture (G_T) as a function of the overall fraction F of low-energy strong boundaries, for intrinsically ductile and brittle materials with different values of (G_R/G_T). It is very interesting to see that the toughness of an intrinsically brittle polycrystalline material (such as refractory metals like molybdenum and ceramics) increases systematically with increasing fraction of low-energy strong boundaries, while in the case of more ductile materials (such as copper and iron)

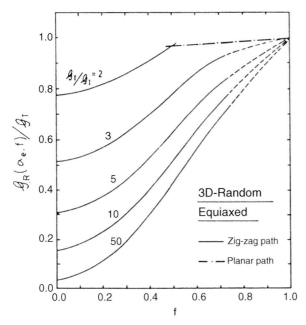

Figure 17.16 The effect of the overall fraction of low-energy boundaries, f, on the toughness of a three-dimensional polycrystal with a random grain boundary character distribution [99].

the toughness increases with the overall fraction of low-energy boundaries, but beyond a certain level of the fraction, the toughness will not increase anymore and will keep almost a constant value, due to the brittle–ductile transition of fracture mode from intergranular brittle fracture to trangranular ductile fracture. It is interesting to see that the change in fracture mode can occur around a specific value of the overall fraction of low-energy boundaries, say 0.5. In fact, this is a good demonstration of GB microstructure-dependent fracture in polycrystals where percolation-controlled crack propagation proceeds and the percolation threshold may exist beyond 0.5 of the fraction of special low-energy boundaries in the 3D-GB network, as confirmed by a recent experimental approach to GBE in metallic materials [68,69,100]. Thus, the possibility of controlling intergranular brittle fracture by GBE for polycrystalline materials has been evidenced by recent theoretical and experimental studies.

17.6.4
Fracture Control Based on Fractal Analysis of GB Microstructure

In the final section, several new findings on GB microstructure-dependent fracture processes in polycrystals are briefly discussed, particularly on the GBE based on fractal analysis for control of segregation-induced intergranular fracture in nickel, performed by Kobayashi et al. [101]. For quantitative discussion and understanding of the effect of microstructure on bulk properties of

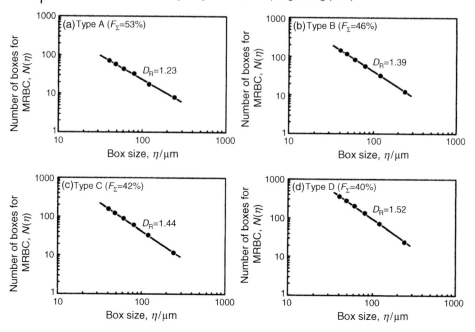

Figure 17.17 Fractal-based relationship between the number of boxes N (η) for complete coverage of the maximum random boundary connectivity (MRBC) and the box size η for the four different types of nickel polycrystalline specimens with different fractions of low-Σ coincidence boundaries, F_Σ [101]. Note that the fractal dimension D_R given by the slope of curve increases systematically with decreasing fraction of strong low-Σ coincidence boundaries, or conversely with increasing fraction of weak random boundaries.

polycrystalline materials, fractal analysis of microstructure is known to be useful and powerful, as demonstrated by Hornbogen [102]. We have recently applied the fractal approach to quantitative analysis of GB microstructure-dependent fracture in order to obtain a new insight and strategy for control of intergranular brittle fracture in polycrystals. We tried to find some correlation between the GBCD and GB character-dependent fracture path observed in sulfur-doped nickel polycrystals. It was found that cracks form preferentially at high-angle random boundaries, and the main crack propagates along connecting random boundaries preferably oriented to the stress axis in the zig-zag path along random boundaries, passing over triple junctions, or changed fracture mode to planner transgranular fracture in the grain interior. The fractal dimension of the GB network was analyzed by using the box-counting method on the basis of the trace of the random boundary network actually obtained from GB microstructure characterized by OIM in real specimens.

Figure 17.17 shows an important finding that the fractal dimension (D_R) for the maximum random boundary connectivity (MRBC), that is the slope of the curve, systematically increases with decreasing fraction of special low-energy strong boundaries (F_Σ), over the range of F_Σ from 53% to 40%, or conversely with increasing fraction of random weak boundaries F_R from 47% to 60%. It is evident that fracture processes and characteristics are definitely controlled by the GBCD and the observed value of the fractal dimension for the maximum random boundary connectivity. This clearly reflects the GB microstructure-dependent fracture processes in extrinsically brittle intergranular fracture in nickel induced by GB segregation of sulfur. Probably this work has first evidenced the applicability of the fractal approach to full understanding and prediction of the brittle–ductile fracture transition based on a key role of GB microstructure quantitatively analyzed by OIM in real polycrystalline materials, as exactly expected and predicted almost 30 years ago by Mandelbrot et al. based on measurements of the impact energy by a standard Charpy impact test and from their fractal analysis of fracture surfaces in maraging steel [103].

17.7
Conclusion

During the last three decades, the concept of GB design and control, or GBE has been successfully developed and almost established through a number of challenges to solve the long pending materials problem, especially "strengthening without loss of ductility" for structural materials. GBE was performed through texturing and found to be powerful for producing optimum grain boundary microstructures for desirable bulk mechanical properties such as higher strength and higher fracture toughness in polycrystalline materials.

Acknowledgments

This chapter was written as a dedication to Prof. Gunter Gottstein on a special occasion of his honorary symposium, with my sincere gratitude for wonderful friendship and collaboration with him for many years. I would like to sincerely thank my co-workers who have been deeply involved in the project on the grain boundary engineering, particularly Prof. S. Tsurekawa and Prof. S. Kobayashi, who have collaborated with me on the recent studies cited in this chapter. My sincere gratitude also goes to Prof. L. Zuo, Prof. X. Zhao of Northeastern University, China, and Prof. C. Esling of University of Metz for their collaboration and constant friendship.

References

1 Gottstein, G. (1978) Recent aspects in the understanding of recrystallization texture development, in *Proc. 5th Intern. Conf. on Textures of Materials (ICOTOM-5)*, Springer, Berlin, pp. 93–109.

2 Gottstein, G. and Shvindlerman, L.S. (2010) *Grain Boundary Migration in Metals – Thermodynamics, Kinetics, Applications*, 2nd edn, CRC Press, Boca Raton, FL, (1999, first edn).

3 Watanabe, T. (1984) An approach to grain boundary design for strong and ductile polycrystals. *Res Mechanica*, **11** (1), 47–84; (1983) Materials design by grain boundary design, *Bulletin of JIM* (in Japanese), 22 (2), 95–102.

4 Watanabe, T. (2011) Grain boundary engineering: Historical perspective and future prospects. *Journal of Materials Science*, **46** (12), 4095–4115.

5 Adams, B.L., Wright, S.I., and Kunze, K. (1993) Orientation imaging: The emergence of a new microscopy. *Metallurgical Transactions*, **A24** (4), 819–831.

6 Smith, C.S. (1975) Metallurgy as a human experience. *Metallurgical Transactions*, **6A** (4), 603–623.

7 Cohen, M. (1983) Materials, materialism and search for meaning – An essay. *Metallurgical Transactions*, **14A** (4), 513–518.

8 Cahn, R.W. (2001) *The Coming of Materials Science*, Pergamon, London.

9 Gensamer, M. (1954) The effect of grain boundaries on mechanical properties, in *Relation of Properties to Microstructure*, ASM, Materials Park, OH, pp. 16–29.

10 Low, J.R. Jr. (1954) The relation of microstructure to brittle fracture, in *Relation of Properties to Microstructure*, ASM, Materials Park, OH, pp. 163–179.

11 Gleiter, H. (1996) Microstructure, in *Physical Metallurgy* (eds R.W. Cahn and P. Haasen), Elsevier Science BV, Amsterdam, pp. 843–942, Chap. 9.

12 Watanabe, T., Tsurekawa, S., Zhao, X., and Zuo, L. (2009) The coming of grain boundary engineering in the 21st century, in *Microstructure and Texture in Steels*, Springer, Berlin, pp. 43–82, Chap. 4.

13 Honeycombe, R.W.K. (1981) *Steels-Microstructure and Properties*, Edward Arnold, London, UK.

14 Tien, J.K. and Ansell, G.S. (eds) (1976) *Alloy and Microstructural Design*, Academic Press, New York.

15 Hondros, E.D. and McLean, D. (1976) Grain boundary fragility, in *Grain Boundary Structure and Properties* (eds G. A. Chadwick and D.A. Smith), Academic Press, New York, pp. 353–381.

16 Baker, T.N. (ed.) (1983) *Yield, Flow and Fracture of Polycrystals*, Applied Science Publisher., London, New York.

17 Hall, E.O. (1951) *Proceedings of the Physical Society of London*, **B64**, 747.

18 Petch, N.J. (1956) The ductile fracture of polycrystalline α-iron. *Philosophical Magazine*, **1**, 186–190.

19 Otooni, M.A., Armstrong, R.W., Grant, N.J., and Ishizaki, K. (eds) (1995) *Grain Size and Mechanical Properties – Fundamentals and Applications, Proc. of MRS Symp*, vol. 362, Materials Research Society, Pittsburgh, PA.

20 Kung, H. and Foecke, T. (eds) (1999) *Mechanical behavior of nanostructured materials, MRS Bulletin*, **24**, (2).

21 Birringer, R., Herr, U., and Gleiter, H. (1986) Nanocrystalline materials – A first report, Proc. 4th JIM Intern Symp. on Grain Boundary Structure and Related Phenomena, Suppl. to Trans, JIM, 27, pp. 43–52.

22 Valiev, R.Z. and Langdon, T.G. (2006) Principles of equal-channel angular pressing as a processing tool for grain refinement. *Progress in Materials Science*, **51**, 881–981.

23 Gleiter, H. (2000) Nanostructured materials: Basic concepts and microstructure. *Acta Materialia*, **48**, 1–29.

24 Gleiter, H. (2009) Are there ways to synthesize materials beyond the limits of today? *Metallurgical and Materials Transactions*, **40A** (7), 1499–1509.

25 Gilman, J.J. (1966) Monocrystals in mechanical technology. *Transaction of ASM*, **59**, Cambell Memorial Lecture, 597–629.

26 Armstrong, R.W. (1974) The strengthening or weakening of polycrystals due to the presence of grain boundaries. *Canadian Metallurgical Quarterly*, **13** (1), 187–202.

27 Watanabe, T. (1983) Grain boundary sliding and stress concentration during creep. *Metallurgical Transactions*, **14A** (4), 531–545.

28 Smith, C.S. (1948) Grins, phases, and interfaces: An interpretation of microstructure. *Transaction of AIME*, **175**, 15–51.

29 Aust, K.T. and Chalmers, B. (1952) Energies and structure of grain boundaries, in *Metal Interfaces*, ASM, Materials Park, OH, pp. 153–178.

30 Gleiter, H. and Chalmers, B. (1972) High-angle boundaries. *Progress in Materials Science*, **16**, Pergamon, London, pp. 1–274.

31 Chadwick, G.A. and Smith, D.A. (eds) (1976) *Grain Boundary Structure and Properties*, Academic Press, London.

32 Aust, K.T. (ed.) (1974) Grain boundaries. *Canadian Metallurgical Quarterly*, **13** (1), pp. 1–316.

33 Biscondi, M. and Goux, C. (eds) (1975) Grain boundaries in metals. *Journal de Physique*, **36**, Colloque C-4, No. 10.

34 Watanabe, T. (1986) Grain boundary design for new materials, in *Proceeding of 4th JIM Intern. Symposium on Grain Boundary Structure and Related Phenomena*, vol. 27 (ed. Y. Ishida), Trans. JIM, pp. 73–87.

35 Watanabe, T. (1993) Toward grain boundary design and control for advanced materials, in *Grain Boundary Engineering*, Proc. K. T. Aust Symp. (eds U. Erb and G. Palumbo), CIMMP, pp. 57–87.

36 Chalmers, B. (1952) Structure of crystal boundaries. *Progress in Metal Physics*, **3**, 293–319.

37 McLean, D. (1957) *Grain Boundaries in Metals*, Clarendon Press, Oxford.

38 Erb, U. and Palumbo, G. (eds) (1993) Grain boundary engineering, in *Proc. K.T. Aust. Symp. held in June 1993*, Canadian Institute of Mining, Metallurgy and Petroleum.

39 Watanabe, T. Tsurekawa, S. Petit, J. Dimitrov, D. and Nagata, N. (eds) (2002) *Interfaces and Related Phenome – The Control of Interfaces, Surface and Environmental Effects on Materials Function and Performance*, Elsevier, Amsterdam.

40 Kumar, M. and Schuh, C.A. (eds) (2006) Viewpoint Set. No. 40: Grain boundary engineering. *Scripta Metallurgica*, **54** (6), 961–1220.

41 Ranganathan, S. and Kaplan, W.D. (eds) (2006) Special issue: Advanced materials and characterization: Proceedings of the Brandon Symposium. *Journal of Materials Science*, **41** (23), 7667–7871.

42 Ikuhara, Y. and Koyama, M. (eds) (2011) Special issue: Intergranular and interphase boundaries in materials. *Journal of Materials Science*, **46** (12), 4093–4434.

43 Sorby, H.C.I. (1887) *Journal of the Iron and Steel Institute*, 255. Cited by Hondros, E.D. (1996) The nature of the grain boundary – an enquiry, in *Structural Materials – Engineering Application through Scientific Insight* (eds E.D. Hondros and M. Mclean), The Institute of Materials, pp. 1–12.

44 King, R. and Chalmers, B. (1949) Crystal boundaries. *Progress in Metal Physics*, **1**, 127–162.

45 Pande, C.S. and Chou, Y.T. (1975) Structure and mechanical behavior of bicrystals, in *Treatise on Materials Science and Technology*, vol. 8 (ed. H Herman), Academic Press, New York, pp. 43–120.

46 Shibata, N., Oba, F., Yamamoto, T., and Ikuhara, Y. (2004) Structure, energy and solute segregation behavior of [110] symmetric tilt grain boundaries in yttria-stabilized cubic zirconia. *Philosophical Magazine*, **84**, 2381–2415.

47 Watanabe, T., Kimura, S., and Karashima, S. (1984) The effect of a grain boundary structural transformation on sliding in $\langle 10\bar{1}0 \rangle$ - tilt zinc bicrystals. *Philosophical Magazine*, **49**, 845–864.

48 Watanabe, T. (1993) Grain boundary design and control for high temperature materials. *Materials Science and Engineering*, **A166**, 11–28.

49 Watanabe, T., Tsurekawa, S., Kobayashi, S., and Yamaura, S. (2005) Structure-dependent grain boundary deformation and fracture at high temperatures. *Materials Science and Engineering A*, **410–411**, 140–147.

50 Watanabe, T. (1994) The impact of grain boundary character distribution on fracture in polycrystals. *Materials Science and Engineering*, **A176**, 39–49.

51 Watanabe, T., Tanaka, M., and Karashima, S. (1984) Intergranular fracture caused by liquid gallium in polycrystalline beta brass with bcc structure, in *Embrittlement by Liquid and Solid Metals* (ed. M.H. Kamdar), TMS, pp. 183–196.

52 Watanabe, T. (1993) Grain boundary design for advanced materials on the basis of the relationship between texture and grain boundary character distribution

(GBCD). *Texture and Microstructures*, **20**, 195–216.

53 Watanabe, T. and Tsurekawa, S. (1999) The control of brittleness and development of desirable mechanical properties in polycrystalline systems by grain boundary engineering. *Acta Materialia*, **47** (15/16), 4171–4185.

54 Watanabe, T., Tsurekawa, S., Zhao, X., Zuo, L., and Esling, C. (2006) A new challenge: grain boundary engineering for advanced materials by magnetic field application. *Journal of Materials Science*, **41**, 7747–7759.

55 Watanabe, T. (1985) Structural effects on grain boundary segregation, hardening and fracture. *Journal de Physique*. Colloque C4, **46**, 555–566.

56 Grabski, M. (1985) Mechanical properties of internal interfaces. *Journal de Physique*. Colloque C4, **46**, 567–579.

57 Wyrzykowski, J.W. and Grabski, M.W. (1986) The Hall–Petch relation in aluminium and its dependence on the grain boundary structure. *Philosophical Magazine*, **A53**, 505–520.

58 Watanabe, T., Fujii, H., Oikawa, H., and Arai, K.-I. (1989) Grain boundaries in rapidly solidified and annealed Fe-6.5 mass%Si polycrystalline ribbons with high ductility. *Acta Metallurgica*, **37** (3), 941–952.

59 Grabski, M.W. and Korski, R. (1970) Grain boundaries as sinks for dislocations. *Philosophical Magazine*, **22**, 707–715.

60 Przetakiewicz, W., Kurzydlowski, K.J., and Grabski, M.W. (1986) Grain boundary energy changes during grain growth in nickel and 316L austenitic steel. *Materials Science and Engineering*, **2**, 106–109.

61 Watanabe, T. (1987) Prediction of change in grain boundary energy during grain growth. *Scripta Metallurgica*, **21** (6), 427–432.

62 Watanabe, T. (1998) Grain boundary architecture for high performance materials, The David A. Smith Symposium on Boundaries and Interfaces in Materials, TMS, pp. 19–28.

63 Kokawa, H., Watanabe, T., and Karashima, S. (1981) Sliding behaviour and dislocation structures in aluminium grain boundaries. *Philosophical Magazine*, **44** (6), 1239–1254.

64 Chaudhari, P. and Mathews, J.W. (1971) Coincidence twist boundaries between crystalline smoke particles. *Journal of Applied Physics*, **42** (8), 3063–3066.

65 Sautter, H., Gleiter, H., and Baro, G. (1977) The effect of solute atoms on the energy and structure of grain boundaries. *Acta Materialia*, **25**, 467–473.

66 Erb, U. and Gleiter, H. (1979) The effect of temperature on the energy and structure of grain boundaries. *Scripta Materialia*, **3** (1), 61–64.

67 Watanabe, T., Yoshikawa, N., and Karashima, S. (1981) Grain boundary character distribution in recrystallization structure of deformed aluminium single crystals, Proc. 6th. Intern. Conf. on Textures of Materials (ICOTOM-6), ISIJ, pp. 609–618.

68 Tsurekawa, S., Nakamichi, S., and Watanabe, T. (2006) Correlation of grain boundary connectivity with grain boundary character distribution in austenitic stainless steel. *Acta Materialia*, **54** (13), 3617–3626.

69 Kobayashi, S., Tsurekawa, S., Watanabe, T., and Kobylanski, A. (2008) Control of grain boundary microstructure in molybdenum polycrystals by thermomechanical processing of single crystals. *Philosophical Magazine*, **88** (4), 489–506.

70 Watanabe, T., Arai, K.-I., Yoshimi, K., and Oikawa, H. (1989) Texture and grain boundary character distribution (GBCD) in rapidly solidified and annealed Fe-6.5mass%Si ribbons. *Philosophical Magazine Letters*, **59** (2), 47–52.

71 Lim, L.C. and Raj, R. (1984) On the distribution of Σ for grain boundaries in polycrystalline nickel prepared by strain annealing technique. *Acta Materialia*, **32** (8), 1177–1181.

72 Aust, K.T. (1981) Comments concerning structure and some properties of grain boundaries, in *Chalmers Anniversary Volume, Progress in Materials Science* (eds J.W. Christian, P. Haasen, and T.B. Massalski), Pergamon Press, London, pp. 27–48.

73 Mackenzie, J.K. (1958) Second paper on statistics associated with the random disorientation of cubes. *Biometrica*, **45**, 229–240.

74 Warrington, D.H. and Boon, M. (1975) Ordered structures in random grain boundaries; some geometrical probabilities. *Acta Materialia*, **23** (5), 599–607.

75 Grimmer, H. (1979) The distribution of disorientation angles if all relative orientations of neighbouring grains are equally probable. *Scripta Materialia*, **13** (2), 161–164.

76 Kobayashi, S., Kamata, A., and Watanabe, T. (2009) Roles of grain boundary microstructure in high-cycle fatigue of electrodeposited nanocrystalline Ni–P alloy. *Scripta Materialia*, **61** (11), 1032–1035.

77 Kobayashi, S., Tsurekawa, S., Watanabe, T., and Palumbo, G. (2010) Grain boundary engineering for control of sulfur segregation-induced embrittlement in ultrafine-grained nickel. *Scripta Materialia*, **62**, 294–297.

78 Kobayashi, S., Fukasawa, R., and Watanabe, T. (2012) Control of grain boundary microstructures in sputtered gold thin films by surface energy-driven grain growth. *Materials Science Forum*, **706–709**, 2880–2885.

79 Bunge, H.J. (ed.) (1993) *Textures and Microstructures*, **20** (1–4), as Proc. of the Symp. on Microscale Textures of Materials (October, 1991).

80 Howell, P.R., Fleet, D.E., Ralph, B., and Welch, P.I. (1976) The relationship between texture and grain boundary structure in commercial tungsten, in *Proc. 4th Intern. Conf. on Texture, Texture and Properties of Materials*, The Metals Society, pp. 121–131.

81 Fleet, D.E., Welch, P.I., and Howell, P.R. (1978) Texture and grain boundary structure in tungsten: I, II and III. *Acta Materialia*, **26** (10), 1479–1489, 1491–1498, 1499–1503.

82 Pumphrey, P.H. (1972) A plane matching theory of high angle grain boundary structure. *Scripta Materialia*, **6** (2), 107–114.

83 Warrington, D.H. (1974) Special grain boundaries in random polycrystalline aggregates. *Journal of Microscopy*, **102** (3), 301–308.

84 Watanabe, T., Obara, K., and Tsurekawa, S. (2004) In-situ observations on interphase boundary migration and grain growth during α/γ phase transformation in iron alloy. *Materials Science Forum*, **467–470**, 819–824.

85 Watanabe, T., Obara, K., Tsurekawa, S., and Gottstein, G. (2005) A mechanism of plane matching boundary-assisted α/γ phase transformation in Fe–Cr alloy based on in-situ observations. *Zeitschrift für Metallkunde*, **96** (10), 1196–1203.

86 Kobayashi, S., Tsurekwa, S., and Watanabe, T. (2006) Structure-dependent triple junction hardening and intergranular fracture in molybdenum. *Philosophical Magazine*, **86** (33–35), 5419–5429.

87 Kobayashi, S., Tsurekawa, S., Watanabe, T., and Palumbo, G. (2010) Grain boundary engineering for control of sulfur segregation-induced embrittlement in ultrafine-grained nickel. *Scripta Materialia*, **62** (5), 294–297.

88 Fortier, P., Aust, K.T., and Miller, W.A. (1995) Effect of symmetry, texture and topology on triple junction character distribution in polycrystalline materials. *Acta Metallurgica et Materialia*, **43** (1), 339–349.

89 Fortier, P., Miller, W.A., and Aust, K.T. (1997) Triple junction and grain boundary character distributions in metallic materials. *Acta Materialia*, **45** (8), 3459–3467.

90 Fray, M. and Schuh, S.A. (2005) Connectivity and percolation behavior of grain boundary networks in three dimensions. *Philosophical Magazine*, **85** (11), 1123–1143.

91 Don, J. and Majumdar, S. (1986) Creep cavitation and grain boundary structure in type 304 stainless steel. *Acta Metallurgica*, **34** (5), 961–967.

92 Garbacz, A. and Grabski, M.W. (1993) The relationship between texture and CSL boundaries distribution in polycrystalline materials: II. *Acta Metallurgica et Materialia*, **41** (2), 475–483.

93 Zhilyaev, A.P., Gertsman, V.Yu., Mishin, O.V., Pshenichnyuk, A.I., Aleksandrov, I.

V., and Variev, R.Z. (1993) Grain boundary misorientation spectra (GBMS) determined by real ODF in FCC-materials susceptible to annealing twinning. *Acta Metallurgica et Materialia*, **41** (9), 2657–2665.

94 Zuo, L., Watanabe, T., and Esling, C. (1994) A theoretical approach to grain boundary character distribution (GBCD) in textured polycrystalline materials. *Zeitschrift für Metallkunde*, **85** (8), 554–558.

95 Garbacz, A., Ralph, B., and Kurzydlowski, K.J. (1995) On the possible correlation between grain size distribution and distribution of CSL boundaries in polycrystals. *Acta Metallurgica et Materialia*, **43** (4), 1541–1547.

96 Romero, D., Martiunez, L., and Fionova, L. (1996) Computer simulation of grain boundary spatial distribution in a three-dimensional polycrystal with cubic structure. *Acta Materialia*, **44** (1), 391–402.

97 Yamaura, S., Furuya, Y., and Watanabe, T. (2001) The effect of grain boundary microstructure on Barkhausen noise in ferromagnetic materials. *Acta Materialia*, **49**, 3019–3027.

98 Watanabe, T. and Tsurekawa, S. (2004) Toughening of brittle materials by grain boundary engineering. *Materials Science and Engineering*, **A387–389**, 447–455.

99 Lim, L.C. and Watanabe, T. (1990) Fracture toughness and brittle-ductile transition controlled by grain boundary character distribution (GBCD) in polycrystals. *Acta Metallurgica et Materialia*, **38** (12), 2507–2516.

100 Michiuchi, M., Kokawa, H., Wang, Z.J., Sato, Y.S., and Sakai, K. (2006) Twin-induced grain boundary engineering for 316 austenitic stainless steel. *Acta Materialia*, **54**, 5179–5184.

101 Kobayashi, S., Maruyama, T., Tsurekawa, S., and Watanabe, T. (2012) Grain boundary engineering based on fractal analysis for control of segregation-induced intergranular fracture in polycrystalline nickel. *Acta Materialia*, **60** (17), 6200–6212.

102 Hornbogen, E. (1989) Fractals in microstructure in metals. *International Materials Reviews*, **34**, 277–296.

103 Mandelbrot, B.B., Passoya, D.E., and Paullay, A.J. (1984) Fractal character of fracture surface of metals. *Nature*, **308**, 721–722.

18
Microstructure and Texture Design of NiAl via Thermomechanical Processing
Werner Skrotzki

18.1
Introduction

NiAl, an intermetallic compound with B2 structure, because of its high melting point, low density, high thermal conductivity, good corrosion, and moderate creep resistance, is considered as a potential material for high temperature structural applications [1–3]. Depending on the application, the strong elastic and plastic anisotropy of NiAl (Figure 18.1) may play a major role [3]. In polycrystalline aggregates, microstructure and texture determine the strength and anisotropy. Therefore, it is the aim of this chapter to give an overview on the design of microstructure and texture in NiAl via thermomechanical processing. Processing consists of different deformation modes including dynamic recrystallization and annealing leading to static recrystallization. The processes presented here, extrusion, compression, rolling, and annealing at different temperatures, may be of technological interest. Other processes, like torsion and equal channel angular pressing, applied to NiAl are described in [4–6]. The results are discussed on the basis of polycrystal deformation modeling considering the special slip system activity in NiAl.

18.2
Experimental

Cast stoichiometric NiAl samples (composition within ±1at.% confirmed by electron beam microprobe analysis) were extruded at different reduction ratios (4:1–7:1) and temperatures (700–1300°C) through a round and rectangular die (Figure 18.2 a) at a strain rate of the order of $1\,s^{-1}$. To diminish friction the samples were sealed using copper and steel cans for extrusions below and above 1000°C, respectively. In general, the extruded rods were air cooled. To study recrystallization and grain growth, slices of the rods about 5 mm thick were annealed in air and air cooled.

The microstructure was investigated on mechanically and electrolytically well-polished sections in the scanning electron microscope (SEM, Zeiss DSM 962) using

Microstructural Design of Advanced Engineering Materials, First Edition. Edited by Dmitri A. Molodov.
© 2013 Wiley-VCH Verlag GmbH & Co. KGaA. Published 2013 by Wiley-VCH Verlag GmbH & Co. KGaA.

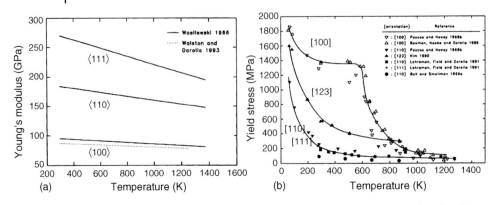

Figure 18.1 Temperature dependence of Young's modulus and the yield stress of NiAl single crystals [3].

back-scatter electron (BSE) contrast. In addition, electron back-scatter diffraction (EBSD) measurements were carried out in order to correlate texture components with the microstructure shown by orientation contrast and to determine the misorientation distribution. The EBSD patterns were analyzed with the software package Channel+ developed by HKL-Software, Hobro, Denmark.

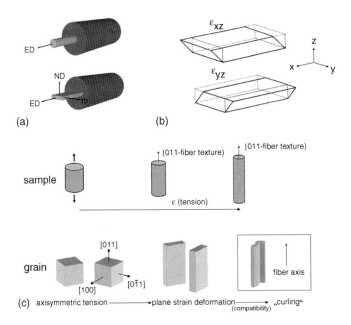

Figure 18.2 (a) Extrusion through a round and rectangular die, with ED = extrusion direction, TD = transverse direction, ND = normal direction. (b) Relaxations used in the "pancake" model of plane strain deformation, with x = ED and rolling direction, y = TD, z = ND and compression direction. (c) Principle of "curling" in axisymmetric tension along the $\langle 110 \rangle$ fiber axis.

Complete pole figures for {200}, {220}, and {111} reflections of the extruded samples were measured by neutron diffraction at GKSS Research Centre Geesthacht, Germany, using cylindrical samples with a height and diameter of 10 mm. Orientation distribution functions (ODFs) were calculated using the series expansion method up to a series expansion degree of 22 and axial or orthorhombic sample symmetry [7]. Generally, the maximum texture intensities are given in multiples of a random orientation distribution (m.r.d.). The volume fraction of texture components with a fixed 15° full width at half maximum were calculated with the component model developed by Helming [8].

Texture simulations were done with the full and relaxed constraints (FC and RC) Taylor theory [9,10] applied on a random distribution of 1000 grains. For extrusion through the rectangular die, rolling and compression the shear components ε_{xz} and ε_{yz} are relaxed ("pancake model"), with x, y, and z being the extrusion (or rolling), transverse, and normal (or compression) direction, respectively (Figure 18.2b). In the case of extrusion through the round die, relaxation of $\varepsilon_{yy}-\varepsilon_{zz}$ and ε_{yz} allows for "curling" around the extrusion axis x (Figure 18.2c) [11]. Curling has to be started after a certain strain in the FC simulation mode when a ⟨110⟩ fiber has developed. In all simulations, the strain was achieved in steps of 5%. The slip systems used are {110}⟨100⟩, {110}⟨110⟩, and {110}⟨111⟩, their activity changing with temperature (Figure 18.3). These slip systems yield three, two, and five independent systems, respectively. Thus, in order to fulfill the von Mises criterion [12], requiring five

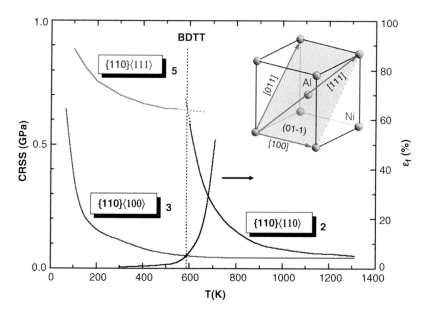

Figure 18.3 Temperature dependence of the critical resolved shear stress (CRSS) for slip on different slip systems in NiAl single crystals combined with the temperature dependence of the strain at fracture ε_f of NiAl polycrsytals deformed at a strain rate of about $10^{-4}\,\mathrm{s}^{-1}$. (BDTT = brittle-to-ductile transition temperature).

independent slip systems for a homogeneous deformation of polycrystals with a random orientation distribution, beside primary slip on $\{110\}\langle 100\rangle$, secondary slip on $\{110\}\langle 111\rangle$ at low temperatures, or slip on $\{110\}\langle 110\rangle$ at high temperatures has to be activated, too. This becomes quite easy above a certain temperature marking the brittle-to-ductile transition temperature (BDTT). Moreover, above BDTT, diffusional processes contribute to the slip of dislocations on $\{110\}\langle 110\rangle$ [13] and facilitate relaxation of constraints.

Uniaxial compression tests under 0.4 GPa gas confining pressure were conducted on differently textured cylindrical samples with a height of 14 mm and a diameter of 7 mm at room temperature and at an initial strain rate of about 10^{-4} s^{-1} in a Paterson rock deformation apparatus [14] with the samples coated with a 0.6 mm copper jacket. The deviatoric stress–strain curves obtained under such conditions were corrected for the strength of the copper jacket assuming ideal plastic behavior above the copper yield stress of 330 MPa. Based on the correction, the accuracy of the NiAl flow stress is better than about 40 MPa. A possible error arising from dilatancy is neglected.

Taylor energies were calculated by $\sum_s \gamma^s \tau_c^s$, with γ^s and τ_c^s being the resolved shear strain and critical resolved shear stress (CRSS) on the active slip system s, respectively. The 18 slip systems used at room temperature (Figure 18.3) are $\{110\}\langle 100\rangle$ with $\tau_c = 1$ (corresponds to 100 MPa) and $\{110\}\langle 111\rangle$ with $\tau_c = 7$ (corresponds to 700 MPa) [15]. The mean Taylor energy of the aggregate was obtained by arithmetic averaging over 4000 single orientations deduced from the initial texture. The mean Taylor energy was calculated for different deformation geometries

$$\begin{pmatrix} q & 0 & 0 \\ 0 & 1-q & 0 \\ 0 & 0 & -1 \end{pmatrix}$$

ranging from axial deformation ($q = 0.5$) to plane strain ($q = 0.0$ and $q = 1.0$).

18.3
Microstructure and Texture Development

The cast microstructure is shown in Figure 18.4 a normal to the ingot axis. It is a coarse grain structure (mean grain area 0.4 mm^2) with grains elongated in radial direction (mean aspect ratio 4) mainly consisting of high angle grain boundaries (maximum misorientation about 45°) (Figure 18.4b). The preferred radial growth direction is $\langle 100\rangle$ resulting in a $\langle 100\rangle$ fiber texture (Figure 18.4c). Consequently, the orientations parallel to the ingot axis are located on the symmetry line between $\langle 100\rangle$ and $\langle 110\rangle$. Similar observations have been made for B2-structured Fe–Al intermetallic alloys [16]. Microstructure and texture are typical for cast cubic metals [17].

In the temperature range investigated up to a reduction ratio of about 5:1 extrusion of NiAl leads to a microstructure that is highly recovered. It consists

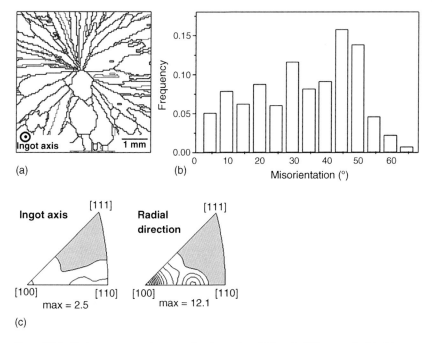

Figure 18.4 Grain structure and texture of cast cylindrical NiAl ingots: (a) grain structure normal to the ingot axis (black and gray lines represent grain boundaries with misorientation angles ≤10° and >15°, respectively), (b) misorientation distribution, (c) inverse pole figures of the ingot axis and radial direction (gray areas are below the level of 1.0 m.r.d.).

of a subgrain structure with a grain size of a few microns (Figure 18.5 a). The frequency of boundaries decreases with increasing misorientation, the majority is in the low angle range up to 15° (Figure 18.5c). For higher reduction ratios above 1100°C, the microstructure is almost completely dynamically recrystallized. The volume fraction of dynamically recrystallized grains exceeds 90%, with the deformed recovered areas mainly existing in the center of the extruded rods. The size of the discontinuously dynamically recrystallized grains is of the order of 100 μm (Figure 18.5b), the misorientation is in the large angle range with a broad maximum at about 30° (Figure 18.5d).

At all temperatures used, extrusion through the round and rectangular die at reduction ratios smaller than about 5:1 leads to a dominant ⟨110⟩ and minor ⟨100⟩ fiber texture and a dominant {111}⟨110⟩ and minor {110}⟨110⟩ texture component, respectively (Figures 18.6, 18.7, 18.10 and 18.11). Dynamic as well as static recrystallization (Figures 18.5b and 18.8) changes the textures to a ⟨111⟩ fiber (Figures 18.6, 18.7 and 18.9) in the axisymmetric case and to a dominant {110}⟨110⟩ component in the near plane strain case (Figures 18.10 and 18.11). The strength of texture is quite weak and slightly increases with temperature and reduction ratio. The grain size after static recrystallization is in the order of 100 μm.

Fischer-Bühner [18] succeeded in rolling (plane strain deforming) NiAl in the lower temperature range of 470–750°C (Figure 18.12). The textures measured by

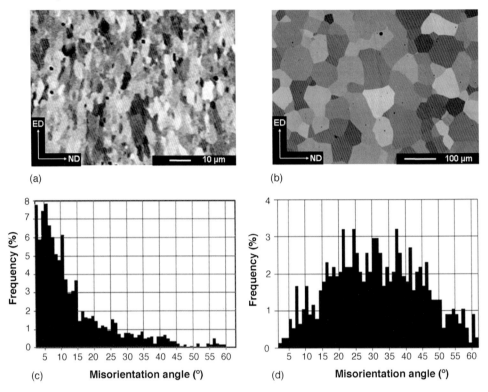

Figure 18.5 Microstructure (BSE images) and misorientation distribution of extruded NiAl: (a) Subgrain structure of a sample extruded through a rectangular die at 1000°C and an extrusion ratio of 5:1, (b) dynamically recrystallized grain structure of a sample extruded through a rectangular die at 1100°C and an extrusion ratio of 6:1, (c) misorientation distribution of a sample extruded through a rectangular die at 950°C and an extrusion ratio of 4:1, (c) misorientation distribution of a sample extruded through a round die at 1100°C and an extrusion ratio of 7:1.

X-ray diffraction follow nicely the results obtained for extrusion through a rectangular die at higher temperatures. However, at the lowest temperatures also the rotated cube component {100}⟨110⟩ develops. Similarly, the static recrystallization leads to the development of the {110}⟨110⟩ component mainly on the cost of {100}⟨110⟩ and slightly {111}⟨110⟩ orientations.

Compression of NiAl at temperatures between room temperature and 1200°C leads to a ⟨111⟩ fiber texture (Figure 18.13) [18]. The texture intensity increases with temperature with the rate of increase being higher above 500°C.

Deformation of NiAl at high temperatures leads to a pronounced texture, in agreement with other investigations [4–6,18–35]. Depending on the deformation mode the textures developed have some similarities to those of base-centered cubic (bcc) metals. In tension, bcc metals develop a ⟨110⟩ fiber texture while in

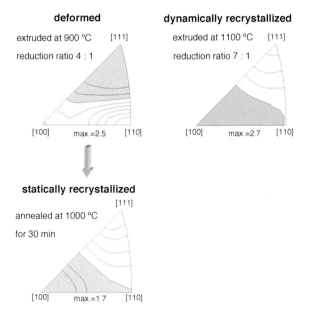

Figure 18.6 Textures of NiAl extruded through a round die in the deformed and dynamically and statically recrystallized state displayed as inverse pole figures of the extrusion direction.

compression it is ⟨111⟩. Plane strain deformation leads to an incomplete covering of the α- and γ-fiber with {112}⟨110⟩ and {111}⟨112⟩ dominating components, respectively. This has been clearly shown for the Fe–Al system [36]. The differences between bcc metals and B2 intermetallic compounds can be accounted for

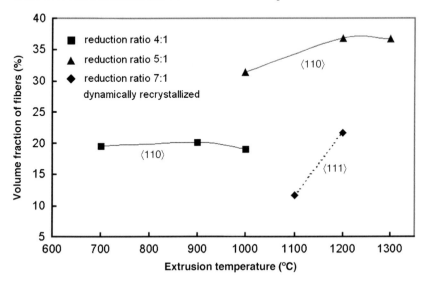

Figure 18.7 Volume fraction of ⟨110⟩ and ⟨111⟩ fibers as a function of temperature for extrusion through a round die.

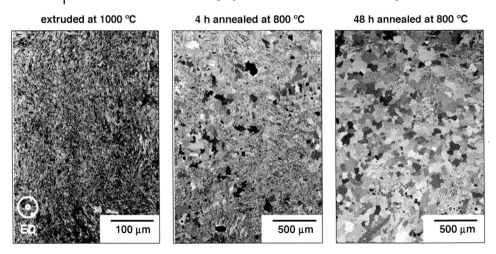

Figure 18.8 Microstructure development during annealing at 800°C of a NiAl sample extruded through a round die at 1000°C.

by the differences in a slip system activity as is demonstrated by texture simulations described in Section 18.3.2.

Dynamic as well as static recrystallization leads to a drastic texture change. In tension, the change is from a predominant ⟨110⟩ to a predominant ⟨111⟩ fiber

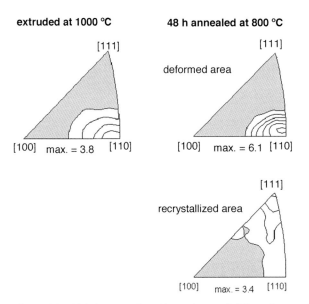

Figure 18.9 Global texture of a NiAl sample extruded through a round die at 1000°C measured by neutron diffraction and textures measured by EBSD in the deformed and recrystallized areas of an annealed sample.

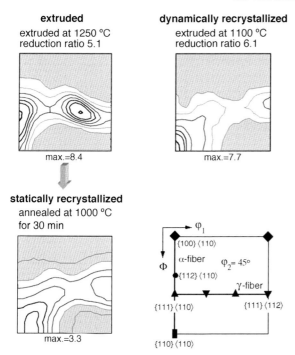

Figure 18.10 Textures of NiAl extruded through a rectangular die in the deformed and dynamically and statically recrystallized state displayed as $\varphi_2 = 45°$ ODF sections showing the main plane strain components given in the key figure.

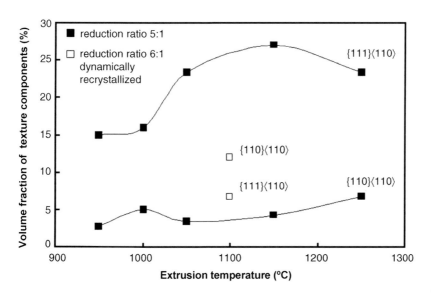

Figure 18.11 Volume fraction of $\{111\}\langle110\rangle$ and $\{110\}\langle110\rangle$ texture components as a function of temperature for extrusion through a rectangular die.

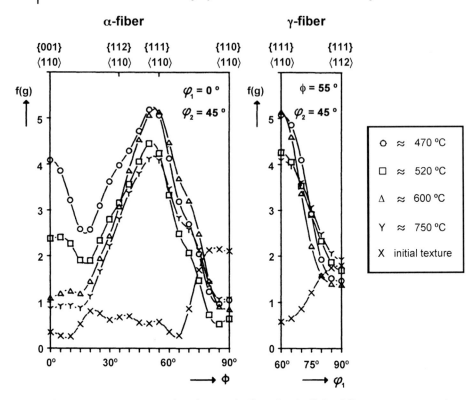

Figure 18.12 ODF intensity along the α- and γ-fiber of NiAl rolled at different temperatures [18].

texture. In a plane strain deformation, there is a change of the partial coverage of the α- and γ-fiber, with the dominant {111}⟨110⟩ component being replaced by {110}⟨110⟩. However, in compression there is no change.

In order to separate the contributions of deformation and recrystallization on texture formation, the orientations of several deformed areas occasionally left in the dynamically recrystallized sample have been measured by means of EBSD [28,29]. There is a tendency for a preferred ⟨110⟩ fiber texture and a {111}⟨110⟩ component for extrusion through a round and rectangular die that corresponds to the deformation texture shown in Figures 18.6 and 18.10, respectively. Likewise, orientations of isolated dynamically recrystallized grains in samples deformed at low strain and high temperature have been measured [31]. However, no trend is observed due to a poor statistics. For partially statically recrystallized samples, a clear separation between deformation and recrystallization contributions could be made (Figure 18.9).

The stored strain energy determines the stability of particular orientations. Highly strained grains have the tendency to nucleate new grains and those with low stored energy (less deformed) have the tendency to grow at the expense of their neighbors. Therefore, the texture developing during recrystallization will depend on the balance

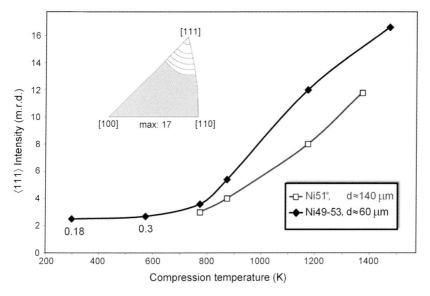

Figure 18.13 Development of the compression texture of NiAl as a function of temperature, after [18]. The compression axis of Ni51* samples, cast and high temperature forged, Ni49–53 samples, high temperature extruded, is normal to the forging and extrusion axis, respectively. The true strain is 1.2, if not indicated. The inverse pole figure belongs to compression of Ni49 at 1473 K.

between nucleation (defined by the probability and "critical strain" for nucleation) and the boundary mobility. NiAl at all temperatures in tension and plane strain deformation develops a major ⟨110⟩ fiber texture and a major {111}⟨110⟩ component, respectively. During recrystallization, there is a change to predominantly ⟨111⟩ and {110}⟨110⟩. The observation of dynamically recrystallized grains in highly strained and recovered regions along grain boundaries in FeAl [36] shows that recrystallization in this B2 structured intermetallic compound is nucleation controlled. The same may be assumed for NiAl in line with the low boundary mobility typical for ordered structures [37,38].

18.4
Texture Simulations

The deformation textures of NiAl can be quite well simulated with the FC and RC Taylor theory assuming major slip on the {110}⟨100⟩ and {110}⟨110⟩ systems and adjusting the critical resolved shear stress (CRSS) ratios (Figures 18.14–18.16). The FC Taylor theory works quite well for the lower temperature range, while for the RC case it is the high temperature range. The RC Taylor model allows shears not producing too large strain incompatibilities (Figure 18.2b and c). At high temperatures, such incompatibilities may be reduced by diffusional processes and grain

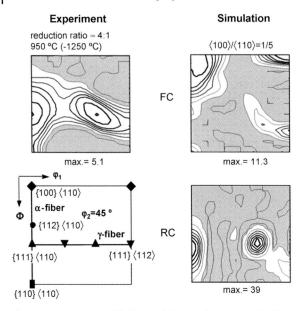

Figure 18.14 Texture of NiAl extruded through a rectangular die represented by $\phi_2 = 45°$ ODF sections in comparison with simulated textures. The main texture components are indicated in the key figure.

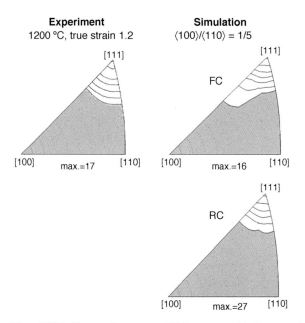

Figure 18.15 Texture of compressed NiAl represented by inverse pole figures of the compression axis in comparison with simulated textures.

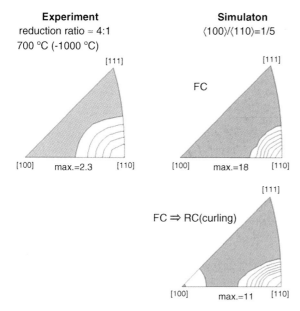

Figure 18.16 Texture of NiAl extruded through a round die represented by inverse pole figures of the extrusion direction in comparison with simulated textures.

boundary sliding. However, compared with experiment, the simulated textures are too strong. This is a generally observed phenomenon in texture simulation that may mainly be attributed to the fact that strain rate sensitivity and grain fragmentation has not been taken into account.

In axisymmetric tension at all temperatures, a major ⟨110⟩ fiber develops. Therefore, grains with a [011] tensile axis have a [100] and [0–11] direction normal to it (Figure 18.2c). In general, these orthogonal directions mechanically behave differently. Thus, grains with a ⟨110⟩ fiber axis will preferentially deform in pure shear, that is, in the extreme case in plane strain. However, as the overall deformation of the aggregate is axisymmetric, in order to relax strain incompatibilities folding around the tensile axis, so-called curling, may take place. The main texture component developing in plane strain deformation is (111)[0–11]. Folding of grains with such an orientation around [0–11] yields a [0–11] fiber texture.

18.5
Mechanical Anisotropy

The room temperature stress–strain curves (Figure 18.17) show that there is a strong plastic anisotropy, that is samples having a ⟨100⟩ crystallographic preferred orientation parallel to the compression axis have a much higher work-hardening rate than those having ⟨110⟩ and ⟨111⟩, with ⟨110⟩ being lowest [33,34]. In all cases, the stress at the onset of deformation is almost the same. The compressive stress of the ⟨100⟩

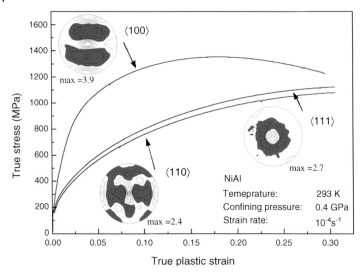

Figure 18.17 True stress–true plastic strain compression curves of differently textured NiAl samples, after [34].

sample goes over a maximum at a strain of about 15%. The reason for this behavior may be the formation of microcracks or shear bands.

The ⟨100⟩ and ⟨110⟩ samples have an asymmetric orientation distribution normal to the compression direction, whereas that of the ⟨111⟩ sample is almost symmetric (Figure 18.17). The asymmetry leads to an inhomogeneous shape change during compression with lateral flow being stronger in the softer direction (Figure 18.18). The strain parameter q was determined for each of the three samples according to the anisotropic shape change measured: ⟨100⟩: $q = 0.79$, ⟨110⟩: $q = 0.70$, ⟨111⟩: $q = 0.53$. The shape change of the samples has been taken to calculate the strain tensor, that is q.

To explain the plastic anisotropy observed on the basis of the measured textures, mean Taylor energies have been calculated for different deformation modes (Figure 18.19). Input data for the calculations are the slip systems as well as their critical resolved shear stresses. As {110}⟨111⟩ slip has been found to be more difficult than {110}⟨100⟩ slip (Figure 18.3), polycrystals with ⟨100⟩ preferred compression axis are expected to be harder than those with ⟨111⟩ and ⟨110⟩. The mean Taylor energy changes in a characteristic manner with the deformation mode. For the samples with orthorhombic texture symmetry, the Taylor energy is minimum for pure shear deformation with $q \approx 0.8$. In the almost axisymmetric ⟨111⟩ sample, it has a minimum at q slightly higher than 0.5 representing axisymmetric compression. Therefore, if not constrained to a certain deformation mode, the samples tend to deform in a mode with minimum Taylor energy, which is observed experimentally. With the q-values derived from the inhomogeneous shape change, the mean Taylor energies can be taken from Figure 18.19. In the series ⟨100⟩, ⟨110⟩, and ⟨111⟩ their ratios are: 1.62 : 1 : 1.02. These ratios are comparable with those of the

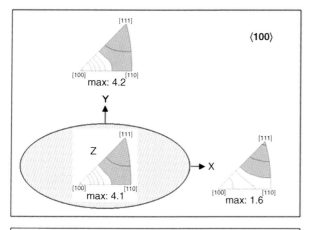

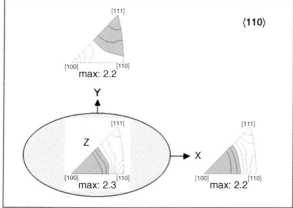

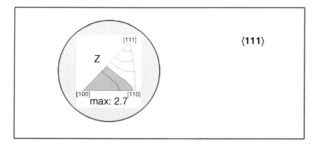

Figure 18.18 Schematic sketch of NiAl sample cross sections after compression (exaggerated for clarity). The crystallographic directions corresponding to X, Y, Z are given by inverse pole figures [34].

flow stresses at 15% strain (1.52 : 1 : 1.06) but not at yielding, where they are close to unity. This discrepancy cannot be explained by the uncertainty in the CRSS ratios resulting from deviations from stoichiometry [1]. It must be due to the fact that strongly plastically anisotropic crystals do not deform homogeneously as required by the Taylor theory, at least at the beginning of deformation. Slip probably starts in

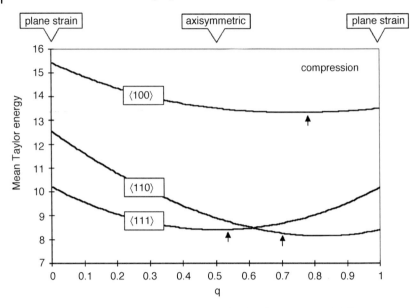

Figure 18.19 Mean Taylor energies as a function of preferred orientation and deformation geometry ($q = \varepsilon_{xx} = 1 - \varepsilon_{yy}$). The actual q-values measured in the middle of the sample length are marked by arrows [34].

crystallites favorably oriented for primary slip. Such crystallites are less frequent in the hard-oriented ($\langle 100 \rangle$) polycrystal and therefore the flow stress rises much faster than in the soft-oriented ($\langle 110 \rangle$, $\langle 111 \rangle$) polycrystals. The internal stresses arising from primary slip alone may be released by microcracking or by the activation of the hard secondary slip systems necessary for fulfilling the von Mises criterion. In polycrystals deformed under atmospheric pressure, microcracking detected by accoustic emission [39] already starts at 0.7% strain. Under confining pressure, microcracking should be suppressed up to higher strains and therefore the activation of secondary slip is more probable. This assumption may be supported by the flow stress ratios at elevated strains as well as the shape change as predicted by the Taylor theory. However, the direct proof by transmission electron microscopy is still missing. So far, in polycrystals deformed under atmospheric pressure essentially only primary dislocations have been observed [40].

Concerning the elastic and plastic properties of single crystals, it should be mentioned that in NiAl Young's modulus and the yield stress are strongly anisotropic, that is, they decrease with orientation in the sequence $\langle 111 \rangle$, $\langle 110 \rangle$, $\langle 100 \rangle$ and $\langle 100 \rangle$, $\langle 110 \rangle / \langle 111 \rangle$ (Figure 18.1). Thus, to increase the elastic stiffness, textures containing the $\langle 111 \rangle$ component in loading direction are desirable. This may be achieved by recrystallization of axisymetrically extruded NiAl. On the other hand such material has the lowest mechanical strength, not only because of texture but also because of large grain size.

18.6
Conclusions

1) Deformation of NiAl leads to specific deformation textures: a ⟨111⟩ fiber texture in compression, a predominant ⟨110⟩ fiber texture in tension, and a predominant {111}⟨110⟩ component in plane strain deformation.
2) Dynamic and static recrystallization may lead to a change of texture type: no change in compression, change from a predominant ⟨110⟩ to a ⟨111⟩ fiber texture in tension and a change in plane strain deformation from a predominant {111}⟨110⟩ to a predominant {110}⟨110⟩ component.
3) The deformation textures with regard to the main texture components can be simulated well with the FC and RC Taylor model at low and high temperatures, respectively.
4) Mechanically, textured NiAl behaves strongly anisotropic with the magnitude of the anisotropy depending on the strength and type of texture. The anisotropy comprises mechanical strength and shape change. This should be taken into account in the design of parts subjected to strain-controlled loading conditions.

Acknowledgments

Contributions of R. Tamm, M. Lemke, O. Perner, M. Scharnweber and discussions with Dr. C.-G. Oertel, Dr. J. Fischer-Bühner, and Dr. B. Beckers are gratefully acknowledged. Prof. Dr. H.-G. Brokmeier, GKSS Forschungszentrum, Geesthacht, and Dr. W. Tirschler, IMPK, TU Dresden, helped in neutron and EBSD texture measurements, respectively. Dr. K.B. Müller, Forschungszentrum Strangpressen, TU Berlin, and Dr. R. Opitz, IFW Dresden, carried out the extrusions. The compression tests under hydrostatic pressure were done by Dr. E. Rybacki, Geoforschungszentrum, Potsdam. Thanks are due to the Deutsche Forschungsgemeinschaft for support through contract Sk 21/14.

References

1 Noebe, R.D., Bowman, R.R., and Nathal, M.V. (1993) Review of the physical and mechanical properties of the B2 compound NiAl. *International Materials Reviews*, **38**, 193–232.
2 Miracle, D.B. (1993) The physical and mechanical properties of NiAl. *Acta Metallurgica et Materialia*, **41**, 649–684.
3 Noebe, R.D., Bowman, R.R., and Nathal, M.V. (1996) The physical and mechanical metallurgy of NiAl, in *Physical Metallurgy and Processing of Intermetallic Compounds* (eds N.S. Stoloff and V.K. Sikka), Chapman & Hall, New York, pp. 212–296.
4 Klöden, B., Oertel, C.-G., Rybacki, E., and Skrotzki, W. (2009) Microstructure development during high strain torsion of NiAl. *Journal of Engineering Materials and Technology*, **131**, 011101-1-9.
5 Klöden, B., Oertel, C.-G., Rybacki, E., and Skrotzki, W. (2009) Texture formation and Swift effect in high strain torsion of NiAL. *Journal of Engineering Materials and Technology*, **131**, 011102.

6 Skrotzki, W., Chulist, R., Beausir, B., and Hockauf, M. (2011) Equal-channel angular pressing of NiAl. *Materials Science Forum*, **667–669**, 39–44.

7 Dahms, M. (1992) The iterative series-expansion method for quantitative texture analysis. 2. Applications. *Journal of Applied Crystallography*, **25**, 258–267.

8 Helming, K. (1995) *Texture Approximation by Model Components*. TU Clausthal Habilitation Thesis.

9 Taylor, G.I. (1938) Plastic strain in metals. *Journal of the Institute of Metals*, **62**, 307–324.

10 van Houtte, P. (1988) A comprehensive mathematical formulation of an extended Taylor–Bishop–Hill model featuring relaxed constraints, the Renouard–Wintenberger theory and a strain rate sensitivity model. *Textures and Microstructures*, **8–9**, 313–350.

11 Wenk, H.-R. (1985) Carbonates, in *Preferred Orientation in Deformed Metals and Rocks: An Introduction to Modern Texture Analysis* (ed. H.-R. Wenk), Academic Press, New York, pp. 361–384.

12 von Mises, R. (1928) Mechanik der plastischen Formänderung von Kristallen. *Zeitschrift für Angewandte Mathematik und Mechanik*, **8**, 161–185.

13 Scharnweber, M., Oertel, C.-G., and Skrotzki, W. (2010) Deformation mechanisms of hard oriented NiAl single crystals. *Journal of Physics: Conference Series*, **240**, 012026.

14 Paterson, M.S. (1990) Rock deformation experimentation, in *The Brittle-Ductile Transition in Rocks* (eds A.G. Duba, W.B. Durham, J.W. Handin, and H.F. Wang), The Heard Volume, American Geophysical Union Geophysical, Monograph, 56. AGU, Washington, DC, pp. 187–194.

15 Noebe, R.D., Bowman, R.R., Cullers, C.L., and Raj, S.V. (1991) Flow and fracture behavior of NiAl in relation to the brittle-to-ductile transition temperature. *Materials Research Society Symposium Proceedings*, **213**, 589–596.

16 Skrotzki, W., Kegler, K., Tamm, R., and Oertel, C.-G. (2005) Grain structure and texture of cast iron aluminides. *Crystal Research and Technology*, **40**, 90–94.

17 Wassermann, G. and Grewen, J. (1962) *Texturen Metallischer Werkstoffe*, Springer, Berlin.

18 Fischer-Bühner, J. (1998) Mechanismen der Mikrostruktur- und Texturentwicklung von polykristallinem NiAl. RWTH Aachen Dissertation.

19 Khadkikar, P.S., Michal, G.M., and Vedula, K. (1990) Preferred orientations in extruded nickel and iron aluminides. *Metallurgical Transactions A*, **21**, 279–288.

20 Bieler, T.R., Noebe, N.D., Whittenberger, J.D., and Luton, M.J. (1992) Extrusion textures in NiAl and reaction milled NiAl/AlN composites. *Materials Research Society Symposium Proceedings*, **273**, 165–170.

21 Dymek, S., Hwang, S.J., Dollar, M., Kallend, J.S., and Nash, P. (1992) Microstructure and texture in hot-extruded NiAl. *Scripta Metallurgica et Materialia*, **27**, 161–166.

22 Bowman, K.J., Jenny, J., Kim, S., and Noebe, R.D. (1993) Texture in hot-worked B2-structure aluminides. *Materials Science and Engineering: A*, **160**, 201–208.

23 Dollar, M., Dymek, S., Hwang, S.J., and Nash, P. (1993) The role of microstructure on strength and ductility of hot-extruded mechanically alloyed NiAl. *Metallurgical Transactions A*, **24**, 1993–2000.

24 Bieler, T.R., Noebe, R.D., Hebsur, M., and Swaminathan, R. (1994) The effects of extrusion parameters on recrystallization kinetics, texture, and corresponding fracture toughness of NiAl, in *Advances in Hot Deformation Textures and Microstructures* (eds J.J. Jonas, T.R. Bieler and K.J. Bowman), The Minerals, Metals & Materials Society, Warrendale, PA, pp. 519–533.

25 Margevicius, R.W. and Lewandowski, J.J. (1993) Deformation texture of hydrostatically extruded polycrystalline NiAl. *Scripta Metallurgica et Materialia*, **29**, 1651–1654.

26 Lee, I.G., Gosh, A.K., Ray, R., and Jha, S. (1994) High-temperature deformation of B2 NiAl-base alloys. *Metallurgical Transactions A*, **25**, 2017–2026.

27 Margevicius, R.W. and Cotton, J.D. (1995) Study of the brittle-to-ductile transition in

NiAl by texture analysis. *Acta Metallurgica et Materialia*, **43**, 645–655.

28 Lemke, M. (1996) Textur- und Mikrostrukturuntersuchungen an stranggepreßtem NiAl. TU Dresden Diploma Thesis.

29 Tamm, R., Lemke, M., Oertel, C.-G., and Skrotzki, W. (1998) Deformation and recrystallization texture of extruded NiAl. *Materials Science Forum*, **273–275**, 411–416.

30 Skrotzki, W., Lemke, M., Oertel, C.-G., and Tamm, R. (1997) Development of microstructure and texture in extruded NiAl. *Materials Science and Engineering: A*, **234–236**, 739–742.

31 Garmestani, H., Harris, K.E., and Ebrahimi, F. (1998) Texture evolution in NiAl. *Materials Science and Engineering: A*, **247**, 187–194.

32 Skrotzki, W., Tamm, R., and Oertel, C.-G. (1999) On the texture change in extruded NiAl, in *Proc. 12th Int. Conf. on Textures of Materials* (ed. J.A. Szpunar), NRC Research Press, Ottawa, pp. 730–735.

33 Perner, O. (2000) Untersuchungen zum Rekristallisationsverhalten von NiAl. TU Dresden Diploma Thesis.

34 Skrotzki, W., Tamm, R., Oertel, C.-G., Beckers, B., Brokmeier, H.-G., and Rybacki, E. (2001) Texture induced plastic anisotropy of NiAl polycrystals. *Materials Science and Engineering: A*, **319–321**, 364–367.

35 Skrotzki, W., Tamm, R., Oertel, C.-G., Beckers, B., Brokmeier, H.-G., and Rybacki, E. (2002) Influence of texture and hydrostatic pressure on the room temperature compression of NiAl polycrystals. *Materials Science and Engineering: A*, **329–331**, 235–240.

36 Skrotzki, W., Tamm, R., Kegler, K., and Oertel, C.-G. (2009) Deformation and recrystallization textures in iron aluminides, in *Microstructure and Texture in Steels* (eds A. Haldar, S. Suwas and D. BhattarcharjeeC), Springer, New York, pp. 379–391.

37 Zhao, X.B. (1998) Grain growth and texture development of NiAl during annealing. *Journal of Materials Science Letters*, **17**, 489–492.

38 Humphreys, F.J. and Hatherly, M. (2005) *Recrystallization and Related Annealing Phenomena*, Elsevier, Oxford.

39 Wanner, A., Schietinger, B., Bidlingmaier, T., Zalkind, H., and Arzt, E. (1995) Monitoring of deformation induced microcracking in polycrystalline NiAl. *Materials Research Society Symposium Proceedings*, **364**, 543–548.

40 Bowman, R.R., Noebe, R.D., Raj, S.V., and Locci, I.E. (1992) Correlation of deformation mechanisms with the tensile and compressive behavior of NiAl and NiAl (Zr) intermetallic alloys. *Metallurgical Transactions A*, **23**, 1493–1508.

19
Development of Novel Metallic High Temperature Materials by Microstructural Design

Martin Heilmaier, Joachim Rösler, Debashis Mukherji, and Manja Krüger

19.1
Introduction

In order to increase the temperature capability of turbine engines beyond the current limit, structural metallic materials need to be developed being able to continuously withstand surface temperatures in excess of 1200 °C in a combustion environment. Currently used Ni-base superalloys have written an over 60 years long success story but have nowadays reached the temperature limit given by their melting point. Hence, it would be attractive not only from industrial, environmental, and socio-economical views but also from a scientific standpoint to develop novel alloy systems with substantially higher melting points than superalloys thus making them attractive candidates for ultrahigh temperature structural applications.

For this purpose, several novel metallic systems have been proposed to yield significant potential such as for instance novel intermetallics (Ru–Al and Ru–Ni–Al) [1,2], γ'-strengthened Co-base superalloys [3–6], and multinary and multiphase Nb–Si alloys [7–9]. While showing promise, none of them have been introduced into industrial service yet. This is simply due to the complexity of the task since it requires a thorough, physical metallurgy-based understanding of the interplay between the following topics essential for any application as high temperature structural material:

1) Processing
2) Strength and ductility at ambient temperatures, damage tolerance
3) Oxidation resistance
4) Creep resistance.

In what follows, we review our current activities in a larger scale research project funded recently by the German Science Foundation DFG focusing on the development of novel high temperature alloys exemplified with the following two alloy systems:

a) Mo–Si–B, in which alloys usually yield melting point of around 2000 °C, thus exceeding those of superalloys by at least 500 K and

Microstructural Design of Advanced Engineering Materials, First Edition. Edited by Dmitri A. Molodov.
© 2013 Wiley-VCH Verlag GmbH & Co. KGaA. Published 2013 by Wiley-VCH Verlag GmbH & Co. KGaA.

b) Co–Re that offers the possibility of continuously tuning-up the melting point of the alloy through Re addition.

This chapter is structured into two main parts that treat the above alloy systems separately. Within these parts, the topics and properties of importance are introduced and discussed. A short conclusion summarizes the currently achieved results and gives an outlook on future development trends.

19.2
Alloy System Mo–Si–B

Refractory metals and alloys (on the basis of Mo, Nb, Ta, and W) currently used as components for ultrahigh temperature applications are limited to protective atmosphere applications because they suffer from severe oxidation in air above around 500 °C, commonly known as *pesting* [10]. Multiphase silicide alloys with Nb or Mo as a base metal have shown promise for high temperature structural applications in air [7,9,11], because they may take advantage of the oxidation resistance of appropriate silicide phases eventually leading to the formation of a protecting silica layer on the surface of the material [12]. Berczik [13] pointed out that an alloy that possesses a reasonable combination of strength and oxidation resistance should lie in the three-phase field Mo solid solution – Mo_3Si–Mo_5SiB_2 (the T2 phase) having the baseline composition Mo-9Si-8B (in at%). In particular, key mechanical properties such as a tensile strength exceeding 700 MPa over the whole temperature range between room temperature (RT) and 1100 °C as well as a brittle-to-ductile transition temperature (BDTT) of around 600 °C have been reported in the patent [13]. The ongoing assessment of several ingot and powder metallurgy approaches [13–19] revealed that a suitable compromise in terms of processing and microstructure needs to be found to optimize simultaneous improvement of the material performance in each of the otherwise mutually exclusive areas, see the scenario in Figure 19.1, schematically depicted from [14]. In what follows we will show one possible way to satisfy this demand by suggesting a unique powder metallurgical approach yielding mechanical alloying (MA) as the crucial processing step [16,19]. We will further describe the impact of processing on microstructure and relevant properties of this alloy system. In addition, the effect of Zr microalloying addition will be assessed comparatively against the ternary baseline alloy composition.

19.2.1
Manufacturing and Microstructure

The alloy compositions chosen for the present chapter were determined to be within the three-phase field between α-Mo, Mo_3Si, and Mo_5SiB_2, as described in the phase diagram in Figure 19.2. Four ternary alloy compositions, namely Mo-6Si-5B, Mo-9Si-8B, Mo-10Si-10B, and Mo-13Si-12B (in at%), were processed by an industrial powder metallurgical (PM) route, in which mechanical alloying was used as the

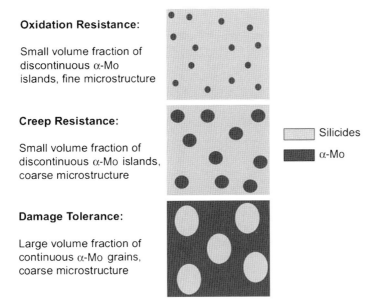

Figure 19.1 Schematic illustrations of ideal microstructures to improve oxidation resistance, creep resistance, and damage tolerance of Mo–Si–B alloys (re-drawn after [20]).

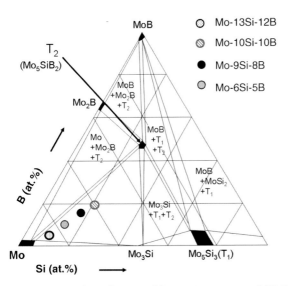

Figure 19.2 Isothermal section of the Mo–Si–B system at 1600 °C (re-drawn after [24]). The compositions (in at%) chosen for this study are marked.

crucial step. This stage was added to an industrial PM fabrication route to achieve fine grained and chemically homogeneous powder particles. As a reference for the matrix phase of the above three-phase Mo–Si–B alloys, high purity Mo(Si) solid solutions with (relevant) Si contents ranging between approximately 0.3 at% and 3 at% were manufactured via a comparable PM route [21,22]. Zr was added (in the form of ZrH_2) to selected compositions, since it was found that Zr improves fracture toughness and grain boundary strength in Mo-base alloys [14,21,23].

In our earlier work [16], we identified milling parameters to find a satisfying compromise between milling progress that is aimed at homogeneous microstructures of powder particles as well as mechanically induced formation of a Mo (Si, B) solid solution together with a low content of impurities. Proceeding on these results, elemental powder mixtures of Mo, Si, and B were mechanically alloyed with tungsten carbide balls in a planetary ball mill (Retsch PM 400) in protective Ar atmosphere at 200 rpm for 20 h utilizing a powder to ball weight ratio of 1:13. Use of these milling parameters yielded powder particles with (i) sizes in the range of 1.8 to 4.6 μm, (ii) domain (grain) sizes as low as 6 nm, and (iii) a substantial amount of Si and B in supersaturated Mo solid solution. The fine scale and chemically homogeneous particle microstructure has been proven beneficial for the subsequent consolidation procedure utilizing sintering [16].

Next, the powders were cold isostatically pressed at 200 MPa and sintered in hydrogen at 1600 °C to reduce the impurity content, mainly oxygen. Finally, residual porosity was minimized during hot isostatic pressing (HIPing) at 1500 °C and 200 MPa to values below 1%. Microstructures of consolidated samples with above compositions are depicted in Figure 19.3. Clearly, the fine scale microstructure has been inherited from the former powder particles. A thorough metallographic analysis demonstrates that the mean sizes of the phases in all alloys lie in the submicron to 2 μm range [17]. Concurrent with increasing Si and B concentrations in Mo–Si–B alloys, the volume fraction of the intermetallic phases (darker gray phases in the SEM micrographs in Figure 19.3) increases, and, simultaneously, the volume fraction of the α-Mo phase (bright phase) decreases. X-ray diffraction measurements (not shown here) verified that all alloys are composed of the three

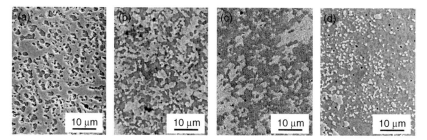

Figure 19.3 SEM micrographs of (a) Mo-6Si-5B, (b) Mo-9Si-8B, (c) Mo-10Si-10B, and (d) Mo-13Si-12B showing the composition and distribution of the α-Mo phase (brightest appearing phase) and the two intermetallic phases Mo_3Si (medium gray phase) and Mo_5SiB_2 (dark gray phase).

phases bcc α-Mo, Mo$_3$Si (A15), and Mo$_5$SiB$_2$ (tetragonal T2 phase) as expected from the ternary phase diagram (Figure 19.2). Rietveld refinement analysis revealed that the amount of Si and B controls the volume fraction of intermetallic phases Mo$_3$Si and Mo$_5$SiB$_2$. Concurrently, the phase arrangement changes: above about 50 vol% of Mo solid solution phase, corresponding to alloy composition Mo–9Si–8B, a microstructure with a continuous α-Mo matrix that surrounds regions of intermetallic phases was fabricated. In contrast to alloys with high-volume fractions of α-Mo phase, Figure 19.3 shows that α-Mo volume fractions lower than 50% invert microstructure morphology. Thus, in these cases either interpenetrating microstructures (Mo-10Si-10B) or intermetallic matrices with embedded island of α-Mo prevail (Mo-13Si-12B). Phase compositions and distributions evaluated on the basis of SEM images could be verified by focused ion-beam (FIB) tomography measurements [25]. Compositions of the investigated alloys fit very well with theoretical phase compositions (see Figure 19.2) evaluated from the isothermal section of the Mo–Si–B phase diagram at 1600 °C [20]. Microalloying with Zr does not significantly alter the phase compositions and arrangement but has a small influence on further promoting the fineness of the microstructure.

19.2.2
Strength and Ductility at Ambient Temperatures

From the above description of the microstructural arrangement of the phases, it can straightforwardly concluded that any ductility and toughness be present in this composite material may stem from the Mo solid solution. Both intermetallic phases are known to be brittle up to temperatures of more than 1200 °C [26,27]. For example, single crystalline Mo$_5$SiB$_2$ possesses at RT an indentation fracture toughness of only 2 MPam$^{1/2}$ [26]. Therefore, in a first step, in Section 19.2.2.1, we report on the temperature dependence of the mechanical behavior of a single phase bcc Mo–Si solid solution with and without the addition of Zr studied utilizing three-point bend tests. In a second step, in Section 19.2.2.2, the mechanical properties of multiphase Mo–Si–B alloys will be reported and the influence of Zr microalloying will be likewise exemplified.

19.2.2.1 Single-phase Mo Solid Solutions
Single-phase Mo–Si alloys with 0, 0.1, 0.5, and 1.0 wt% (0, 0.34, 1.69, and 3.34 at%) Si have been fabricated yielding comparable grain sizes of around 50 μm utilizing a standard industrial PM route, for details see [22].

Fracture toughness values were determined utilizing three-point bending experiments on bars with rectangular cross-section prenotched by electrodischarge machining. Figure 19.4 shows that the room temperature fracture toughness of Mo–Si decreases precipitously with the Si concentration. In view of the pronounced strengthening for the higher Si concentrations, see [28], this is not unexpected – often the fracture toughness scales inversely with the yield strength. What is unexpected is the fact that at room temperature the Mo-0.1 wt% Si material is significantly more brittle than pure Mo, although its yield strength is lower [22]. This is believed to be due to a reduction in the grain boundary strength as suggested by

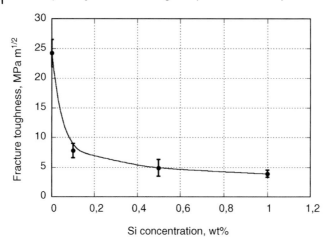

Figure 19.4 Room temperature fracture toughness of binary Mo–Si solid solutions as a function of the Si concentration [22].

the transition from transgranular to intergranular fracture, see Figure 19.5. In other words, the observed fracture toughness values are a lower limit to those that might be observed if fracture were transgranular. If the transition in the fracture mode could be eliminated, the fracture toughness might not be degraded as much by Si. In arc-cast and annealed Mo-0.75 wt% Si, for example, room temperature fracture is mostly transgranular [29]. This may be due to differences in the type and amount of grain boundary segregants, predominantly oxygen, as compared to the PM material.

As pointed out earlier, microalloying additions such as C, B, or Zr are known to segregate to grain boundaries in Mo [23] and may alleviate intergranular fracture. Therefore, in a next step Zr has been added to a binary Mo-1.5 at% Si solid solution [21]. This composition was chosen because (i) it already showed brittle deformation behavior in PM-processed material (Figure 19.4) and (ii) it is a relevant Si concentration in the Mo–Si solid solution in multiphase Mo–Si–B alloys. In an alternative second approach, a small amount of incoherent nanoscale Yttria particles has been

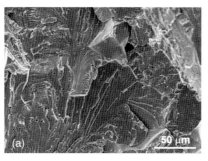 

Figure 19.5 SEM micrographs of room temperature fracture surfaces of (a) Mo and (b) Mo-0.1 wt% (0.34 at%) Si [21].

incorporated into the same material composition. In analogy to the manufacturing of multiphase Mo–Si–B alloys (see Section 19.2.1) MA has been employed to achieve chemically homogeneous, fine grained, and thus, high strength Mo–Si solid solutions.

In bcc metals and alloys ductility can be improved by reducing the grain size since this reduces the effective oxygen concentration at the grain boundaries [29]. Both, the MA approach together with the alloying additions substantially reduce the mean grain size towards around 1 µm (as compared to above binary Mo–Si solid solutions). While lowering the grain size due to Yttria dispersoid pinning raises the maximum strength to about 2 GPa, the material still behaves totally brittle in three-point bending tests up to temperatures of 538 °C (corresponding to 1000 °F), see Figure 19.6. The addition of Zr, however, improves both the maximum strength and

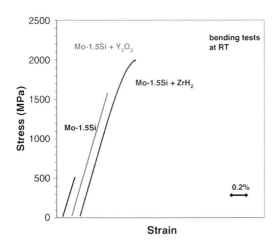

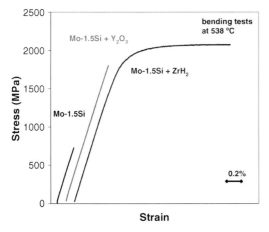

Figure 19.6 Stress–strain curves (derived with elastic beam theory from the three-point bend tests for the outer tensile fiber, over- and underestimating stresses and strains as plasticity sets in) of pure Mo-1.5Si and Mo-1.5Si with Y_2O_3 and ZrH_2 addition at (a) room temperature, (b) at 538 °C [21].

the ductility of the Mo–Si solid solution: even at RT the stress–strain curve of Mo-1.5Si-1Zr in Figure 19.6a indicates an onset of plastic deformability at a stress level of more than 2 GPa. At 538 °C (1000 °F), Figure 19.6b, the binary- and the yttria-doped Mo–Si solid solution still behave brittle, whereas the Zr-microalloyed material can be deformed plastically without failure at a still extremely high stress level of around 2 GPa. Therefore, the concept of reducing the concentration of interstitial impurities at the grain boundaries to ductilise the alloys may be ruled out here.

Fractured surfaces of the bend samples deformed at RT have been analyzed by SEM and Auger analysis. As compared to both, the binary Mo–Si solid solution and the yttria-doped alloy, fracture in the Zr containing alloy occurs to a much lesser extent by intergranular failure. In other words, a substantially higher amount of transgranular failure indicates already an increased grain boundary cohesion caused by the presence of Zr. Therefore, Auger spectra of the grain boundary regions (intergranular fracture areas) and the grain interior (transgranular) have been taken for MA Mo-1.5Si and MA Mo-1.5Si-1Zr solid solution. The only obvious difference of the spectra refers to the size of the Si peak, whereas the Zr peak could not be analyzed due to overlapping with the much stronger Mo peaks (not shown here, see [21]). While this peak is most pronounced for the grain boundaries in the MA Mo-1.5Si alloy, it decreases for the MA Mo-1.5Si-1Zr grain boundaries and becomes minimal for the transgranular fracture area. The quantitative analyses of the Si concentration on the basis of Auger spectra on several different grain boundaries and interiors, Figure 19.7, confirm that the Si concentration is reduced at the grain boundaries of the Zr containing Mo-1.5Si solid solution. Hence, Si is strongly segregated at the grain boundaries in all binary Mo–Si alloys but to a much lesser extent in the Zr containing alloy, indicating that the presence of Zr leads to depletion of Si at the grain boundaries.

From above observations, the following conclusions can be drawn with respect to the observed deformation behavior: although the portion of transgranular fracture

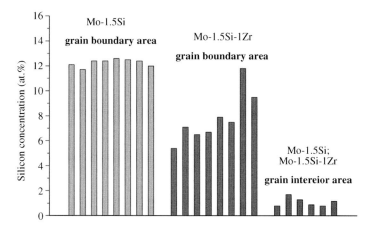

Figure 19.7 Quantitative analysis of the Si concentration of Auger spectra taken at several different grain boundaries and in the grain interior of Mo–Si solid solution alloys [21].

increases with the addition of Zr, there is still intergranular fracture in these specimens [21]. This suggests that further improvement of the ductility by further strengthening of the grain boundaries is possible until the fracture mode may become completely transgranular. This might be reached by (i) further increase of the amount of Zr addition, (ii) lowering the oxygen concentration by optimization of the sintering treatment and (iii) further alloying additions that might lead to an increased depletion of Si and O at the grain boundaries. Carbon, titanium, and boron are among the possible candidates for the latter two approaches [23,30].

19.2.2.2 Multiphase Mo–Si–B Alloys

Figure 19.8 proves the above outlined concept of (a) the establishment of a continuous Mo matrix and (b) the addition of Zr with the three-point bend tests of the multiphase Mo–Si–B alloys. Here, the outer fiber tensile strain (employing classical elastic beam theory) is plotted in dependence on test temperature revealing the BDTT. For all alloys, the data points reflect the average of three independent tests. They indicate failure unless having reached the arbitrary value of 8% strain or stated otherwise. As anticipated already from the mechanical properties of the silicide phases [26,27], the material with an intermetallic matrix exhibits the highest BDTT by far. More than 1100 °C are needed to observe significant (plastic) straining in the three-point bending. This value is in nice coincidence with the one determined for a differently processed, namely gas atomized Mo-8.9Si-7.7B alloy with eutectic matrix and embedded Mo_{ss} particles [19]. Reducing the amount of Si and B to 10% each leads to an interpenetrating microstructure (see Section 19.2.1) and already to a decrease of BDTT to about 1050 °C. The effect of a continuous Mo_{ss} matrix is reflected by the two overlapping curves for Mo-9Si-8B and Mo-6Si-5B, see the full squares and gray triangles respectively, leading to a BDTT of roughly 950 °C.

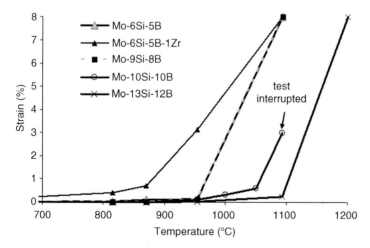

Figure 19.8 Outer fiber tensile strain versus temperature for the three-point bend tests of various Mo–Si–B alloys. Back-extrapolation to the onset of strain defines the brittle-to-ductile transition temperature BDTT.

Finally, the curve for the Zr-microalloyed Mo-6Si-5B alloy is given as black triangles in Figure 19.8 for comparison. Clearly, a further impact on BDTT can be noted as compared to the both reference alloys with continuous Mo_{ss} matrices. However, its effect on reducing BDTT is a mere 150 K that is much less pronounced as in the case of the Mo solid solution where clear signs of plasticity at temperatures as low as 500 °C have been observed, see Figure 19.6. Why the Zr segregation towards grain and phase boundaries, with the aim of increasing their cohesion, in multiphase Mo–Si–B alloys does not show a similarly pronounced effect as in the Mo solid solution is unclear at the moment. Further work with atom probe tomography and transmission electron microscopy is ongoing to shed light on this phenomenon and to fully exploit its potential in reducing BDTT.

19.2.3
Oxidation Resistance

Since single phase Mo alloys cannot withstand temperatures beyond 700 °C in air due to volatile Mo-oxide formation [28], protection can only be achieved by adding Al or Si [31]. The sources of oxidation protection in multiphase Mo–Si–B are the intermetallic phases Mo_3Si and Mo_5SiB_2 (T2) that provide the Si and B reservoir necessary for borosilicate scale formation. Generally, oxidation experiments, utilizing, for example, thermogravimetric analysis (TGA), of Mo–Si–B alloys first exhibit a pronounced transient mass loss that can be attributed to the formation of volatile MoO_3. Steady-state oxidation begins once a continuous and slow-growing borosilicate layer (SiO_2 with B_2O_3 in solid solution) has formed, see Figure 19.9 for a test temperature of 1300 °C, which protects the alloy from further rapid oxidation [18,32,33]. SiO_2–B_2O_3 exists as a continuous outer layer, whereas beneath the layer SiO_2 precipitates form an internal oxidation zone with a Mo-rich matrix depleted in intermetallic phases If, however, the temperature is below 800 °C, ternary Mo–Si–B exhibit a rather poor oxidation resistance, since a single and highly porous oxide layer consisting of MoO_3 and SiO_2–B_2O_3 develops, which

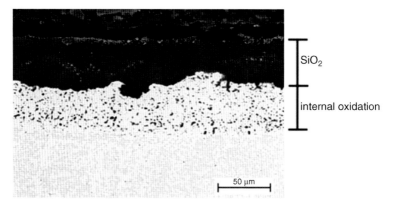

Figure 19.9 Cross-section of Mo-9Si-8B alloy oxidized in air at 1300 °C for 72h (SE image) [33].

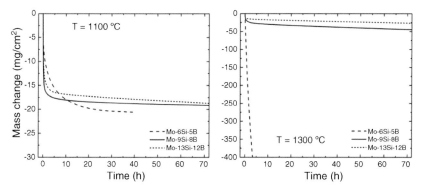

Figure 19.10 TGA results for Mo–Si–B with varying Si and B content, (a) 1100 °C, (b) 1300 °C.

yields only moderate oxidation protection [33]. This phenomenon is commonly called *pesting* in literature [10,33].

A steady oxidation state is clearly observed in the temperature range between 1000 °C and 1300 °C. This is exemplified in Figure 19.10 a and b at 1100 °C and 1300 °C, respectively, for all Mo–Si–B alloys with contents of Si > 9% and B > 8%, respectively. This behavior is found to be *independent* on microstructural morphology. Surprisingly, also the alloy with the least Si and B content and continuous Mo matrix (Mo-6Si-5B) shows comparable self-protection capability at 1100 °C, see Figure 19.10a, indicating the importance of a judicious choice of the B–Si ratio [18]. Lowering the B–Si ratio yields a decrease of the high temperature oxidation resistance because silica forms at a very slow rate at low temperatures (750 °C) [34]. However, at 1300 °C, Figure 19.10b, this alloy exhibits a dramatic weight loss indicating that no protective silica scale is formed. Obviously, this is caused by the heavy sublimation of Mo oxides together with the well known volatilization of B_2O_3 [35,36]. Therefore, also in the two other, self-protecting alloys no Boron was found in the silica scale via Raman Spectroscopy [33].

Adding small amounts of reactive elements (e.g., Y, La, Zr) may further help to reduce the initial weight loss [37]. Since they usually become incorporated into the forming surface scale, they can also act as helpful fluxing agents. The positive effect of Zr on the oxidation behavior is most impressive between 1000–1150 °C, cf. Figure 19.11b with Figure 19.10a (no Zr). Generally, Mo–9Si–8B–1Zr experiences less weight loss than the reference composition Mo–9Si–8B. This is the consequence of the grain and phase refinement caused by Zr microalloying [21] that allows a faster sealing of the surface by silica. In particular, it exhibits the best oxidation performance at 1100 °C, where the weight loss after 72 h is almost identical to the initial value ($-6\,\text{mg/cm}^2$). However, the situation is reversed at temperatures of 1200 °C and higher. The reason for that peculiar behavior is sketched in Figure 19.11a: monoclinic Zirconia incorporated into the forming scale undergoes a phase transformation towards tetragonal crystal structure at temperatures beyond 1150 °C accompanied with a distinct volume reduction. This causes

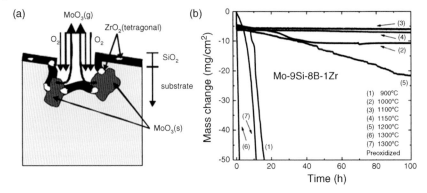

Figure 19.11 Mo-9Si-8B-1Zr: (a) schematic representation of the initial oxidation mechanism at $T > 1150\,°C$, (b) TGA results for $T = 900-1300\,°C$ [33].

damage to the scale and gives access to the substrate beyond 1200 °C. As a consequence, heavy bubbling of MoO_3 increases formation of pores. With increasing oxidation time volatilization of B_2O_3 hinders fluxing of silica and sealing of the pores.

Application of a preoxidation treatment under reducing conditions aiming to stabilize a dense silica layer and thus, minimize or completely avoid the sample weight loss due to volatile Mo-oxide formation has been proven partly successful. While a preoxidation treatment at 1200 °C and oxygen partial pressures below 10^{-12} bar for 50 h was not successful to establish a dense and protective silica scale, see curve (7) showing catastrophic oxidation in Figure 19.11b, a prolonged annealing time to 100 h or an increase of annealing temperature to 1300 °C verified this concept with a measured marginal weight loss [12,36].

19.2.4
Creep Resistance

Any application as a high temperature structural material requires sufficient creep resistance, that is, resistance against plastic deformation at elevated temperatures due to mechanical loading. Usually Mo base alloys such as TZM (Mo-0.5Ti-0.08Zr) intrinsically possess an outstanding creep resistance up to temperatures of 1300 °C due to their high melting point [38–40]. Therefore, one may anticipate a similar finding even for the fine grained Mo–Si–B alloys discussed in this manuscript. Both, compressive and tensile creep testing was carried out at constant true stresses or constant loads. Testing was performed in vacuum in the range of 10^{-3} Pa to exclude effects from ongoing oxidation during creep. The temperature was chosen to be 1200 °C that is already about 100 K beyond the commonly accepted capability of Ni-base superalloys.

Figure 19.12 exemplifies the (minimum) strain rate versus stress data that were taken and plotted in double logarithmic representation. It is obvious

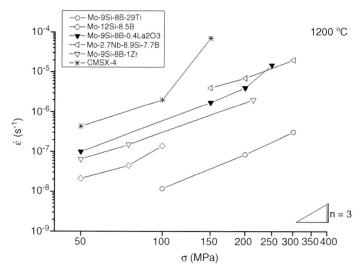

Figure 19.12 Log minimum strain rate–log applied stress relationship at 1200 °C for a number of Mo–Si–B alloys, compositions, and symbols as given in the insert. For reference, a state-of-the-art nickel-base single crystalline alloy CMSX-4 is plotted as stars.

(i) that all Mo–Si–B alloys (triangles and squares in Figure 19.12) exhibit similar slopes, that is, stress exponents (assuming power law creep)

$$n = \Delta \log \dot{\varepsilon}/\Delta \log \sigma \qquad (19.1)$$

with n values between 2 and 3 and (ii) that they are clearly superior to a state-of-the-art Ni-base single crystal CMSX-4 (stars in Figure 19.12) that creeps about one order of magnitude faster than its Mo–Si–B counterparts. A similar trend was also noted at 1100 °C (not shown here) [40]. Summarizing the detailed discussion on the acting creep mechanism given in [38,40] yields that creep is clearly matrix-controlled with an activation energy comparable to that of Mo self diffusion (about 400 kJ/mole). Finite element analysis and TEM investigations on deformed samples illustrate strain localization in the matrix (and partly in the Mo_3Si intermetallic compound), whereas the T2 phase particles are elastically highly stressed and show only scarce dislocation activity [38]. Higher temperatures or lower strain rates/stresses are required to plastically deform the T2 phase, thus, leading to a more homogeneous deformation of the composite.

As an outlook to future work, we have added in Figure 19.12 the creep characteristics of a Ti-containing alloy as open circles. This alloy with composition Mo-9Si-8B-29Ti is based on a thermodynamic approach by Yang et al. [41] who recently found out that heavily alloying with Ti enables to substitute the intermetallic compound Mo_3Si for $(Mo,Ti)_5Si_3$ in three-phase Mo–Si–B alloys. More recently, the latter intermetallic compound was found to be much more oxidation resistant than the former one [42,43]. While in-depth, investigations of the oxidation behavior are open to future studies our first creep tests (open circles in Figure 19.12) on Mo–9Si–8B–29Ti material manufactured via arc melting show promise with minimum creep rates being an additional order of magnitude lower than their Ti-free counterparts.

19.3
Alloy System Co–Re–Cr

Rhenium has been added to Ni-base superalloys starting in the early 1980s [44,45] as it not only causes strengthening via solid solution hardening and retardation of γ'-growth. It also belongs to the few alloying elements that increase rather than decrease the melting range of superalloys. Unfortunately, solubility of rhenium in nickel is limited and so is the beneficial effect of increasing the melting range. This can already be appreciated by inspecting the binary Ni–Re phase diagram (Figure 19.13 a). In contrast, complete miscibility exists in the binary Co–Re phase diagram (Figure 19.13b), offering the possibility to continuously increase the melting range with rhenium addition from that of cobalt (1495 °C) to that or rhenium (3186 °C). In 2007, this led to the idea to develop Co–Re-based alloys with melting temperatures beyond the capability of Ni-base superalloys [47]. A particular beauty of this system is that the character of the material can be adjusted continuously from that of a contemporary Co-based alloy to that of a high melting point material, depending on the rhenium content. This provides the opportunity to tune properties in a wide range and find a suitable balance between requirements for ductility, toughness, and producibility on the one hand and strength at temperatures beyond the capability of Ni-base superalloys on the other hand.

Obviously, further alloying elements besides rhenium are needed to attain useful properties for high temperature service and it is the intent here to review the current alloy development status since the idea for Co–Re-based high temperature alloys was born 5 years ago [47]. The alloys discussed in the following contain 17% to 31% rhenium (all compositions are given in at% unless otherwise stated), having a Re–Co ratio of about 0.28 to 0.67. They were manufactured by vacuum arc melting.

19.3.1
Oxidation

Aluminum, chromium, and silicon are the three fundamental alloying elements to be considered for oxidation protection via formation of a protective oxide scale. For the first phase of alloy development, chromium was selected as primary alloying element because of its good solubility in cobalt, neutral effect regarding the melting range, and proven ability to provide oxidation protection in contemporary cobalt alloys [48,49]. Yet, the unknown was the effect of rhenium to the overall oxidation behavior.

Results of initial oxidation studies [50,51] are shown in Figure 19.14 for a ternary Co–Re–Cr alloys. In analogy to conventional Co–Cr alloys devoid of rhenium [48], the oxide scale essentially consists of three layers: CoO and $CoCr_2O_4$ spinel as fast

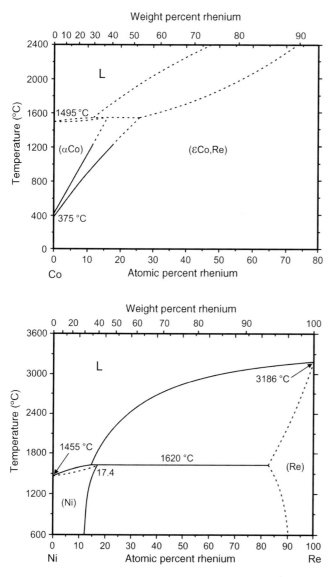

Figure 19.13 Binary phase diagrams for Ni–Re (a) und Co–Re (b), according to [46]. Note complete solubility of rhenium in cobalt in contrast to limited solubility in nickel. The ε- and α-phases in the Co–Re diagram are hexagonal close packed and face centered cubic, respectively.

growing, nonprotective layers followed by a thin, innermost Cr_2O_3 scale, expected to provide oxidation protection. Despite this seemingly favorable microstructure, the results obtained after exposure in air at 1000 °C were devastating. Irrespective of the added chromium content (23% and 30%), all Co–Re–Cr alloys showed substantial mass losses, presumably due to evaporation of volatile Re-oxides, after exposure

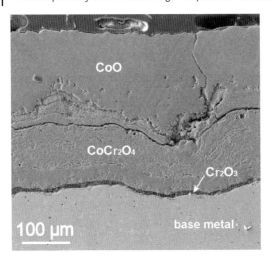

Figure 19.14 Co-17Re-23Cr after air exposure at 1000 °C/8h. Clearly visible is a layered oxide scale consisting of outermost CoO followed by $CoCr_2O_4$ and Cr_2O_3.

times as small as 15 min while Co-23Cr, Co-23Cr-2.6C, and the commercial cobalt alloy X-40 (composition: Co-29Cr-10Ni-2.4W-1.5Si-2.4C) exhibited small mass gains as expected when a protective oxide scale is formed [50]. This demonstrates the adverse effect of rhenium on the oxidation properties and may shed some doubt on the ability to develop Co–Re-based alloys with balanced properties. However, it is also noteworthy that the alloy X-40 was performing significantly better than the Re-free binary and ternary alloys, which may be due to its Si-content. Thus, the effect of silicon on oxidation properties of Co–Re–Cr alloys was explored in a second step [52,53], using silicon contents of 1% to 3% in combination with 23% to 30% chromium. Interestingly, the oxidation behavior is much improved by this measure: Figure 19.15 demonstrates a parabolic mass gain of the Co–Re–Cr–Si alloys

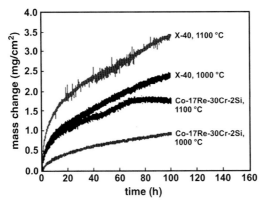

Figure 19.15 Oxide growth kinetics measured as mass change versus exposure time in air (data reproduced from [53]).

compared to that of X-40 instead of the mass loss observed beforehand. Inspection of the oxide scale revealed internal oxidation of silicon but no silica-scale beneath the Cr_2O_3 scale. Thus, the beneficial effect of silicon in the concentrations studied here is merely to promote formation of a dense, that is, protective, Cr_2O_3 scale. According to Stringer et al. [48], this comes about because the internally formed SiO_2 oxide particles serve as nucleation sites for the Cr_2O_3 scale. As a result, more Cr_2O_3 nuclei are formed, making it easier to connect to a dense scale by lateral growth. Similarly, the beneficial effect of Y_2O_3 particles on the oxidation resistance of ODS-alloys can be explained [54], lending credit to this interpretation.

The results demonstrate that Co–Re-based alloys can attain an oxidation resistance comparable to contemporary Co alloys despite their high rhenium content when chromium is added in combination with silicon. Noting that the latter materials are routinely used to manufacture coated first stage turbine vanes, that is, particularly hot gas turbine components, it is anticipated that suitably coated Co–Re–Cr–Si alloys provide sufficient oxidation resistance for high temperature applications as well.

19.3.2
Strength and Ductility at Ambient Temperatures

The results presented in the previous section have demonstrated the beneficial effect of chromium on the oxidation resistance when combined with small amounts of silicon. The critical question is then, whether about 20% to 30% chromium can be tolerated from the perspective of mechanical properties. Addition of chromium in these quantities has a distinct effect on the microstructure of Co–Re-based alloys. Instead of being single phase, a Cr_2Re_3-type σ-phase appears as expected from the ternary Co–Re–Cr phase diagram [55]. Its volume fraction not only increases with the chromium but also with the rhenium content (see Figure 19.16).

The σ-phase is very strong, having a micro-hardness of about 1400 HV 0.005 [47]. Thus, composite type strengthening by load transfer from the softer matrix onto the harder phase is anticipated on the one hand [56,57]. On the other hand, the σ-phase may also cause embrittlement. Bending tests on ternary Co–Re–Cr alloys [58] showed, in fact, little plastic deformation (Figure 19.17) seemingly fitting to that expectation. Yet, inspection of the fracture surface did not reveal cleavage failure of the σ-phase particles and/or debonding between particle and matrix as dominant failure mode. Instead, it mainly showed intercrystalline failure (Figure 19.18a). Such intercrystalline failure modes are neither unusual for high strength nickel alloys nor for the Mo–Si–B alloys discussed in Section 19.2 and indicative of grain boundary embrittlement due to environmental or segregational effects.

A remedy well established for these materials is to add boron and zirconium as grain boundary strengtheners. While the latter element was discussed in Section 19.2 on the example of the Mo–Si–B alloys, the former was explored on Co–Re–Cr alloys containing 200–1000 wt-ppm (i.e., 0.15 to 0.72 at%) boron [58]. As shown in Figure 19.17, the result on mechanical behavior is remarkable. All doped alloys displayed substantial plastic deformation and the failure mode changed from intercrystalline to transcrystalline (Figure 19.18), demonstrating the profound effect

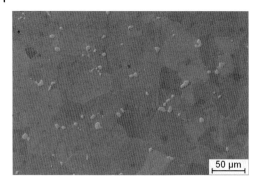

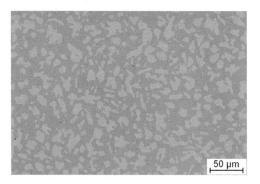

Figure 19.16 Microstructure of alloy Co-17Re-23Cr (a) and Co-31Re-23Cr (b) after solution heat treatment at 1350 °C/5h + 1400 °C/5h + 1450 °C/5h. The volume fraction of the σ-phase (bright) is increasing with the Re-content.

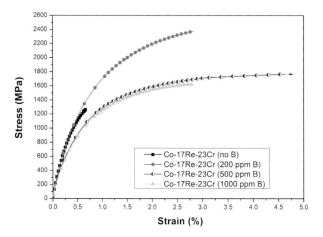

Figure 19.17 Three-point bend test results obtained at ambient temperature for Co-17Re-23Cr-B alloys [58]. The test of the alloy containing 500 wt.-ppm boron was discontinued prior to failure. Note that stresses and strains have been calculated by elastic beam theory for the outermost tensile fiber, overestimating stresses and underestimating strains as soon as plasticity takes place.

 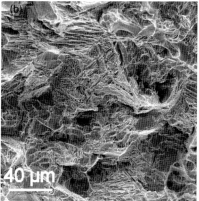

Figure 19.18 Fracture surface of Co-17Re-23Cr without (a) and with addition of 500 wt.-ppm boron (b) after tensile testing at ambient temperature [58].

of boron additions on grain boundary cohesion. Cleavage fracture of the σ-phase particles is now apparent (Figure 19.18b), showing that the weakest link in the microstructure has shifted from the grain boundaries to the brittle σ-phase. Yet, the ductility of the Co–Re-rich matrix ensures sufficient plastic deformation of the two-phase alloy prior to fracture.

Boron addition not only improves grain boundary cohesion and, by that, ductility. It also causes (i) formation of $(Cr,Re)_2B$ borides situated along grain boundaries [59] and affects the amount of σ-phase precipitation in a complex fashion [60]. As a result, there is also a remarkable interdependence between flow strength and boron content (compare with Figure 19.17), the former being highest at about 200 ppm boron.

Summarizing these results, it is noted that chromium not only causes oxidation protection (in combination with silicon) but it also imparts particle strengthening (in combination with rhenium) by formation of the σ-phase. Provided boron is added, ductile behavior can be ensured even though cleavage of the σ-phase particles initiates fracture. Furthermore, *in situ* neutron diffraction experiments [61,62] as well as heat treatment studies have shown that the σ-phase is stable up to very high temperatures. In fact, it was not possible to dissolve the σ-phase during heat treatments up to 1450 °C. While this is desirable to ensure composite-type strengthening at very high temperatures, it hinders refinement of the microstructure by controlled dissolution and reprecipitation of the σ-phase. In this respect, an interesting effect was noticed: if nickel is added to Co–Re–Cr alloys, the solvus temperature of the σ-phase is suppressed to such an extent that complete dissolution becomes possible. Subsequent precipitation heat treatment leads then to a substantially refined morphology of the σ-phase [63]. This is shown on the example of alloy Co-17Re-23Cr-25Ni in Figure 19.19b. The coarse σ-phase could be completely dissolved during solution heat treatment at 1350 °C/5 h + 1400 °C/5 h + 1450 °C/5 h and reprecipitated in fine lamellar form at 1225 °C/3 h. The effect of this microstructural change on ductility is immediately apparent from Figure 19.19 :

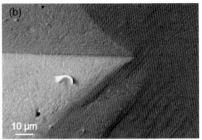

Figure 19.19 Microstructure of alloy Co-17Re-23Cr + 1000 wt.-ppm B after solution heat treatment at 1350 °C/7.5h + 1400 °C/7.5h (a) and Co-17Re-23Cr-25Ni after solution heat treatment at 1350 °C/5h + 1400 °C/5h + 1450 °C/5h and aging at 1225 °C/3h (b). Hardness indentations are visible in (a) and (b) in the middle and at the left side, respectively. While the coarse σ-phase (bright) in (a) is fracturing due to the hardness indentation, the refined σ-phase in (b) is not.

while the large σ-phase particles of the Ni-free alloy were fractured due to hardness indentation (Figure 19.19a), the fine σ-phase particles of alloy Co-17Re-23Cr-25Ni stayed intact. The reason for this size effect is the lack of critical flaws in the refined σ-phase particles. Note also the presence of slip lines in the matrix, demonstrating again the intrinsic ductility of the Re-rich matrix.

Despite the availability of the σ-phase for particle strengthening and substantial solid solution hardening of the matrix due to its high re-content, it is sensible to explore further strengthening concepts. One avenue is to utilize carbide strengthening much as it is done in contemporary cobalt alloys. For this reason, Co–Re–Cr–C– (Ta) alloys were produced, containing 1% to 3% carbon. This adds considerable complexity to the microstructure of the alloys and its evolution during heat treatment. First, $Cr_{23}C_6$-type carbides are formed on two length scales, that is, blocky carbides with dimensions of a few micrometers and fine lamellar carbides with thicknesses and spacings on the submicron scale [47]. Heat treatment studies at temperatures between 1000 °C and 1200 °C showed significant coarsening and dissolution so that these precipitates can be ruled out for high temperature applications. In contrast, addition of 1% to 2% tantalum led to the formation of very fine and stable TaC carbides (Figure 19.20). Their initial size is about 30 nm [62] and there appears to be little coarsening during elevated temperature exposure. Furthermore, *in situ* neutron diffraction experiments have demonstrated a constant volume fraction between 1000 °C and 1300 °C [62], that is, thermodynamic stability up to very high temperatures. Thus, particle strengthening by TaC carbides appears to be another attractive strengthening mechanism.

19.3.3
Creep Resistance

The above discussion has highlighted particle hardening concepts as attractive means to achieve sufficient creep strength. TaC carbides appear to be attractive because of their stability. Furthermore, TEM investigations after creep deformation

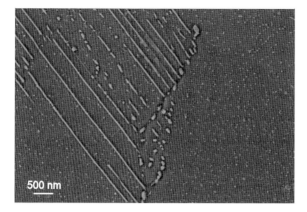

Figure 19.20 Microstructure of alloy Co-17Re-23Cr-1.2Ta-2.6C after solution heat treatment (1350 °C/7.5h + 1400 °C/7.5h) and aging at 1100 °C/0.5h. Visible are finely dispersed TaC particles as well as lamellar $Cr_{23}C_6$-carbides, which start to break up due to the exposure at 1100 °C.

indicate pinning of dislocations at the TaC particles due to an attractive particle–dislocation interaction (Figure 19.21) [64]. The attraction stems from partial relaxation of the dislocation stress field at the particle–matrix interface [65]. Consequently, the relaxed energy has to be expended when the dislocation is detaching itself from the particle under the action of the applied stress and the activation energy for that process is given by [66]

$$E_d = Gb^2 r[(1-k)(1-\sigma/\sigma_d)]^{3/2} \tag{19.2}$$

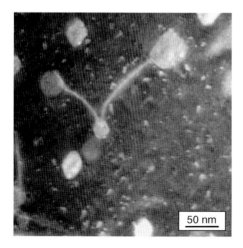

Figure 19.21 Dislocation pinning at TaC particles after creep deformation of alloy Co-17Re-23Cr-1.2Ta-2.6C at 1100 °C/50 MPa [64].

where G is shear modulus, b is the Burgers vector, r is the particle radius with the relaxation parameter

$$k = T_p/T_m \tag{19.3}$$

where T_p, T_m are dislocation line energies at the particle–matrix interface and in the matrix, respectively, and athermal detachment stress [67]

$$\sigma_d = \sigma_o(1 - k^2)^{1/2} \tag{19.4}$$

where σ_o is the Orowan stress. It can be easily appreciated from Eq. (19.2) that the activation energy increases with increasing athermal detachment stress and decreasing relaxation parameter k. Typically, k is about 0.75 to 0.95 leading to stress exponents $n > 10$ for $\sigma < \sigma_d$, see Eq. (19.1) for definition of n. In these cases (and assuming small particles), the time required to overcome the energy barrier associated with the detachment process is much larger than the time required for the climb step over the particle [68]. For this reason, the observation of an attractive particle dislocation interaction after creep deformation is a good indicator of a powerful particle strengthening mechanism at elevated temperatures.

The creep strength of Co–Re-based alloys was measured and compared to a conventional cobalt alloy for high temperature applications (X-40) in [64]. All materials were subjected to the same manufacturing route, namely vacuum arc melting in a laboratory furnace. The Larson–Miller plot in [64] shows a temperature benefit of the Co–Re-based alloys relative to lab scale X-40 of about 120 K. This is a promising result, noting that the temperature benefit of polycrystalline superalloys such as MM002 relative to X-40 is significantly less, namely about 40 K. However, on closer inspection of the results, there are also critical questions. First of all, addition of tantalum in case of the alloy Co-17Re-23Cr-1.2Ta-2.6C did not lead to strengthening relative to the Ta-free counterparts (Co-17Re-23Cr and Co-17Re-23Cr-2.6C) devoid of TaC particles. Second, the measured stress exponents n were very low (about 1.4 to 3.0) irrespective of the tantalum content. Both observations are surprising in view of the above discussion, which suggests large stress exponents and significant strengthening when TaC particles are introduced. Third, the creep strength of lab scale X-40 is far inferior to the data measured on industrial scale X-40 [64]. In this context, it is important to note that the lab scale materials were fine grained with typical grain sizes $\leq 100\,\mu m$. As the stress exponent is also suspiciously small, this leads to the conclusion that diffusional creep accommodated by grain boundary sliding rather than dislocation creep is rate controlling. Accepting this interpretation, the lack of strengthening by the TaC particles and the inferior creep strength of lab scale X-40 compared to coarse grained industrial scale material becomes understandable. Investigations by Depka (T. Depka, 2012, unpublished research), who placed a grid of parallel lines by focused ion beam milling on the specimen of a Co–Re–Cr alloy prior to creep testing and observed substantial displacement of the lines at grain boundaries due to grain boundary sliding after creep testing, lend further credit to this interpretation.

In summary, one has to state that the creep test results are not yet conclusive. Tests on large-grained materials have to be conducted in order to separate diffusional creep from dislocation creep mechanisms and assess the intrinsic creep strength of the Co–Re-based alloys.

19.4 Conclusions

Mo–Si–B and Co–Re-based alloys offer significant potential for high temperature applications beyond the scope of Ni-base superalloys. The current research on Mo–Si–B alloys has demonstrated that three-phase fine-grained alloys offer an acceptable compromise between creep strength and oxidation resistance (both being superior to current single crystal Ni-base superalloys) on the one hand and ambient temperature toughness on the other hand. A suitable industrial manufacturing route comprises powder metallurgy with mechanical alloying as the crucial processing step. However, three drawbacks have to be mentioned: first, all alloys developed do not show tensile ductility at room temperature, rather the BDTT is on the order of 800 °C. Second, these alloys still suffer from significant pesting in the temperature range between 700 and 900 °C unless a silica oxide scale has been applied by preoxidation at higher temperatures and lower oxygen partial pressures [36]. Finally, the density of these alloys exceeds those of Ni-base superalloys considerably that makes applications in aviation questionable. Due to promising creep strength and favorably low densities of below 8 g/cm^3 Ti-containing Mo–Si–B alloys with titanium contents above 20% may offer a significant step forward towards a new generation of high temperature structural materials beyond Ni-base superalloys. This will be subject of future studies, in particular with respect to so far unknown properties like room temperature toughness and oxidation resistance.

The current research on Co–Re-based alloys has demonstrated that ductility, oxidation resistance, and strengthening for high temperature applications can be attained by suitable addition of alloying elements. One, but possibly not the only way to oxidation protection is addition of chromium in combination with silicon. Ductility can be ensured by addition of grain boundary active elements, in particular boron. Particle hardening on different length scales can be attained through (i) the σ-phase, when chromium is added to the solid solution-strengthened Co–Re matrix, and (ii) MC-carbides, particularly TaC. Yet, there are still many issues to be tackled, which is not surprising in view of the short time span since the idea for the development of Co–Re-based high temperature alloys was introduced in 2007 [47]. Understanding (i) the role of grain boundaries and diffusional creep along with measures to prevent this fast creep deformation mechanism is certainly one area of utmost interest. Another one is the understanding of phase transformations and kinetics in complex alloys along with the interaction of different alloying elements. Even though Co–Re-based alloys appear to be producible much as cobalt alloys and superalloys (which is, of course, a particular attraction of this material system), in-depth investigation and understanding of the entire processing chain, particularly the casting process, is also urgently needed. Only then will it be possible to eventually introduce Co–Re-based alloys for applications beyond those of Ni-base superalloys.

Acknowledgments

Financial support of the research unit 727 "Beyond Nickelbase Superalloys" by the Deutsche Forschungsgemeinschaft (DFG) is acknowledged gratefully. The authors are indebted to the further members of this research unit (S. Burk, B. Gorr, M. Azim, H.-J. Christ, D. Schliephake, T. Depka, Ch. Somsen, G. Eggeler, M.-C. Bölitz, M. Brunner, R. Hüttner, R. Völkl, U. Glatzel) who have contributed to this work.

References

1 Cao, F. and Pollock, T.M. (2008) Creep deformation mechanisms in Ru-Ni–Al ternary B2 alloys. *Metallurgical and Materials Transactions*, **39**, 39–49.

2 Mücklich, F., Ilic, N., and Woll, K. (2008) RuAl and its alloys, Part II: Mechanical properties, environmental resistance and applications. *Intermetallics*, **16**, 593–608.

3 Sato, J., Omori, T., Oikawa, K., Ohnuma, I., Kainuma, R., and Ishida, K. (2006) Cobalt-base high-temperature alloys. *Science*, **312**, 90–91.

4 Suzuki, A. and Pollock, T.M. (2008) High-temperature strength and deformation of γ/γ' two-phase Co–Al–W-base alloys. *Acta Materialia*, **56**, 1288–1297.

5 Titus, M.S., Suzuki, A., and Pollock, T.M. (2012) Creep and directional coarsening in single crystals of new γ/γ' cobalt-base alloys. *Scripta Materialia*, **66**, 574–577.

6 Klein, L., Bauer, A., Neumeier, S., Göken, M., and Virtanen, S. (2011) High temperature oxidation of γ/γ'-strengthened Co-base superalloys. *Corrosion Science*, **53**, 2027–2034.

7 Bewlay, B.P., Jackson, M.R., Zhao, J.-C., Subramanian, P.R., Mendiratta, M.G., and Lewandowski, J.J. (2003) Ultrahigh-temperature Nb-silicide-based composites. *MRS Bulletin*, **28**, 646–653.

8 Sekido, N., Kimura, Y., Miura, S., Wie, F.-G., and Mishima, Y. (2006) Fracture toughness and high temperature strength of unidirectionally solidified Nb–Si binary and Nb–Ti–Si ternary alloys. *Journal of Alloys and Compounds*, **425**, 223–229.

9 Jéhanno, P., Heilmaier, M., Kestler, H., Böning, M., Venskutonis, A., Bewlay, B.P., and Jackson, M.R. (2005) Assessment of a powder metallurgical processing route for refractory metal silicide alloys. *Metallurgical and Materials Transactions*, **36**, 515–523.

10 Parthasarathy, T.A., Mendiratta, M., and Dimiduk, D.M. (2002) Oxidation mechanisms in Mo-reinforced Mo_5SiB_2(T2) –Mo_3Si alloys. *Acta Materialia*, **50**, 1857–1866.

11 Dimiduk, D.M. and Perepezko, J.H. (2003) Mo–Si–B alloys: Developing a revolutionary turbine-engine material. *MRS Bulletin*, **28**, 639–645.

12 Burk, S., Gorr, B., Krüger, M., Heilmaier, M., and Christ, H.-J. (2011) Oxidation behavior of Mo–Si–B– (X) alloys: Macro- and microalloying (X = Cr, Zr, La_2O_3). *Journal of the Minerals Metals and Materials Society*, **63**, 32–36.

13 Berczik, D.M. (1997) Oxidation resistant molybdenum alloy. US Patent 5,693,156.

14 Schneibel, J.H., Tortorelli, P.F., Ritchie, R.O., and Kruzic, J.J. (2005) Optimization of Mo–Si–B intermetallic alloys. *Metallurgical and Materials Transactions*, **36**, 525–531.

15 Yoshimi, K., Nakatani, S., Nomura, N., and Hanada, S. (2003) Thermal expansion, strength and oxidation resistance of Mo/Mo_5SiB_2 in-situ composites at elevated temperatures. *Intermetallics*, **11**, 787–794.

16 Krüger, M., Franz, S., Saage, H., Heilmaier, M., Schneibel, J.H., Jéhanno, P., Böning, M., and Kestler, H. (2008) Mechanically alloyed Mo–Si–B alloys with a continuous α-Mo matrix and improved mechanical properties. *Intermetallics*, **16**, 933–941.

17 Jéhanno, P., Heilmaier, M., Saage, H., Heyse, H., Böning, M., Kestler, H., and Schneibel, J.H. (2006) Superplasticity of a multiphase refractory Mo–Si–B alloy. *Scripta Materialia*, **55**, 525–531.

18 Mendiratta, M., Parthasarathy, T.A., and Dimiduk, D.M. (2002) Oxidation behavior of α Mo-Mo$_3$Si-Mo$_5$SiB$_2$ (T2) three phase system. *Intermetallics*, **10**, 225–232.

19 Jéhanno, P., Heilmaier, M., and Kestler, H. (2004) Characterization of an industrially processed Mo-based silicide alloy. *Intermetallics*, **12**, 1005–1009.

20 Lemberg, J.A. and Ritchie, R.O. (2012) Mo–Si–B alloys for ultrahigh-temperature structural applications. *Advanced Materials*, **24**, 3445–3480.

21 Saage, H., Krüger, M., Sturm, D., Heilmaier, M., Schneibel, J.H., George, E.P., Heatherly, L., Somsen, C., Eggeler, G., and Yang, Y. (2009) Ductilization of Mo–Si solid solutions manufactured by powder metallurgy. *Acta Materialia*, **55**, 525–531.

22 Sturm, D., Heilmaier, M., Schneibel, J.H., Jéhanno, P., Skrotzki, B., and Saage, H. (2007) The influence of silicon on the strength and fracture toughness of molybdenum. *Materials Science and Engineering*, **A463**, 107–114.

23 Miller, M.K. and Bryhan, A.J. (2002) Effect of Zr, B and C additions on the ductility of molybdenum. *Materials Science and Engineering*, **A327**, 80–83.

24 Sakidja, R., Park, J.S., Hamann, J., and Perepezko, J.H. (2005) Synthesis of oxidation resistant silicide coatings on Mo–Si–B alloys. *Scripta Materialia*, **53**, 723–728.

25 Hassomeris, O., Schumacher, G., Krüger, M., Heilmaier, M., and Banhart, J. (2011) Phase continuity in high temperature Mo–Si–B alloys: A FIB-tomography study. *Intermetallics*, **19**, 470–475.

26 Ihara, K., Ito, K., Tanaka, K., and Yamaguchi, M. (2002) Mechanical properties of Mo$_5$SiB$_2$ single crystals. *Materials Science and Engineering*, **A329–331**, 222–227.

27 Rosales, I., Schneibel, J.H., Heatherly, L., Horton, J.A., Martinez, L., and Campillo, B. (2003) High temperature deformation of A15 Mo$_3$Si single crystals. *Scripta Materialia*, **48**, 185–190.

28 Northcott, L. (1956) *Molybdenum*, Academic Press, New York.

29 Wadsworth, J., Nieh, T.G., and Stephens, J.J. (1986) Dilute Mo–Re alloys – a critical evaluation of their comparative mechanical properties. *Scripta Materialia*, **20**, 637–642.

30 Geller, C.B., Smith, R.W., Hack, J.E., Saxe, P., and Wimmer, E. (2005) A computational search for ductilizing additives to Mo. *Scripta Materialia*, **52**, 205–210.

31 Birks, N. and Meier, G.H. (eds) (1983) *Introduction to High Temperature Oxidation of Metals*, E. Arnolds Ltd., London.

32 Woodard, S.R., Raban, R., Myers, J.F., and Berczik, D.M. (2008) Improved oxidation resistant Molybdenum alloy. EU Patent EP 1 382 700 B1.

33 Burk, S., Gorr, B., Trindade, B.V., and Christ, H.-J. (2010) Effect of Zr addition on the high-temperature oxidation behaviour of Mo–Si–B alloys. *Oxidation of Metals*, **73**, 163–181.

34 Supatarawanich, V., Johnson, D.R., and Liu, C.T. (2003) Effects of microstructure on the oxidation behavior of multiphase Mo–Si–B alloys. *Materials Science and Engineering*, **A344**, 328–339.

35 Weimer, A.W., Roach, R.P., Haney, C.N., Moore, W.G., and Rafaniello, W. (1991) Rapid carbothermal reduction of boron-oxide in a graphite transport reactor. *AICHE Journal*, **37**, 759–768.

36 Burk, S., Gorr, B., and Christ, H.-J. (2010) High temperature oxidation of Mo–Si–B alloys: Effect of low and very low oxygen partial pressures. *Acta Materialia*, **58**, 6154–6165.

37 Pint, B.A. (2003) Optimization of reactive-element additions to improve oxidation performance of alumina-forming alloys. *Journal of the American Ceramic Society*, **86**, 686–695.

38 Alur, A.P., Chollacoop, N., and Kumar, K.S. (2004) Creep effects on crack growth in a Mo–Si–B alloy. *Acta Materialia*, **52**, 5571–5587.

39 Jain, P. and Kumar, K.S. (2010) Tensile creep of Mo–Si–B alloys. *Acta Materialia*, **58**, 2124–2142.

40 Jéhanno, P., Heilmaier, M., Saage, H., Böning, M., Kestler, H., Freudenberger, J., and Drawin, S. (2007) Assessment of the high temperature deformation behavior of molybdenum silicide alloys. *Materials Science and Engineering*, **A463**, 216–223.

41 Yang, Y., Bei, H., Chen, S., George, E.P., Tiley, J., and Chang, Y.A. (2010) Effects of Ti, Zr, and Hf on the phase stability of Mo$_{ss}$

+Mo$_3$Si +Mo$_5$SiB$_2$ alloys at 1600 °C. *Acta Materialia*, **58**, 541–548.

42 Burk, S., Gorr, B., Christ, H.-J., Schliephake, D., Heilmaier, M., Hochmuth, C., and Glatzel, U. (2012) High-temperature oxidation behaviour of a single-phase (Mo,Ti)$_5$Si$_3$ (Mo–Si–Ti) alloy. *Scripta Materialia*, **66**, 223–226.

43 Rioult, F., Imhoff, S.D., Sakidja, R., and Perepezko, J.H. (2009) Transient oxidation of Mo-Si-B alloys: Effect of the microstructure size scale. *Acta Materialia*, **57**, 4600–4613.

44 Schweizer, F.A. and Duhl, D.N. (1980) Single crystal nickel superalloy. US Patent US 4 222 794.

45 Henry, M.F. (1983) Cyclic oxidation-hot corrosion resistant nickel-base superalloys. US Patent 4 388 124.

46 Massalski, T.B. (1986) *Binary Alloy Phase Diagrams*, ASM, Materials Park, OH

47 Rösler, J., Mukherji, D., and Baranski, T. (2007) Co–Re-based alloys: A new class of high temperature materials? *Advanced Engineering Materials*, **9**, 876–881.

48 Jones, D.E. and Stringer, J. (1975) Effect of small amounts of silicon on the oxidation of high purity Co-25 wt.% Cr at elevated temperatures. *Oxidation of Metals*, **9**, 409–413.

49 Irving, G.N., Stringer, J., and Whittle, D.P. (1974) Effect of possible fcc stabilizers Mn, Fe, and Ni on high temperature oxidation of cobalt–chromium alloys. *Oxidation of Metals*, **8**, 393–407.

50 Gorr, B., Trinidade, B.V., Burk, S., Christ, H.-J., Klauke, M., Mukherji, D., and Rösler, J. (2009) Oxidation behaviour of model cobalt–rhenium alloys during short-term exposure to laboratory air at elevated temperature. *Oxidation of Metals*, **71**, 157–172.

51 Klauke, M., Mukherji, D., Gorr, B., Da Trindade Filho, B.V., Rösler, J., and Christ, H.-J. (2009) Oxidation behaviour of experimental Co–Re-base alloys in laboratory air at 1000 °C. *International Journal of Materials Research*, **100**, 104–111.

52 Gorr, B., Burk, S., Depka, T., Somsen, C., Abu-Samara, H., Christ, H.-J., and Eggeler, G. (2012) Effect of Si addition on the oxidation resistance of Co–Re-Cr-alloys: Recent attainments in the development of novel alloys. *International Journal of Materials Research*, **103**, 24–30.

53 Gorr, B. (2011) Ph.D. Dissertation, University of Siegen.

54 Stringer, J. and Wright, I.G. (1972) High temperature oxidation of cobalt – 21 wt.% chromium –3 vol. % Y$_2$O$_3$ alloys. *Oxidation of Metals*, **5**, 59–84.

55 Sokolovskaya, E.M., Tuganbaev, M.L., Stepanova, G.I., Kazakova, E.F., and Sokolova, I.G. (1986) Interaction of cobalt with chromium and rRhenium. *Journal of the Less Common Metals*, **124**, L5–L7.

56 Christman, T., Needleman, A., and Suresh, S. (1989) An experimental and numerical study of deformation in metal ceramic composites. *Acta Metall*, **37**, 3029–3050.

57 Rösler, J. (2003) Particle strengthened alloys for high temperature applications: strengthening mechanisms and fundamentals of design. *International Journal of Materials and Product Technology*, **18**, 70–90.

58 Mukherji, D., Rösler, J., Krüger, M., Heilmaier, M., Bölitz, M.-C., Völkl, R., Glatzel, U., and Szentmiklosi, L. (2012) The effects of boron addition on the microstructure and mechanical properties of Co-Re-based high-temperature alloys. *Scripta Materialia*, **66**, 60–63.

59 Bölitz, M.-C., Brunner, M., Völkl, R., Mukherji, D., Rösler, J., and Glatzel, U. (2012) Microstructural study of boron-doped Co–Re–Cr alloys by means of transmission electron microscopy and electron energy-loss spectroscopy. *International Journal of Materials Research*, **103**, 554–558.

60 Wehrs, J. (2012) Diploma Thesis, Technical University of Braunschweig.

61 Mukherji, D., Strunz, P., Gilles, R., Hofmann, M., Schmitz, F., and Rösler, J. (2010) Investigation of phase transformations by in-situ neutron diffraction in a Co–Re-based high temperature alloy. *Materials Letters*, **64**, 2608–2611.

62 Mukherji, D., Strunz, P., Piegert, S., Gilles, R., Hofmann, M., Hölzel, M., and Rösler, J. (2012) The hexagonal close-packed (HCP) a double dagger dagger face-centered cubic (FCC) transition in Co–Re-based experimental alloys investigated by neutron scattering. *Metallurgical and Materials Transactions*, **43**, 1834–1844.

63 Mukherji, D., Rösler, J., Strunz, P., Gilles, R., Schumacher, G., and Piegert, S. (2011) Beyond Ni-based superalloys: Development of CoRe-based alloys for gas turbine applications at very high temperatures. *International Journal of Materials Research*, **102**, 1125–1132.

64 Brunner, M., Hüttner, R., Bölitz, M.-C., Völkl, R., Mukherji, D., Rösler, R., Depka, T., Somsen, C., Eggeler, G., and Glatzel, U. (2010) Creep properties beyond 1100 °C and microstructure of Co–Re–Cr alloys. *Materials Science and Engineering*, **528**, 650–656.

65 Srolovitz, D.J., Petkovic-Luton, R.A., and Luton, M.J. (1984) Diffusionally modified dislocation particle elastic interactions. *Acta Metallurgica*, **32**, 1079–1088.

66 Rösler, J. and Arzt, E. (1990) A new model-based creep equation for dispersion strengthened materials. *Acta Metallurgica et Materialia*, **38**, 671–683.

67 Arzt, E. and Wilkinson, D.S. (1986) Threshold stresses for dislocation climb over hard particles – the effect of an attractive interaction. *Acta Metallurgica*, **34**, 1893–1898.

68 Rösler, J. and Arzt, E. (1988) The kinetics of dislocation climb over hard particles: 1. Climb without an attractive particle–dislocation interaction. *Acta Metallurgica*, **36**, 1043–1051.

Index

a

ab-initio quantum mechanical studies 4
abnormal grain growth (AGG) 310
absolute grain boundary mobility, measurements 239–243
absorption contrast tomography 371
activation energy 81
activity coefficients 271, 273
advanced photon source (APS) 370
age-hardenable alloy ZM61 95
ALAMEL model 31–38
Allen–Cahn kinetic formulation 276
Al–Li alloy 390
alloy systems, types 289
aluminum alloys 9, 397
aluminum bicrystals 242
aluminum polycrystals 437
aluminum processing, main texture components 389
aluminum wire 391
anisotropy 91, 162, 400
– affected by aging treatments 96
– energy 162, 163
– inclination-based 174
– mesoscopic simulation approaches 164–179
– mobility 163, 164, 166–170, 179
– plastic anisotropy 19, 27, 33, 49, 98, 447
– reduced 97, 103
– slip-induced 100
– stress and strain 92–103
annealing 8, 10, 91, 166, 192, 238, 247, 249, 250, 255, 259, 285, 310, 368, 388, 399, 420, 422, 426, 478
– annihilation length 73
– fcc metals and alloys 432
– grain growth during 258
– induced boundaries 368
– magnetic 252
– polycrystal by means of magnetic 249
– recrystallization during magnetic 261
– texture dominated by the cube component 166
– tilt grain boundary during 242
– time most appropriate revealed by 250, 378
– undergone spheroidization 285
APT. See atom probe tomography (APT)
arc zone melting technique 414
Arrhenius equation 79
as-quenched sample 292
asymmetrical grain boundaries
– coupled motion of 216–221
atomic bonds decohesion 269
atomic force micrograph, of surface of AZ31 tensile specimen 99
atomic-scale chemical analysis 292
atom probe tomography (APT) 280–282
– disadvantage of preparing 281
Auger spectra 474
– Si concentration, quantitative analysis 474
austenitic stainless steel
– Wilkinson (small strain) method 355
austenitic stainless steel (SUS304L) 424
axisymmetric tension 448, 459

b

basal slip 111
base-centered cubic (bcc) metals 452
$BaTiO_3$ polycrystals 314, 315
bicrystals, magnetically driven grain boundary motion 237–246
– absolute grain boundary mobility, measurements 239–243
– magnetically induced motion 237–239
– misorientation dependence of 243–245
– temperature dependence 243, 246
– tilt boundary mobility, boundary plane inclination effect 245, 246

bismuth bicrystals 240
– geometry of 238
Boltzmann factor 165
Boolean' model 131
boundary conditions 25, 26, 42
boundary energy 172
boundary migration process 374
boundary velocity 313
box-counting method 440
Bragg's law simulations 345
brittle–ductile fracture transition 441
brittle-to-ductile transition temperature (BDTT) 450, 468, 475, 476, 489
Brownian regime 214
bulk diffusion 303
Burgers vector 209

c

CA codes 8
Cahn–Hilliard formulation 174
Cahn–Hoffman formalism 301
Cahn's expression 144
capillarity-mediated branching 329–333
– capillary-mediated energy 331–333
– Gibbs–Thomson–Herring interface potential 329, 330
– local equilibrium 329
– tangential gradients and fluxes 330, 331
capillarity vector 163
capillary driving force 307
capillary energy density 332
capillary energy fields 324–328
– capillarity on melting, influence 327, 328
– melting experiments 325
– self-similar melting 326, 327
capillary length 324
capillary-mediated fluxes 331, 333
capillary-mediated interface energy fields
– branching 333, 334
– capillarity-mediated branching 329–333
– capillary energy fields 324–328
– dynamic solver results 334–336
– zeros of surface Laplacian 333, 334
capillary-mediated interface rotation 335
capillary potential 330
capillary vector 301
carbon atoms 281, 284, 286, 288, 293
– map 284, 287
carbon concentration gradient 283
carbon-enriched austenite layer 293
carbon-enriched ferrite
– dislocation substructure 285
carbon steels
– austenite reversion in 278

– modeling grain boundary segregation 278
Cartesian coordinates 329
casting process 489
causal cone notion 124–126
– geometric interpretation 125
CCD chip 347
cellular automata (CA) 3, 9, 169
– modeling of recrystallization 8
Chalmers' technique 414
charge coupled device (CCD) detectors 369
chemically decorated grain boundaries
– phase transformation, and nucleation on 288–294
– – at grain boundaries 288–290
– – in martensitic Fe-C steels 290–294
chemical potential 271–273
– gradient 313
classic nucleation model 379
cobalt alloys 480
Co-base superalloys 467
coincidence axial direction (CAD) 428
coincidence boundaries 432
– frequency 431, 432, 436
coincident site lattice (CSL) 204
– boundaries 433
– condition 428
– role of 162
cold-rolled titanium 249, 258
cold-rolled Zr sheet, orientation density 259
columnar polycrystal aggregate 44
commercial purity (CP) 73
compression test 59, 93, 97, 98
– uniaxial 450
computational model 327
computer-assisted SEM-EBSD/OIM technique 411, 419, 427, 434
continuum simulations 42
convergent beam electron diffraction (CBED) 362
copper, grain boundary motion 194
Co–Re-based alloys 482, 483, 489
– creep strength 488
Co-17Re-23Cr alloy
– fracture surface 485
– microstructure 484, 486, 487
Co-Re-Cr alloy system 480–489
– creep resistance 486–489
– oxidation 480–483
– strength and ductility at ambient temperatures 483–486
correlated electron microscopy 280–282
CPFE methods, success of 45
CPFE simulations 42

critical resolved shear stress (CRSS) 21, 104–106, 109, 449, 450
critical shear stress for prismatic slip 99
cross-correlation method 345, 348, 360
cross-slip of screw dislocations 71
CRSS. *See* critical resolved shear stress; critical resolved shear stress (CRSS)
crystal deformation models 27
crystal elastoplasticity, modeling 41
crystal lattice 21, 25
crystallographic-geometrical analyses 426
crystallographic lattices of grains 19
crystallographic plane 20, 83, 389
crystallographic textures 7, 19, 51, 167, 399
crystallographic textures and magnifying glass
– aluminum alloys
– – oriented nucleation, oriented growth and *in situ* recrystallization 397, 398
– – recrystallization in particle containing Al alloys 399
– casting 389
– deformation mechanisms 389–396
– – stacking fault energy analysis in rolling texture formation 395
– – surface shearing and lubrication effects 395, 396
– – textures revealing 391–395
– information in metal processing
– – texture evolution and exploitation 388–399
– to investigate materials 387–400
– recrystallization 397–399
crystal–melt surface energy density 329
crystal orientation 6, 20, 26, 33, 41, 414, 423
crystal plasticity 5, 20, 21, 41
– concepts 7
crystal plasticity modeling 41–45, 49
– boundary value solvers 49
– constitutive models 46, 47
– – dislocation slip 47
– – displacive transformations 47, 48
– decomposition, of deformation gradient 45, 46
– effective material properties 55
– – direct transfer of microstructures 55, 56
– – representative volume elements 57–59
– – virtual laboratory 59–61
– homogenization 48
– texture, and anisotrop 49
– – optimization of earing behavior 51–55
– – prediction of earing behavior 50, 51
– – prediction of texture evolution 49, 50
– total deformation gradient, 44, 45
– velocity gradient 46
crystal-scale micromechanics 43

Cu Al alloy 379
Cube component 399
curling 285, 448, 449, 459
curvature-driven grain boundary motion 9
cyclic softening 73
cyclic stress–strain curve 72, 74

d

damage-tolerant steels 267
data analysis 282, 381, 382
3D computer simulation, of microstructures 120
3D/4D techniques 381
Debye frequency 75
de-convolution 75
defectant theory 270
defect densities 20
deformation 11, 93, 94
– behavior of ferritic steels 74
– behavior of recrystallized Zircaloy-4 76
– induced dislocation boundaries 368
– mechanism 395
– modes in magnesium alloys 92
– by shear banding 91
– temperature 97
– texture components
– – volume fraction of 260
– twinning 96
deformed state 8, 284, 373, 420
dendritic growth
– conventional noise-based theories 324
– kinetics 323
– noise-based theories 323
dendritic morphogenesis 323
deviation angle 433
diffraction contrast tomography (DCT) 370, 371, 379
diffraction patterns 370
diffusion-bonding method 413
diffusion coefficient 303
diffusion kinetics 306
diffusion mechanism 306, 307
dislocation-assisted (DA) mechanism 299
dislocation boundaries 83, 211, 374
– analyses 363
– deformation-induced 368, 374
dislocations 20, 21
– arrays 226, 285, 363
– creep mechanisms 489
– densities 6, 47, 70, 71, 74, 76, 79, 80, 82, 114, 356, 359, 378
– discrete 205
– on discrete slip systems 42
– glide 20, 70, 76, 114, 203, 216, 389, 400

– gliding 206
– mobile 37, 47, 73
– related stress 74
– screw 226, 299, 309
displacement gradient tensor 342
double kink formation energy 75
3D polychromatic X-ray microdiffraction technique 375, 381
driving force 302, 309, 315
2D simulation, using interface phase-field method 173
dual-beam focused ion beam (FIB) microscope 281
ductility 95, 97, 100, 101, 270, 407, 441, 485, 486, 489
3DXRD method 370, 372
– examples 372
– limitations 381, 382
– mapping mode 376
– microscope
– – permanent installation 368
– – principal components 369
dynamic recovery processes 79

e

EBSPs
– many-beam dynamical simulations 362
EBSP technique 380
ECAP grain refinement treatment 101
edge dislocation 21
effective flux 313
elastic beam theory 473
elastic deformation 197
elastic–plastic deformation 41
elastic–plastic self-consistent (EPSC) models 103
elastic strain measurement 353
electrochemical polishing sequence 282
electro-deposition methods 420
electron backscatter diffraction (EBSD) 252, 281
– data analysis 254
– grain microstructures 260
– high angular resolution (See high angular resolution EBSD)
– maps 291
– measurements 448
– pattern 343
– – formation 343
– technique 405
electron back-scatter electron (BSE) contrast 448
electron back scattering pattern (EBSP) techniques 380
– simulated 346
electron beam microprobe analysis 447
electron channeling patterns (ECPs) technique 417
electron energy loss spectroscopy (EELS) 411
electron microscopy 279–281, 295, 355, 367, 376, 411, 413, 462, 476
ellipsoidal interface 327
empiricism 104
energy anisotropy 167, 168, 277
energy dispersive spectroscopy (EDS) 280, 291
enthalpy 163, 215, 243, 244, 245, 274
entropy 79–81, 80, 278, 279, 316, 416
equal channel angular pressing (ECAP) 82–86
equilibrium grain boundary segregation theory 271–280
– Gibbs adsorption isotherm 271, 272
– interface complexions at 278–280
– Langmuir–McLean isotherm equations 272–274
– phase-field modeling 274–278
equilibrium segregation phenomena 270
equilibrium shape of a crystal (ECS) 300
European Synchrotron Radiation Facility (ESRF) 368
evolution equation for back stress 75
experimental validation, prediction
– cup drawing textures 36, 37
– rolling textures 34–36

f

fabricating processes 400
face centered cubic (FCC) metals 390
fatigue limit 74
fcc metals 162, 167
– rolling texture in 392, 393
FC Taylor model 393, 394
Fe–Al system 453
Fe-based Si alloys 432
Fe–Cr–C sample 294
Fe–Cr steels 267
Fe–C system 275
Fe–Mn steels 267
Fe–Mn systems 289
FEM simulations 11, 13, 352
– of deep drawn cup 13
– of materials processing 5
ferrite nucleation rate 379
ferrite subgrain boundaries 286
field emission gun scanning electron microscope (FEGSEM) 351
field ion microscopy (FIM) 411
finite element method (FEM) 3, 42, 54, 58
floating zone technique 414

flow stress 94
focused ion-beam (FIB) tomography 368, 471
Fokker–Planck equation 83
fourfold anisotropy 329
four-point bending tests 437
fractal analysis 439
fractal-based relationship 440
fracture processes 413, 419
fracture stress 408
Frank–Bilby equation 209
free energy density 275
friction 25, 92, 212, 395, 396
front-tracking algorithms 9
Fuchizaki's model 176

g

GBE. *See* grain boundary engineering (GBE)
GB microstructure-dependent fracture processes 438
GB phenomena, characteristic features 410
GBs. *See* grain boundaries (GBs)
generalized Schmid law 22, 23, 25, 26
geometrically necessary dislocations (GNDs) 354
Ge single crystal 345
GIA model 393
Gibbs adsorption isotherm 270–272, 288
Gibbs–Duhem equation 271
Gibbs formulation 288
Gibbs free energy 272, 273, 303
Gibbs–Thomson equilibrium 195, 304, 308, 327
Gibbs–Thomson–Herring (GTH) equation 330, 331
Ginzburg–Landau equation 172
GND analyses 361
GND density distribution 356
GND density maps 357
gradient-weighted moving finite element (GWMFE) method 176, 177, 179
grain-averaged activation energies
– distribution 377
grain boundaries (GBs) 73
– Al bicrystal 226
–– surface topography 203, 207, 220, 227
– Al, stress-driven migration of 226
– anisotropy (*See* anisotropy)
– atomic mechanisms 209
– beneficial effect 411
– crystal symmetry analysis 217
– Cu-bicrystal, AFM topography measurement 192
– curvature 9
– displacement and shear stress 213
– energy 9, 191, 258
– engineering (*See* grain boundary engineering (GBE))
– enrichment factor 274
– experimental migration rate 214
– free surface line tension 192
– gliding dislocations, traces
–– MD simulations 206
– groove
–– cubicrystal, AFM topography measurement 192
– junctions 189–198
– lattice misorientation angle, time dependence 224
– microstructures 435
– migration (*See also* grain boundary migration)
–– control 235
–– in metals 163
– migration and shear strain
–– coupling modes 208
– mobility 8, 260
–– measurements 215
–– in Zn 244
– molecular dynamics 204
– motion (*See also* grain boundary motion)
–– recrystallization 235
– network 9
– particle
–– schematic view of 193
– periodic boundary conditions 204
– plastic deformation (*See* plastic deformation)
– plot of population (area) of 173
– polycrystalline materials 216
– properties 278
–– calculation 178
– role 268
– schematic view of 195
– segregation coefficient 273
– segregation phenomenon 268, 271
– shearing 100
– shrinking grain 223
– simulation geometry 204
– sliding 99
– stick–slip process 212, 213
– stress-induced 206
–– migration 215, 221
–– motion 203
– structure and properties 412
– structure-dependent fracture processes 418
– surface tension 192
– three-dimensional 205
– tilt angle, schematic dependence 217
– triple junctions 189

– – line tension 192
– types 424
– velocity 8
grain boundary character distribution (GBCD) 162, 407
grain boundary engineering (GBE)
– basic concept 411–413
– beneficial, and detrimental effects 410, 411
– for fracture control through texturing 434–441
– – control of microscale texture, and microstructure 434–436
– – fracture control based on fractal analysis 439–441
– – microstructure-dependent brittle fracture in molybdenum 437, 438
– – percolation-controlled and microstructure-dependent fracture 438, 439
– historical background 406–410
– long pending materials problem, and dilemma 407–409
– microstructures
– – change in energy during grain growth 420, 421
– – characteristic features 420–426
– – characterization parameters 419, 420
– – grain size dependence 420
– – relation between character distribution and grain size 421–426
– – relation between texture and grain 426–436
– – in textured polycrystals, characteristic features 427–434
– – in textured polycrystals, theoretical basis 426, 427
– origin of heterogeneity 409, 410
– progress in 405
– strong and tough structural materials, demand for 406, 407
– structure-dependent fracture processes in polycrystals 417–419
– structure-dependent grain boundary properties in bicrystals 413–417
grain boundary-free surface 197
– absolute values of 192
grain boundary migration 235, 236
– AA3103 261
– aluminum alloy sheet, pole figures 249
– control of 235
– conventional curvature driving force 236
– driving forces for 236, 237
– forces acting 255
– magnetic driving forces 248, 255
– in magnetic field 237
– – opposite directions 240

– ND–TD plane, pole figures 250
– nonferromagnetic materials 248
– Ti sheet sample, pole figures 252
– titanium and zirconium 251
– Young's modulus 236
grain boundary motion
– computer simulation methodology 204–206
– coupling effect 202, 203
– – dynamics of 212–216
– – and grain rotation 221–227
– coupling factors, multiplicity of 208–212
– experimental methodology 206–208
– with local migration rate 9
grain boundary triple line
– absolute values of 192
– grain growth, driving force 192, 193
grain cluster model GIA 7
grain growth 7, 9
– in $BaTiO_3$, abnormal 119
– ClaNG model 10
– model 300
– simulation 9
– three-dimensional vertex modeling 9
grain interaction model (GIA) 7, 31, 35, 38, 393
grain microstructure evolution 235
grain refinement 94, 104
– of wrought magnesium alloys 95
grain sizes 113
– effect 420
GTH potential 333, 336

h

Hall–Petch behavior 109, 110, 112, 201, 437
Hall–Petch effect 270, 282, 438
Hall–Petch plots, magnesium alloy AZ31 from compression 94
hardening coefficient 357
hardening rate 94
hcp crystal structures 91
heat flux 326, 328
Henry's law 273
Hertzian attraction load 198
hexagonal closed packed (HCP) 391
hexagonal materials, challenges to constitutive modeling 75
high-angle grain boundaries (HAGBs) 167
high angular resolution EBSD (HR-EBSD) 341, 353, 359
– analyses 346
– applications 351–359
– crystal defects
– – and elastic strains 359
– – and lattice misorientations, GND analysis 354–359

– EBSP numerical simulations 345
– elastic strain analysis 360, 361
– – application 361
– evolution, and validation 359–362
– geometry 342–345
– history 341, 342
– lens distortion 346, 347
– low-angle grain boundaries 351
– materials applications 341
– orientation gradients 347–349
– pattern center 349–351
– practical problems 345–351
– pure elastic strains 351–354
– sensitivity 350
High Energy Materials Science beamline (HEMS) 370
high-performance steels 268
– grain and phase boundaries analyais in 267–271
high-temperature processing 270
homogeneous Poisson point process 119, 120
HR-EBSD. *See* high angular resolution EBSD (HR-EBSD)
hydrostatic pressure 79
hypereutectoid pearlitic steel 286
– wires
– – 3D carbon atom maps 284
– – TEM images 286
hysteresis loop 72

i

inclination angles 172
incoherent nanoscale Yttria particles 472
in situ neutron diffraction experiments 485, 486
in-situ recrystallization 398
integrated computational materials engineering (ICME) 5
intensity measure 122
interface potential 335
interface reaction 306
interface reaction-controlled crystal growth 303–306
– growth from a re-entrant edge 305, 306
– growth from flat surface atom 303, 304
– growth from screw dislocation 304, 305
interface segregation
– in advanced steels studied at atomic scale 267
– atom probe tomography 280–282
– chemically decorated grain boundaries
– – phase transformation and nucleation on 288–294
– correlated electron microscopy 280–282

– equilibrium grain boundary segregation theory 271–280
– ferrite phase of pearlitic steel, atomic-scale experimental observation 282–288
– grain and phase boundaries analyais in high-strength steels 267–271
interfacial energy 174, 178, 272, 324, 332
interfacial flux 332
interfacial stiffness 174
interfacial tension 271
intermetallic particles 289
interphase-boundary-related phenomena 406
intrinsic hardness 93
inverse strength-ductility
– problem 267
– relationship 269
iron–tin alloy 409
isothermal annealing 376

j

Johnson, Kendall, and Roberts (JKR) model 197, 198
Johnson-Mehl-Avrami-Kolmogorov (JMAK)
– analysis 170
– theory 119–121, 126–129

k

Kikuchi band edges 342
kinetic freezing 292
Kocks-Mecking-Estrin (KME) model 6, 71
– coefficients, re-interpreted 81
– dislocation density evolution 79
– modification, predicting deformational behavior 76
– total entropy flux 79
– for UFG titanium 74

l

LAMEL model 7, 35, 48, 50
Langmuir–McLean type approaches 272, 274
Langmuir–McLean-type monolayer adsorption films 279
Laplace heat field 327
Laplace pressure 191
– for 3D-curved grain boundary surfaces 191
Larson–Miller plot 488
lattice orientations 20
lattice rotation 23
304L-austenitic stainless steel 425
lens distortion calibration
– schematic presentation 347
"lightening rod" effect 326
linear stability analysis 86
linear stress–velocity relation 214

liquid-gallium-induced fracture 418
liquid matrix 303, 306
local chemical compositions 293
local electrode atom probe (LEAP™) 282, 291
lost-evidence problem 373
low-angle grain boundaries (LAGBs) 167
low-energy boundaries 439
LSW theory 308, 309

m

macroscopic gradients 333
macroscopic properties 5
magnesium alloy, basal pole for 98
magnetically anisotropic polycrystals, magnetic driving force impact
– texture/grain structure development
– – grain growth, texture evolution 248–253
– – growth kinetics 253–258
– – microstructure evolution 253–258
magnetic annealing. See annealing
magnetic energy density 249
magnetic field, rolling direction (RD) 253
magnetic susceptibilities 237
mark distribution 123
mark space 122
martensite matrix 293
martensite-to-austenite reverse transformation 289
martensitic stainless steel 290
material point method (MPM) 87, 120
materials science, advantages for 341
materials science and engineering (MSE) 406
Matérn cluster process 130, 131
– computer simulation 131
matrix material 288
maximum random boundary connectivity (MRBC) 440
mean densities
– associated to a birth-and-growth process 123, 124
mean extended surface density 124
mean extended volume density 124
mean surface density 124
mean surface measure 124
mean Taylor energies 462
mean volume density 123
mechanical alloying (MA) 468
mechanical anisotropy 459–462
mechanical behavior of TRIP steel 75
mechanical characterization 295
metallic systems 467
metal microstructures, characterization 367–382
– applications 372–379

– challenges and suggestions for 379–381
– 4D characterizations by 3DXRD 368–371
– 3D mapping of crystallographic contrast microstructure by synchrotron x-rays 371
– 4D studies 377–381
– – grain growth 378
– – phase transformation 379
– – plastic deformation 377, 378
– 3DXRD microscope 369–371
– recrystallization studies 373–377
– – grain boundary migration 374, 375
– – nucleation 373, 374
– – recrystallization kinetics 376, 377
metal production processes 388
MgO particles 421
microstructural changes
– along the process chain of tube fabrication 4
microstructural path concept 120
microstructural path method (MPM) 120
microstructural processes 6
microstructure, and atomistic processes 5
microstructure builder 166
microstructure evolution 5, 387
– continuum approach 5
microstructure–property relationships 5
microstructures, types 315
migration mechanism
– of faceted solid–liquid interface 314
Mises criterion 449
misorientation angle 167
misorientation function 428
Mn-rich austenite 290
mobility anisotropy 167
modeling anisotropy
– stress and strain 103–113
modeling birth-and-growth processes 121–123
model validation 21
molecular dynamics (MD) simulations 3, 4, 204
– block 222
molybdenum single crystals 428
monochromatic beam 369
Monte Carlo simulations 3, 6, 8, 9, 166
Mo-oxide formation 478
Mo–Si alloys 474
Mo-Si-B alloy 468–480, 469, 470, 475, 476, 479, 483
– creep resistance 478–480
– cross-section 476
– isothermal section 469
– manufacturing and microstructure 468–471
– multiphase Mo-Si–B alloys 475, 476
– outer fiber tensile strain vs. temperature 475

– oxidation resistance 476–478
– single-phase Mo solid solutions 471–475
– strength and ductility at ambient temperatures 471–476
– TGA results 477
Mo–Si–B phase diagram 471
Mo-9Si-8B-1Zr 478
multiphase field approach 275
multiscale modeling 5
multislip situation in polycrystal 43

n

nanocrystalline material 408
nanoscale impurity-containing intergranular films 279
nanostructured metallic alloys 283
Neyman–Scott processes 130
NGG. *See* normal grain growth (NGG)
NiAl compound
– anisotropy 447
– compression 452
– compression texture 457
– deformation 452
– deformation textures 457
– sample 454
– sample, schematic sketch 461
– textures 455, 458, 459
– Young's modulus 462
NiAl flow stress 450
Ni-based superalloy single crystal 351, 352
Ni-base superalloys 467, 480
Ni–P alloy 427
Ni–Re phase diagram 480
Ni single crystal
– GND density maps 358
– orientation maps 358
nitial cubic cell
– elastic deformation 344
noise-based theories 323
"noise-free" dendrites 335
noisy fluctuations 323
nonzero lattice rotation 25
normal grain growth (NGG) 299, 310, 317
normal growth model 123
normalized recrystallization kinetic curves 378
novel metallic high temperature materials development
– Co-Re-Cr alloy system 480–489
– by microstructural design 467–490
– Mo–Si–B alloy system 468–480
nucleation 119, 375
– barriers 277
– in clusters 130

– – evaluation of integral 131–133
– – influence of cluster radius 133–135
– – influence of number of nuclei per cluster 135, 136
– – Matérn cluster process 130, 131
– – numerical examples 133
– frequency 8
– and growth 119
– kinetics 8
– on lower dimensional surfaces 136
– – bulk nucleation 137, 138–140
– – derivation of expressions for surface and bulk nucleation 136
– – numerical examples 138
– – simultaneous bulk and surface nucleation 140, 141
– – surface nucleation 136, 137, 138
– rates 9, 137
– sites exhausted 120
numerical validation tests 360
Nye's dislocation density tensor 354
Nye's tensor 356

o

ODF intensity 456
ODFs. *See* orientation distribution functions (ODFs)
OIM technique 434
optical micrograph, of AZ31 after elongation in tension 109
orientation distribution functions (ODFs) 31, 35, 37, 38, 250, 254, 449
– cold-rolled Zr sheet 259
– (0002) pole figures 259
– titanium and zirconium 254
Orientation Gradients 347–349
orientation imaging microscopy (OIM) 406
orthorhombic texture symmetry 460
oxidation resistance 468

p

pancake model 449
parameters used in modeling flow stress and r-values 109
partial differential equations (PDEs) 3
particle–matrix interface 487
particle-stimulated nucleation (PSN) 399
Peach–Kohler forces 226
pearlitic steel
– ferrite phase, atomic-scale experimental observation 282–288
Peierls–Nabarro stresses 211
Peierls stress 74
percolation-controlled crack propagation 439

phase field crystal (PFC)
- boundaries, simulations 218
- MD simulations 218, 219
- methodology 205
phase-field modeling 9, 171, 280
- advantage 275
phase transformations 10
- energy 325
- process 326
phase transitions 288
pivalic anhydride (PVA) crystallites 325
- progressive melting 326
plane-matching-type boundary 428
plastic anisotropy 27, 41, 42, 460
- of polycrystalline materials 33, 34
- in pure magnesium and the traditional alloys 100
plastic deformation 235
- grain boundaries (GBs) 201
- heterogeneity of 44
- of metallic crystals 20
- of polycrystalline materials 19
plastic incompatibilities 75
plastic strain 268
- rates 24
- ratios 103
Plateau's rule 313
Poisson's ratio 107, 197
polar-to-equatorial fluxes 328
polychromatic X-ray microdiffraction technique 371
polycrystal deformation modeling 447
polycrystal deformation software 24
polycrystalline aluminum alloy 242
polycrystalline cold-rolled zinc 236
polycrystalline materials 405, 421, 426, 441
- bulk properties 419
- mechanical properties 407
- properties 413
polycrystalline zinc 258
polycrystal plasticity models, for single-phase materials 27, 28
- grain interaction models 30–33
- relaxed constraints Taylor theory 29, 30
- Sachs model 28
- Taylor theory 28, 29
polycrystals
- equilibrium boundary shape 301
- equilibrium crystal shape 300, 301
- fundamentals, equilibrium shape of interface 300, 301
- interface structure-dependent grain growth behavior in 299–318

- solid–liquid two-phase systems, grain growth in 302–312
- solid-state single-phase systems, grain growth in 312–317
- system 175
- texture/microstructure evolution
-- magnetic field influence 258–261
polymer crystallization 119
potential energy 171
Potts algorithms 165
power–law relationship 374
processing–property relationships 4
proportionality constant 314
pseudonormal grain growth (PNGG) 310
pyramidal slip 111

q
Q-state Ising model 164
quasistatic melting 327
quenched systems, types 289

r
Raman spectroscopy 477
Random velocity 145
- cases 147, 148
- computer simulation 148–150
- time-dependent 145–147
rare earth (RE) elements 100
rate insensitive methods 25
rate sensitive methods 25
Read–Shockley model 176
real-space simulations 8
recrystallization 7, 8, 9, 10, 91, 119, 121, 169, 261
- mechanism 399
- modeling 8
- nucleation 373
- static/dynamic 92
- stored energy level in 165
- textures 397, 398, 399
-- components 398
- of warm rolled aluminum 1050 166
regions of interest (ROI) 342
ReNuc, nucleation model 8
residual shifts 348
resolved shear stress 20, 21
ribbon-like lamellar structure 285
Rietveld refinement analysis 471
rigid body rotation 352
Robo-Met.3D technique 380
rolled/annealed FCC sheet
- orientations in 392
rotated reference technique (RRT) 349, 361
rotation angle, function 348

rotation matrix 280
round-edged polyhedral 307
r-values 98, 99, 107, 108, 111, 112

s

scale factor 329, 344
Schmid factors 93, 98, 108
Schmid's law 103
Schmid tensors 24
Schmidt factors 211
screw dislocation-assisted mechanism 305
secondary ions mass spectrometry (SIMS) 280
segregation coefficient 274
segregation-induced embrittlement 427
selected area channeling pattern (SACP) technique 423
selectively grown grains, relative frequency 247
SEM-EBSD/OIM technique 422, 423
severe plastic deformation (SPD) 69, 70
– numerical simulations of SPD processes 82–86
– one-internal variable models 70–77
– three-internal variable models 81, 82
– two-internal variable models 77–81
SFE stainless steel crystal 363
shadow casting technique 351
shear modulus 488
shear strain 25, 30, 73, 80, 208, 352, 353
– components 353
– through an expression of the Voce type 103
shear stress 27, 78, 79, 91, 98, 103
shear zone 396
shift error distributions 349
shift errors 349
SiGe–Si sample
– HR-EBSD analysis 354
silver alloys
– vapor-deposited samples 421
simple beam theory 352
simulation approaches 161, 162, 164
– cellular automata 169, 170
– cusps in grain boundary energy 174
– level set 174, 175
– moving finite element 176–179
– particle pinning of boundaries 179
– phase field 170–174
– Potts model 164–169
– vertex 175, 176
single crystal approach 408
site-saturation 121
site-specific joint scanning transmission electron microscopy (STEM)–APT analysis 281

skew-symmetrical velocity gradient tensor 26
slip direction 20
– families 21
slip domination, factors 91
slip plane families 21
slip planes 20, 27
slip rates 23, 25, 26
slip systems 20, 21, 25, 395, 449, 454
– geometry 389
smart cellular automata code CORe 8
Smith–Zener equation 179
smooth particle hydrodynamics (SPH) code 87
softening phenomena 7
solidification 169, 323
solidification process 325, 389
solid–liquid interface 304, 330
solid-liquid two-phase systems 299
– diffusion-controlled crystal growth 302, 303
– grain growth 302–312
– – behavior 307–312
– interface reaction-controlled crystal growth 303–306
– mixed controlled growth of faceted crystal 306, 307
– nonstationary grain growth, in systems with faceted grains 309–312
– single crystal in liquid, growth mechanisms and kinetics 302–307
– stationary grain growth in systems with spherical grains 308, 309
solid oxide fuel cell (SOFC) 353
solid-state single-phase systems
– grain boundary, migration mechanisms and kinetics 312–315
– grain growth 312–317
– – behavior 315–317
solid–vapor system 305
solid volume fraction 309
S-orientation 168
SPD. See severe plastic deformation (SPD)
spherical system 308
spheroidization 286
spiral growth mechanism 304
stable cyclic deformation. 73
stable hysteresis loop 74
stacking fault energy (SFE) 359, 393
– evaluation 396
stagnant grain growth (SGG) 310
standard electron backscatter diffraction (EBSD) 341
standard optimization techniques 343, 375
static recrystallization 451
Stefan balance 336
strain 26

– components 342
– deformation 453
– errors 348
– free cells 378
– measurement errors 347, 350
– rate tensor 21, 23
– ratio 108
stress-dependent energy 75
stress–strain behavior
– in pure magnesium and a magnesium–gadolinium alloy 101
stress–strain curves 356, 359, 437, 450, 459, 473
– for AZ31 110
– for bars of magnesium alloy WE54 in tension and compression along 102
stress–true strain, behavior evaluation 107
stress vs. strain 110
structure-dependent fracture processes 417
structure-dependent grain boundary properties 415, 416
subpixel registration techniques 345
superposition principle 140
surface nucleation 136, 138
surface tension 165

t

tangential divergence operator 332
tangential gradient operator 331
Taylor energy 450, 460
Taylor theory 7, 23, 26, 27, 73, 461
Taylor-type condition 78
Taylor-type models, applied to magnesium 103, 392
temperature
– and degrees of GBS 99
– fluctuations 14
– influencing plastic anisotropy of magnesium alloys 99
– influencing r-values for AZ31 sheets 100
tensile creep testing 478
tensile strength 96, 283
tensile stress–strain curves, for AZ31 alloy 95
tensile test 94, 98, 107
tension tests 94
tension twinning 91
texture development 9
texture/grain structure development
– grain growth, texture evolution 248–253
– growth kinetics 253–258
– microstructure evolution 253–258
texture index (TI) 37
texture ODF intensity 397
thermal stability 416

thermogravimetric analysis (TGA) 476
thermomechanical processing 422, 424, 429, 430, 435, 447
– experimental 447–450
– mechanical anisotropy 459–462
– microstructure and texture development 450–457
– NiAl microstructure and texture design 447–463
– texture simulations 457–459
three-dimensional (3D) Hough transformation 342
three-dimensional X-ray diffraction (3DXRD) 368
through-process modeling 10–14
– aluminum sheet fabrication 11
through-process modeling (TPM) 4
Ti-containing alloy 479
TiC–WC–Co system 312
tilt boundary mobility, boundary plane inclination effect 245–246
tilt grain boundaries
– in bismuth 246
– bismuth bicrystal, during annealing 240
– images 239
– migration activation 244
– migration of 240
– orientation/inclination 246
time integration 26
torque force 301
total deformation tensor 349
total dislocation density 80, 81
total free energy 171
transformation-induced plasticity (TRIP) 269
transformations 119, 150, 151
– application to recrystallization of IF steel 153–157
– kinetics, cluster radius on 133
– mean volume density transformed as a function of 134
– objectives 150
– sequential transformations 152
– simultaneous transformations 151, 152
transformations, nucleated on random planes
– analytical expressions for 141
– – behavior at origin as model for behavior in "unbounded" specimen 142
– – computer simulation results 145
– – nucleation on random parallel planes 143, 144
– – results for nucleation on random planes 141, 142
transition probability 165
triple junction line tension

– AFM topography measurement 191
– crystalline nanoparticles agglomerate 196–198
– on Gibbs–Thompson relation, effect of 195, 196
– Zener force, effect of 193–195
triple line energy 190
– AFM topography measurement 191
– schematic three-dimensional (3D) view 190
triple line tension
– on thermodynamics, and kinetics 192
true stress 108
true stress–true plastic strain compression curves 460
turbine engines
– temperature capability 467
twinning 93
– reduced in RE-alloys 102
twinning fraction 112
twinning-induced plasticity (TWIP) 269
two-dimensional atomic nucleation (2DN) 299
two-dimensional nucleation and growth (2DNG) 309, 310
two-dimensional nucleus 303
two-phase systems 315

u
uniaxial deformation 390

v
vapor-deposition methods 420
vector function 301
velocity gradient 24
– tensor 21, 26
visco-plastic methods 25
viscoplastic self-consistent (VPSC) 103, 106
– calculations 110
– effect of strengthening by grain refinement on *r*-value 112
– fraction of strain accommodated by 111
Voce-type expressions 106

volume fractions 123, 124
– of major texture components 12
von Mises criterion 462
von Mises equivalent strain rate 27
von Mises yield condition 102

w
WC grains 312
WE54 alloy, tension and compression curves 101
weighted essentially nonoscillatory (WENO) approach 172
work hardening model 13
wrought magnesium alloys
– mechanical properties 91
Wulff construction 300
wurzite–zincblende phase boundary 314

x
X-ray diffraction 291, 388, 427, 470
X-ray line broadening experiments 377
X-ray texture data 387
X-ray tomography 370
X-ray tomography techniques 368

y
yield stress 448
Young's modulus 107, 197, 448, 462
yttrium alloy 103

z
Zener force 193
zinc bicrystals
– magnetic driving force 241, 243
– specimens
– – geometry of 238
zinc single crystal 246
– under magnetic driving force 246–248
zirconium 249
– annealing, in magnetic field 258